Hybrid Vehicles
and the Future of Personal Transportation

Hybrid Vehicles

and the Future of Personal Transportation

Allen Fuhs

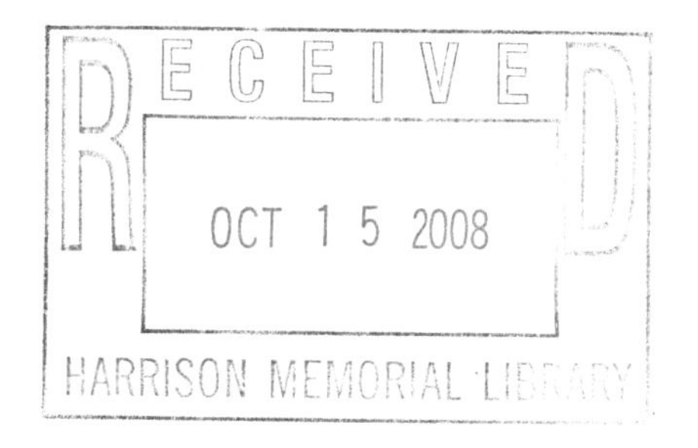

CRC Press is an imprint of the
Taylor & Francis Group, an **informa** business

CRC Press
Taylor & Francis Group
6000 Broken Sound Parkway NW, Suite 300
Boca Raton, FL 33487-2742

Printed in the United States of America on acid-free paper
10 9 8 7 6 5 4 3 2 1

International Standard Book Number-13: 978-1-4200-7534-2 (Softcover)

Library of Congress Cataloging-in-Publication Data

Fuhs, Allen E.
Hybrid vehicles and the future of personal transportation / by Allen Fuhs.
p. cm.
Includes bibliographical references and index.
ISBN-13: 978-1-4200-7534-2
ISBN-10: 1-4200-7534-9
1. Hybrid electric cars. I. Title.

TL221.15.F84 2009
629.22'9--dc22 2008013028

Visit the Taylor & Francis Web site at
http://www.taylorandfrancis.com

and the CRC Press Web site at
http://www.crcpress.com

Dedication

To Maxine, my companion
through trials and triumphs

Contents

Preface

The preface is an opportunity for the author to have a conversation with the reader concerning various aspects of the book. Many readers skip the preface. One author was so determined that his preface be read, that he inserted it between the first and second chapters.

The intended audience includes the buyers or would-be buyers of hybrid vehicles. This book provides reasons why or why not to buy hybrids, as well as the facts necessary to guide the buyer's choice of a hybrid. The reader who is interested in the impact of hybrids on the environment will find useful concepts herein. The future of personal transportation vis-à-vis global warming and the finite limit of petroleum is addressed in great detail and questions such as "What's new on the horizon?" and "When will these advances appear in the showroom?" are answered.

More specifically, the intended readers are potential hybrid customers; automobile enthusiasts who subscribe to *AutoWeek*, *Motor Trend*, *Car and Driver*, *Automobile*, etc.; automobile dealership sales and repair personnel; clients of both educational and public libraries; government workers from the Environmental Protection Agency, Department of Energy, California Air Resources Board, state departments of transportation, etc.; rescue personnel from fire departments, ambulances, and highway patrols; and members and staff of Congress. Some business executives need a reliable source of information about hybrids to make intelligent business decisions.

One goal of the book is to provide information for the educated citizen. Important decisions are to be made in the coming years. Driving a car is fun, but the cumulative consequences of 1,000,000,000 cars on the road are serious. (That is right, in the near future, 1 billion cars will be on the roads of the world.) The potential for wars being fought over oil and the subsequent economic disaster is great. Intelligent decisions may avoid the rocks and shoals of the future.

The book is itself a hybrid of material primarily for the intelligent general reader with added insight for the equation-confident reader. The level of writing is similar to that of *Scientific American*. The book combines descriptions in nontechnical English with the accuracy and insight of equations. With regard to equations, approach them with an open mind. If you balance your own check book, you are qualified for this book. If you prepare your own income taxes (using TurboTax or TaxCut, even), then you are overqualified for the contents of this book. Equations provide transparency; all the assumptions are clearly presented. The technology discussed is neither shallow nor overwhelming.

For the engineer, the book provides a starting point for additional research. Many engineers in the scope of their technical knowledge do not cover all the different topics presented in this book. An engineer specializing in structures and the finite element method for computation is unlikely to be familiar with electrochemistry. An expert in control theory will likely not be conversant with advances in internal combustion engines. For the various engineering disciplines, the book provides a nontrivial introduction to each of the technologies essential for hybrids.

To really understand this subject, one must have some feeling for the magnitude of numbers and what a certain magnitude means. Most of the car-buying and car-driving public recognizes that 50 hp is really small for a car while 300 hp is quite powerful. For hybrids, electrical units, such as those for power (kW) and energy (kWh), are being force-fed to the public. The hybrid enthusiast develops an understanding of the implications of a 2 kWh battery. This book nurtures and expands your understanding of numbers related to hybrids.

This book is written by a retired engineer and college professor. Engineers love to learn how gadgets work. Around two years ago, the author read an article that listed the components in a hybrid. From the list, the functioning of hybrids was apparent. This stimulated two years of research on hybrids followed by 18 months, more or less, writing the book.

Besides a career teaching at the college level and a background in engineering, the author was exposed to automobiles while still a teenager. During World War II, he managed a Shell service station on old Highway 66 in Gallup, New Mexico. Later in college, he worked for $1/h in a service station to help pay college expenses. He has owned cars (from the United States, Germany, Italy, Japan), pickups (from the United States), sports cars (from Italy, Britain), and minivans (from Japan, the United States). He has overhauled Ford Model A four-cylinder engines, Ford V-8s, and Rio pick-ups; and also differentials and cooling systems. He later became a fellow of the Society of Automotive Engineers (SAE). Four wheels and an open road still remain a strong attraction.

This may be the tenth and final book the author has written or edited over a span of 40 years. Approximately three months were devoted to Chapter 12. Considerable material was assembled, and numerous equations were derived. Enough information was gathered to write another book!

Parts of the contents have formed the basis for several luncheon and dinner lectures. At the end of the talks, the author came to know the interests of the audiences. Also, he was made aware of the weak spots in the presentations.

One feature of the book is the information placed at your fingertips on the inside front and back covers. This includes data on the energy density of gasoline, conversion factors from horsepower to kilowatt, etc.

The data for this book have been gathered from the Internet, SAE technical publications, auto enthusiast magazines, and government reports. With regard to the Internet, two types of postings are available. One type of Internet listing is called "buoyant nonsense," buoyant because of an optimistic (unrealistic) view and nonsense because of the huge disconnect from the realities of the technical world. Another type is called "accurate" and "informative." This book uses the latter type to its advantage and discards all that is unnecessary. The SAE reports are written by and written for automotive engineers working on hybrids and other vehicles. These reports have been dissected and presented herein.

Julie Fleer and Donna Aikins, who are professional graphics artists, translated most of my hand drawn sketches into exceptionally fine figures. Figures in Chapter 14 were drawn by Larry Omoto of the Hana Group, Monterey, California. A few of the figures in Chapter 17 were drawn by Thomas Blackwell while he was completing his degree in graphic arts.

1 Historical Roots of Hybrid Automobiles

Surprise: The hybrid was invented more than 100 years ago.

A definition of a hybrid seems appropriate as the first paragraph of the first chapter. What is a hybrid? A hybrid combines two methods for the propulsion of a vehicle. Possible combinations (all discussed later) include diesel/electric, gasoline/flywheel, and fuel cell/battery. Typically, one energy source is storage, and the other is conversion of a fuel to energy. As discussed in more detail in Chapter 4, the combination of two power sources may support two separate propulsion systems. This is true for the parallel design hybrid.

Historians recognize that certain events are milestones. These events can change the course of history. Milestone events divide history into periods. Names are assigned to the periods of history. An example of a milestone event is 9/11, a watershed event in the history of global terrorism. Yet note that 50–60 major terrorist attacks had occurred before 9/11.

What are the significant events and periods with regard to hybrid automobiles?

Three periods can be defined: ancient, modern, and current. Two events are significant. The first event, which occurred a century ago, was the production of the world's first hybrid automobile. The OPEC-induced gas shortages of 1973 and again in 1979 constitute the second event. The period between the first and second milestones is considered "ancient." Ancient history is ancient only relative to hybrids and cars. The period following the gas shortages is considered "modern." The modern period, of course, ends with the present and ushers in the current period.

ANCIENT HISTORY: CROSSROADS

In 1900, cities were urban and not suburban as today. Ninety percent of the population dwelled within 5–8 mi of the heart of most cities. Cities were not spread out which was a favorable factor for cars with limited range such as electric cars.

Existing roads in the United States were mainly stage coach trails or freight wagon byways. These roads were dusty in dry weather and muddy barriers in wet weather. Paved roads, although found exclusively in cities, were few. For travel between cities, railroads were the only practical means of transport for either passengers or freight. Canals, where present, offered economical but slow transportation.

Table 1.1 shows the 1900 automobile census. The left-hand column has three different types of automobiles: steam, electric, and gasoline. The estimated number of carriages is shown for perspective. The middle column provides the number of vehicles in descending order. The right-hand column presents the major limitation for vehicle range. Range is the distance that the car can travel without stopping for service.

STEAM CARS

In 1900, steam technology was advanced. On May 10, 1869, the transcontinental railroad had been completed with the tying of rails at Promontory, Utah. Samuel Langley was experimenting with a steam-powered airplane to be flown from a barge on the Potomac River. Engraved in the marble on the front of a large building on the campus of the University of Illinois was "Railroad Engineering." Large electrical generators driven by steam engines generated the electricity in many towns and cities. Hence, the application of steam to cars was natural.

TABLE 1.1
Automobile Census by Type in 1900

Automobile Census in 1900—New York, Boston, and Chicago

Automobile Type	Number	Limiting Factor for Range
Steam	1,170	Feedwater for boiler
Electric	800	Energy storage in battery
Gasoline	400	Size of gas tank
Carriages	294,689	Endurance of horse

Source: From SAE Historical Committee, *The Automobile: A Century of Progress*, Society of Automotive Engineers, Warrendale, PA, 1997, 292 pp. With permission.

The advantages of steam-powered cars included high performance in terms of power and speed. Steam engines and electrical motors share a common feature; both have high static torque, that is, zero-rpm (revolutions per minute) torque. Steam cars had boundless torque from rest, and hence had excellent acceleration. A clutch and a gearshift were not required for a steam car. Similar to electric cars, steam-driven cars were quiet.

The disadvantages of steam-powered cars included poor fuel economy and the need to "fire up the boiler" before driving. A trip must be anticipated. As shown in Table 1.1, range was limited by the necessity to replenish feedwater. As in railroad locomotives, feedwater was lost in the exhaust along with the smoke from fuel combustion. Steam powered ships could not tolerate the loss of fresh water. Steam condensers were developed, and a closed cycle became possible. Later, the same technology was applied to the steam car to solve the feedwater problem; however, the timing was too late. Gasoline cars had won the marketing battle.

In a competition for speed among gasoline, steam, and electric cars, a 1906 Stanley Model K Racer stunned the automotive world with a speed of 128 mph [8]. Compare with the electric-cars below. See Reference [1–8], *Autoweek*, February 27, 2006. In 2008, interest in speedy steam powered cars continues. The British Steam Car Challenge is producing a vehicle with a target speed of 320 km/h (200 mph).

ELECTRIC CARS

In comparison, electric cars were comfortable, quiet, clean, and fashionable. Electric cars did not need to be cranked, a feature especially attractive to women. Ease of control was also a desirable feature. However, as shown in Table 1.1, the range was limited by energy storage in the battery. For city use, the range was adequate. After every trip, the battery required recharging. Lead acid batteries were used in 1900. Lead acid batteries are still used in modern cars. Hence lead acid batteries have a long history (since 1881) of use as a viable energy storage device [1]. See Figure 1.1.

Chapter 6 discusses the relationship between the state of charge (SOC) and battery life. A battery retaining 60% of its charge has SOC of 60%. A fully charged battery has SOC of 100%. Repeated deep discharges, that is, SOC near zero, reduce battery life. To achieve maximum range, the early electric cars usually had cycles with deep discharges. Battery life was short, incurring high costs.

An electric car does not imply a plodding vehicle! The first car to exceed a speed of 60 mph was an electric car. In fact, the speed record of 65.8 mph was set in April 1899 [5]. At the Bonneville Salt Flat in Utah along I-80, in October 2004 the electric race car, *Buckeye Bullet*, achieved 271 mph. The *Bluebird Electric BE3* is aiming for over 300 mph. In April 2007, an electric powered motorcycle, which used Li-ion batteries, set the quarter mile speed record of 156 mph. Building on this heritage of speed, two start-up car companies in Silicon Valley are building pure electric sports cars using

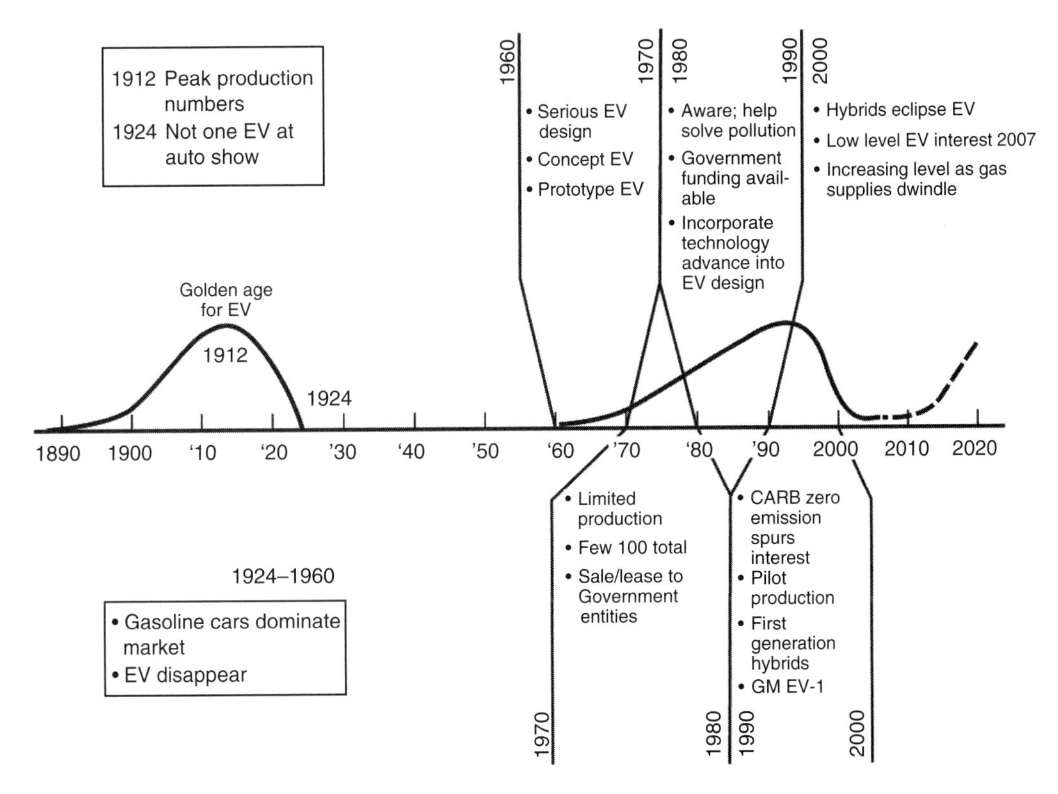

FIGURE 1.1 Curves are qualitative and provide a feel for the level of interest and activity in the EV from 1890 to present day. EV merged into HEV (hybrid electric vehicle). In the future, the HEV may become EV.

Li-ion batteries to improve range. The electric motors produce enormous torque for amazing performance. While on the subject of speed records, what is the record for a fuel-cell-powered car? In August 2007, a fuel-cell-powered Ford Fusion, known as Hydrogen 999, set the record of 207 mph at Bonneville Salt Flat [6,9]. Drag coefficient for the Hydrogen 999 was $C_D = 0.21$.

What is the world record for the longest distance traveled by an electric vehicle (EV) on a single charge? A special 1999 Mitsubishi coupe using Li-ion batteries covered 2124 km (1330 mi) on a single charge. This is the distance from San Diego to Vancouver, BC or along the east coast the distance from Miami to Boston. Going west from Washington, DC, it is almost the distance to Denver. Going east from San Francisco, the Mitsubishi coupe could almost be driven to Omaha. Mitsubishi has been an ardent supporter of EVs (Figure 1.2).

At the 1924 automobile show, no electric cars were on display. This announced the end of the Golden Age of electric-powered cars.

GASOLINE CARS

Gasoline cars of 1900 were noisy, dirty, smelly, cantankerous, and unreliable. Cranking of engine produced a kickback that typically resulted in broken thumbs. Flat tires were common and most cars carried multiple spares. However, the gasoline car had one dominant feature; it used gasoline as a fuel. The range of a gasoline car was far superior to that of either a steam or an electric car.

The superior range of a gasoline car can be explained by the energy density of gasoline compared to a battery. A quantity of 1 kg, 2.2 lb, of gasoline has 360 times the energy of a 1 kg, fully charged, lead acid battery. Electric cars are more efficient than gasoline cars. Accounting for the relative drive efficiencies of electric and gasoline cars, gasoline still enjoys a 70 to 80 times advantage.

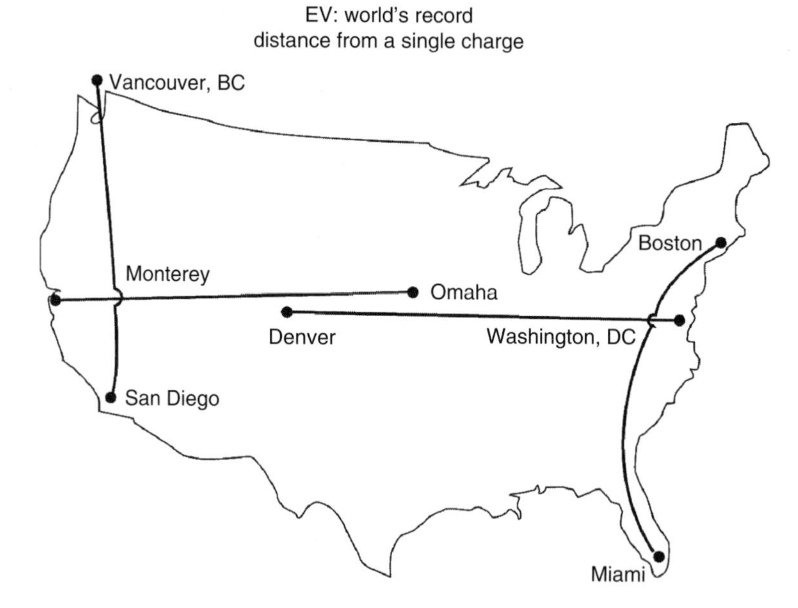

FIGURE 1.2 World's record for the greatest distance on a single charge for an EV.

CARRIAGES

The number of carriages, 294,689.4, is the author's estimate; we know for sure it is not 294,689.4 carriages. Automobiles were displacing carriages, although in 1900, the penetration of the personal mobility market was very limited. If a car is parked in a garage, no maintenance is necessary. If a horse is lodged in a stable, feed and water must be provided.

CROSSROADS

Obviously, the high-tech automobile of 1900 would ultimately replace horses and carriages. Which car type would dominate or if any car type would dominate at all was not apparent in 1900. Personal mobility transportation was at the crossroads. As stated later, personal mobility transportation is again at the crossroads today. The uncertainty is the same, but circumstances are very different today as compared to 1900. This uncertainty is the theme of Ref. [10].

INVENTION OF HYBRID AUTOMOBILE

Customers realized that despite its many advantages, the limited range of the electric car was a big letdown. Moreover, the inconvenience of recharging and the long recharge times reduced its appeal.

Engineers recognized that the good features of the gasoline engine could be combined with those of the electric motor to produce a superior car. The gasoline engine contributes the favorable range capability. The electric car provides quiet comfort and ease of control. A marriage of the two yields the hybrid automobile. Most early hybrids could be started with the simple motion of pushing a button; this was a major advantage. As is the case for today's hybrids, the electric motor can spin the gas engine at high rpm for starting.

On occasion, in the early days of the gasoline car, the driver and the passengers would be forced to walk home. Engine failures were not unknown. The electric car was more reliable. A hybrid automobile offered the no-walk-home insurance of an electric motor.

TABLE 1.2
Key Features of Hybrid Automobiles; Then and Now. The Only Thing Not Invented a Century Ago Was the Word *Hybrid*

Key Feature of Hybrid	Woods Gas-Electric Car	2007 Toyota, Ford, Honda
Downsized gasoline engine	X	X
Dual-purpose motor/generator	X	X
Regenerative braking	X	X

KEY FEATURES OF HYBRID AUTOMOBILE

By combining the outputs of the electrical motor and gasoline engine a smaller gasoline engine can be used. A smaller gasoline engine can operate under conditions yielding higher efficiency. Likewise, compared to a pure electric car, for the same performance, a smaller electrical motor can be employed while retaining performance levels.

Although not mentioned in Table 1.2, the electrical battery, for the same performance, can be downsized compared to a pure electric car. The point is often made that a hybrid has two propulsion systems, which is true. However, the detrimental effects of having two systems are partially mitigated by the fact that many components can be downsized.

A dual-purpose motor/generator (M/G) facilitates regenerative braking and, perhaps as significant, regenerative "coasting" downhill.

EARLY HYBRID VEHICLES

As evident from Table 1.3, numerous hybrids were designed at the turn of the century. Some were concept cars for display at the Paris Salon. Others were put into production, albeit on a small scale. Reference [1] provides further details. Here we note that the benchmark for the start of the ancient period for history of hybrids was the turn of the century.

TABLE 1.3
Early Hybrids in Europe and United States
Early Hybrid Vehicles

Manufacturer or Engineer	Country	Year
Pieper	France	1898[a]
Vendovelli & Priestly	France	1899[a]
Jenatzy	?	1901[a]
Krieger	France	1902
Lohner-Porsche	Germany	1903
Auto-Mixie	Germany	1906
Mercedes-Mixie	Germany	1907
Pope	United States	1902[b]
Baker	United States	1917
Woods	United States	1917

[a] Concept vehicle for Paris Automobile Salon.
[b] Prototype caught fire and burned on first test run.

WOODS GAS-ELECTRIC CAR

The first hybrids to enter production in the United States were the Baker and Woods vehicles. Both Baker and Woods were graduate electrical engineers. Both were manufacturers of pure electric cars before designing and producing hybrids. The 1902 Woods Phaeton EV, which was little more than a carriage with an electric motor, had a range of 18 mi, a top speed of 14 mph, and sold for $2000. We will now focus briefly on the 1916–1919 Woods gas-electric car.

The body style was vintage 1910s but the technology was advanced (Figure 1.3). The Woods gas-electric car was surprisingly modern. An article in *Scientific American* describing this car [2] reads like the sales brochure for a modern hybrid car. The Woods gas-electric car was expensive and could not compete with the pure electric cars of the day and in particular with the Ford Model T (Figure 1.4). Sales were relegated to small numbers with low profits. Inevitably, the Woods gas-electric car vanished into oblivion.

CONTROL SYSTEMS FOR EARLY HYBRIDS

Chapter 8 emphasizes the crucial importance of the overall control of the hybrid propulsion system. During the ancient period of hybrid history, both the control hardware and the control theory were in their infancy. Therefore, control of early hybrids was rudimentary, the onus for control totally on the driver. One hybrid had a three-position switch:

1. Battery connected to motor only
2. Generator connected to battery only
3. Short circuit motor acting as a dynamic brake

Number 3 was regenerative braking at 0% efficiency.

The lines of gasoline and electric cars are attractively combined in the Woods dual power car

FIGURE 1.3 Lines of gasoline and electric cars are attractively combined in the Woods dual power car also known as Woods gas-electric coupe. (This drawing appeared in *Motor Age*, 31 August 1916, page 48. Reproduced with permission of *Motor Age*.)

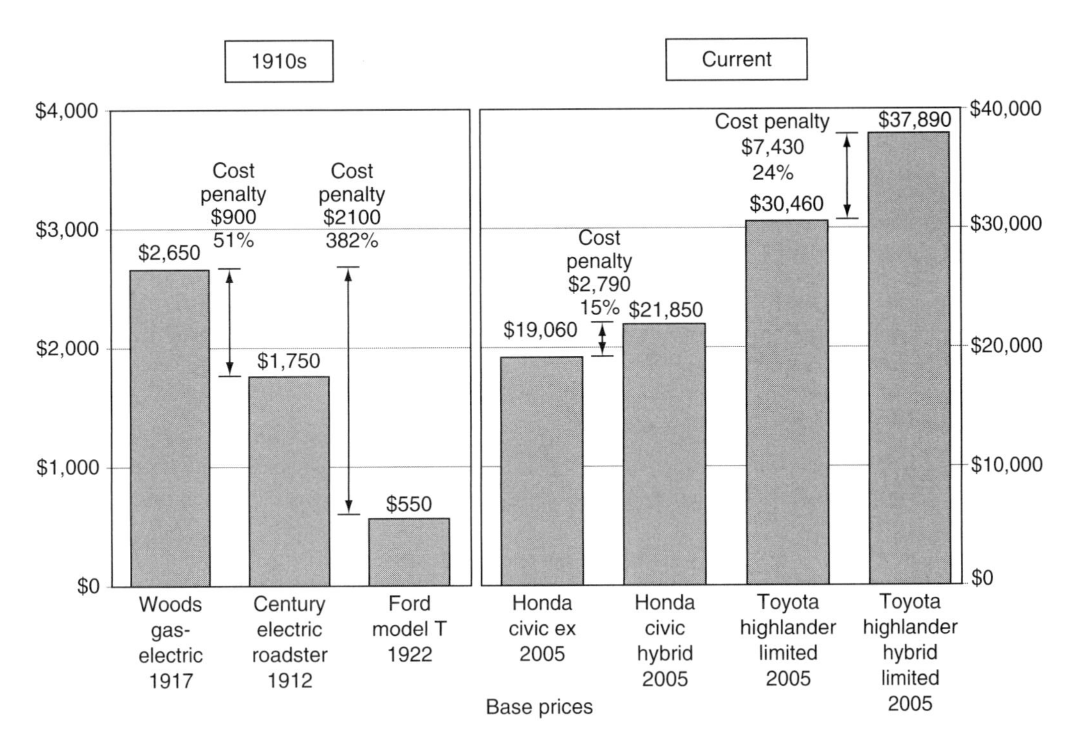

FIGURE 1.4 Comparison of prices for gas-electric, pure electric, and pure gasoline cars in 1910s. Also a comparison of prices for two HEV with their pure gasoline counterparts.

MODERN PERIOD OF HYBRID HISTORY

The modern period starts with the oil embargoes and the gasoline shortages during the 1970s which created long lines at gas stations. To avoid waiting, entrepreneurs formed a new business. Instead of you waiting in line to refuel your car, they would do it for you for a fee. After waiting in line, only 8 gal could be bought. The long lines focused the public's attention on energy.

Congress responded to the gasoline shortages of the 1970s by passing the Corporate Average Fuel Economy (CAFE) legislation. By the 1980s, concern over fuel availability had faded. Gasoline was plentiful at a low price. Car buyers demanded more performance without regard to miles per gallon (mpg). Interest in CAFE waned. Legislatively driven mpg and market-driven mpg were not in sync. From today's viewpoint, the decline in interest can be assessed as a failure on the part of the federal government. However, Congress responds to the voters who wanted bigger cars; so the failure to increase mpg by means of CAFE can be traced back to you, the voter.

Chapter 2 asks the question "What are the reasons for poor gas mileage?". Here we provide an introduction to the question posed in Chapter 2 based inpart on discussions with a retired California highway patrol officer. Sports utility vehicles (SUVs) contribute to the gas crisis. CAFE specifies minimum mileage for different classes of vehicles. One loophole in CAFE was the definition of cars and trucks. Large SUVs do not need to conform to passenger car CAFE, which was 27.5 mpg in 2007. SUVs are rated as trucks (by the federal government) and meet only the CAFE required of trucks.

Is an SUV a car or a truck? It depends on whom you ask. As stated above, according to CAFE, an SUV is a truck. According to licensing and traffic enforcement, an SUV may be a truck or a car. The classification varies from state to state. For example, in California, a Lexus RX 400 h and the Hummer H-3 are SUV-cars while the Hummer H-1 and H-2 are SUV-trucks.

Here we classify different vehicles and continue our observations on the questions posed in Chapter 2. California has posted different speed limits for cars and trucks at 65 and 55 mph, respectively.

Since some SUVs are trucks, the SUV-trucks at speeds over 55 mph should be ticketed. Further, trucks must drive in the right-hand two lanes only; the SUV-trucks legally are so restricted. These are examples of the disparity between law and actuality. Finally, consider this peculiar situation without clear legal precedence. Trucks cannot drive in commuter lanes. Can a hybrid SUV-truck legally drive in commuter lanes with only the driver on board?

HYBRID AUTOMOBILES: STIRRING OF INTEREST IN THE 1990s

Let us begin the discussion of hybrids with an introduction to a modern, milestone setting EV. Obviously, the technology of hybrid electric vehicles (HEVs) and EVs is interwoven.

GM EV-1 Electric Car

The GM EV-1 electric car, which was introduced in 1997, is discussed in Chapter 6 and Appendix A. As background for later chapters, consider the quote from the GM advertisement appearing in Ref. [7].

> "There will be a day when cars won't make any sound, won't run on any gas, won't do anything to harm our health or environment. There will be a day when cars run by alternate fuel will be commonplace, when every highway will be filled with them, when every kid will be driven to soccer practice in one."
>
> "Yes, there will be a day when EV-1 will be no big deal but that day has not yet come."

The advertisement continues with additional comments and the following questions interspersed:

"Are you a missionary?"
"Are you a maverick?"
"Do you want to change the world?"

Read these quotes from two points of view. First, compare the issues that GM considered to be of importance and interest two decades ago with our current concerns and interests (the same concerns). It is clear that GM was 20 years ahead of its time. Second, what are the psychological appeals being exploited in this advertisement? (Figure 1.5) (As you are undoubtedly aware, almost all advertising attempts to exploit your psychological "hot buttons.")

This discussion on GM EV-1 leads us directly to "Reasons to Buy a Hybrid" in Chapter 3. The motivations for buying a hybrid fall into three categories. Of the three, two are objective reasons while the third is subjective, that is, hybrids give special concessions that some drivers value highly.

Subjective reasons to buy a hybrid depend on each individual's personal evaluations. Environmental awareness is a determining factor for many people. Having more playthings than other cars, the unique experience of driving a hybrid, augmented prestige, being a pioneer, and the recognition of advanced technology are other subjective reasons. As more hybrids hit the road, the prestige of ownership becomes smaller. As part of the unique hybrid driving experience, to achieve good mpg, special driving techniques are required. Some owners enjoy achieving good mpg more than video games.

The movie, *Who Killed the Electric Car?*, focused attention on GM EV-1 and created controversy over EV-1.

Renewed Interest in Hybrids

Global warming and energy security for the United States generated renewed interest in hybrids. In the 90 years since 1900, many new and relevant technologies have matured. The convergence of these technologies indicated advanced hybrids might be successfully designed. Improved batteries

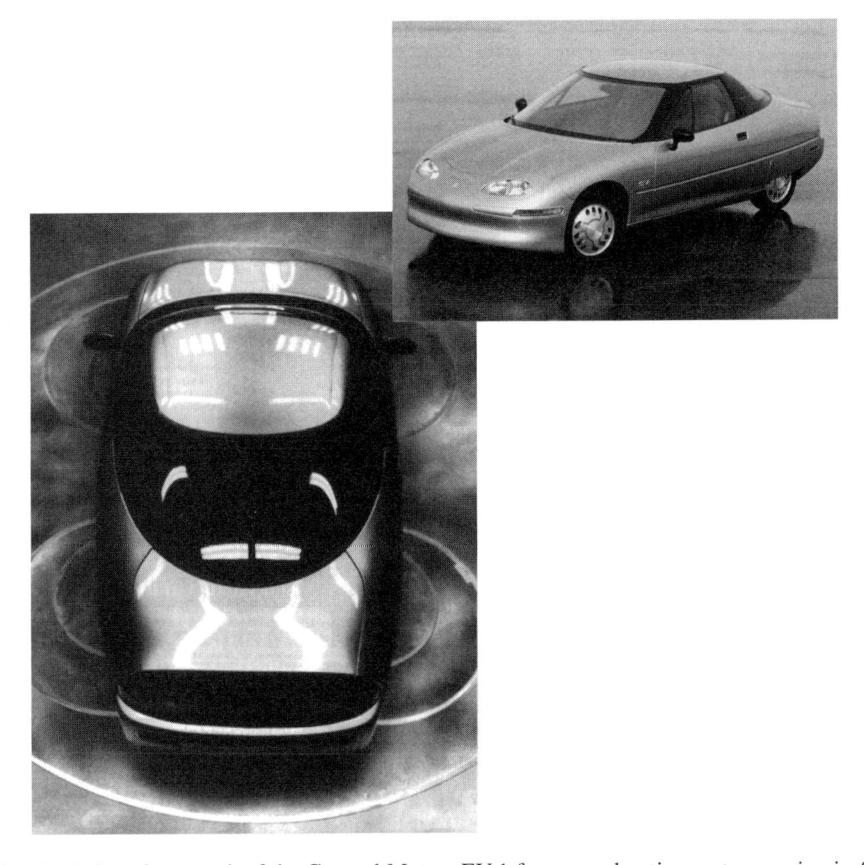

FIGURE 1.5 Debut photograph of the General Motors EV-1 from an advertisement appearing in *Scientific American*, October 1998, page 44. In this view, you are looking at the rear of EV-1. (Reproduced with permission of General Motors Corporation.)

such as nickel metal hydride (NiMH) became available. Major advances in computer hardware and software permitted use of complex control algorithms. The key elements of a hybrid are the motor and generator, or as frequently designed, a M/G device. The development of permanent magnets permits higher efficiency motors and generators (Table 1.4).

TABLE 1.4
Hybrid Solutions to Urban and Global Problems

	Need for Supercar
Urban Problems	
Air pollution	Decreased emissions
Traffic congestion	Little contribution to solution
Limited parking	Shorter vehicles (8 ft long)
Global Problems	
Global warming	Decreased emissions
Exhaustion of petroleum	Increased mpg delays time
Total number of vehicles	Impact of each vehicle on environment

One deficiency was disinterest in electromechanics, which, as the name suggests, is the science and engineering of electromechanical devices such as motors. Digital electronics, solid-state electronics, photonics, computers, and computational solutions to the equations for electricity and magnetism were the most popular topics. Hybrids have revived interest in electromechanics. Universities have once again added the course to their syllabus.

Advantages of the hybrid to national energy security were recognized. Hybrids offered a way to decrease the emission of gases causing global warming. Pollutants could be reduced by lower gasoline consumption resulting in less pollution from refineries, from distribution of gasoline, and during refueling by the customer at the gas station.

PARTNERSHIP FOR A NEW GENERATION OF VEHICLES

"Supercar" was the unofficial name for the Partnership for a New Generation of Vehicles (PNGV) program, which brought the federal government together with the domestic car manufacturers, Ford, GM, and Chrysler. Supercar was keenly supported by President Clinton and particularly the "green" Vice President Al Gore.

The lead department on Supercar was the Department of Commerce (DOC) because the program involved cooperation with industry. Funding was shared jointly by government and industry. Other departments that were involved were the Department of Transportation (DOT), the Department of Energy (DOE), and the Department of Defense (DOD).

The goal of Supercar was not to build a car that was nearly ready for production. The goal was to build prototypes and demonstrators that industry could then translate into versions suitable for production. Even though one of the goals was to build only prototypes, none of the designs showed any clear potential for production. Thus, in a sense, Supercar can be faulted for not providing a ready path to a production car.

An arbitrary goal of 80 mpg was set for Supercar. This goal was never questioned in the 9 years of the program. An overly ambitious goal led to the failure of the program. A car with 80 mpg could not be produced at a price the public was willing to pay. A price of $50,000–$100,000 far exceeded the rates for the new car market. A goal of 80 mpg, which was impossible with the expertise available in the 1990s, misdirected R&D efforts. The focus was on expensive lightweight materials and extreme aerodynamics. Also, a PNGV goal was to reduce the 10 kW cruise power at 60 mph to 7 kW by introducing

- Belly pan to reduce C_D
- Double-acting brake calipers (eliminating disk brake drag)
- Low rolling resistance tires

As discussed in Chapter 7, car weight is a perennial problem. For Supercar, composite materials were used to reduce weight. Composite materials typically have high strength fibers buried in a plastic matrix and are widely used in aircraft.

If you were to see the underside of your car in the new car showroom, you would change your decision to purchase a car. The bottom of your car is an untidy, even ugly, array of mufflers, catalysts, parking brake cables, coil springs, suspension parts, oil pans, etc. This jumbled mess of parts does not hinder component performance, but it does create high aerodynamic drag. To achieve low aerodynamic drag and high mpg, a belly pan can be added. The belly pan reduces drag but adds cost. Yet another means of reducing aerodynamic drag is to reduce the frontal area of the car. The expression, "like a barn door" is relevant here.

Supercar attracted some competent and brilliant participants. Under contract to GM, Dr. Elbert Rutan built a Supercar, the Ultralite, which is illustrated in Figure 1.6 and is discussed in Chapter 3. See also "Aerodynamic Drag; Wheel Wells, Hub Caps, and Wheels" in Chapter 12. One reason GM selected Rutan's company is because it is named Scaled Composites; as the name indicates, they

FIGURE 1.6 Burt Rutan's Supercar built under contract to GM. The hybrid by which all other hybrids should be judged. (Reproduced with permission of General Motors Corporation.)

have extensive experience with composite materials. Burt Rutan is a famous designer of aerospace vehicles including his first aircraft the VariViggen. His brother, Dick Rutan, and Jeana Yeager flew the Burt Rutan–designed, piston-powered Voyager around the world without refueling. Voyager was the first aircraft to accomplish this feat. He also designed the jet-powered Virgin GlobalFlyer that enabled a solo, unrefueled, flight around the world. He later won the X Prize for the first commercial entry into space.

Supercar did have successes. Public interest was spurred by the various concept cars. Supercar also did R&D on HEV and EV technology. Although it was acknowledged that an 80 mpg vehicle could not be produced, GM, Ford, and Chrysler made significant advances. For example, the roots of the Chevrolet Tahoe SUV Hybrid, GMC Yukon SUV Hybrid, Saturn Aura Hybrid Sedan, Chevrolet Silverado Full Hybrid Pickup, and the Saturn Vue Green Line SUV Hybrid are found in the GM concept vehicles such as Precept and Ultralite.

As an overall summary of the Supercar program, consider the following quote (from Ref. [4]):

> "Arbitrary adherence to an arbitrary goal, 80 MPG, ensured Supercar would not succeed in putting a high-mileage vehicle into production."

An unseen aspect of the Supercar program was the stimulus it provided to the European and Japanese car manufacturers to get on to the hybrid bandwagon. Toyota was able to scoop the domestic car manufacturers. The United States announced Supercar in the 1990s. Toyota's management, very perceptively, realized that 80 mpg was an unrealistic goal for a production car (Table 1.5). Toyota's goal was an achievable 55 mpg—twice (and not thrice as was the case for the Supercar) the mileage of existing cars. This was the foundation for the Prius hybrid.

The first generation Prius, which initially was sold only in Japan, received tax exemption from the Japanese government. Although the performance was not adequate for the U.S. market, the first Prius was an ingenious and realistic solution unlike the Supercar. The first generation Prius, which benefited from the experience in the Japanese market, was introduced successfully in the U.S. market. It proved to be very reliable. The second generation Prius is a superbly engineered

TABLE 1.5
Typical Supercar Specifications and Design Goals
Design Specification for a Typical Supercar

Performance Factor	Supercar	Hybrid Supercar
Mileage, mpg	80	80
Gas engine, kW (hp)	22.5 (30)	22.5 (30)
Electric motor, kW (hp)	—	45 (60)
Maximum speed, mph	80	80
Acceleration, 0–60 s	20	8
Empty weight, lb	1650	1900
Payload, lb	880	880
Payload/empty weight, %	53	46

vehicle. Toyota and the Prius benefited immensely from the price spike to $3/gal of gasoline in the United States. When gasoline increased from an astonishing $3/gal to an excessive $4 gal, the Prius became even more alluring.

As a follow-on to PNGV, DOE has a new program called FreedomCAR and Vehicle Technologies (FCVT). Much like PNGV, it is in partnership with industry. The goals are to develop advanced technology for personal vehicles as well as commercial vehicles. Long-term goals are to greatly lessen the dependence on petroleum as a fuel. Also, zero emission or near-zero emission vehicles are to be developed. Further, through reduced fuel consumption or by use of hydrogen as a fuel, CO_2 emissions are reduced. Supercar and FreedomCAR developed NiMH batteries under Advanced Battery Consortium using PNGV funding. They also developed a composite carbon fiber 60% lighter than steel with the same strength, and a new motor oil filtration method which cuts oil changes by half.

CROSSROADS 2008: SNEAK PREVIEW OF THE FUTURE

This brief discussion is an overview for Chapter 22. The discussion continues with a focus on Figure 1.7. The basis for the sketch is the fact that petroleum is limited and will someday run out of supply. Additional information concerning petroleum reserves appears in Chapter 2, which discusses the gradual decrease in oil production after peaking and the question of when oil will run out.

The fuzzy area in Figure 1.7 separates the pre-petroleum era from the postpetroleum era. Even with a clear crystal ball, events within the fuzzy area are even more fog shrouded. The years 2037 and 2042 are arbitrary and are unknown. Speculative events both going into and coming out of the fuzzy area are shown.

In the arbitrary year 2037, an estimated 1,000,000,000 (that is right, 1 billion) petroleum-fueled vehicles will be on the world's roads. See Ref. [3] which predicts 1 billion cars by 2020. These cars with gas tanks will need conversion kits to salvage the body and chassis. The kit replaces the gas engine with a battery and M/G. Downstream of 2042, the kit cars fade away. Do you remember Y2K: the scare over computer malfunctions at the turn of the new millennium? The 1 billion cars make Y2K look trivial. Governments of the world need to have solutions for the 400 million otherwise useless cars.

In 2037, gasoline will become prohibitively expensive. To reiterate, 2037 is a completely arbitrary choice. What provisions will be necessary for the citizens to cope with this crisis? The year "gasoline runs out" means petroleum will no longer be used for personal mobility. The remaining petroleum will be used for producing plastics, medicines, and similar applications.

For EV_B, central power stations are necessary. In a future with central power stations, the energy source can be nuclear, wind, solar, or wave motion. For EV_H, a vast infrastructure is necessary, which requires a long time to create the network of fueling stations, etc.

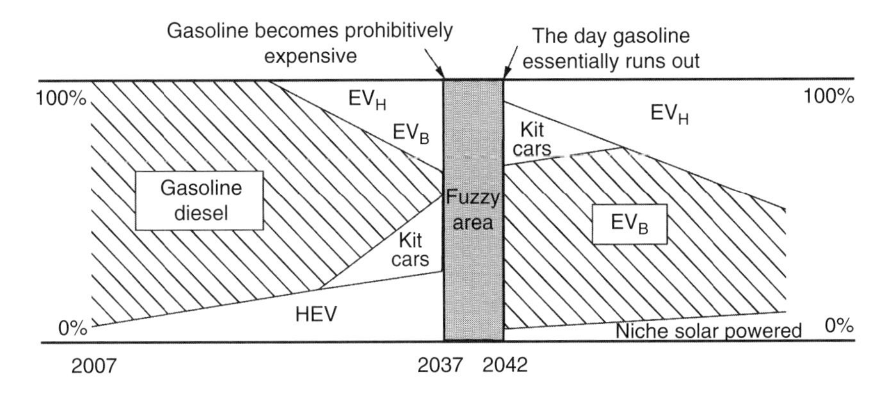

FIGURE 1.7 Shift in types of vehicles in the future as petroleum supplies diminish.

A market may develop for solar-powered EVs of the size of a scooter or golf cart. While having limited speed and payload, the niche vehicle provides personal transportation for small distances.

Any discussion of the future of ground transportation must include buses and trains. Buses and trains likely will be more important. Since hybrid technology applies to heavy vehicles, hybrid buses and hybrid trains will be more significant.

The future of oil extraction holds many surprises. As an example, what is the connection between U.S. Navy torpedo warhead research and the day oil runs out? The connection is very slight, but there is some connection nevertheless. Torpedo warheads may use a device called shaped charge that creates a jet of very fast metal—a pencil of metal moving at incredible speeds. This jet can penetrate deep into steel armor plating. Hard rock can be drilled several inches, or even a few feet, with a single shaped charge. When drilling for oil, very hard rocks are occasionally encountered; these rocks may force a decision to stop drilling because of the great expense involved. However, use of shaped charge warheads may reduce costs and allow drilling to proceed. If the oil well is not a dry hole, it adds to the pool of available oil and extends the day oil runs out.

REFERENCES

1. E. H. Wakefield, *History of the Electric Automobile*, Society of Automotive Engineers, Warrendale, PA, 1998, p. 81.
2. Staff, Wood's dual power car. *Scientific American*, 6 January 1917, p. 23.
3. J. B. White, One billion cars. *The Wall Street Journal*, 17 April 2006, p. R1.
4. J. M. German, (Ed.), *Hybrid Gasoline-Electric Vehicle Development*, PT-117, Society of Automotive Engineers, Warrendale, PA, 2005.
5. SAE Historical Committee, *The Automobile: A Century of Progress*, Society of Automotive Engineers, Warrendale, PA, 1997, pp. 292.
6. Staff, Ford Earns Green. Fast. *Motor Trend*, November 2007, p. 17.
7. General Motors Corporation Advertisement, This may not be the perfect car. It may be a critical step in creating a more perfect world. *Scientific American*, October, 1998, p. 44.
8. *Autoweek*, 27 February, 2006, p. 23.
9. R. Gehm, Ford in a hydrogen-fueled rush, *Automotive Engineering International*, Society of Automotive Engineers, Warrendale, PA, October 2007, pp. 60–61.
10. S. Birch, Spoilt of choice, the auto industry's options list includes biofuels, hydrogen, hybrids, fuel cells, and batteries—but will it be all, or nothing at all? *Automotive Engineering International*, Society of Automotive Engineers, Warrendale, PA, June 2007, p. 21.

ADDITIONAL REFERENCES

References from National Automobile Museum, Reno, NV

Woods dual power, *ATJ*, May 6, 1916, p. 231.
Woods gas electric, *Horseless*, May 15, 1916, p. 397.
Woods gas electric, *Motor Age*, May 11, 1916, p. 25.
Woods dual power, *Motor Age*, August 31, 1916, p. 48.
Woods dual power, *American Chauffeur*, July 1916, p. 324.
Woods gas electric, *Motor Age*, December 28, 1916, p. 10.
Woods dual power, *MoTor*, May 1916, p. 93.
Woods gas electric, *Auto Topics*, May 13, 1916, p. 24.
Woods gas electric going out of business, *Motor Age*, January 23, 1919, p. 9.

2 Why the Crisis? Hybrid Vehicles as a Mitigation Measure

Trust everyone, but always cut the cards.

—Mark Twain

It wasn't raining when Noah built the ark.

—Howard Ruff

INTRODUCTION

Today, automobile transportation is in a crisis with regard to the high price of gasoline. In the future, the crisis will take on unprecedented proportions. Not just apparent; overwhelming. The supply of oil will diminish and hybrid vehicles will play a major role in diffusing the situation.

In this chapter, we intend to provide a balanced view on petroleum. A balanced portrayal is necessary for intelligent decisions when faced with many alternatives. An unbalanced or biased story runs contrary to the goals stated herein. If the reader agrees with a slanted version and finds the statements in harmony with his/her feelings, then agreement is strong. If the reader does not agree, then the material is dismissed as propaganda. A broad and balanced discussion of energy appears in Ref. [1]. Relative to oil production one extreme is the alarmist position. The "sky is falling" or "cry wolf" syndromes prevail. Past predictions of depletion of oil supplies have proven false; hence impending doom seems unlikely. The other extreme is one of nonchalance. More oil can always be found. Anyhow, if new petroleum reservoirs are not found, then alternate fuels will replace oil. In the short term, the hybrid electric vehicle (HEV) (gasoline or diesel) that relies more on compressed natural gas will start the transition. Beyond natural gas is the introduction of biofuels on a larger scale with a sizeable percentage decrease (20%–30%) in petroleum usage. In midterm, fuel cells and the hydrogen infrastructure look likely. (See Figures 1.7 and 22.9. See also Ref. [21].)

A balanced discussion incorporates arguments and facts from both ends of the spectrum. The basic uncertainty in predicting the time for the peak in oil production is recognized. The unknowns, that might be called wildcards, are not dismissed. These unknowns can be favorable or unfavorable.

New buzzwords will enter the vocabulary of the average citizen. Among these are Hubbert's curve and mitigation measures. As you would expect, Hubbert's curve, which predicts the peak and decline of petroleum production, is discussed in the following paragraphs. As a tribute to his curve, it is used by both ends of the spectrum. The alarmists consider the peak as a thing of the past and thereby declare doom. The optimists consider the peak as a distant possibility and remain nonchalant. A middle path is to accept that oil production is decreasing and to think of ways to compensate for that loss. Hybrid vehicles can be one mitigation measure.

In the political arena, 30-second TV spots are used to sway voters. The spots are intended to push a hot button and stimulate the desired response. The voter is not swayed by facts. Let us consider the next four sentences in the boxes as themes for TV spots. As an informed citizen, what should be your response? This chapter does not tell you what your response should be, but provides sufficient facts for you to form an appropriate opinion.

Oil is finite and someday will run out.

When oil production peaks, the economic consequences may be much more severe than the Great Depression of the 1930s.

Global oil production is already in a state of decline—we just do not recognize it!

New technology will allow us to find new oil reservoirs as demand increases.

Two words, unprecedented and inevitable, describe the situation with regard to petroleum production. Unprecedented is correct since past oil problems in the 1970s and the early 1980s did not involve lack of supplies to meet demand. These crises were because of the withholding of oil from the market. The oil was available but was not for sale. Inevitable is correct since the amount of global oil is finite. At some future date, oil will not be available at any price. The change to an oil-free economy is not optional but mandatory.

STORY

In the early 1960s, newly elected President John F. Kennedy (JFK) directed that an interdepartmental energy study be undertaken. JFK was on the Senate Energy Committee when still a senator. When the study idea was presented to the senate for approval, one senator had three or four stacks, each stack 3–4 ft tall, of previous energy studies placed on a groaning table. The senator said "What is the Interdepartmental Energy Study going to show that hasn't already been studied?" Thus, the government has been studying energy for decades and to date has no coherent program. This is the feeling of some citizens.

In line with the previous paragraph, the gas crisis of the 1970s spawned many books on energy. One such book is *Energy, a Crisis—A Dilemma or Just Another Problem?* published in 1977 (Ref. [22]). This 30-year-old book could have been written in present times with little change. The vocabulary, the table of contents, the theme, the sense of urgency, and the recommendations for an intelligent energy policy have not changed in the last 30 years.

SETTING THE STAGE

As background consider the history and developments in railroads and in particular in locomotives. Initially, up to the 1840s or so, steam locomotives used wood as fuel. The number of freight cars for a locomotive was about four. One limit was on the amount of wood that one man, the fireman, could provide the furnace. In the mid-1800s and early 1900s, wood was replaced by coal. The source of wood, the forests, was rapidly being depleted, and this could readily be seen and comprehended. Today, the source of oil, the petroleum fields, is rapidly being depleted and this fact cannot readily be seen or comprehended. In a growing country, wood was more valuable for buildings than as a fuel. The railroads were competing among themselves. Huge quantities of wood were needed for the ties as the expansion spread across the nation. Wood burned as fuel was not available for ties. With the greater energy from coal, the fireman could supply enough energy for the locomotive to pull eight cars. Water was still needed to make steam.

The Archimedes screw was introduced to feed coal to the boiler automatically. A single steam locomotive could now pull 16 cars. Locomotives grew in size to accommodate more power and more water.

From the early 1900s to the mid-1900s, coal gave way to oil. Oil could be readily pumped into the boiler firebox. With the convenience of oil, locomotives continued to grow in size. As many as 32 freight cars could now be pulled by a single steam locomotive.

In the 1930s, diesel/electric locomotives were introduced, thus eliminating the need for water to be used. Compared to steam locomotives, efficiency was increased allowing for 64 freight cars to be pulled. Diesel fuel is absolutely essential.

The world economy will suffer when there is a mismatch between oil production and demand. Globalization will be severely handicapped by the loss of economical and efficient transportation. The domestic economy will not be immune. Think of a train stalled (without diesel fuel) in the middle of the desert with a load of perishable food.

The recounting of the technology for locomotives introduces three transitions.

$$\text{Wood} \rightarrow \text{Coal}$$
$$\text{Coal} \rightarrow \text{Oil}$$
$$\text{Steam} \rightarrow \text{Diesel}$$

Depletion of oil will introduce another transition.

$$\text{Oil} \rightarrow ?????$$

There is a vast difference between the first three transitions and the last. In the first three cases, change was optional. The change involved going from one known quantity to another known quantity. Substitution of the new for the old was motivated by economics (cheaper) and efficiency. The initial fuel had not been depleted. Three major transitions were made over a span of 150 years. In contrast, the transition from oil to the unknown is not optional but will be forced upon us. The transition is from a known fuel to an unknown substitute. Another 10–20 years are needed to define and put into production a suitable substitute. The models of the first three transitions cannot be followed. New policies and methods are essential and must be developed. The script for the latter transition is yet to be written.

What should be done about the lack of diesel fuel? Go back to steam and coal? Develop alternate liquid fuels for diesel locomotives. The knee-jerk reaction and response is biodiesel fuel.

As another example of the impact of expensive fuel ($20/gal) on economy and our normal daily lives, consider grapes from Chile. In the winter, the grapes are flown by Boeing 747 Freighter to the United States. Even with cheap jet fuel, the grapes are a luxury item. With expensive fuel costs, importing grapes from Chile becomes uneconomical. This has a ripple effect on the economy. Some of the grape pickers in Chile are left unemployed; the aircrew and ground crew of the aircraft are fired. The laundry where the B-747 captain had his/her shirts/blouses done has a drop in business. The cascading effects are obvious. That is a microscopic view.

For a macroscopic (large-scale) view, consider the fact that the world's food supply is heavily dependent on petroleum. No, we do not eat oil. With modern food production, a few percent of the population provides 100% of the food. In the early days of the United States, the population consisted mainly of farmers. At all stages, modern food production consumes energy in the form of oil. From planting, irrigation, feeding, harvesting, processing, packaging, and finally distribution to consumers, oil is needed. In addition, fertilizers, plastic wrapping, and pesticides are made from oil. The seriousness of the mismatch between oil supply and demand is, hopefully, evident [6].

The economic consequences of the demand for oil exceeding supply are discussed in Stephen Leeb's book *The Coming Economic Collapse* [17]. The title of Chapter 3, *The Collapse of Civilizations: Causes and Solutions*, along with the title of the book are a fair indicator as to the message conveyed by this book.

DEPLETION OF PETROLEUM

Globally, with growing populations and growing economies, the demand for oil is increasing [2]. J. B. Heywood [16] reports that in 2006, the worldwide use of petroleum was 80 mbd (million barrels per day), of which 53 mbd is overall transportation and 29 mbd is land transport of people. By 2025, the demand for oil is estimated to go up to 105 mbd, which is a 60% rise from 2007 [9]. With regard to supply, the U.S. Energy Information Agency states global production might reach 119 mbd [8]. About 2025, the supply of conventional oil will no longer satisfy global demand. The supply of oil available to meet demand is described in terms of a bell-shaped curve known as the Hubbert curve. A typical Hubbert curve is shown in Figure 2.1.

The Hubbert curve has a peak, or maximum, for the rate of oil extraction. The time of peaking is important. After peaking

- Rate of oil production can never increase.
- Production will decrease over time.

Peaking for the United States occurred in 1970. It is still uncertain when global peaking will occur. Peaking is likely to cause unprecedented worldwide economic problems. Because of potential major problems, the nations of the world should prepare now for the eventual peaking. Today, cooperation between nations is possible. For example, information about alternate fuels can be candidly exchanged at international conferences. The decline after peaking will see intense competition. Each country will vie for the remaining oil. Being prepared can mitigate the situation.

Peaking has been predicted for more than 100 years. It is claimed that peaking will occur in 10–20 years. Obviously, these claims have been false. Earlier predictions were no more than uninformed guesses. Today, in spite of the improved quality of geological data and the better understanding, significant uncertainty exists. This does not mean newer predictions are also wrong and can be ignored. Incorrect predictions have anesthetized the public and the governments. Peaking is like death; you do not know when it will occur, but death and taxes are certain.

Prediction of oil peaking has improved for several reasons. First, massive databases from drilling are available; however, some nations do not provide information. Second, exploration techniques

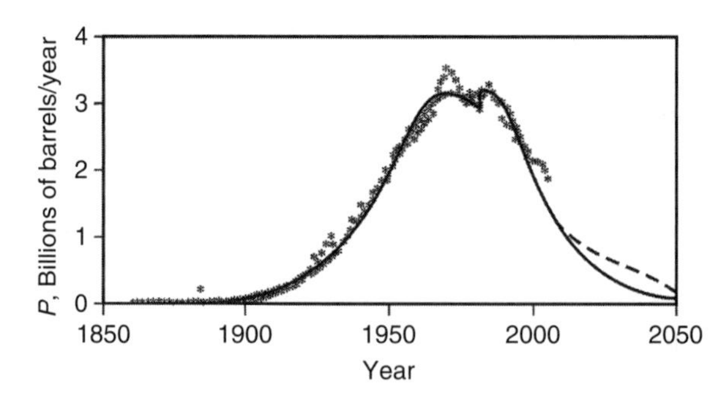

FIGURE 2.1 U.S. annual production of oil in billions of barrels from 1859 to 2005. The solid line is a fit to the data projected to 2050. The dashed line after 2010 is an estimate of total U.S. oil production if oil is taken from the Arctic National Wildlife Reserve (ANWR). (Reproduced with permission from R.J. Wiener, Forum on Physics & Society, *American Physical Society*, 35(3), July 2006.)

have advanced. (See *Introduction to Geophysical Prospecting* [5].) This is the fourth edition of the book (1988). The first edition was in 1960. Consequently, the authors of this book cover the time of new exploration methods based on ever-increasing capability of instruments and computers. Even though advances have been made in exploration, less and less oil is being found. Third, credible analysts have become more pessimistic about peaking and oil production. Even optimistic forecasts see global peaking within the next 20 years. Fourth, in the current energy and economic environment, peaking might create a disruption on a much larger scale than the previous oil crises in the 1970s and the 1980s.

The stock markets [19] believe that cheap, easily extracted, conventional oil is finite, and the day is approaching when one half will have been consumed. Several mutual funds have been created that focus on alternate energy sources such as solar, wind, hydrodynamic, biofuels, and hydrogen.

ESTIMATING OIL PRODUCTION

Estimation of oil depletion involves two different but interrelated questions. The first question is how much oil will eventually be produced? The answer is the estimated ultimate recoverable (EUR) oil and is found by assessing all oil fields in the world. Not all hydrocarbons found in the earth are the same. The answer depends on the kinds of oils that are included and those that are excluded. The second question is regarding the year that oil production rate will peak. Analysts are frustrated by the quality of data from certain regions of the world such as the Middle East and Russia.

Production from three different sources is considered: oil reservoirs in production, oil reservoirs that have been found but that are not in production, and yet-to-be-discovered oil reservoirs. The complexity can cause significant errors.

Except for abiotic oil [9,10], all petroleum is considered to be the result of geological processes millions of years ago. The word abiotic can be broken down into "a" (not), "bio" (biological source), and "tic" (for pleasing the grammarians). When the planet Earth was formed, in the swirling soup of atoms and molecules were hydrocarbons such as methane (CH_4). That is abiotic oil.

Gold's book [10] focuses on methane. Reference [18] discusses methane trapped in clathrate hydrates (see Chapter 22 for further details). The tremendous amount of trapped methane residing within continental shelves and in polar regions forms a possibly vast natural resource. Consistent with Gold's theme, methane may be an even more important source of hydrocarbons than envisioned today.

The differing means of creating petroleum and the different geology throughout the earth cause wide variations in oil reservoirs. The differences encompass depth of oil, size of oil reservoir, and the quality of the oil. Supergiant oil reservoirs (reservoirs with more than 5 billion barrels) are the easiest to find and the most economical to develop. The giant oil reservoirs have the longest producing life upwards of 50 years. The world's 370 known giant oil reservoirs contained 75% of all oil discovered up to 1985. Since 1985, few giant oil fields have been discovered [11]. There are very few giant oil reservoirs in reserve to replace those that are in decline of production.

Extensive data are available showing the increasing production, peak, and decline for individual oil reservoirs. By summing the individual oil reservoirs, the production for a region or country can be estimated. Adding together the production from the various regions or countries worldwide yields a global Hubbert curve.

Since the early days when oil seepages revealed oil below, oil has not been easy to find. A surface signature for oil has not been used for decades. Current oil reservoirs are found at great depth or in deep water. Once found an exploratory well is drilled. If successful, many production wells are drilled. With several wells, the information about the reservoir can be refined.

OIL RESERVES AND RESOURCES

Reserves are the amount of oil that can be extracted at an assumed cost. A higher price for oil means more oil can be removed. Reserves increase, however, a limit is imposed by the geology of the oil field. The increase in reserves due price is 10%–20%. Even if oil is priced at $1000/barrel, a limit exists. Reserve estimation is an ongoing process. As the reservoir is developed, new data allow refined estimates. From the same data, different analysts will estimate different reserves.

Reserves do not equal future production. Other factors influence production such as production history, position of the peak (before or after), local geology, available technology, and oil prices.

DEMAND FOR PETROLEUM

To meet demand, new reservoirs are needed to compensate for declining production from depleted oil fields (for data on demand, see Ref. [7]). The DOE Energy Information Administration states global demand will be up 50% by 2025. Without new discoveries, world production cannot meet demand. Demand for imported oil by the United States was 33% in 1973, 60% in 2006, and is estimated to be 70% by 2025.

Oil is classified as conventional or unconventional. Conventional oil is of the highest quality, is the lightest, and can be pumped from underground reservoirs with relative ease. Simply put, conventional oil is cheap pumpable petroleum.

According to Ref. [12], 95% of all oil produced so far has been conventional. Unconventional oil includes tar sands, oil shales, and oil not recoverable by today's technology. There may be unconventional oil yet to be found in Greenland, Canada, and Antarctica but not at today's prices using today's technology. The cost of oil from unconventional petroleum is likely to be astronomical [15].

Oil is limited by geology, and higher prices cannot create more oil. Recent high prices have stimulated extensive exploration. Unfortunately, more and more exploration finds less and less oil. Exploration is less successful. Figure 2.2 is particularly sobering. More oil is being used than is being found. Figure 2.2 indicates the world enjoyed a long period (1941–1984) when more oil was found than was consumed. This trend was reversed from 1985. Figure 2.2 is but one of numerous trends that suggest oil is fast approaching the inevitable peaking of production.

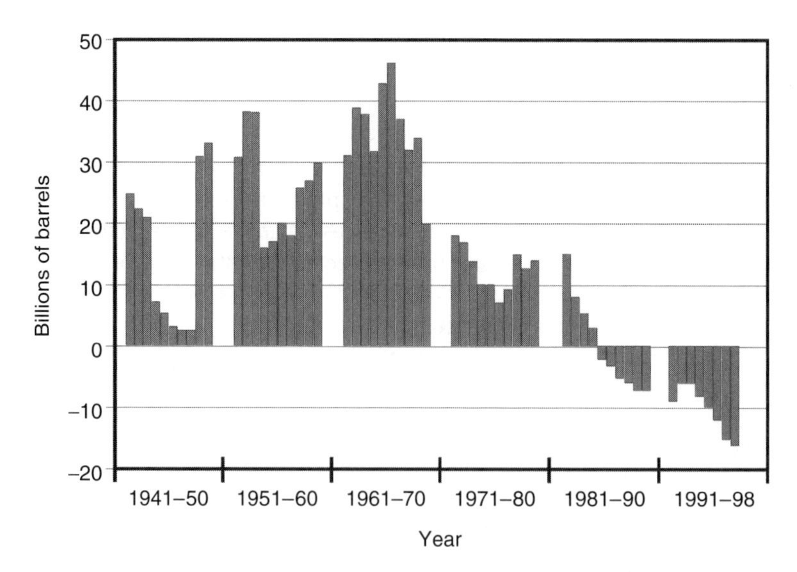

FIGURE 2.2 Net difference between annual additions to global oil reserves and annual consumption. (Reproduced from R.L. Hirsch, R.H. Bezdek, and R.M. Wendling, Peaking oil production: Sooner rather than later? *Issues in Science and Technology*, Spring, 2005, http://www.issues.org/21.3/hirsch.html. This reference written under contract to Department of Energy.)

Synergy: Gasoline/Asphalt

The combination that almost defines the word synergy is gasoline/asphalt from a barrel of petroleum. Energy (gasoline) is needed to move an automobile. Roads (asphalt) are needed for the automobile to drive upon. Both are obtained from petroleum. If petroleum usage is greatly reduced, a shortage of asphalt will likely occur.

The black sticky stuff that holds aggregates together is asphalt. The asphaltic concrete forms more than 90% of roads in the United States. Although natural deposits of asphalt occur, the major fraction is derived from petroleum. Research is under way to find adequate sources of nonpetroleum-based asphalt.

OIL IN THE UNITED STATES

Figure 2.1 is the sum of the lower-48 oil production with the giant oil field at Prudhoe Bay, Alaska. The dots are actual data. The secondary peak in the 1980s is because of North Slope coming on line. Even with Alaska oil, production has been declining for 35 years since 1970. A small temporary reversal occurred because of Alaska oil. The dashed line shows an estimate of the effect of ANWR oil on the production in the United States. About 2030, the oil production is almost doubled. However, the oil production is very small at less than 1 mbd.

The 1980s and 1990s were an era where exploration technology was at its pinnacle. Three-dimensional (3-D) seismology was practical. Horizontal drilling became economically feasible. Geological understanding was dramatically improved [5]. What could and what did high oil prices and advanced technology do for U.S. oil production? Based on the data in Figure 2.1, the answer is not much. The oil crises of 1973 and 1979–1980 caused only minor blips in production in spite of very large increases in oil prices. From 1970 to 1981, oil crude prices increased fivefold [7].

More recently, ultra-low frequency electromagnetic (EM) waves have been used to find petroleum deposits. Because of ultra-low frequency, the EM waves can penetrate to great depths. The EM waves supplement the usual 3-D seismic surveys [20].

The lower-48 can serve as a surrogate for global prediction of peaking. The United States at one time had giant oil reservoirs with the richest fields. The geology was widely varied. The oil fields were very productive. Extrapolating the U.S. experience to the global peaking question is reasonable. The dots in Figure 2.1 show how well the theoretical curve tracks the data.

The bottom line is that there is disagreement when predicting the amount of the EUR oil and the year of peaking. Some of the uncertainty has been discussed above. Other factors are summarized in Table 2.1. Even then, the following areas of broad agreement exist:

- Global oil is finite.
- Daily oil production has possibility of more growth.
- Most of the world's oil is in OPEC's control.
- Possible new oil might be found in the deep waters of the Gulf of Mexico and South Atlantic.
- More giant fields might be found in Iraq.
- Parts of the world are currently off-limits for exploration.
- Back side (negative slope) of Hubbert's curve may be more shallow than the front side (positive slope).
- Demand is key to predicting peaking.
- Production over the next 10–15 years may be constrained more by social, political, economic, and technical issues than by geological limit.

TABLE 2.1
Uncertainty of the Unknowns

Factor	Favorable	Unfavorable
Year of peaking	Predictions of early peaking are wrong	Early peaking occurs as predicted
Supergiant oil fields	A number of new are found and produce before global peaking	None are found
Middle East oil reserves	Higher than reported	Lower than reported
Oil prices	Sustained high prices stimulate efficiency and investment in alternate fuels before peaking	Prices misread; complacency and complaints prevail but without action
Fuel efficiency, the United States and Europe	New, tough standards are mandated not only for cars but for homes, aircraft, etc.	Consumers demand continuation of status quo with heavier and more hp in vehicles
Fuel efficiency, China and India	High standards are set	Efficiency ignored
Natural gas	Huge new fields found	None or little discovered
Scientific Breakthroughs	Out of laboratory and into economy, demand reduced	Misguided R&D efforts on wrong topics, money dissipated
Terrorism	Abates	Destructive to oil fields, pipelines, refineries, etc.
Political	None occurs that affects oil production	Changes hinder oil production.
Oil markets	Market signals clearly indicate need for action	Erratic signals confuse market and governments
Natural disasters	Absence or no increase	Increasing hurricanes, earthquakes, etc.
Weather	Normal or less severe	Unusual cold in winter, unusual heat in summer

HUBBERT'S PEAKING THEORY

SWIMMING POOL MODEL

As an introduction, an analogy between a swimming pool and an oil field is discussed. Pumping water from a swimming pool is extreme simplicity. Pumping oil from an oil field is extreme complexity.

Think about pumping water from a swimming pool. Consider a constant flow pump with a rate of 10,000 gal/day. In 7 days, a pool containing 70,000 gal can be emptied. The top curve in Figure 2.3, which is a swimming pool model of oil field production and reserves, shows the amount of water in the pool. The bottom curve is the pumping rate which is constant.

For an oil field, similar curves apply. Figure 2.4 has the Hubbert curves for oil remaining in the reservoirs, Q_R, and the production rate, P. Compare with the swimming pool model shown in Figure 2.3. For an oil field, days are replaced by years. Gallons of water are replaced by billions of barrels of oil. Note the curve for the remaining oil, Q_R, is rounded. Also the production rate, P, has a bell shape and a peak value. P is the negative of the slope of the Q_R curve. At the time for the peak, the Q_R curve has an inflection point. The curves for both Q_R and P have obvious symmetry, which will be exploited in later derivations. Also the curves for Q_R and P are data for oil production in the United States. Time zero corresponds to 1970. Frequently used is the cumulative oil production, Q, which is related to Q_R by

$$Q_R = Q_T - Q \tag{2.1}$$

where Q_T is the total oil in the reservoir. Normally, Q, and not Q_R, is used in discussions of oil supply.

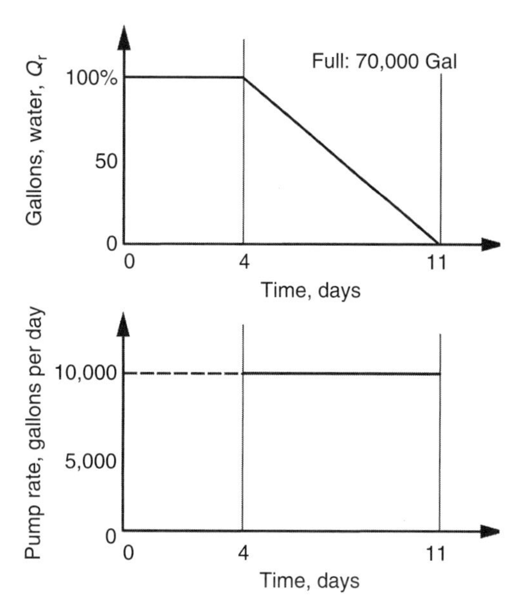

FIGURE 2.3 Swimming pool model of oil field production and reserves.

The swimming pool model is a case in contrast to Hubbert's curves. For a swimming pool, the extent of the water is evident whereas for an oil field the size and boundaries are only vaguely known in the early days after discovery. The number of pumps for the swimming pool requires little thought. For the oil field, a few hundred oil wells may ultimately be needed. A test well is drilled. If it is successful, then well number 2 is drilled, and the process continues. The number and rate of new oil wells drilled is dictated by economics; typically new wells are drilled at a gradual rate.

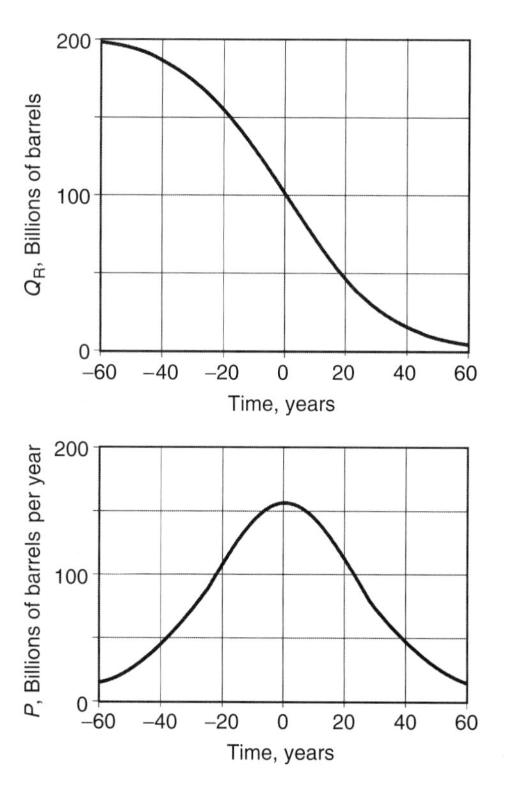

FIGURE 2.4 Oil field production and cumulative production.

The size of a pump for a swimming pool can be ascertained by knowing how many days are required to empty the pool. For an oil well, the oil must migrate to the sucking point deep in the earth. The flow of oil to the sucking point depends on geology and determines the pump rate. The spacing of oil wells depends on how big a circle a single oil well draws upon. Easy flow means a large circle and fewer oil wells to be drilled and bought.

The location of the suction hose for a swimming pool is at the deep end of the pool. For an oil field, the boundaries are not precisely known. The best locations for oil wells are gradually determined as the oil field is developed. Acquired geological knowledge is necessary for the best use of capital.

The total number of pumps is easily determined for a swimming pool. For an oil field, the determination of the total number of oil wells requires detailed geological information that slowly evolves.

Based on the comments concerning pump size, locations, and number of oil wells, the rounded nature of the curves in Figure 2.4 is evident.

Standard Presentation of Hubbert's Curves

Hubbert's curves are based on the logistics equation. An excellent discussion appears in Ref. [4]. The logistics differential equation is

$$P = \frac{dQ}{dt} = r_0\left(1 - \frac{Q}{Q_T}\right)Q \tag{2.2}$$

where

Q_T is the total amount of oil in the reservoir
Q is the amount of oil pumped from the reservoir at time t
r_0 is the initial rate factor

The ratio P/Q varies linearly with Q/Q_T as is seen in Equation 2.2.

$$\frac{P}{Q} = r_0\left(1 - \frac{Q}{Q_T}\right) \tag{2.3}$$

The linear variation is important for finding values of r_0 and Q_T. Figure 2.5 illustrates the use of Equation 2.3. When Q is zero, the ordinate-intercept is r_0. When P/Q is zero, the abscissa-intercept is $Q = Q_T$.

Integration of Equation 2.2 yields

$$Q(t') = \frac{Q_T}{1 + \exp\left[r_0\left(t_m - t'\right)\right]} \tag{2.4}$$

The symbol, t_m, is the time at which the peak in the P curve occurs. At t_m, the midpoint in the Q curve occurs. Time, t', is an arbitrary time.

$$P(t_m) = P_P = \text{peak value of } P$$

$$Q(t_m) = \tfrac{1}{2}Q_T$$

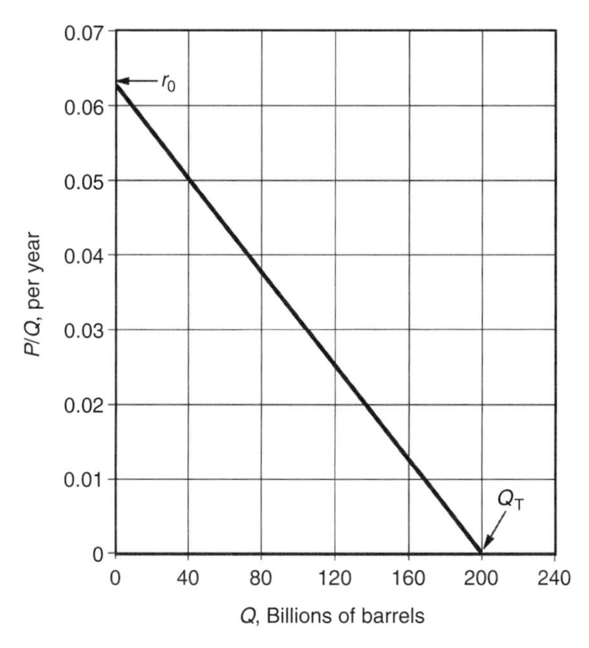

FIGURE 2.5 Illustration of the linear relationship between P/Q and Q.

The production curve, P, is obtained from Equations 2.2 and 2.4. Differentiate Equation 2.4. The result is

$$P(t') = \frac{r_0 Q_T \exp[r_0(t_m - t')]}{\{1 + \exp[r_0(t_m - t')]\}^2} \tag{2.5}$$

The use of the time, t', does not allow exploitation of the symmetry which is evident in the curves of Figure 2.4.

New Analytical Representation

Of interest is $\delta\tau = t_{peaking} - t_{today}$ which is not directly provided by use of t_m. Prediction of the year of the peak in P, known as peaking, is of great interest. $\delta\tau$ is the number of years to peaking measured from today. Since t' and t_m are arbitrary without physical significance, eliminate both by defining a new time $t = t_m - t'$. Insert t into Equations 2.4 and 2.5. Define a new variable, Q'

$$Q' = Q - \frac{1}{2}Q_T \tag{2.6}$$

Figure 2.6 shows a plot of Q'. Use of the function, Q', makes the cumulative curve an odd function. Also P is an even function as a result of putting time equal to zero at the peak. The time interval, $\delta\tau$, is also shown in Figure 2.6.

From Equation 2.4 with $t = t_m - t'$ the equation for Q' becomes

$$Q' = \frac{1}{2}Q_T \tanh\left(\frac{1}{2}r_0 t\right) \tag{2.7}$$

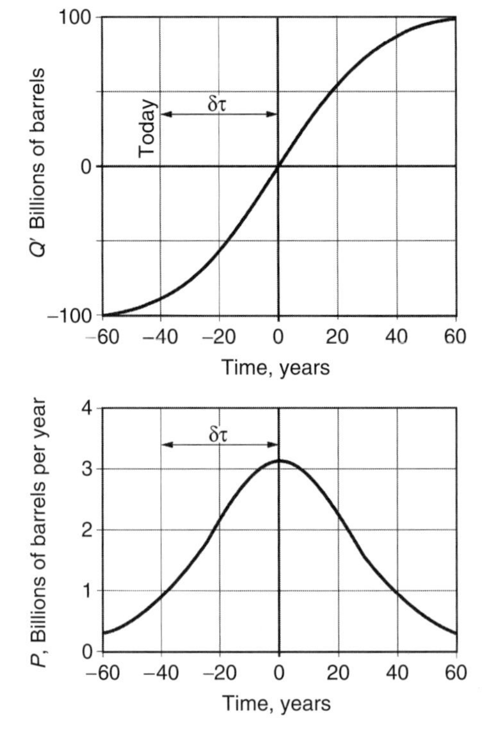

FIGURE 2.6 Symmetry of curves due to use of the function Q' and time $t = 0$ at the peak.

Combining Equations 2.6 and 2.7 gives

$$Q = \frac{1}{2} Q_T \left[1 + \tanh\left(\frac{1}{2} r_0 t \right) \right] \tag{2.8}$$

Since P is the derivative of Q or Q'

$$P = \frac{dQ}{dt} = \frac{dQ'}{dt} \tag{2.9}$$

the value of P is found using Equation 2.9 as

$$P = \frac{1}{4} r_0 \, Q_T \, \text{sech}^2 \left(\frac{1}{2} r_0 t \right) \tag{2.10}$$

The differentiation of Equation 2.9 is simple because of known properties of the hyperbolic functions. Also compare the simplicity of Equation 2.10 with the complexity of Equation 2.5.

Variation of the Production Rate, *r*(*t*)

The production rate is a function of time, t, and the defining equation is

$$r(t) = \frac{P(t)}{Q(t)} \tag{2.11}$$

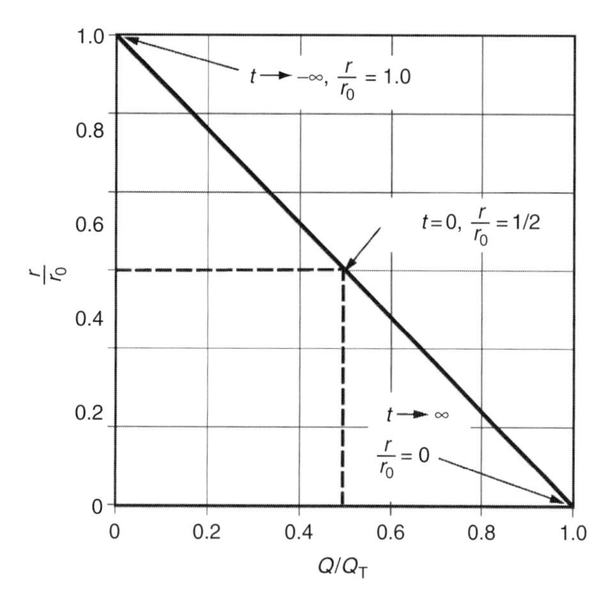

FIGURE 2.7 Plot of the production rate, $r(t)$, as a function of the cumulative oil production, Q/Q_T.

The properties of $r(t)$ are

$$\text{at } t \to -\infty,\ r(-\infty) = r_0$$

$$\text{at } t = 0,\ r(0) = \frac{1}{2}\ r_0$$

$$\text{at } t \to +\infty,\ r(+\infty) = 0$$

These values are shown in Figure 2.7. Also note that

$$\frac{r(t)}{r_0} = \frac{P(t)}{Q(t)r_0} \tag{2.12}$$

The ratio, $r(t)/r_0$, is used in Figure 2.7 as the ordinate variable. At any time

$$P(t) = r(t)Q(t) \tag{2.13}$$

As $Q(t)$ approaches Q_T late in production, $r(t)$ approaches zero. As a result, $P(t)$ falls.

Limit of P/Q as Time Approaches Negative Infinity

Figure 2.7 shows that as Q/Q_T tends to zero, P/Q becomes equal to r_0. Graphically, the result is easily accepted. A limiting process is implied. As an application of the hyperbolic functions above, the limiting process is now accomplished. From Equations 2.8 and 2.10,

$$\frac{P}{Q} = \frac{1}{2} r_0 \frac{\operatorname{sech}^2\left(\frac{1}{2} r_0 t\right)}{1 + \tanh\left(\frac{1}{2} r_0 t\right)} \tag{2.14}$$

The first step in the limiting process yields

$$\lim_{t \to -\infty} \frac{P}{Q} \to \frac{0}{0} \tag{2.15}$$

The usual procedure for indeterminate forms leads to

$$\lim_{t \to -\infty} \frac{1}{2} r_0 \frac{\operatorname{sech}^2\left(\frac{1}{2} r_0 t\right)}{1 + \tanh\left(\frac{1}{2} r_0 t\right)} = \frac{r_0}{2} 2 = r_0 \tag{2.16}$$

The mathematics yields the correct answer.

Characteristic Production Time, τ

Nothing inherent in the physics of the problem defines a unique average production rate. A reasonable assumption is to use $P_{AVG} = ½ P_P$.

$$\tau = \frac{Q_T}{\frac{1}{2} P_P} = \frac{2Q_T}{P_P} \tag{2.17}$$

Introducing an equation for P_P gives

$$P_P = r(t)Q(0) = \left(\frac{r_0}{2}\right)\left(\frac{Q_T}{2}\right) = \frac{r_0 Q_T}{4} \tag{2.18}$$

Combining Equations 2.17 and 2.18 gives the final result

$$\tau = \frac{8}{r_0} \tag{2.19}$$

Figure 2.8 illustrates the average production rate and the characteristic time, τ. For the case illustrated, $\tau = 127$ years. Within that time, 96% of all oil will have been pumped.

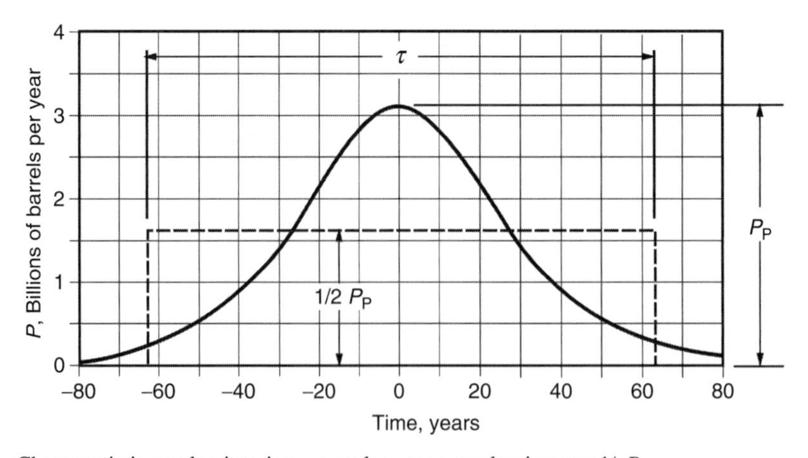

FIGURE 2.8 Characteristic production time, τ, and average production rate ½ P_P.

TABLE 2.2
Sample Data Used to Obtain r_0 and Q_T

Year	*P*, Billions Barrels/Year	*Q*, Billions Barrels	*P/Q*, per Year	Hubbert's Time, *t*, Years
1930	0.876	15.1	0.0581	−40
1940	1.448	26.54	0.0546	−30
1950	2.180	44.60	0.0489	−20

Finding r_0 and Q_T

As oil is pumped, data for Q and P are accumulated. Table 2.2 provides some sample numbers for the application here. These data are used to find values for r_0 and Q_T.

The three data points for the years 1930, 1940, and 1950 from Table 2.2 are plotted in Figure 2.9. A straight line is drawn through the three points and extended until the line intercepts the abscissa and the ordinate. The ordinate-intercept provides r_0. A value of 0.063 was obtained. The abscissa-intercept gives Q_T. A value of 203 was found, which has an error less than 2%. The values for r_0 and Q_T can be used in Equations 2.7 and 2.9 to determine $P(t)$ and $Q(t)$. Obviously, more accurate techniques such as a linear least squares curve fit to the data could be used.

Of interest is the time to peaking. This is found by solving Equation 2.8 for the time to peak, t_P. The result is

$$t_P = \frac{2}{r_0}\tanh^{-1}\left(\frac{2Q}{Q_T} - 1\right) = 4\tau\tanh^{-1}\left(\frac{2Q}{Q_T} - 1\right) \tag{2.20}$$

Subscript p suggests peaking. Further, the value for P_P can be estimated from

$$P_P = \frac{1}{4} r_0 Q_T \tag{2.21}$$

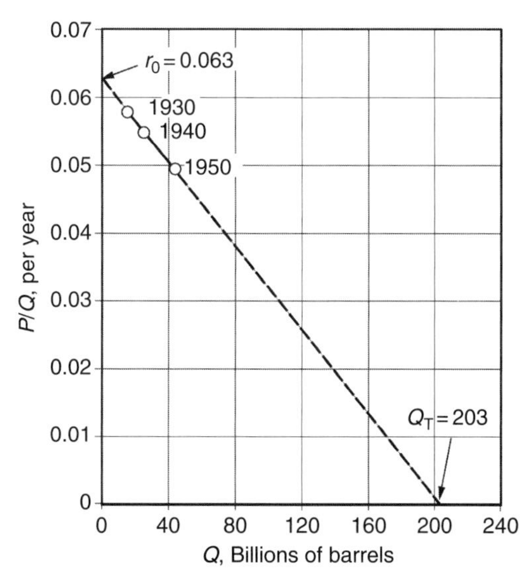

FIGURE 2.9 Plot of P/Q as a function of Q using early, partial, data for P and Q. From the plot, r_0 and Q_T are obtained.

Equations 2.20 and 2.21 complete the tools for an overall assessment of an oil field.

As the years go by and additional data are accumulated, the estimates for r_0 and Q_T are refined giving a better estimate for the time to peak, t_P. The period considered in Table 2.2 extends from 1930 to 1950, which seems like a long time. The data are for the domestic production of all oil in the United States.

When Do Hubbert's Curves Apply?

It is highly unlikely that the *P*s and *Q*s of oil production follow precisely the pure mathematics and perfect symmetry of Hubbert's curves. However, it is quite likely that oil production rate, *P*, follows a bell-shaped curve. Initially, *P* increases and attains a peak. After the peak, *P* eventually decays to zero. When the goal is to extract the oil as rapidly as possible subject to constraints, Hubbert's curves apply. The developers of the oil field intend to pump as much oil as possible in the shortest time. Their intentions are subject to geological and economical constraints. This means action on the part of the developers once the oil field has been discovered.

The pumping rate, *P*, is affected by geology of the oil field and economics. Economics implies the wise investment of capital. Geology affects the number of oil wells that can be usefully employed. Economics affects the number and timing of drilling new oil wells.

Prediction of the peak depends on knowledge of r_0 and Q_T. Any factor that affects either r_0 or Q_T alters Hubbert's curves.

When Do Hubbert's Curves Not Apply?

Obviously, if all power is turned off to an oil field the pumps stop, and *P* drops to zero. The cumulative curve, *Q*, becomes a horizontal straight line. For this extreme case, Hubbert's curves do not apply. Hubbert's curves are not some irrefutable or unavoidable law of nature that must always apply. Factors affecting the application of Hubbert's peaking theory include

- Growth of the oil reserves
- New technology
- Basic commercial factors; supply and demand for oil
- Impact of war and geopolitics

When Hubbert's peaking theory applies, as it did with domestic production in the United States, the future output can be predicted.

NEAR-TERM SITUATION

Transportable Fuels Crisis

Currently, for transportation, be it cars, trucks, trains, ships, or aircraft, liquid hydrocarbon fuels are used. These fuels offer ease of handling coupled with high energy density. When oil production peaks, alternate fuels become essential. The alternate fuels should preferably be in liquid form, but this is not absolutely essential. Note the title of this chapter is not "Why the Energy Crisis?" The title is "Why the Crisis?" Relative to transportation, today's predicament is not an energy crisis. It is a transportable (liquid) fuels problem.

Current Fleet of Low Miles per Gallon Vehicles

The fleet of vehicles that is on the road today yields poor fuel economy when compared to European and Japanese vehicles. Compared to what might have been had technology been applied differently, poor is once again accurate.

TABLE 2.3
Characteristics of Light-Duty Vehicles, for Four Model Years

	1975	1987	1997	2006
Adjusted fuel economy	13.1	22.1	20.9	21.0
Weight (lb)	4060	3220	3727	4142
Horsepower	137	118	169	219
0–60 time (s)	14.1	13.1	11.0	9.7
Percent truck and SUV sales	19%	28%	42%	50%
Percent 4WD sales	3%	10%	42%	50%
Percent manual transmission	23%	29%	14%	8%

Starting from today, if every new vehicle sold gave 40 mpg (miles per gallon), it would still require 10–15 years for the full benefit to be felt. The 10–15 years are required for cars currently on the road to be replaced. Appendix 12.2 provides an equation for calculating the fleet average mpg. Using that equation, assume that today's fleet average is 25 mpg, and that all new vehicles being sold henceforth give 40 mpg. Assume a linear transition from old cars (25 mpg) to new cars (40 mpg) occurs over 16 years. The fleet average is at 4 years (28.75 mpg), at 8 years (32.50 mpg), at 12 years (36.25 mpg), and at 16 years (40.00 mpg).

Table 2.3 provides information that is relevant to later discussion. Over a span of 30 years, significant changes have been made in vehicles.

Application of Increased Engine Efficiency

The two major factors forcing change are: (1) need for increased fuel economy and (2) reduced emissions. Great uncertainty reigns within the car makers. Research and development must include three areas: (1) gasoline engine, (2) diesel engine, and (3) hybrid technology as an ally. Hybrid technology involves electrochemistry, motors, power electronics, and control. Strides in technology come with a high unit cost.

Engine advances that improve efficiency include variable valve timing, computerized engine control, reduced friction, and various forms of fuel injection (see Chapter 17). Not only has engine efficiency been improved, the level of emissions has also been greatly reduced.

Two choices for application of the advanced engine technology are shown in Figure 2.10. One choice was to stress fuel economy by decreasing vehicle size and power. Even maintaining

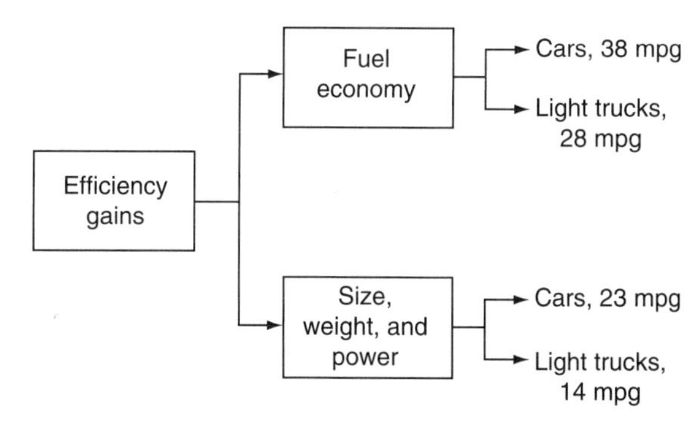

FIGURE 2.10 Two choices for application of efficiency gains with the effect on fuel economy. The car buying public wanted size, weight, and power.

vehicle size and power without growth would have been desirable [14]. The other choice was to manufacture what the car buyers wanted. According to some accounts, the automobile lobby stressed the public (voters) wanted more power, greater weight, and larger size. (Actually, lobbyists are a topic beyond the scope of this book.) Congressmen are told that an increase in Corporate Average Fuel Economy (CAFE) will cause loss of jobs for the automakers. Increased mpg is too much of a technical challenge. The Congressmen are reminded that an increase in CAFE likely will make cars more expensive and that the voters will be angry. Customer choice must be preserved. The gas crises of the 1970s were forgotten. The mpg shown in Figure 2.10 are real life values (see Chapter 12).

Vehicle Power

The general population loves raw horsepower (hp), or so it would seem. Yates [13] looked at the power for the cars on the ten best list during 20 years. Here is what he found: 1985, 130–141 hp; 1995, 192–207 hp; and 2005, 239–298 hp. In keeping with this trend, the Dodge Caravan was introduced in 1984 with a four-cylinder 101–104 hp engine. In 2007, the engine choices include a V-6 rated at 200 hp, which is almost double. Compare a Chevrolet Corvette and a Honda Accord. In 1985, the Corvette had 230 hp and in 2005 it has augmented to 400 hp. In 1985, the Honda Accord had 80–101 hp while in 2005 it offered 160–240 hp. One model of the 2005 Honda Accord had more power than the Corvette of 1985! Table 2.3 shows an increase in average power from 137 hp in 1975 to 219 hp in 2006.

Chapter 12 focuses on loss of fuel economy with increased power. Chapter 4 discusses engine downsizing and the major effect on fuel economy. Added power does lessen mpg.

Vehicle Weight

Fuel economy, mpg, can be correlated with vehicle weight. This correlation is so important that an entire chapter (Chapter 7) is devoted to the topic. Chapter 12 also provides equations for estimating loss of mpg because of added weight.

Table 2.3 indicates that average weight increased from 3220 lb in 1987 to 4142 lb in 2006—a 29% increase. Roughly, a 10% decrease in weight causes a 10% gain in mpg. If the weight had remained at 3220 lb and if new technology had been applied to gain fuel economy, the adjusted fuel economy in 2006 would have been

$$22.1 \text{ mpg } (1 + 0.29) = 28.5 \text{ mpg}$$

Looking at the adjusted fuel economy, a lower value is expected for 2006. The fact that it is not lower is because of advanced vehicle efficiency.

Table 2.3 provides information on four-wheel drive (4WD) vehicles. The percentage of 4WD vehicles has increased from 3% in 1975 to 50% in 2006. Addition of 4WD increases vehicle weight compared to 2WD. Further, the powertrain is less efficient because of added components. The mpg suffers significantly because of the combined effects of increased weight and lost powertrain efficiency.

Vehicle Size

Size and weight usually go hand in hand. In Chapter 12, the effect of aerodynamic drag on fuel economy is presented. Aerodynamic drag depends on the frontal area of the vehicle. Increased frontal-areas means increased drag, which also means fewer mpg. Sitting up high above the road (and other cars), which is one appeal of SUVs and light trucks, increases frontal area.

Volume for cargo is desirable; however, more volume usually means larger size. Large size typically results in fewer mpg.

Vehicle Powertrain

Vehicle powertrain includes those components that deliver torque from the engine to the drive wheels. Powertrain is the torque converter (automatic transmissions only), transmission, driveshaft, differential, and axle shafts.

In earlier times, automatic transmissions have been inferior to manual transmissions with regard to efficiency. The torque converter is a major culprit. In recent times, the number of speeds for automatic transmissions has increased. Six speed automatics have become common. Some luxury brands are offering seven or eight speed automatics. With increasing number of speeds, the efficiencies have also increased. Automatic transmissions rival manual transmissions. Table 2.3 depicts the decline in the number of manual transmissions at 23% in 1975 to only 8% in 2006.

A new type of transmission is the direct sequential gearbox (DSG), also known as the direct shift gearbox (DSG). This transmission is mentioned in Chapter 3. DSG has the efficiency of a manual transmission with the ease of operation of an automatic transmission.

Options and Accessories

Chapter 5 lists many automobile auxiliary and accessory components that are new or are more widely used since 1973 and that affect fuel economy. Auxiliaries in this category are

Entertainment systems	Electric valve actuation (forthcoming)
Window defrost	Electric windows
Electric brakes (coming)	Headlights and running lights
Electrically heated catalyst	Adaptive cruise control (uses radar)
Electrically heated seats and mirrors	Navigation system
Power door locks	Power sunroof

A typical luxury car has alternator power of 2–4 kW. Because of the options and accessories, mileage ratings decrease by 10%–15%.

Chapter 8 discusses air-conditioning (A/C). For some hybrids the A/C compressor is driven by an electrical motor. In other hybrids, the A/C compressor is engine driven. In heavy stop-and-go traffic, the A/C ECU can request the hybrid ECU to start the engine to obtain power for driving the A/C compressor.

Chapter 12 develops equations for the prediction of the loss of fuel economy because of use of A/C. EPA is revising their tests. The new tests and standards will be introduced in a two-phase program. In 2008, the effects of A/C will be included.

Posted Speed Limits

In the 1970s a federal speed limit was imposed at 55 mph. This limit, which was known as the double nickel, was not popular and was almost universally exceeded. Even so, fuel was saved. For either a V-6 mid-sized sedan or a V-8 large sedan, approximately 22% fuel savings resulted at 55 mph compared to 75 mph.

As a nation, the population has selected speed, power, weight, and size over fuel economy. Will oil peaking alter this trend?

Energy Conservation

Conservation can be applied to transportation. By habit or custom, some trips are made by car when walking would be better. You do not need to drive your car to visit your next-door neighbor. Over the past 50 years, sidewalks have disappeared in most California towns. Bicycles are a productive form of energy conservation. On a hot day, the engine is left running with the A/C on. Presumably human comfort overrides fuel conservation. On a cold day, the engine is left running with the heater on. Once again, human comfort overrides fuel conservation. Getting good gas mpg is a form of conservation and is discussed in Chapter 12. For a short trip, when walking is unreasonable (rainy day), is the vehicle providing the best mpg? As energy costs increase, conservation becomes more popular. How much are you willing to pay for energy?

SUMMARY

In the near term, 10–20 years, the United States is stuck with a fleet of gas guzzlers. Even if mandated by Congress, a shift to higher mpg will not be timely.

Hubbert's curves identify the existence of a bell-shaped curve for oil production. His curves provide a tool for analysts to predict when oil production will peak. The time of peaking is important. After peaking

- Rate of oil production can never increase.
- Production will decrease over time.

A new formulation for Hubbert's curves uses hyperbolic functions.

In the longer term, 20–60 years, oil production will peak. Petroleum is finite. When oil production peaks, alternate fuels become essential (see Chapter 22). The alternate fuels should preferably be in liquid form, but this is not absolutely essential. Relative to transportation, today's predicament is not an energy crisis. It is a transportable (liquid) fuels problem.

REFERENCES

1. Special Issue, Energy's future beyond carbon; how to power the economy and still fight global warming, *Scientific American*, September 2006.
2. R. B. Stobaugh and D. Yergin, Eds., *Energy Future*, Random House, New York, 1979, p. 353.
3. R. L. Hirsch, R. H. Bezdek, and R. M. Wendling, Peaking oil production: Sooner rather than later? *Issues in Science and Technology*, Spring, 2005, http://www.issues.org/21.3/hirsch.html.
4. R. J. Wiener, Drilling for oil in the Artic National Wildlife Reserve, Forum on Physics and Society, *American Physical Society*, 35(3), July 2006.
5. M. B. Dobrin and C. H. Savit, *Introduction to Geophysical Prospecting*, 4th ed., McGraw-Hill, New York, 1988, p. 867.
6. N. Church, Why our food is so dependent on oil, 1 April 2005. Available at http://www.energybulletin.net/print,php?id=5045.
7. C. Glover and C. E. Behrens, Energy useful facts and numbers, *Congressional Research Service, Library of Congress*, November 2004.
8. S. Andrews and R. Udall, Oil prophets: Looking at world oil studies over time, Association for the Study of Peak Oil, ASPO Conference, Paris, France, May 2003.
9. Hubbert Peak Theory, Wikipedia, Last modified December 2006. Available at http://www.wikipedia.org/wiki/Peak_oil.
10. T. Gold, *The Hot Deep Biosphere*: *The Myth of Fossil Fuels*, Copernicus Books, Springer-Verlag, New York, 2001, p. 243.
11. L. F. Ivanhoe, Exports—the critical part of global oil supplies, Hubbert Center, Colorado School of Mines, *Newsletter* # 2000/4–1, October 2000.
12. Staff, The Hubbert peak for world oil, December 2006. Available at http://www.hubbertpeak.com/summary.htm.
13. B. Yates, H-bombs, WMP, and girls in motorsports, *Car and Driver*, March 2005, p. 28.
14. Staff, Fuel economy; why you're not getting the MPG that you expect, *Consumer Reports*, October 2005, pp. 20–23.
15. C. Hoffman, Oil sands tapping North America's vast reserves, *Popular Mechanics*, March 2007, p. 70.
16. J. B. Heywood, Fueling our transportation future, *Scientific American*, September 2006, pp. 60–63.

17. S. Leeb, *The Coming Economic Collapse*, Warner Business Books, New York, 2006, p. 211.
18. W. L. Mao, C. A. Koh, and E. D. Sloan, Clathrate hydrogen under pressure, Feature Article, *Physics Today*, October 2007, pp. 42–47.
19. N. Bullock, Betting on the sun, *Smart Money*, April 2007, pp. 80–83.
20. R. Cunningham, Wildcats in the Orphan Basin, *The Lamp*, Exxon Mobile, 88(2), 2006, pp. 1–4.
21. S. Birch, Searching for fossil fuel alternatives, *Automotive Engineering International*, Society of Automotive Engineers, March 2007, pp. 56–58.
22. J. S. Doolittle, *Energy, A Crisis—A Dilemma or Just Another Problem?* Matrix Publishers, Champaign, IL, 1977, p. 313.

3 Overview of Hybrid Vehicles

The future of hybrid vehicles rests on the twin pillars of enhanced fuel economy and lower emissions.

INTRODUCTION

Throughout this book, "motor" always refers to the electric motor or motor/generator (M/G); "engine" always refers to the gasoline or diesel engine and never the M/G; and "hybrid" refers to any hybrid vehicle.

What is a hybrid? A hybrid combines two methods for the propulsion of a vehicle. Possible combinations include diesel/electric, gasoline/flywheel, and fuel cell (FC)/battery. Typically, one energy source is storage, and the other is conversion of a fuel to energy. As discussed in more detail in Chapter 4, the combination of two power sources may support two separate propulsion systems. This is true for the parallel design hybrid.

APPROACHES TO BEAT THE HIGH COST OF GASOLINE

Hybrids are only one approach. The alternatives to hybrids are better than expected. Currently, the hybrids on the road are hybrid electric vehicles (HEV). Other vehicles that offer high miles per gallon (mpg), either now or in the immediate future, are clean diesel engines and higher-efficiency gasoline engines. Chapter 17 discusses clean diesel engines and higher-efficiency gasoline engines. In the more distant future, vehicles using hydrogen in an internal combustion engine (ICE) and hydrogen in FCs will be available. For driving hints for better mileage, refer to Chapters 5 and 12.

FCs are high on the priority lists of the government as well as car manufacturers. FCs signal the independence from petroleum as an energy source. The fuel cell vehicle (FCV) will have zero CO_2 emission, which is important to reduce global warming. Other emissions will also be low. The FCV will likely be a hybrid with a fusion of FC and battery.

Although ethanol and other alternate fuels have an influence on mpg, the change is not as great as the switch to hybrids. Since diesels are likely to play a major role in future personal and noncommercial transportation, biodiesel is important. Biodiesel is a product of soybeans and other organic materials collectively known as biomass. The leftover material from the soybeans can be used as animal feed. To generate methane, biomass relies on the decomposition of organic material. Methane from manure is an example. A process known as gas-to-liquid can also create fuel for diesels. Natural gas is converted to a liquid and is a temporary solution.

Alternate fuels are important for energy availability. The energy security of the United States will likely depend on alternate fuels. Brazil has developed alternate fuels and does not import petroleum.

SAVING GASOLINE

The national energy goal is to reduce both oil imports and CO_2 emissions. Does hybrid technology applied to gas-hogs yield greater reductions than when applied to fuel-sippers? The answer is gas-hogs. The annual saving of gasoline in gallons, S, is given by the following equation:

$$S = \frac{M}{G_C}\left(\frac{f}{1+f}\right) \tag{3.1}$$

where

M is the miles per year driven
f is the fractional gain in mpg because of hybrid
G_C is the mpg of the conventional vehicle
G_H is the mpg of the hybrid

This equation is derived in Appendix 3.1.

For a sample calculation, the same value for $M = 12{,}000$ mi/year is used. Values for the SUV and pickup truck (the very definition of gas-hog) are $G_C = 18$ mpg and $f = 20\%$. For the gas-hogs, $G_H = 21.6$ mpg. The gasoline saved annually is 111 gal. Values for the small fuel-sipper are $G_C = 36$ mpg and the same $f = 20\%$. The gasoline saved annually is 56 gal. Hence, the gas-hog saves 55 gal more than the fuel-sipper.

Even if $f = 40\%$ for the fuel-sipper, which is twice as large and twice as favorable as for the gas-hog, the gas-hog saves more gasoline compared to the fuel-sipper. For a small fuel-sipper, $G_H = 50.4$ mpg. For the fuel-sipper, the gasoline saved annually is 95.2 gal as compared to 111 gal for the gas-hog—a considerable difference of 15 gal. Although gas-hogs may not be in favor, for these vehicles, a small improvement (even with $f = 20\%$, not 40%) in mpg yields a big increment in gasoline saved. This information is relevant to the cases of the Chevrolet Silverado mild hybrid pickup and the GMC Sierra mild hybrid pickup discussed later.

Hybrid Performance

Performance is an all-encompassing word that includes many factors such as vehicle dynamics, acceleration time from 0 to 60 mph, ride comfort, the braking distance to stop from 60 mph, and the turning radius. For hybrids, performance is measured by mileage and distance covered in electric-only mode. Another important factor is reduced emissions. In this chapter, performance is merely stated or reported. It is later elaborated upon in the remainder of this book. For information on the technical basis of mpg, refer to Chapter 5.

Weight is a major impediment to performance. Chapter 7 discusses loss of mileage because of extra weight. Hybrids, by their very nature, are more heavy than conventional vehicles. Because of this, hybrids start at the bottom of an "mpg hole." To improve mileage, the hybrid must first come out of the mpg hole. Certain benchmarks are used to specify performance. Two measures of acceleration are the time taken from 0 to 60 mph and the speed at the quarter mile distance. On the international scene, the reported acceleration time is 0–100 km/h (0–62 mph).

Allure of Plug-In Hybrids

The beauty of the gasoline/electric or diesel/electric hybrids is that the battery is charged as the vehicle is driven. The gasoline/electric and diesel/electric hybrids are jointly designated as HEVs. The irksome chore of recharging the battery is avoided. The driving public is fickle. A decade or so ago, pure electric vehicles (EVs) were shunned for the very reason that the EV needed to be recharged.

Now that the HEV is gaining popularity, people want to add a larger battery to an HEV to get 80, 90, 100, or 200 mpg. The more energetic battery creates a class of hybrids known as the plug-in hybrid electric vehicle (PHEV). A plug-in hybrid, as the name implies, requires plugging in to an electrical outlet for recharging. Without recharging from an external source, the promise of 80, 90, 100, or 200 mpg is not realized.

How far can a plug-in go in electric-only mode with a large battery? The answer can be found in data for the GM EV-1 which had an E_B (battery energy) of 16.2 kWh. With this battery, a distance of 145 km can be covered. From these numbers the energy per kilometer equals 16,200 kWh/145 km = 112 W h/km. For a PHEV, assume the allowable change in state of charge (SOC) is 25%. Assume that

E_B = 10kWh for the PHEV. For a PHEV the same weight and size as the GM EV-1, the electric-only range is R = (10kWh)(0.25)/(112W h/km) = 22km = 13mi. If SOC is 100%, R = 88km = 52mi.

PHEV will most likely use lithium-ion (Li-ion) batteries to lessen the weight gain of a large battery. However, Li-ion batteries are expensive. Toyota and Honda feel that plug-ins are troublesome. In 2006, Toyota indicated a PHEV was being developed. The vehicle must be plugged in, which is a big chore. The much larger battery pack adds weight and significant cost ($8,000–$12,000). To deliver performance that matches the promises, more discharge must be allowed, which reduces battery life. Aftermarket plug-in kits are available (Chapter 20). Daimler/Chrysler is working on a small fleet of Dodge Sprinter van plug-ins; the fleet is for demonstration purposes only. For more information on plug-in hybrids, refer to Chapter 11. Web sites, such as www.news.com, contain considerable information on plug-in hybrids. At press time, a commercial Web site, www.edrivesystems.com, provides information on the modification of Prius II from HEV to a PHEV.

Conversion or Clean Sheet Design

Only three production hybrids start with a "clean sheet of paper" design. These are the Honda Insight and the Toyota Prius I and II. All other production hybrids are conversions (modifications) of an existing product line. An example is the Honda Civic and the Honda Civic hybrid. When hybrid production starts with an existing car design, hybrid components may increase braking distance, decrease storage space, increase weight, and affect handling characteristics, when compared to non-hybrid cars. During the conversion to a hybrid, design changes are necessary to offset the effects of added weight. The potential customer should check and verify these design changes.

Conversion or Clean Sheet Design: Impact on AWD and 4WD

As discussed in Chapter 11, the torque split between the front and rear wheels depends on the conventional production all-wheel drive (AWD) or four-wheel drive (4WD) vehicle being converted to a hybrid AWD. The torque split front/rear (F/R) is usually best with 50/50 F/R or with a bias of 30/70 F/R on the rear wheels. Too much torque to the front wheels may overload front tires which affects handling capability.

As an example of the torque split between F/R wheels, the Kia KCD-II Mesa concept SUV shifts torque from 100% rear drive to 50/50F/R under certain road conditions. In line with the discussion above, the maximum torque to the front wheels is limited to less than 50% [1].

Making AWD is relatively easy once a commitment is made to create a hybrid. One set of wheels, front or rear, can be driven by an M/G. The drive shaft, differentials, and transfer cases are eliminated. A bigger battery, at least one more M/G, and a more complex control system are added.

The customer should ask when buying an SUV hybrid or other AWD hybrid the proportion of torque split front to rear. Details are given in Chapter 11.

Current View on Hybrids

Hybrids are a somewhat unanticipated commercial success spurred by $3–$4/gal cost of gas. Some car manufacturers are in a speed-up, catch-up, mode. Hybrids on the market span from micro, mild, to full hybridness. A micro hybrid has a very small motor compared to the gasoline engine. A mild hybrid has a motor with power about 15%–25% the power of the engine. A full hybrid has a motor with power about 40%–50% the power of the engine.

Mild hybrids are gaining in popularity with extra features offered such as 110V outlets for running power tools. Mild hybrids yield slightly reduced fuel consumption with lower emissions. Compared to the conventional vehicle (CV), mild hybrids have a small price premium. To further decrease costs of the mild hybrid, the possibility of using ultracapacitors (UC) to replace either nickel metal hydride (NiMH) or Li-ion batteries is being investigated. Most of the mild hybrid capability (regenerative braking, idle-off start–stop) would be retained. For cost reasons, the mild hybrid-based UC uses existing lower voltages, that is, 14 V [18].

In 2007, in the United States, 1 out of 50 cars sold was a hybrid. Annual sales were about 330,000. By 2011, 1 out of 20 cars sold will be a hybrid. In 2011, estimated sales of hybrids will be 750,000 and the number of different hybrids on sale will be 75. Other estimates state that hybrid sales will be 900,000 in 2013. Reference [13] provides yet another set of estimates for global sales. In 2007, global sales of hybrids were 385,000. By 2010, this will reach 1,100,000 and will touch 2,000,000 by 2015. In the next 15–20 years, hybrids face stiff competition from advanced ICE. Clean diesels will compete with hybrids. Likewise, newly developed engines such as the homogeneous charge compression ignition (HCCI) engines put gasoline-powered cars on par with HEVs. For more information on the HCCI see Chapter 17.

Reasons to Buy a Hybrid

The first buyers of hybrids were enamored by the novelty of the new technology. Today's hybrid customers are more interested in the tangible benefits such as higher fuel efficiency, reduction in emissions, and reduced dependence on foreign oil. Apart from this, the intangible benefits described in the following paragraphs influence today's hybrid customer. The motivations for buying a hybrid fall into three categories. Of the three, two are objective (tangible) reasons, and the third is subjective (intangible), that is, hybrids give special concessions that some drivers value highly.

Objective reasons are those that can be quantified by a number. Fuel economy, that is, mpg and the savings on gasoline, is a leading factor. Higher mpg provides greater range on one tank of gas. Lower fuel reserve requirements are possible because of better fuel economy. When the red fuel warning light comes on, the vehicle is beginning to use its fuel reserves. The fuel reserve is determined from

$$\text{Fuel reserve} = (\text{comfortable distance to next gas station})/\text{mpg}$$

For a hybrid, typical fuel reserve numbers are 80 mi/40 mpg = 2 gal. For an SUV, reasonable numbers are 80 mi/20 mpg = 4 gal reserve. The reserve fuel is added weight, which is hauled around only to prevent embarrassment and inconvenience.

Other objective reasons include lower emissions, that is, driving green. Some hybrids offer improved acceleration. The hybrids emphasizing mpg are usually smaller and easier to park.

Subjective reasons to buy a hybrid depend on each individual's personal evaluations. Environmental awareness is a determining factor for many people, but also, having more playthings than others, the unique experience of driving a hybrid, augmented prestige, being a pioneer, and the recognition of advanced technology are other subjective reasons. As more hybrids hit the road, the prestige of ownership becomes smaller. As part of the unique hybrid driving experience, to achieve good mpg, special driving techniques are required. Some owners enjoy achieving good mpg more than video games.

Buying a hybrid promotes something meaningful in this confused world. A hybrid is an anti-establishment symbol, the ultimate antistatus symbol, and a political expression. A hybrid is the ultimate feel-good vehicle.

Special concessions are not to be ignored. In many states, the hybrid is car-pool-lane legal even when there is only one driver. For some drivers, accessibility at metered exits and on-ramps is even more valuable. At some hotels and municipal parking lots, the parking fee is waived. Of course, having tax exemptions from both the state and the federal government is another incentive to buy. If you think income tax rules are complicated then—no surprise—hybrid tax credits are no less complex and daunting.

Hybrids are good vehicles for commuting. The typical driving cycle (stop-and-go, crawl-and-accelerate) favors hybrid. If the primary reason to buy a hybrid is commuting, consider a full hybrid which provides electric-only propulsion.

"More Playthings than You"

A powerful marketing tool is the materialistic appeal of having more playthings than anyone else. Examples include having more buttons on the steering wheel, having bigger (22 in. diameter) wheels with more sparkle (lots of chrome), and having more horsepower (hp). Compare the Bugatti Veyron at 1001 hp (expensive $1,300,000) and the Prius at 110 hp (sensible $25,000). In the same vein as more hp is greater acceleration resulting in lower 0–60 mph acceleration time. Compare once again the acceleration time of 2.5 s from 0 to 60 mph for the Bugatti Veyron with the acceleration time of 10.5 s from 0 to 60 mph for the Prius.

Reasons Not to Buy a Hybrid

For superior mpg, special driving techniques are necessary. Avoid speeding and drive in the slow lane. Avoid jackrabbit starts, apply light pressure on the gas pedal. Do not step on gas to get through a green light just about to turn red. Anticipate stops and deceleration to put more juice back in battery by regenerative braking (see Chapter 9).

Hybrids have a price premium. For the extra costs, see Chapter 15. There is a long waiting time for some hybrids; the time is shorter if you are willing to accept the color and set of options of a car that a dealer might have in stock. The logic with regard to mpg is that small cars, CV or HEV, have better fuel economy. For this reason, many high mpg hybrids are small. Tall people may be uncomfortable in a small hybrid. People with lots of luggage may find a small hybrid inconvenient.

Converting an existing production vehicle to a hybrid may add 100–200 lb (45–90 kg). The effect of the extra weight on braking distances was discussed earlier in this chapter. The brake pedal feel of regenerative braking is good but usually not as good as the non-hybrid version.

Besides adding weight, the battery usually reduces space in a hybrid. A safe place for the battery is between the rear wheels. Approximately 3 to 5 cu ft of luggage space can be lost to accommodate the battery. Some sedans have flip-down rear seats or pass-through doors that allow hauling long objects in the cabin. Because of the battery, the flip-down rear seats or pass-through doors may be lost.

Diesel/Electric Hybrids

The diesel/electric hybrid offers better fuel economy with reduced emissions compared to the gasoline/electric hybrid. Diesel engines are more expensive. The electrical components of a hybrid add even more to cost. Consequently, diesel/electric hybrids will be really expensive. Cost is a serious drawback. The torque of a diesel is usually high at fewer revolutions per minute (rpm), which does not match the motor torque as well as a gasoline engine does. Making the diesel/electric hybrid affordable is an R&D goal.

Role of Design in Buying Decision

Advanced technology does not automatically exclude compelling good looks or a persuasive driving experience. A winning hybrid must have a combination of looks, superior drivability, and advanced technology. As the number of available hybrids increases, the car buying public will have a wide array to choose from. Some outstanding elements are needed to distinguish one hybrid from another. Styling and design will provide the necessary "wow" factor.

Vehicle design affects subjective feelings even when buying a hybrid. The Toyota Prius hybrid shouts "I'm a hybrid!" for the Honda Civic hybrid, and many other hybrids, you need to look for a chrome badge on the trunk that states hybrid. Design involves many other facets other than identity as a hybrid.

CONCEPT HYBRIDS AND ELECTRIC VEHICLES

Different Types of Concept Hybrids

Concept cars are made to illustrate a manufacturer's prowess, to advertise, and to assess public response. Some concept vehicles are duds and others win the best-of-show award. Concept vehicles include gas/electric, HEV; diesel/electric, HEV; and hybrids other than gas or diesel. The last category of vehicles use flywheels instead of batteries (see Chapter 16).

The Partnership for Generation of Vehicles (PNGV) program, also known as Supercar, spawned several concept cars (see Chapter 1). A few of the concept cars from PNGV are presented below. Also PNGV technology such as the R&D leading to the GM/Chrysler/Mercedes Benz two-mode full hybrid architecture is discussed.

Audi Metroproject Quattro

This plug-in hybrid concept car uses the Audi A1 platform. A turbocharged four-cylinder engine drives the front wheels. The engine is teamed with a 30 kW (40 hp) motor which drives the rear wheels. Compared to a similar AWD vehicle, the hybrid is expected to yield a 15% increase in mpg. A lithium-ion battery provides a maximum range on electric-only of 100 km (62 miles). The battery can be recharged by plug-in or by the gasoline engine. Of course, regenerative braking is used. Performance is good with 0 to 100 km/h (62 mph) of 7.8 seconds and a maximum speed equal to 200 km/h (124 mph).

BMW Connected Drive X5 Hybrid SUV

This concept vehicle is a mild hybrid. The V-8 engine provides torque of 1000 N·m (740 ft·lb_f) at 1000 rpm. The electric motor gives 660 N·m (485 ft·lb_f) torque at the same rpm. The combined torque provides a decrease of 30%–40% in the acceleration time from 0 to 60 mph with a 15% increase in mpg. Instead of batteries, capacitors are used to store energy. The capacitors have a storage energy of 200 W h. Prius II batteries store 10 times more energy. Maximum transient power from the capacitors is 350 kW (467 hp) with short durations (5–6 s) of 35 kW (47 hp). The capacitors weigh 65 kg (143 lb).

BMW X6 ActiveHybrid Sport AWD Coupe

The X6 has a body shape of a coupe but has four doors. Also this vehicle has features found in an SUV, namely ground clearance and AWD. The Global Hybrid Corporation sponsored jointly by BMW, GMC, and DaimlerChrysler developed the dual mode hybrid system. The dual mode hybrid system is used with a V-8 engine. Fuel economy is improved by an estimated 20%. The key feature of the dual mode hybrid system, which is lacking in competitive systems, is the four fixed gear ratios overlaid on the electronically variable transmission (EVT).

Cadillac Provoq FC Hybrid

This is a concept car—the Cadillac Provoq (as in provocative). It is a cross over vehicle and is a fuel cell hybrid. The hybrid uses a lithium-ion battery with 9 kWh energy and 60 kW (80 hp) power. The fuel cell is a GM fifth generation with continous power of 88 kW (117 hp). Power to the front wheels is 70 kW (93 hp), and power to the rear wheels is by means of two 40 kW, in-hub, rotors giving a total of 80 kW (107 hp). Acceleration of is 0–60 time of 8.5 sec. Maximum speed is 100 mph (160 km/h). Range due solely to stored H_2 is 450 km (280 miles), and the range is 32 km (20 miles) from the energy in the battery. Two tanks store H_2 at 10,000 psi (680 bar, 69 MPa) with mass of 6 kg (13.2 lb) hydrogen. A solar panel on the roof powers audio and interior lights.

Chinese GAC A-HEV Sedan

The sedan, which was shown at the 2008 Beijing Auto Show, is reportedly a diesel-electric hybrid in one version. Another version uses fuel cells. With advanced electronics, the concept cars have steer-by-wire, drive-by-wire, and brake-by-wire.

Chinese Great Wall Gwkulla Concept Electric Car

This pure electric vehicle is small being approximately 3 feet (1.0 m) wide and is similar in appearance to the Smart ForTwo cars [20]. Intended for urban transportation, the Gwkulla has a top speed of 40 mph (66 km/h).

Chinese Plan 863 for Hybrids and Alternate Fuels

Plan 863 is a 5-year, high-technology, plan started in 2001 to focus the Chinese automobile companies on alternate powertrains and alternate fuels. The plan is now producing numerous concept cars. Remarkable local know-how is emerging in full hybrids, fuel cells, Li-ion batteries, compressed natural gas, and hydrogen as fuel [19].

Chrysler Citadel

The Citadel is for the market segment that demands ample cargo room while providing the driving passion of the Chrysler 300 M. A V-6 powers the rear wheels. An M/G powers the front wheels. The result is a vehicle with AWD. From the data on engine and motor power, the hybridness (H) is 19%. This value makes the Citadel a mild hybrid.

Chrysler Eco Voyager FCV

Hydrogen is feed to a PEM fuel cell to provide power for a 200 kW (267 hp) motor. The powertrain is stored beneath the cabin floor. Vehicle mass is 2750 lbs (1250 kg). Hydrogen storage is at 10150 psi (700 bar). Design goals are a range of 300 miles and a 0–60 time of less than 8 s. The car is a hybrid since a battery is used. The body is a one-box design; another description is an egg-shaped minivan.

Citroën Airscape Hybrid

This concept convertible uses a diesel engine with a reversible starter/alternator which is an M/G. Of special interest is the use of supercapacitors to store electrical energy. The mild hybrid design allows regenerative braking. The air-conditioning (A/C) compressor is driven by an electrical motor.

Citroën4 Hybrid Sedan

The Citroën4 hybrid sedan is designed to be superefficient. It is a diesel/electric hybrid with an estimated 82 mpg.

Citroën C-Cactus Hybrid Sedan

The diesel engine provides 52 kW (70 hp) and the motor gives 22 kW (29 hp). Hybridness is 30%. Anticipated fuel consumption is 2.0 L/100 km or, equivalently, 81 mpg. The vehicle has a ZEV (electric-only) mode for city driving. Maximum speed is 150 km/h (93 mph).

Diahatsu Hybrid Vehicle Sports: Sports Car

The Diahatsu hybrid vehicle sports (HVS) has a 1.5 L, 103 hp (77 kW) engine. The front M/G is 50 hp (38 kW) while the rear M/G is 27 hp (20 kW). The vehicle is projected to deliver 60 mpg. The HVS can cover the quarter mile in less than 17 s.

Dodge ESX3 PNGV Sedan

The Dodge ESX3 integrated almost all the PNGV technology developed by Chrysler into one concept vehicle. The body had advanced PNGV body technology that uses plastics. The vehicle incorporated PNGV chassis technology with new PNGV aerodynamics technology. The drag coefficient, C_D, was 0.22. Typical C_D ranges from 0.26 to over 0.34. A periscope mirror was used. External mirrors cause aerodynamic drag as well as noise. The engine was a 1.5 L diesel. The battery was a 150 V Li-ion. The motor size made this diesel/EV a mild hybrid. Dodge used the catchy word *mybrid* to replace mild hybrid.

Dodge Powerbox SUV Hybrid

The Powerbox evolved from the 1999 Dodge Power Wagon concept. The vehicle is powered by a supercharged V-6 187 kW with an estimated 25 mpg, which is 60% better than a Dodge Durango. Acceleration time from 0 to 60 mph is 7 s. The front wheels are driven by an M/G 52 kW (69 hp). The electric traction motor has no connection electrically or mechanically with the V-6 engine. So, it is a through-the-road hybrid. The road effectively transmits torque between rear and front wheels.

Dodge Zeo EV

This is a 2 + 2 sport wagon. An advanced Li–ion battery supplies energy for 250 mi of driving on a single charge. The range of 250 mi seems optimistic. Acceleration time from 0 to 60 mph is 6 s. The sports wagon body has four doors for four passengers.

Fisker Hybrid Luxury Sedan

The four-door sedan is a plug-in hybrid with 50 mi (80 km) of electric-only driving. The estimated mpg is 100 with a total range of 620 mi (992 km). The engine for range extension is either diesel or gasoline. The anticipated price is $80,000. The sedan will go on sale in late 2009 or early 2010.

Ford Edge with HySeries Drive

The Ford Edge is a cross-over vehicle which has been modified as a concept vehicle offering a fuel cell/battery hybrid configuration. The unusual aspect is that it is a plug-in hybrid. Just as the plug-in HEV has increased range and fuel economy due to the plug-in feature, likewise this Edge also has increased range and fuel economy for the same reason.

Ford Fiesta Hybrid

The Fiesta belongs in the same size category as Honda Fit, Nissan Versa, and Chevrolet Aveo. The CV version is ready for production. A mild hybrid (regenerative braking, motor assist, and integrated starter/alternator) forms part of the future plans for Fiesta.

Ford Focus FCV

This modification of a production car combines the FC with an NiMH battery creating a hybrid vehicle.

Ford Focus Hydrogen–ICE Conventional Vehicle

This vehicle is not a hybrid. Some features (state-of-the-art 2008) of the Ford Focus ICE, hydrogen-fueled vehicle include a 2.3 L, four-cylinder, supercharged, intercooled, engine with a four-speed automatic. The car mass is 1452 kg (3200 lb). Favorable aerodynamics yield drag area $A = 2.06\,m^2$ and drag coefficient, $C_D = 0.31$. These values are used to calculate aerodynamic drag using the formula

$$D = \frac{1}{2}\rho V^2 C_D A$$

The rolling resistance coefficient is a very low $f = 0.008$. Rolling resistance drag is

$$\text{Drag} = 0.008 \times 3200\ \text{lb} = 25.6\ \text{lb}$$

$$\text{Drag} = 0.008 \times 1452\ \text{kg} = 114\ \text{N}$$

The power to overcome rolling resistance at 30 m/s (67 mph) is only

$$(114\ \text{N})(30\ \text{m/s}) = 3.4\ \text{kW because of rolling resistance}$$

Using the data above, the power to overcome aerodynamic drag is calculated to be 10.3 kW giving total power to cruise at 30 m/s (67 mph) as 13.7 kW (18.3 hp).

Ford Reflex or Refl3x

The Ford Reflex and Refl3x is the same car with different cute names. The car has seats for three passengers, which is the origin of the name, Refl3x. A 1.4 L turbocharged diesel 41 kW (55 hp) at 6000 rpm is used. For a diesel, 6000 rpm is a high speed. Diesel torque is 175 N·m (120 ft·lb_f). The Reflex has an acceleration time of 7 s from 0 to 60 mph. Traction M/G is 30 kW for the front wheels and 15 kW for the rear wheels. The two M/G provide AWD. Hybridness, *H*, is 52%. Estimated fuel economy is 65 mpg. Solar cells power a fan to help cool the cabin during hot-weather parking. (Solar cells are much too feeble to contribute to vehicle propulsion.)

Ford HySeries Edge: An FC/Battery Plug-In Hybrid

Figure 11.2 relates the series hybrid and its derivatives to hybridness, *H*. Using Equation 6.15, the hybridness, *H*, can be calculated. The Li-ion battery has maximum power of 130 kW (174 hp), and the FC provides 35 kW (48 hp). Hybridness, *H*, is 80% which accounts for the "plug-in" in the title. Range extender could have been used as a name. Range solely from the battery is 25 mi (40 km). With the FC activated to charge the battery, the range is 225 mi (363 km). The range of 225 mi is limited by the amount of hydrogen for the FC.

GM Chevrolet Volt (Cadillac Volt)

The Volt is a plug-in hybrid offering greater electric-only range than an HEV. Up to 40 mi (64 km) electric-only range is possible. The Volt is one of a family of modular plug-in drive systems which will also be used by the Opel Flextreme. The battery can be recharged with the common 110 V power in your home. A 1.0 L engine coupled to a 53 kW (71 hp) generator keeps the 136 kW (180 hp) battery charged. Battery energy is 16 kWh which is about eight times the energy of the battery in a Prius II. Total range is 640 mi (1060 km).

Production of the Volt will begin at Hamtramck, Michigan in 2010 for 2011 model year (MY). The Volt will use the Cobalt platform. The car is identified as the Global Delta Volt and not Chevrolet Volt. By not using the Chevrolet name, GM's first plug-in may be sold as a Cadillac. This possibility is because of cost [7]. Chevrolet is considering leasing the Li-ion batteries. Leasing suggests low volume production and enables production to start in 2010.

GM Graphyte SUV

As is expected for an SUV, the Graphyte has AWD. The vehicle employs the dual-mode full hybrid system which has two M/Gs. The dual-mode hybrid system has been used in vehicles of widely varying size ranging from big trucks and buses to small sedans such as the Opel Astra described next [2]. The Graphyte, which was highlighted in GM's 2004 Annual Report, is designed without compromise on performance. The dual-mode hybrid system combines a V-8 engine, which has displacement on demand (DoD) with NiMH batteries for the M/G.

GM Precept: Product of PNGV Program

The Precept was a product of PNGV. PNGV is discussed in Chapter 1. Two versions of the concept car were investigated: hybrid and FC. The two Precept versions have certain technologies in common. The mass was low because of aluminum construction. A belly pan was used to reduce aerodynamic drag. See Chapters 1, 5, 8, and 10, which discuss drag. A TV camera replaced outside mirrors to reduce aerodynamic drag. Low rolling resistance tires helped improve mpg. Double-acting brake pistons in the calipers reduced brake drag by retracting the brake pads from the rotors.

GM Precept: Diesel Hybrid

The Precept diesel design goals were a Supercar, four-door sedan, with fuel economy of 80 mpg. The M/G for front-wheel drive (FWD) had power of 30 kW (40 hp). The compression ignition, direct injection, 1.3 L, rear mounted diesel produced 40 kW (53 hp). The diesel compression ratio was 18.3. Hybridness, *H*, was 47% which makes it a full hybrid. The Precept had a multipurpose M/G with a power of 14 kW (19 hp).

GM Precept: FC Version

A proton exchange membrane (PEM) FC provided a power of 70 kW (93 hp) continuously from a package of 1.3 cu ft. Loaded, the FC voltage was 95 V with a current of 750 A. Current density in the FC was 2 A/cm^2. Other features include

Ultralow weight aluminum	Belly pan to reduce C_D
Low rolling resistance tires	Stack volume 1.3 cu ft
TV camera replaces rear view mirrors	$P = 70$ kW continuously
Open Circuit Voltage (OCV) = 95 V at 750 A	$J = 2$ A/cm^2
Double-acting brake calipers (eliminate disk brake drag)	

GM Sequel

The GM Sequel reinvents the automobile. (Don't all concepts claim to do that?) The Sequel forms the bridge for the long-term solution to two dominant problems: emissions and use of petroleum fuels. Its FCV design removes the automobile from the environmental debate: it is a zero emissions vehicle (ZEV). The Sequel is the first FCV to have a 300 mi range between fill-ups. Range is limited by the size of the hydrogen tank. The storage uses high-pressure hydrogen gas in a carbon composite tank [2].

The goals are to design and validate an FCV by 2010. The Sequel is an example of an FCV. The vehicle should be competitive with a CV in terms of performance, durability, and cost. An affordable price is one of the keys to success. Further, the FCV must be such that it can be built affordably in high volume.

FC have problems that need to be overcome, such as cold weather freezing of water in the FC, cold start, sagging voltage with increasing current, sagging efficiency with increasing current, and slow transient response to power demands. To compensate for some of these limitations, an Li-ion battery adds to performance and range.

Sequel has a packaging concept called "skateboard." The front wheels are driven by a single M/G. The rear wheels have a hub M/G for each rear wheel. See Ref. [11] for more information.

GM Ultralite: Another PNGV Supercar

The GM Ultralite was aptly named. The structure was monocoque, which means all structural loads are carried by the skin. Monocoque structures are common in aircraft. Most of the vehicle was made of composite material. Vehicle mass (weight) was 625 kg (1400 lb) which is very light (see Chapter 1 and Figure 1.4).

The body was aerodynamically optimized yielding a very low C_D of 0.192. Combined with the low frontal area $A = 1.71\,m^2$ (18.7 sq ft), aerodynamic drag was impressively low. The road-speed rolling resistance coefficient was 0.007.

Honda Clarity FC Hybrid

This vehicle is based on technology from the Honda FC Hybrid, FCX. Low-volume production commences in 2008 with leasing of the clarity to pioneering customers.

Honda CR-V Crossover Hybrid

The decision to apply hybrid technology to the CR-V, which was completely redesigned for the 2007 MY, is still pending.

Honda CR-Z Hybrid

CR-Z (compact renaissance-zero) is a two-seater similar in concept to the Honda CR-X of a few years ago. The next generation hybrid technology (after the Civic Hybrid) of HEV will be used.

Honda FC Hybrid, FCX

Initially Honda developed an EV that was reworked into the first generation FC car. The FCX is the second generation, all-new, FC car. Some features are

- Start-up and power generation at −20°C (−4°F)
- 80 kW
- UC + FC combined to give higher initial acceleration and overtaking acceleration
- Higher efficiency, η, for FC
- Higher overall efficiency 55%; twice that of an HEV, and three times that of a CV
- Driving range 430 km (258 mi)

Honda Hybrid Sedan

The car is being designed to surpass the Prius. With five seats, the small economy hybrid will sell for less than $22,000. Fuel economy is anticipated to exceed 60 mpg. The sales goal is 100,000 per year, three times that of the Civic Hybrid.

Honda Small Hybrid Sports Concept

This two-seat coupe was introduced at Geneva Auto Show in 2007. It is based on the 2007 Civic platform and will use a 1.4 L gasoline engine.

Honda Puyo FCV

The body is a rounded bubble with soft-to-the-touch exterior surfaces. It is powered by a hydrogen FC.

Hyundai i-Blue FCV

The i-Blue vehicle will be test driven in 2008 with production planned for 2012. The FC stack produces 100 kW (133 hp) with a 100 kW (133 hp) motor for the front wheels and 20 kW (27 hp) motors for each rear wheel. Hydrogen is stored in two high pressure tanks at 700 atm (10,200 psi, 70 MPa). Anticipated range is 600 km (370 mi).

Jeep Renegade Diesel–Electric Hybrid

Vehicle weight is 3150 lb (1430 kg). The diesel, which is a Bluetec engine with a Mercedes Benz parentage, provides a power of 115 hp (86 kW). The diesel has three cylinders with a displacement of 1.5 L. The diesel fuel tank holds 10 gal (38 L). The diesel is teamed with the motors to create a hybrid. The diesel provides range extension beyond the 40 mile (64 km) electric-only range. Total range is 400 miles (645 km). The motors provide the 4WD that you would expect in a jeep. Combined power of the motors is 85 kW (113 hp). The fuel economy goal is 110 mpg. Aluminum, is used in the body to reduce weight. Features include a cut-down windshield, roll bar, openings in the door panels, and a hose-out interior.

Kia FCV

The 100 kW (133 hp) FC is packaged into an SUV. The motor for the front wheels is 100 kW (133 hp) while the motor driving the rear wheels is 20 kW (27 hp). A pair of hydrogen tanks are pressurized to 10,100 psi (70 MPa). Range is stated to be 380 mi (610 km).

Land Rover LRX SUV

This hybrid is being designed to provide real off-road capability. The M/G augments the mechanical rear drive to gain low-speed control and low-end torque. New technologies include hybrid design features of a turbodiesel battery, use of biodiesel fuel, and exhaust heat recovery. The electric drive rear axle allows electric-only propulsion at low speeds and provides AWD.

Land Rover "E" SUV

The Land Rover "E" SUV is a skeleton with all the parts hanging out; it is not a styling demonstrator. The goal is 30% better mpg using a hybrid powertrain. Production is a goal. The vehicle uses an auto-connecting rear drive shaft that has less drag thereby providing better mpg.

Range Rover Hybrid

The hybrid will start with the new aluminum body. Lower emissions and better fuel economy are promised. The result will be a luxurious, expensive, vehicle.

Lexus LF-Xh

The Lexus LF-Xh is the next generation of the Lexus RX SUV hybrid. Layout of powertrain is similar to the old model, but the body size has increased.

Loremo Diesel 2 + 2 Subcompact

The Loremo—short for low resistance mobile—is not a hybrid but its fuel consumption can compete with a hybrid. Three factors contribute to the very low fuel consumption: (1) efficient, common rail, diesel; (2) low weight of 600 kg (1320 lb); and (3) low aerodynamic drag with $C_D = 0.22$. Fuel consumption is 2.0 L/100 km (118 mpg).

Loremo Diesel GT Subcompact

The GT is an updated version of the 2 + 2 stated above with a larger diesel engine. The 37 kW (50 hp) diesel provides a maximum speed of 220 km/h (137 mph). Fuel consumption is higher than for the 2 + 2 at 3.0 L/100 km (79 mpg).

Mazda Premacy Hydrogen RE Hybrid

The Premacy is a van or crossover vehicle using a series-hybrid configuration. The motor is driven by a hydrogen-fueled rotary (Wankel) engine. Driving range is 125 mi.

Mazda *Senku* HEV Sports Car

Senku means "pioneer." Features of this rotary-powered sports car include a mid-engine and direct-injection engine which is marketed to a maturing audience. It is a four-seater with electric sliding doors. The combination of a rotary engine with electrical motors is quite exclusive among concept cars.

Mercedes Benz C-Class

This was an early, mid-1990s, concept four-door sedan to evaluate the merits of series versus parallel hybrid architecture. Different heat engines were compared including gasoline, diesel, gas turbines, and the Stirling cycle (see Chapter 17). Besides improved mpg, low emissions were a design goal.

A matrix of engines was studied. Another matrix of M/G was added to the study. The motors considered were DC, DC permanent magnet, transversal field, switched reluctance, asynchronous, and synchronous. See Chapter 14 for a discussion of various M/Gs. Combining the two matrices, the mpg and emission levels for various combinations of engines, motors, and series or parallel hybrid architecture were evaluated.

Mercedes Benz F-Class HCCI Hybrid

This vehicle, F700 concept sedan, is discussed in more detail in Chapter 17 along with the Mercedes Benz DiesOtto engine which is also known as the HCCI engine. The vehicle teams HCCI with hybrid technology. The F700 has a fuel economy of 44 mpg, the highest for any luxury sedan in the history of the automobile.

Mercedes Benz S-Class Bluetec Hybrid

This other version of an S-Class hybrid is close to production but is also still a concept car. The propulsion is a 3.0 L, V-6 diesel, with M/G. The motor is 20 hp (15 kW).

Combined power is 243 hp (182 kW) with combined torque of 424 ft·lb$_f$ (565 N·m). The key performance factors are an acceleration time of 7.2 s from 0 to 62 mph while yielding a combined 40.5 mpg. The concept car evolved into the Mercedes Benz S-400 Bluetec Hybrid.

Mercedes Benz S-Class Direct Hybrid

This S-Class direct hybrid is close to production but is still a concept car. The propulsion for the hybrid consists of 3.5 L, V-6, gasoline engine with M/G. Combined power is 300 hp (225 kW) with combined torque of 291 ft·lb$_f$ (388 N·m). The key performance factors are an acceleration time of 7.5 s from 0 to 62 mph while yielding a combined 34 mpg.

Mercedes Benz Vision Grand Sports Tourer Hybrid

The engine is V-8 with 184 kW (247 hp) and the motor is 50 kW (67 hp). Hybridness, H, is 21% making this a mild hybrid. The NiMH battery stores energy 1.5 kWh with terminal voltage of 270 V. The electrical aspects of this vehicle are viewed as an introduction to an FC version.

Mercury Meta One Crossover

This diesel/electric hybrid is designed to meet the emission standards for partial ZEV and will be the first in the world to do so. The goal is to demonstrate that diesels can meet the strictest emission standards using a modern downsized diesel with exhaust after treatment.

The Mercury Meta One showcases cutting-edge safety, environmental, and electronic technology. The fuel for this concept car can be biodiesel that is pure and sulfur-free. However, it is waxy and usually is blended with 30% petroleum diesel fuel.

Mitsubishi Concept CT MIEV Hybrid

MIEV stands for motor in-wheel EV. Each wheel has a power of 20 kW (27 hp) giving a total power of 80 kW (108 hp). Because of a motor at each wheel, the hybrid is AWD. The rotor is outside the stator which helps with unsprung mass. Propulsion components are a gasoline engine 50 kW (62 hp) and two M/Gs with combined power of 40 kW (53 hp). This concept vehicle uses an Li-ion battery with power of 50 kW (62 hp). Hybridness, H, is 62%.

Mitsubishi Pure EV

This is a lightweight EV intended for production; for this reason, it may not properly be called a hybrid concept vehicle. Mitsubishi intends to launch the EV in 2010. Two paths exist for the transition to the FCV. The first path is HEV → FCV. The second path is EV → FCV, which is the Mitsubishi path. The EV will use the Li-ion battery and wheel-in-rotor as was done in the concept MIEV hybrid above. According to Mitsubishi the Li-ion battery has six times the energy density (kWh/kg) as lead acid, and the Li-ion has twice the energy density (kWh/kg) as NiMH. The range using a 150 kg Li-ion battery was 150 km in 2006 with a projected range of 250 km by launch time in 2010. In 2006, the fleet tests using 150 kg lithium battery were giving 150 km range. For more details, see Ref. [6].

Mitsubishi iMiEV Sport Pure EV

This version of the Mitsubishi small EV has three motors. Two 20 kW motors are in-wheel driving the front wheels. A single 47 kW motor drives the rear wheels. Active yaw control is incorporated. Extensive use of aluminum reduces weight and improves performance. Vehicle mass is 970 kg (2140 lb). The batteries are located in a mid-vehicle position for favorable vehicle balance.

Nissan Mixim EV

This EV uses motors in the wheels which allow a flat floor in the passenger compartment. The tunnel down the middle of the interior is eliminated. Li-ion batteries are used to attain maximum range.

Nissan Pivo 2

Designed as a very small city car, electric motors are installed in the wheels. The electrically driven wheels can steer in any direction. The vehicle is a pure EV. Li-ion batteries supply the power.

Nissan Tino Hybrid Sedan

The Tino four-door sedan is propelled by a four-cylinder gasoline engine, dual overhead cam (DOHC), 1.8 L, providing power of 73 kW (97 hp) and a traction motor giving 17 kW (23 hp). The motor is permanent magnet synchronous. The M/G, which is also permanent magnet synchronous, provides 13 kW (17 hp). The transmission is a belt-cones continuously variable transmission (CVT) with an electromagnetic clutch. An Li-ion battery stores energy.

Opel (GM) Astra Diesel Electric

The Astra two-door coupe demonstrates that the dual-mode hybrid architecture can be scaled down into a compact, FWD, diesel/EV. The goal is a combined 59 mpg. The dual-mode hybrid system combines a 1.7 L, 93 kW (125 hp) diesel with two electric M/Gs having power of 30 kW (40 hp) and 40 kW (53 hp). The two M/Gs are integrated into the dual-mode transmission. NiMH batteries are used for energy storage.

Opel Corsa Hybrid

The mild hybrid design allows engine-off at traffic lights, torque boosting (motor helps engine), and regenerative braking. The diesel is 1.3 L yielding 55 kW (74 hp); it is the world's smallest four-cylinder, common rail diesel. The available torque is 179 N·m (125 ft·lb_f) over rpm range from 1750 to 2500. The starter/alternator, which is an M/G, is belt driven. Fuel consumption is 3.6 L/100 km (65 mpg).

Opel Flextreme

A novel feature of the Flextreme is storage space for a pair of electric scooters. See "Intercity Congestion Forces Trend" for motivation for this arrangement. Using a diesel, plug-in, series, hybrid architecture, fuel consumption of 1.5 L/100 km (157 mpg) is achieved. The series hybrid means that the diesel is not connected to the wheels but drives the generator which charges the Li-ion battery. The battery has energy of 16 kWh which is about eight times more energy than a Prius II. Peak battery power output is 136 kW (181 hp), and the motor has peak power of 120 kW (160 hp). Electric-only mode has range of 55 km (33 mi). The diesel uses closed loop control of the fuel injection profile. See Appendix 17.1 for more information. General Motors is considering leasing the Li-ion batteries. Leasing suggests low volume production and enables production to start in 2010.

Peugeot 307 Hybrid Sedan

The Peugeot sedan is a diesel/electric hybrid. The estimated fuel economy is 82 mpg. Finally, this is a hybrid that matches the Supercar goals of yore (see Chapter 1).

Phoenix Electric Sport Utility Truck

The Phoenix sport utility truck is not a hybrid but is a pure EV. An Li-ion battery with energy of 30 kWh yields a range in excess of 100 mi. Prius II has battery energy of 2 kWh. Phoenix has 15 times more energy. The body is from SsangYong Motors, South Korea. Sales price is $45,000 of which a significant fraction is the cost of the battery.

Porsche Panamera Sports Sedan Hybrid

The front-engine, four-door, four-seat sedan will be launched in 2009 as a CV. In 2012, a hybrid version will be offered. The power of the motor is 100hp (75kW) and the V-6 engine is 300hp (225kW). Top speed is 175mph (280km/h).

Saturn Vue Green Line Plug-In Hybrid

Li-ion batteries from two different manufacturers are under test. The batteries have sufficient energy for meaningful range. A time frame for production of the plug-in version has not been announced by GM. Timing will depend on battery development.

Subaru B5-TPH Two-Seat GT

TPH stands for turbo parallel hybrid, and parallel refers to hybrid architecture. This sporty two-seater uses a 2.0L turbocharged boxer engine. The boxer engine, which is a Subaru trademark engine, gives 191kW (256hp) with torque of 342N·m (252ft·lb_f). The engine uses the Miller cycle (see Chapter 17). The small M/G of 10kW (13hp) fits in the torque converter and provides torque of 150N·m (111ft·lb_f). The battery is Li-ion. Hybridness, H, is 5%.

Subaru G4e

Green for earth (G4e) is a small-car-size (five passengers) EV using Li-ion batteries. Driving range is 125mi (200km). The battery can be fully charged at home in 8h. An 80% charge is possible in 15min. The aerodynamic drag coefficient, C_D, is 0.276 which helps increase range and performance. (The reported $C_D = 0.276$ should be questioned; accuracy to three decimal places is difficult to obtain.)

Subaru R1e

The R1e is a mini-EV for use in metropolitan areas. The front wheels are driven by a 40kW motor. Electrical energy is from Li-ion batteries.

Suzuki Kizashi 2

This is a crossover sport wagon with a hybrid version.

Suzuki Twin-Turbo

This is a small two-seater powered by a 660cm^3 (0.66L) tiny gas engine. The engine qualifies for a class of vehicles in Japan that receives special tax exemptions. The single M/G is for motor assist and regenerative braking.

Tesla Electric Sports Car

The Tesla has generated considerable attention in the press [5,15–17]. It is a sleek roadster based on the Lotus Elise sports car. By adapting an existing body, development time and expense are reduced. The Tesla, which is an expensive ($98,000–$108,000 fully loaded) pure electric car, has outstanding performance. The acceleration time is 4 s from 0 to 60 mph with a top speed of 135 mph (225 km/h). The lightweight car weighs 2690 lb (1222 kg) with an anticipated range of 240 mi (400 km) on a single charge. Being an EV, the Tesla is more environmentally friendly than either a CV or even an HEV.

The initial design for Tesla battery used commercially available Li-ion cells. The cells are the same as those found in laptops. The 950 lb (431 kg) battery consists of (initial design) 6831 Li-ion individual cells. This vast number of cells requires a sizeable battery management system. Stated battery life is adequate for 100,000 mi. Battery cost is $20,000 which is about 20% the sales price for Tesla. Charging time is 3.5 h at high voltage and 7 h for plug-in at home with 110 V. The battery is encased in fire retardant material. See Appendix 21.1 for more information on Li-ion battery safety. Because of the tremendous torque from the motors, a two-speed transmission is satisfactory. Speeds up to 60 mph (100 km/h) in first gear are possible.

The business plan calls for a relatively small capitalization. Usually car companies need billions to introduce a new car. Some say it has delusions of grandeur. The business plan is to sell to the rich and not the budget-minded consumers. The first year's production of 600 cars was sold out. Introduction of Tesla was delayed by tricky transmission control and safety (collision) testing. An initial public offering was planned when the sports car hit the market. The Toyota (plug-in Prius with Li-ion batteries) and the GM Volt provide indirect competition to the Tesla. Tesla has opened an office in Detroit and may well pave the way for future EVs.

Tesla Motors Blue Star Economy EV

The Blue Star is an economy, five-passenger, pure electric sedan. The base price will be $30,000 with production following the White Star sedan. The long-term goal is to become the plugged-in Ford for the masses.

Tesla Motors White Star Luxury EV

The White Star is a five-passenger, pure electric sedan with a projected base price of $50,000. The goal is to start production in 2010.

Toyota 1/X Plug-In Hybrid

The 1/X, which is also known as the one-Xth, is Prius-size with a huge reduction of mass to 420 kg (925 lb). Extensive use of carbon fiber composites (expensive) helped reduce the mass. The Prius II has a body mass (weight) of 1312 kg (2890 lb), which is somewhat less than that of a Prius I at 1515 kg (3337 lb). The large reduction in mass yields improved fuel economy (see Chapter 7). Using the mass data here and in Chapter 7, estimate the improved mpg. Low mass also means low engine power. A 500 cc (0.5 L) engine helps to power this plug-in hybrid. Mathematically, 1/X is the reciprocal of X. What does this mean with regard to this vehicle?

Toyota A-BAT Hybrid Pickup

A-BAT is advanced breakthough aero truck. The unibody has four doors and is in the same size category as the Ford Ranger. Some goals were to provide an utilitarian, compact design but to have a small-but-tough look. The design borrows from the Honda Ridgeline (cargo bed being part of cabin) and Chevrolet Avalanche (mid-gate opening into cabin for long length). Some styling cues are from Prius. The cab is lengthened to get maximum cargo bed length which is 48 in. (1220 mm). The Toyota hybrid system is used in this small pickup truck, which has AWD. This HEV pickup may use lithium-ion batteries.

Toyota FT-HS Sports Car Hybrid

The combined motor and 3.5 L V-6 engine produce 400 hp (300 kW). Since the concept is a sports car, the styling is original, tough looking, and trendy. It is a four-seat, rear drive car. The planned acceleration time is 4 s from 0 to 60 mph, which is impressive even for a sports car.

Toyota Hi-CT Plug-In Hybrid

This is a small vehicle of the box-shaped camp. It has three doors, and two full size seats and two smaller seats. The vehicle uses Toyota's plug-in hybrid technology.

Toyota Hybrid X

The vehicle is more of a styling exercise than advanced hybrid concepts.

Volkswagen Pure Diesel "Ecoracer"

This vehicle is not a hybrid; it is shown for comparison. The propulsion is solely by means of a common rail, direct injection, four-cylinder, turbocharged, diesel with a power of 101 kW (134 hp) and a torque of 245 N·m (184 ft·lb_f) over 1900–3750 rpm. A six-speed, twin clutch, direct shift gearbox (DSG) transmission puts the diesel engine power to the wheels. The layout is a mid-engine design. DSG is also short for direct sequential gearbox, which more accurately describes the transmission.

As discussed later in Chapter 7, car weight is an energy of good mpg. This vehicle has taken that thought to an extremely low weight. The body is carbon-fiber monocoque chassis with a weight (mass) of 1874 pounds (851 kg). That is really lightweight.

The performance parameters are extraordinary. The fuel economy is 80 mpg. Acceleration time from 0 to 60 mph is an impressive 6.3 s. Top speed is 143 mph. The Ecoracer is a sports car which accounts for some of the driving impressions. The ride is stiff, the cockpit has limited room, and it is noisy.

Volkswagen Space Up! Blue FC Hybrids

The Space Up! Blue is a small van which is 2 ft (60 cm) shorter than a microbus. A 12 kW FC mounted in the front charges 12 Li-ion batteries at the rear. The 40 kW (53 hp) motor is located at the rear. The electric-only range is 65 mi (108 km), which is exceptional for a hybrid. Starting with a fully charged battery and 3.3 kg (7.3 lb) of compressed H_2, the anticipated total range is 220 mi (365 km), top speed is 75 mph (36 m/s or 125 km/h), and the acceleration time is 13.7 s from 0 to 100 km/h (62 mph).

Volkswagen Up! and Space Up! Plug-In Hybrids

The Up! and Space Up! are rear-engined, entry-level, small cars. The diesel or gasoline-powered versions will be converted to plug-in hybrids.

Volvo 3CC

Volvo 3CC has unique 2+1 seating to accommodate three people. A double floor allows use of whatever is the most appropriate powertrain of the future: gasoline, diesel, biofuel, gasoline/electric, or diesel/electric. Space within the double floor provides for a versatile design and makes for a future-proof design [3]. The Battery has 3000 Li-ion cells arranged to give a voltage of 330–420 V.

Battery current is up to 250 A. Battery power can be calculated from these data since power, P, is

$$P = (\text{volts})(\text{amperes}) = 82.5 - 105\ \text{kW}\ (110 - 140\ \text{hp})$$

Plug-in charging from 110–220V outlets is managed by the hybrid controller.

Volvo ReCharge

The ReCharge concept is a plug-in, series, hybrid that uses lithium polymer batteries. The on-board 1.6L, four-cylinder, engine provides recharge when battery energy has been reduced to SOC below 30%. Electric-only range is 100km (62mi). For a 150km (93mi) trip, the fuel economy is 124mpg (1.4L/100km). Four electric wheel motors are used to provide AWD.

CONCEPT HYBRIDS ON TWO (OR THREE) WHEELS

As reported in Chapter 1, an electric powered motorcycle, which used Li-ion batteries, set a record of 156mph for the quarter mile. This record was set in April 2007. Motorcycles (and scooters) have received attention in the form of HEV, EV, plug-in, and FC.

Suzuki Crosscage FC Motorcycle

An FC is used and includes an Li-ion battery to create a hybrid. "Hybrids Using Electrochemical Storage" in Chapter 6 discusses the FC/battery combination for a hybrid. The technology is similar to that used for the Suzuki Crosscage FC motorcycle.

Twike

This human/electric hybrid tricycle is made by Fine Mobile in Germany and sells for $26,000. The two-seat tricycle has a range of 60mi at speeds up to 45mph. When the battery is low, human energy keeps the 500lb Twike moving.

Vectrix Electric Maxi-Scooter

For urban transportation, the $11,000 electric scooter may just be the answer. Peak battery power is 21kW (28hp) with continuous power up to 7kW (9.3hp). Dry weight is 462lb (210kg). Acceleration time is 7s from 0 to 50mph—lively performance indeed. The NiMH battery operates at 125V. Brushless DC hub-mounted, air-gap motors provide the propulsion. See Chapter 14 for motor details. Missing from the data is the range on a single charge. The one essential data point for assessing an EV is range.

Yamaha Luxair Motorcycle Hybrid

A parallel hybrid configuration is used along with regenerative braking to enhance performance.

PARK THE CAR

Intercity Congestion Forces Trend

London has imposed a sizeable fee to drive into the inner city. Because of the success (reduced congestion and new source of revenue) in London, other large cities are considering a similar fee. Daily commuting by car becomes financially prohibitive. To solve the problem, several automakers are designing concept vehicles that have storage space for an electric scooter. On sunny days, the owner parks his/her car on the edge of the fee zone and proceeds efficiently into the city on his/her scooter, thus avoiding the fee. One such concept vehicle is the Opel Flextreme discussed earlier in this chapter. For hybrids, the congestion tax is usually forgiven.

DESCRIPTION OF A FEW SELECT HYBRIDS

CHEVROLET TAHOE SUV 2008 AND GMC YUKON SUV 2008: FULL DUAL-MODE HYBRID

These two vehicles, which were introduced in 2007, are identical except for grills, wheels, and badges. The dual mode hybrid system was developed jointly by GM, BMW, and Chrysler (formerly known as DaimlerChrysler). The dual-mode hybrid system has been thoroughly tested in buses (see "Gillig-GMC Hybrid Bus" this chapter). The first mode, called the input-split mode, allows electric-only, engine-only, or combined motor/engine. The second mode, called the compound-split mode, normally uses the power from the V-8 engine alone, but motor assist is possible. The V-8 has 332 hp (249 kW) power, and two motors each having 60 kW (80 hp) power. Hybridness is 33% indicating it is a full hybrid. Trailer towing (6200 lb RWD) is not hindered by being a hybrid. The estimated improvement in mpg for city driving and highway driving is 40% and 25%, respectively. The vehicles are offered as AWD or RWD. Accessories are driven electrically.

FORD ESCAPE SUV

The Ford Escape is the world's first SUV hybrid. Ford had some help from Toyota having licensed technology from that company. Ford's hybrid is not a clone of a Toyota vehicle. The Escape hybrid has significant Ford exclusive features as evidenced by more than 100 patents based on Ford Escape hybrid technology. In spite of the PNGV Supercar program, Ford had less hybrid development experience than Toyota.

The propulsion consists of the engine and the electrical components. The engine is a 2.3 L, four-cylinder, Atkinson cycle yielding 99 kW (133 hp). Engine torque is 175 N·m (129 ft·lb_f). The traction motor is rated at 65 kW (87 hp). Combined power, engine plus motor, is 116 kW (155 hp). The Escape has a planetary gear set like the Prius and operates as a CVT. The hybridness, H, is 40% that makes the Escape a full hybrid. In regard to emissions, the Escape qualifies for advanced technology partial ZEV. The Escape is being considered as a hybrid taxi in San Francisco and Chicago.

With regard to performance, this SUV provides Environmental Protection Agency (EPA) mileage of 34/40 mpg with V-6 acceleration from a four cylinder. The acceleration time from 0 to 60 mph for hybrids (8.9 s) and conventional V-6 Escape (8.5 s) is comparable. Cruise range is an impressive 419 mi.

For an AWD SUV, the vehicle mass is a reasonable 1634 kg (3600 lb). The only external identifying sign is "Hybrid" on the back. As an option, 110 V outlets are available (Table 3.1).

TABLE 3.1
Year of Introduction for Production of Hybrid Vehicles. A Few Pure EV and FC Hybrids Are Included

Year of Introduction	Vehicle
1997–2005	Ford Escape hybrid SUV
	Toyota Prius I hybrid sedan
	Toyota Prius II hybrid sedan
	Toyota Highlander hybrid SUV
	Honda Insight hybrid sedan
	Honda Civic hybrid sedan
	Honda Accord hybrid sedan
	Lexus RX 440h SUV
	Chevrolet Silverado mild hybrid pickup

TABLE 3.1 (continued)
Year of Introduction for Production of Hybrid Vehicles. A Few Pure EV and FC Hybrids Are Included

Year of Introduction	Vehicle
	GMC Sierra mild hybrid pickup
	Dodge Ram contractor special hybrid
2006	Lexus GS 450 h hybrid sedan
	Toyota Camry hybrid sedan
	Nissan Altima hybrid sedan
	Saturn VUE Green Line mild hybrid SUV
2007	Lexus 600 h hybrid sedan
	Saturn Aura hybrid sedan
	Subaru Legacy hybrid sedan
2008	Chevrolet Malibu mild hybrid sedan
	Chevrolet Silverado full dual mode hybrid pickup
	Chevrolet Tahoe SUV full dual mode hybrid
	GMC Sierra full dual mode hybrid pickup
	GMC Yukon SUV full dual mode hybrid
	Honda FC hybrid, FCX (leasing only)
	Dodge Durango full dual mode hybrid
	Ford Edge hybrid CUV
	Ford Explorer hybrid (unibody)
	Ford Fusion hybrid sedan CUV
	Ford Fusion hybrid sedan
	Mercury Milan hybrid CUV
	Ford 500 hybrid sedan
	Mazda Tribute hybrid SUV
	Phoenix Electric sport utility truck (pure electric)
	Tesla sports car (pure electric)
	Toyota Highlander hybrid SUV second generation
2009–2010	Audi Metroproject hybrid
	BMW X3 SUV hybrid
	Fisker plug-in hybrid luxury sedan
	Honda sports car hybrid
	Honda CR-Z hybrid two-seater coupe
	Honda affordable four-seat hybrid sedan
	Lexus LF-Xh (next generation of Lexus RX)
	Lincoln MKX hybrid CUV
	Mercedes Benz S-400 Bluetec hybrid
	Mercury Montego hybrid sedan
	Nissan SUV hybrid
	Porsche Cayenne hybrid SUV
	Saturn Vue plug-in hybrid (Li-ion batteries)
2011	Chevrolet Volt (Cadillac Volt)
	Tesla Motors White Star Luxury sedan EV
2012	Porsche Panamera sports sedan hybrid
2013	Nissan EV

GM Chevrolet Malibu Mild Hybrid

Using the mild hybrid system from the Saturn Vue Green Line (see section above), the improvement is 2 mpg over the four-cylinder Malibu. The mpg for the hybrid is 26 in the city and 34 on the highway. Manufacturers suggested retail price (MSRP) is $22,790 for the Malibu hybrid.

GM Saturn VUE Green Line Mild Hybrid SUV

The Saturn VUE Green Line is GM's third hybrid on the market. Green Line designates the hybrid version of Vue. The same mild hybrid technology appears in the Chevrolet Malibu. The parts for the mild hybrid add $2000 to the base price. As of 2007, the base price is $23,000 which is $4,000 less than the Ford Escape hybrid SUV and $10,000 less than Toyota Highlander hybrid SUV. For this mild hybrid, relatively little modification is needed to produce a Green Line from the conventional Vue. Using belt drive, the design approach is economical.

The engine is a 2.4 L, 16 valve, DOHC, 128 kW (170 hp) named Ecotec by GM. The Vue has an electronically controlled four-speed transmission. The power of the battery mounted between the rear wheels is a modest 10 kW (13 hp). Mass (weight) of the NiMH battery is 75 kg (165 lb). The battery pack provides 36 V. The M/G, which is belt driven, replaces the alternator. Disturbances such as rapid fuel cutoff can be damped by the M/G. The initial start uses the 12 V starter while all other starts use the M/G. The M/G in M-mode serves as the starter after engine-off. Estimated hybridness, H, is 7.3%, which corresponds to a very mild hybrid.

The Green Line Vue yields 20% better mpg than the base Vue. Mileage for the base Vue is 22/27 mpg. The fuel economy for the Green Vue according to EPA is 27/32 mpg with a combined value of 29 mpg. GM's estimates were identical to EPA values for mpg. The mass removed from the hybrid was 20 kg giving a net weight penalty of 165 – 44 = 121 lb. The Green Vue SUV has a mass (weight) of 1700 kg (3750 lb). By reducing the road clearance, the aerodynamic drag is reduced slightly for the hybrid (from the regular Vue $C_D = 0.38$ to $C_D = 0.36$).

The hybrid modes for Green Line Vue are

Accelerating		
Auto start	Fuel on	Motor assist
Cruising		
Charge battery (NiMH)	Fuel on	
Decelerating		
Early fuel off	Torque smoothing	Regenerative braking
Full stop		
Fuel off	Engine off (obviously since fuel is off)	
Smooth engine off at stop		

Note: The hybrid has both motor assist and regenerative braking.

Honda Civic Hybrid 2006

The 2006 model of the Honda Civic Hybrid is quite different from the 2005 model. See Chapter 14 for information about modifications to the engine and motor. The hybrid system used by Honda is called integrated motor assist (IMA). As is common with HEV, the propulsion consists of an engine and an M/G. The engine is a 1.3 L, four-cylinder, intelligent variable valve timing and lift electronic control (iVTEC) engine, which has three stages with regard to the valves. The stages are low/medium speed cam profiles, high speed cam profiles, and variable cylinder management (VCM).

VCM deactivates all four cylinders by closing all valves. See Chapter 17 and Figure 17.4 for more information concerning the closing of all valves. The engine has a compression ratio of 10.8 with port (manifold) fuel injection. Red line is 6300 rpm. The motor is mounted on the flywheel, a location that has advantages but also significant disadvantages.

The EPA mileage for the 2005 Civic hybrid is 47/48 mpg while the EPA mileage for the 2006 Civic hybrid is 50/50 mpg. In actual driving, the observed fuel economy ranges from 38 to 40 mpg. Mileage in actual driving situations depends on many variables including the aggressiveness of the driver, headwind, tailwind, ambient temperature, traffic, hills, and driving speeds.

The Civic has fully electronic brake assist controlling the friction and regenerative braking. Vacuum brake assist has been removed. A partially charged battery has an effect on performance. Acceleration numbers for a fully and partially charged battery are

From 0 to 60 mph: 10.8 and 12.3 s

Standing quarter mile: 18.3 and 19.1 s

Speed at quarter mile: 78 and 75 mph

Mass (weight) of the Civic hybrid is 1317 kg (2900 lb). Low-resistance tires and special low aerodynamic-drag wheels add to mpg. The Civic lacks the hatchback of the Prius II. Because of the space occupied by the battery, the fold down rear seats are lost.

The Civic has hill-start assist. However, the vehicle cannot launch from zero speed on motor alone. Electric-only operation in speed range 15–20 mph is possible if the battery SOC is near 100%.

Honda Accord Hybrid 2007

Not all hybrids are a success. Production of the 2007 Accord hybrid ceased because of poor sales. This hybrid will be replaced by a diesel Accord. A performance-oriented hybrid was not commercially successful even though the performance was amazing. The emphasis was on high acceleration instead of better mpg. The acceleration time from 0 to 60 mph time was 6.9 s. The fuel economy according to EPA (2007 test procedures) was 29/37 mpg with a combined value of 32 mpg. In on-the-road tests, the observed mileage varied from 27 to 35 mpg.

Toyota Highlander SUV/Lexus RX 400 h

For drawings of the Toyota Highlander SUV with all the internal hybrid parts, see Figures 21.1 and 21.2. The emphasis is on acceleration and achieving V-8 performance from a V-6. Fuel economy is not as important. Impressive propulsion uses a 3.3 L V-6 giving 155 kW (207 hp). The engine torque is 288 N·m (216 ft·lb_f). The SUV is AWD with an M/G at the rear wheels having 50 kW (67 hp). Torque of the rear wheel M/G is 130 N·m (98 ft·lb_f). The AWD has an M/G for the rear wheels and mainly mechanical drive for the front wheels.

Buried in the transmission within the two planetary gear sets are two M/Gs for splitting of the torques. Each M/G has a power of 123 kW (164 hp). The M/Gs and the gear sets serve as a CVT. The torque of each M/G in the gear set is 333 N·m (250 ft·lb_f). The M/G operates at 650 V while battery voltage is 288 V. The combined power from the engine and motors is 205 kW (268 hp).

The Lexus CVT has two planetary gear sets while the Prius has only one. Each M/G must balance torques, thereby forcing electricity to be generated by the generator, which can then be used to drive the motor. The motor in turn helps drive the front wheels. Torque split between front and rear wheels can be changed in a few milliseconds.

The battery, which is under the third row seats, is 288 V NiMH. The battery is air cooled. The power rating of the battery is short-term 45 kW and long-term 36 kW. The SUV has 1.7 times more battery power than the Prius II. Hybridness, H, is 59%. This large value is because of AWD.

The acceleration time from 0 to 60 mph is an amazing 7.4 s. The fuel economy is EPA 31/27 mpg while the conventional Lexus RX 400 is 18/24 mpg. The two SUVs, Toyota and Lexus, meet the California SULEV emission standards. The vehicles have good brake pedal feel, which is tough to achieve with combined friction and regenerative braking. Both the Highlander and the Lexus RX 400 h have relatively long braking distances to stop. Mass (weight) of the Highlander and the Lexus RX 400 h is 2011 kg (4430 lb) and 2062 kg (4540 lb), respectively.

Toyota Prius I Hybrid Sedan

Prius I is the first generation hybrid from Toyota. The Toyota hybrid system, THS, was introduced in Prius I. This is a clean sheet design and is not a modification of a production car. Prius I was introduced in the U.S. market in December 1997.

For more details on the propulsion, see Table 3.2. The electric M/G is water cooled. The engine speed is limited to 4500 rpm, which puts less stress on the engine parts. Various parts of the engine can be made lighter. Prius I hybridness, *H*, is 41% that makes this a full hybrid. In some tests, the observed fuel economy was 41 mpg. The body mass (weight) is 1515 kg (3337 lb).

Toyota Prius II Hybrid Sedan

Prius II is the second generation hybrid from Toyota. The Prius II, which is an excellent commuter car, introduced Toyota synergy drive (TSD). Both Prius I and Prius II are superbly engineered vehicles. Competitive acceleration is a Prius II trait. For more technical details see Table 3.3. The Prius II has a roomy interior with space for five passengers. The hatchback design is practical and versatile. Prius II, like Prius I, is a clean sheet design and not a modification of a production car [4]. Prius II was introduced in the U.S. market in 2004.

For details on the engine, see Table 3.3. The CVT is similar to Prius I, which has two M/G buried inside the single planetary gear set. Each electrical M/G is an AC permanent magnet synchronous machine. M/G power is 30 kW (40 hp) at 940–2000 rpm. Torque is 305 N·m (229 ft·lb_f) from 0 to 940 rpm. The M/Gs are water cooled. The hybridness, *H*, is 47% indicating that it is a full hybrid. If the 12 V battery is dead the car will not start even if NiMH battery is charged. In 2004, Toyota switched from low rolling resistance tires to the tires used in the Corolla. Comfort prevails over the marginal gain in mpg. The body mass (weight) of 1312 kg (2890 lb) is somewhat less than that of Prius I at 1515 kg (3337 lb) (see Figure 3.1).

Discussion of Prius Configuration

The hybrid transmission consists of a planetary gear set that has three shafts. The planetary gear set is identified as power split device in Figure 3.1. Planetary gear sets are not new. About 10 million sets were made for the Ford Model T. Of the three shafts, one is for the engine, another for the generator, and the third for the motor that drives the front wheels making it an FWD. The torques for all

TABLE 3.2
Torque and Power for Combined Motor Plus Engine for the 2006 Honda Civic Hybrid

	Engine	Motor	Engine + Motor
Torque (N·m)	134	76	167
Power (kW)	70	15	83

TABLE 3.3
Specifications for Prius I and Prius II

Item	Prius I	Prius II
Introduction	1997	2004
Battery	NiMH	NiMH
Cells/volts	240/288 V	168/202 V
M/G volts: inverter out	274 V	500 V
Battery weight	45 kg (100 lb)	45 kg (100 lb)
Battery energy (kWh)	1.8	1.3
Battery power (kW)	21	21
M/G	DC brushless	DC brushless
Motor torque (N·m)	350	400
Motor power	33 kW at 1040–5600 rpm	50 kW at 1200–1540 rpm
Engine power	53 kW at 4500 rpm	57 kW at 5000 rpm
Engine torque	115 N·m at 4200 rpm	115 N·m at 4200 rpm
Engine cycle	Atkinson	Atkinson
Compression ratio	13.5	13.0
Motor + engine		
Maximum power (kW) at 85 km/h	65	82
Maximum torque (N·m) at 22 km/h	378	478

three shafts must be balanced one against another. The engine operates at its optimum throttle setting establishing a certain engine torque. The vehicle speed and rolling resistance establish the motor torque. For these values of engine torque and motor torque, the generator must provide the exact torque to balance the overall transmission torques. This forces the generator to operate at a

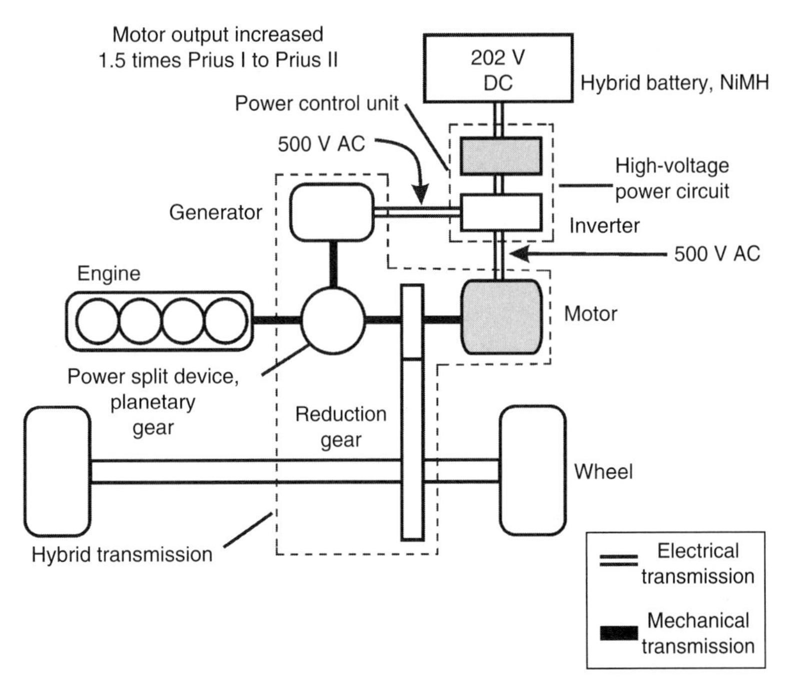

FIGURE 3.1 Prius I and Prius II system configuration. (Reprinted with permission from SAE Paper *Hybrid Gasoline-Electric Vehicle Development*, Edited by John M. German, SAE PT-117, 2005 SAE International.)

certain power and torque level. In turn, this requires a fraction of the engine torque to be diverted to the generator on its way to the wheels. Rather than planetary gear set that forms the electrical *virtual* CVT of the Prius, some hybrid designers prefer the mechanical CVT.

The most efficient path for engine torque is mechanically and directly to the drive wheels. Because of the need to balance torques in the transmission, the path of a portion of the energy is

$$\text{Engine} \rightarrow \text{generator} \rightarrow \text{motor} \rightarrow \text{drive wheels}$$

This path is less efficient because of the numerous transfers of energy.

The Prius transmission has both positive and negative aspects. The design, which is clever and well executed, provides an electronically controlled CVT. The cost is low because of the simple design and the small number of parts. However, the design forces some of the engine power to take a low-efficiency path to the wheels. Also, under certain operating conditions, recirculating electrical currents can occur. These currents rob energy but do not contribute to propulsion. Since the Prius yields very good mpg, the deficiencies noted here are minor but can be improved upon.

Chevrolet Silverado Mild Hybrid Pickup

The design goals were to retain performance and retain towing capability. As design proceeded, the costs/benefits of various hybrid features were considered. For example, the value of motor assist was assessed relative to the bigger batteries, bigger M/G, and integration of bigger components. Motor assist was not added. In the final design, the conversion to hybrid cost $2500 or 8% of the base price.

The hybrid uses integrated starter alternator damper (ISAD) with a power of 15 kW (20 hp). The ISAD is flywheel mounted within the torque converter. The 5.3 L Vortec V-8 is retained and provides 221 kW (295 hp). The M/G is an AC induction machine. The stator is water cooled. The 42 V battery is lead acid and is located under the rear seats. To increase mpg, the electrohydraulic steering pump replaces the belt-driven pump. The hybridness, H, is 6.4%, which is a mild or micro-hybrid.

Introduction of a mild hybrid with seemingly small gain in mpg does save considerable gasoline. So, do not sneer at a small gain. See the discussion of gasoline savings under "Saving Gasoline". The mild hybrid yields a benefit of 1.5 mpg only. The conventional Silverado has EPA rating of 15.5 mpg while the hybrid gives EPA 17/19 mpg. The Silverado saves almost as much gasoline as a lightweight full hybrid HEV. Acceleration time from 0 to 60 mph is a brisk 8.2 s. Towing ratings were retained. Big pickups are heavy with a mass (weight) of 2383 kg (5259 lb).

The Silverado hybrid has regenerative braking and engine-off at stop, but it does not have motor assist. Hybrids provide an opportunity for synergism with other technologies to offer features that the customer deems valuable and useful. In this category of synergistic exploitation, the Silverado hybrid provides 4 AC 110 V outlets with total power of 2.4 kW. The initial source of the power for the outlets is the battery. When battery SOC is too low, the engine automatically starts to drive the M/G in G-mode.

From engine-off at a stop, release of the brake or a touch of the throttle restarts the engine. At times, quick foot movement from the brake to the throttle causes a hard slam. Also the Silverado has an uphill problem. When the Silverado is stopped and facing uphill, the engine is not running. Releasing the brake initiates restart; however, before engine starts, the pickup truck wants to roll back. As the engine starts, a hard slam is experienced. The driver cannot prevent this slam. If he/she hits the brake again, the engine goes into the idle-off mode.

As stated in Chapter 5, for a small car, the A/C requires 30%–50% of cruise power. Of course, for a pickup, the A/C requires a smaller fraction of vehicle power. In any event, A/C is not a trivial problem when the goal is to improve mpg. For the Silverado hybrid, on hot days the engine stays on to drive the A/C; idle-off is overridden. On milder days, the engine stops, but the cooling is reduced.

Toyota 2007 Camry Hybrid Sedan

The design emphasis is on mpg and not on acceleration. The Camry hybrid sips fuel and assures a pleasant ride. The cost hovers around $30,000 with a price premium of about $3000 over the base four-cylinder Camry. The hybrid Camry is less expensive than the Honda Accord. It has room for five passengers. The Camry is FWD with a CVT. The transmission differs little in concept from the successful Prius I and Prius II. The transmission has M/G buried within a planetary gear set. The engine is 2.4 L, four-cylinder, 110 kW (147 hp), Atkinson cycle, with DOHC. Engine torque is 184 N·m (138 ft·lb_f).

The two M/Gs are AC permanent magnet and each produces a power of 106 kW (141 hp). The motor torque is 265 N·m (199 ft·lb_f). Combined power, that is, engine plus motor, is 144 kW (192 hp). The mass of the battery is 68 kg (150 lb). The battery maximum power is 34 kW (45 hp). Electric-only operation is possible at launch and low speeds. Hybridness, *H*, is 49%. The Camry is a full hybrid.

Even though the hybrid acceleration time from 0 to 60 mph is 7.7 s some drivers feel the Camry needs a catapult "C" button instead of an engine-braking "B" button. Brake pedal feel is always a bugaboo on hybrids. The Camry hybrid is better than most other hybrids, but is not as good as a non-hybrid Camry (Table 3.4).

The 2007 EPA fuel economy is 39 mpg combined, and the 2008 EPA fuel economy is 34 mpg. Observed mpg when driving easy on throttle gives a combined mileage on city/highway of 33–34 mpg. When driving heavy on throttle, the combined mileage on city/highway is 24–25 mpg. The base Camry non-hybrid four cylinder gives a realistic combined 31 mpg. The Camry hybrid can travel 600 mi on one tank of gasoline with less than 2 gal reserve. The hybrid mass (weight) is 1635 kg (3600 lb). Cargo space for a conventional Camry is 15 cu ft as compared to the 10.6 cu ft for the hybrid Camry. Because of the battery, 4.4 cu ft of space is lost.

Mass to Power Ratio

Mass to power ratio has a major influence on the acceleration capability of the vehicle. Table 3.5 gives values of mass, *M*, to power, *P*, ratio for several hybrids. A sluggish vehicle typically will have a large *M*/*P* ratio. Prius I has the largest value in the table. It also has the longest acceleration time from 0 to 60 mph. Honda Accord has the smallest *M*/*P* ratio at 8.3 making the Accord a hot rod among sedans. The Accord's acceleration time from 0 to 60 mph is the lowest in the table at 6.9 s. The Prius is designed to be fuel efficient and has very good mpg. The Accord's design emphasizes acceleration instead of mpg.

TABLE 3.4
Comparison of Conventional and Hybrid Camry

Engines	Power (hp)	2007 EPA City/Highway	Automatic Transmission Gears
2.4 L, four cylinder	158	25/34	Five-speed
3.5 L, V-6	268	—	Six-speed
Hybrid, four cylinder	147	40/38	CVT
M/G power	141	—	—
Battery's maximum power	45	—	—
Combined power	192	—	—

TABLE 3.5
Mass to Power Ratio for Several Hybrid Vehicles

Hybrid Vehicle	Mass (kg)	Power (kW)	Mass to Power Ratio	t_{60}, 0–60 Time, (s)	M/(P t_{60}) kg/kW s
Honda Insight	835	56	15.0	12	1.25
Honda Civic	1317	82.5	16.0	10.3	1.55
Honda Accord	1583	190	8.3	6.9	1.21
Prius I	1515	65	23.3	12.6	1.85
Prius II	1312	82	16.0	10.5	1.52
Lexus RX 450 h	2011	201	10.0	7.4	1.35
2007 Camry	1653	144	11.5	7.7	1.50
Ford Escape	1634	116	14.1	8.9	1.58
VW Jetta TDI	1501	75	20.0	10.3	1.94

Note: All are hybrids except the VW Jetta TDI, which is diesel only and is shown for comparison.

HYBRID TRUCK AND BUS TECHNOLOGY

Economics of Bus and Truck

Whereas hybrid cars may sell at a premium without hope of full recovery of the added cost from savings on fuel, in the commercial world this is not the case. Car owners will buy for reasons other than saving money. For a business, the total cost of ownership dictates purchase or none; the *only issue is the effect on the bottom line.*

Comparison Lifetimes or Cycles of Operation

The vast difference in lifetime between a light duty vehicle (LDV) (passenger car) and a truck or bus is given in Table 3.6. Because of the differences, a design for the LDV has a separate focus. Two factors separate trucks and buses from LDV: economics and lifetime.

Attractiveness of Supercapacitor for Trucks and Buses

FC and batteries have chemical reactions during the charge/discharge cycle; these reactions limit lifetime to a few 10,000 cycles. To obtain long life, batteries also are limited to about 25% SOC change during discharge. Capacitors do not have chemical reactions and consequently have a much longer lifetime (>500,000 cycles) expressed as number of cycles before failure. Also the supercapacitors can be operated between SOCs of 0% and 100% without degrading the lifetime. Because of the very large number of charge/discharge cycles, supercapacitors may be selected for most future hybrid truck and bus applications.

TABLE 3.6
Comparison of Lifetimes and Cycles during Lifetime Operation

Hybrid Vehicle	Miles (km)	Braking Cycles	Charge/Discharge Cycles
Light duty vehicle (LDV)	124,000 (200,000)	800,000	Few 10,000
Truck or bus	370,000 (592,000)	2,400,000	Few 100,000

Supercapacitors have poor energy density, that is, energy per volume with units kWh/m^3. This means supercapacitors take up a lot of room. Supercapacitors can be placed almost anywhere in the truck or bus; for example, one possible location is under the fenders. The failure of electrical storage, batteries versus supercapacitors, adversely affects the maintenance costs and utility of the hybrid concept. For trucks and buses, batteries are seen as a high failure item.

Duty Cycle and Application of Hybrid

Duty cycle or driving cycle affects the viability of hybrids for trucks and buses. An ideal duty cycle involves many stops and starts and has long periods of idling. The highway, long distance heavy haulers, such as 18-wheelers, do not fit the ideal duty cycle for hybrids. Other features may make hybrids attractive for 18-wheeler long distance carriers (see Chapter 17).

The trucks and buses that have a duty cycle favorable to hybrid design include trash trucks, door-to-door delivery trucks and vans, just-in-time delivery and supply, Internet-based home shopping, delivery to urban areas, shuttle buses, and transit buses. The first vehicle segment to adopt capacitor technology is buses [10].

Emissions

Meeting the regulations for emissions is nonnegotiable and ranks high with the economics of operation. Trucks and buses favor diesel engines. Diesel exhaust is dirty without extensive after-treatment. For more information on diesels, see Chapter 17.

Future for Diesel/Electric Heavy-Duty Hybrids

As of 2008, the Japanese manufacturers lead in HEV technology whereas the European manufacturers lead in diesel engine technology. Where does the United States fit into the diesel/electric heavy-duty hybrids? The question is answered below. The United States is a major consumer of trucks and buses.

Organizations Considering Hybrid Trucks

Big fleet operators have the most to gain by adapting hybrids. For example, FedEx, Purolator, and the U.S. Post Office operate sizeable fleets. These organizations are considering hybrids.

Truck Original Equipment Manufacturers Working on Hybrids

The major original equipment manufacturers (OEM), Kenworth, International, Freightliner, Volvo, and AM General have hybrid projects. Because of the duty cycle, the future of hybrid technology for largest heavy-duty trucks is uncertain and even questionable.

CONCEPT AND PRODUCTION HYBRID TRUCKS

FedEx Hybrid Delivery Truck

The goals of the FedEx hybrid delivery truck are to reduce pollution by 90%, reduce fuel consumption by 50%, demonstrate economical and functional viability, and accelerate the time-to-market of a hybrid delivery truck. FedEx bought 20 demonstrator hybrid-diesels.

The Li-ion battery for the FedEx hybrid delivery truck has a voltage of 173 V with an energy of 1.07 kWh. The mass of the battery is 20 kg.

GM/Chrysler Hybrid Light Trucks

GM and Chrysler jointly have hybrid light trucks under development.

Diahatsu Hybrid Small Light Delivery Truck

The Diahatsu light truck is a tall delivery van with the powertrain under the floor. The vehicle is for the Japanese home market. It is a rear-wheel drive (RWD). The engine is a small three-cylinder gasoline engine with 0.66L displacement. The M/G is a DC synchronous motor. Operational modes are idle-stop, engine plus motor assist, and regenerative braking.

Ford Hytrans Light Commercial Vehicle

The Hytrans is a micro hybrid which means it has a small value for hybridness, *H*. Hytrans will give 20% better mpg which translates to a 3–4 year recovery of the price premium through fuel savings. The operational modes are idle-stop, engine only with no motor assist, and regenerative braking. On the engine side of the hybrid vehicle, a diesel engine is used. On the electrical side of the hybrid, the M/G is 4kW (5hp) with 42V lead acid batteries. The starter/alternator is belt driven.

Dodge Sprinter Plug-In Hybrid Van

The Dodge Sprinter is a plug-in hybrid. Dodge, which is part of Daimler/Chrysler, is working on a small fleet of Sprinter van plug-ins; the fleet is for demonstration only.

Kalmar Hybrid Terminal Tractors

Marine shipping terminals have tractors (trucks) to move cargo at ports. The operation of these tractors is ideal for hybrid; tractor operation includes lots of stops and starts with has long periods of idling. Engine-off during idling, made feasible by hybrid, will significantly reduce emissions. The "stop" part of stop and start allows regenerative braking while the performance during the "start" part is enhanced by either electric-only mode or motor-assist mode. Kalmar is developing two types of hybrid tractors: diesel-electric and diesel-pneumatic. The better hybrid design will be selected based on actual performance.

Nissan Condor Medium-Size Hybrid Truck

The Condor has a 4 ton payload and is a parallel diesel/electric hybrid. Supercapacitors are used for energy storage. The engine is a six-cylinder diesel producing 152kW (203hp). The M/G develops 55kW (73hp). Hybridness, *H*, for the Condor Hybrid is 27%.

HYDROGEN ELECTRIC RACING FEDERATION

According to Ref. [12], the proposal by the Hydrogen Electric Racing Federation (HERF) is for racing to serve as the ultimate proving ground for the electrification of the automobile. Closed wheel race cars would weigh less than 900kg (1984lb) and feature a minimum power of 300kW (400hp) from FC driving electric motors. The anticipated speed of the race cars is 185mph (300km/h). The title for the race is the Hydrogen 500. The first race is expected to be held in 2010, which is an optimistic schedule.

PRODUCTION HYBRID BUSES

Gillig-GMC Hybrid Bus

These buses use GM's dual-mode hybrid technology, which is an EVT and provides step gear ratios in addition. The two M/Gs are integrated into the transmission. The operational modes have diesel plus motor assist or diesel only. The NiMH battery has a voltage of 660 V. The diesel has a displacement of 8.9 L and replaces the 11.0 L diesel used on non-hybrid buses. In addition to improved mpg, the hybrid buses reduce pollution. The Gillig hybrid bus is 40 ft long and uses GM Allison hybrid technology. The hybrid provides a 90% reduction in particulates, HC, and CO. The improvement in mpg is 20%–55%. The bus has better acceleration than a non-hybrid bus. The prices for the hybrid and non-hybrid buses are $530,000 and $340,000, respectively.

Sales of the Gillig-GMC hybrid bus have been brisk. Yosemite National Park uses GM hybrid buses (see Figure 3.2). Seattle/King County has 235 GM hybrid buses. Eighteen other cities have bought the Gillig-GMC hybrid buses.

Mercedes Benz Hybrid Bus

The Mercedes hybrid bus is RWD with Siemens M/G driving the rear wheels. Each of the two M/Gs provides 150 kW (200 hp). To reduce M/G speed, reduction gears are used from the M/G shaft; however, a step-by-step transmission is not used.

Mitsubishi Aerostar Hybrid Bus

The Aerostar hybrid bus is diesel/electric with a six-cylinder diesel displacing 8.2 L. Li-ion batteries provide the storage.

FIGURE 3.2 Free Yosemite hybrid shuttle, a gillig bus with GMC hybrid propulsion. (Photograph by the author in Yosemite National Park, 30 April, 2007.)

Orion VII Hybrid Bus (BAE Systems)

A single traction motor propels the Orion VII hybrid bus. The electrical power is supplied by a diesel/generator. The diesel is a Cummins 5.9 L, the same as that used in the Dodge Ram pickup. New York City (NYC) operates a fleet of Orion buses. In 2006, NYC added 500 buses swelling their fleet to 825 buses by mid-2007. NYC has an option for 389 more buses.

MILITARY APPLICATIONS OF HYBRIDS

Hybrids and Military Interest

Hybrids offer many advantages for military vehicles [8]. Stealth is of prime importance; hybrids have lower signatures. Less heat because of higher efficiency means a reduced infrared signature. A hybrid is very quiet in the electric-only mode of propulsion and yields a reduced acoustical signature. Because of the electrical components, a military hybrid may have a higher radiated electromagnetic signature.

New military equipment need much more electrical power; hybrids possess plenty of electrical power [9]. Electrical power is required to direct energy weapons such as high power lasers and high power microwaves. The abundant power produced enable deployment of electrothermal guns, electromagnetic guns, and powered armor. The number of electrical generators in trailers can be reduced or eliminated.

Hybrids improve range by up to 35%. The logistics chain is reduced, requiring less transport, ships or aircraft, to position ground forces far away from the United States. Less fuel is needed. Fewer tanker trucks are needed in the battlefield. Fewer drivers of tanker trucks means more manpower is available for other duties. Apart from the better fuel economy, the military requires quick-reaction, dash speeds for vehicles that go in harm's way.

Vehicle maintenance is a critically important aspect for adopting hybrid technology. Hopefully, on-going research proves maintenance demands will be reduced. Hybrids might increase maintenance problems. The required level of maintenance is vital to making a decision.

DARPA Advanced (Military) Vehicle

The Department of Defense (DoD) in the late 1990s sponsored work on military EV and HEV through the Defense Advanced Research Projects Agency (DARPA). DoD was interested in this advanced military vehicle for many obvious reasons. Stealth, reduced logistics supply chain, etc. Also included in the work was the demonstration of a 5 min charging station.

Shadow: A Reconnaissance, Surveillance, and Tracking Vehicle

The objective of the RST-V (reconnaissance, surveillance, and tracking vehicle) is explained in the name itself. This vehicle is sponsored by the U.S. Marine Corps. and Special Operations Command. Shadow gets three times more mpg than a military Hummer and twice the mpg as a civilian Hummer (Figure 3.3).

PARTING PERSPECTIVE

Alternate Vehicles

Within a decade, vehicles having impressive fuel economy will be available. Once a fuel economy of 50 mpg is attained, technology should focus on making the vehicle more affordable. Table 3.7 reaffirms the earlier statement that hybrids are but one approach. The alternatives to hybrids are *better than expected*. As the last set of thoughts before the summary, Table 3.7 is most appropriate.

FIGURE 3.3 Shadow reconnaissance, surveillance, targeting vehicle (RST-V) developed for USMC and special forces. RST-V is a full hybrid. www.army-technology.com/project_printable.asp?ProjectID=2451.

SUMMARY

One sentence summarizes this chapter. The future of hybrid vehicles rests on the twin pillars of enhanced fuel economy and lower emissions. The early discussion focuses on how to save gasoline and where to look for savings. Look for modest improvements in gas-hogs. Hybrid performance encompasses not only enhanced fuel economy and lower emissions but also acceleration, handling, and driving pleasure. Plug-in hybrids have a certain allure. Once the novelty of the hybrid wears off, the customer will revert to the technology (fuel) that offers the most convenience with the least cost.

TABLE 3.7
Mileage Estimates for a 2800 lb (1270 kg) Vehicle Using Technology Available within the Decade, Except for the FC/Battery Hybrid Which May Require More Time

Alternate Vehicle	Mileage (mpg)	Chapter
Diesel/electric	80	17
Gasoline/electric	75	17
Clean diesel	55	17
High-efficiency gasoline	50	17
Hydrogen internal combustion	50[a]	17
FC/battery hybrid	60[a]	22

[a] Gallon in mpg is either gasoline or diesel fuel. For hydrogen-powered vehicles, an adjustment has been made accounting for the different energy content of a gallon of gasoline and a gallon of hydrogen. The numbers are gas equivalent values.

A hybrid may be a clean sheet of paper design or a conversion of a production vehicle. If the hybrid is a conversion, the customer should be aware of certain possible pitfalls. The reasons to buy a hybrid are discussed as well as the reasons not to buy one. The rationale for the larger price premium of a diesel/electric hybrid is presented.

Considerable space is devoted to a catalog of hybrid concept vehicles. The list includes recent and decade-old concepts. A pure diesel concept sports car is given to assist comparison. A few of the important hybrids that were launched recently are described in brief. The mass-to-power ratio is introduced, and its effect on acceleration is examined.

Hybrids are not limited to SUVs and passenger cars. Hybrid technology can be applied and incorporated into trucks. Some truck duty cycles or driving cycles are favorable for hybrids, while some others are not. The economics of hybrid ownership is much more important in the commercial world. Lifetimes and operating cycles are different for cars and trucks and affect design choices. Buses come next on the agenda. Finally, the considerable interest of the military in hybrids is reviewed.

APPENDIX 3.1

This appendix derives the equation for gallons of fuel saved annually. The symbols have been defined under "Saving Gasoline."

$$S = \frac{M}{G_C} - \frac{M}{G_H} \tag{A.3.1}$$

Since

$$G_H = G_C(1 + f) \tag{A.3.2}$$

Combining Equations A.3.1 and A.3.2 and performing the algebraic manipulations leads to the desired Equation 3.1.

$$S = \frac{M}{G_C}\left(\frac{f}{1+f}\right) \tag{3.1}$$

The equation highlights the statements made earlier with regard to gas-hogs and fuel-sippers.

REFERENCES

1. P. Ponticel, Global vehicles, *Automotive Engineering International*, Society of Automotive Engineers, Warrendale, PA, February 2005, p. 32.
2. K. Jost, Global vehicles, *Automotive Engineering International*, Society of Automotive Engineers, Warrendale, PA, February 2005, pp. 12–14.
3. D. Alexander, Global vehicles, *Automotive Engineering International*, Society of Automotive Engineers, Warrendale, PA, February 2005, p. 22.
4. K. Muta, M. Yamazaki, and J. Tokieda, Development of new-generation hybrid system THS II – Drastic improvement of power performance and fuel economy, *Hybrid Gasoline-Electric Vehicle Development*, John M. German (Ed.), Society of Automotive Engineers PT-117, Warrendale, PA, 2005, p. 47.
5. J. Kwon, P. Sharer, and A. Rousseau, Impacts of combining hydrogen ICE with fuel cell system using PSAT, Society of Automotive Engineers, *Applications of Fuel Cells in Vehicles 2006*, SP-2006, Warrendale, PA, Organized by R. K. Stobart and S. Kuman, p. 13.
6. D. Schneider, Who's resuscitating the electric car? *American Scientist*, September–October, 2007, pp. 403–404.
7. T. Lassa, GM-The future car, *Motor Trend*, December 2007, pp. 14–15.

8. D. Axe, Engine check: Technology limitations stall military hybrids, *National Defense*, September 2006, pp. 30–33.
9. S. Magnuson, Where's the juice? Army explores alternate ways to add power on the battlefield, *National Defense*, September 2006, pp. 34–35.
10. K. Jost, Ultracapacitors charge ahead, *Automotive Engineering International*, Society of Automotive Engineers, Warrendale, PA, September 2004, pp. 46–53.
11. A. Taylor, The new fuel thing, *Fortune*, October 2, 2006, pp. 47–48.
12. Competition, Hydrogen fuel-cell racing might be on the way, *AutoWeek*, January 22, 2007, pp. 44–45.
13. D. Carney, Electricity drives forward; continuous advances in component design and integration are helping to overcome cost and performance obstacles, *Automotive Engineering International*, Society of Automotive Engineers, Warrendale, PA, July 2007, pp. 48–51.
14. Web site www.army-technology.com/project_printable.asp?ProjectID=2451.
15. M. Vaughn, Not dead yet: The electric car makes a comeback, *AutoWeek*, July 12, 2006, p. 5.
16. Staff, Electric cars; not so shocking, *The Economist*, July, 29, 2006, p. 73.
17. Staff, Tesla sports car, *Newsweek*, October 29, 2007, pp. E6–E10.
18. S. Birch, Powertrain: Ultracapacitors meet mild hybrids, *Automotive Engineering International*, Society of Automotive Engineers, Warrendale, PA, January 2008, p. 36.
19. M. Wilkinson, Plan 863 spurs flood of hybrid, alt-fuel vehicles, *Automotive Engineering International*, Society of Automotive Engineers, June 2008, pp. 16–17.
20 P. Ponticel, Light-truck maker Great Wall gears up for cars, *Automotive Engineering International*, Society of Automotive Engineers, June 2008, pp. 18–19.

4 Hybrid Automobile: What Is It?

From the broad definition of hybrid, the narrower scope of the hybrid electric vehicle is emphasized.

HYBRID DEFINITION

A hybrid vehicle combines any two power (energy) sources. Possible combinations include diesel/electric, gasoline/flywheel, and fuel cell/battery. Typically one energy source is storage and the other is conversion of a fuel to energy. As discussed below, the combination of two power sources may support two separate propulsion systems. This is true for the parallel design hybrid.

An alternate definition for hybrid depends on one mode of operation. To be a hybrid, the vehicle must have at least two modes of propulsion. Recall the several modes of operation for the Saturn Vue Green Line in Chapter 3. A truck that uses a diesel to drive a generator, which in turn drives several electrical motors for all-wheel drive is *not a hybrid*. If the truck has electrical energy storage to provide a second mode, which is electrical assist, then it is a hybrid.

The broad definition of hybrid encompasses some strange vehicles. Some years ago, a kit consisting of a small, 1-cylinder, 2-cycle, gasoline engine could be bought. Installing the kit on a bicycle created a hybrid vehicle. One energy source was the bike rider; the other was the gasoline engine. With regard to conversion of energy, the rider converted ham and eggs into revolutions per minute (rpm) at the pedals. The engine converted fuel into mechanical energy at the bike wheel. Subsequent discussion will enable you to answer two questions posed here. First, is this hybrid vehicle classified as a series or parallel design? Second, what would need to be added to the kit to allow regenerative braking by the bicycle? Both answers are given in the summary at the end of the chapter. Although not of interest here, another hybrid (two forms of lift) is an autogiro which has a rotor like a helicopter and wings and tail like an airplane.

NEW IDEA FOR A HYBRID

A small company in the San Francisco Bay Area is developing a hybrid outboard motor for use in boats. The idea applies to inboard motors also. The propeller acts as a turbine or water wheel when the boat is slowing down. This is the same concept as regenerative braking for a hybrid car. Does a boat operational cycle involve enough stop-go to make hybrid attractive? Probably not. If the boat were *anchored* in a flowing stream, energy could be extracted just like a windmill or wind turbine.

HYBRID ELECTRIC VEHICLE

Consistent with the definition of hybrid above, the hybrid electric vehicle (HEV) combines a gasoline engine with an electric motor. An alternate arrangement is a diesel engine and an electric motor. Throughout this book, the word "engine" is used to denote the gasoline or diesel engine. Further the word "motor" signifies an electric motor or the motor/generator (M/G). As shown in Figure 4.1, a hybrid vehicle may consist of a gasoline engine combined with an M/G. An HEV is formed by merging components from a pure electrical vehicle and a pure gasoline vehicle. The EV has an M/G which allows regenerative braking for an EV; the M/G installed in the HEV enables regenerative

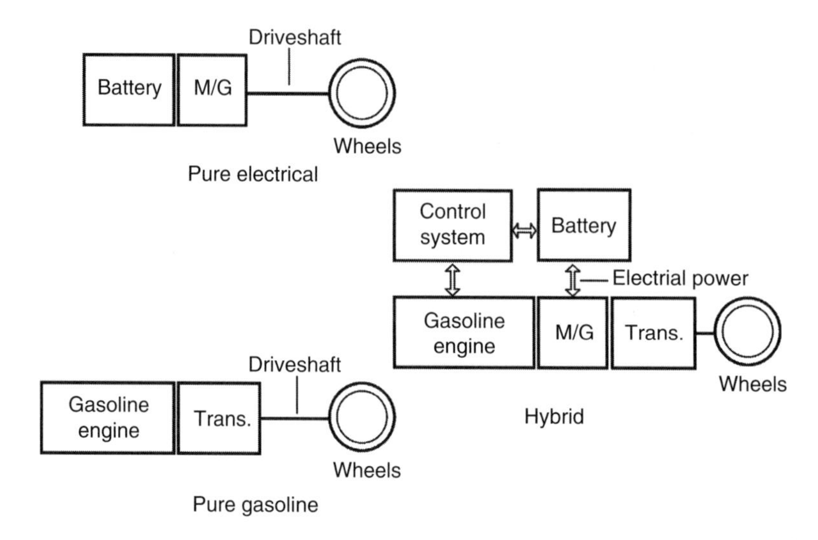

FIGURE 4.1 Components of a hybrid car. The hybrid is formed by combining a gasoline CV with a pure EV.

braking. For the HEV, the M/G is tucked directly behind the engine. In Honda hybrids, the M/G is connected directly to the engine. The transmission appears next in line. This arrangement has two torque producers; the M/G in motor mode, M-mode, and the gasoline engine. The battery and M/G are connected electrically.

One new component, the control system, has been added. The function of the control system, which is discussed in Chapter 8, can be likened to that of a conductor of a symphony orchestra. A good conductor can inspire pleasant music from the symphony orchestra; a good control system can enhance driving experience. Mixing propulsion from the engine and M/G opens many new control dilemmas to be solved and decisions to be made. Major functions are to maximize miles per gallon (mpg) and minimize exhaust emissions, assuming, of course, that maximum mpg is the design goal. Integration of mechanical components, electrical components, and the control software is as important as hardware, if not more so. Integration and software challenges are enormous. Components must engage and disengage smoothly.

Numerous minor functions of the control system include component protection by monitoring battery state of charge (SOC), battery temperature, electric motor overheating, and engine overheating. The control system provides the fail-safe failure modes that yield the limp home capability. Also on-board diagnostics (OBD) is a feature of most control systems.

An interplay exists between the engine and motor; this interplay depends on driver demands and operational conditions. Figure 4.2 shows three different operating conditions: accelerating from a stop sign, cruising on the freeway, and climbing a hill. (This figure is also discussed in Chapter 5.) When large power demands are made on the hybrid, such as climbing a hill, the motor supplements the engine. This is known as motor assist. The gray shaded areas in Figure 4.2 are motor assist. The HEV allows use of a smaller gasoline (or diesel) engine while retaining performance.

A driving cycle consists of the elements of the operational conditions, that is, accelerating from a stop sign, cruising on the freeway, climbing a hill, idle time, etc. The data from driving cycle are extremely important for hybrid design.

To have regenerative braking, which is examined in Chapter 9, a dual purpose M/G is needed. The same electrical device provides, on command, either motor or generator. Regenerative braking returns a portion of the kinetic energy of HEV motion to the battery. Regenerative coasting, either downhill or slowing speed, also uses the M/G in G-mode to charge the battery. The battery pack is usually high voltage.

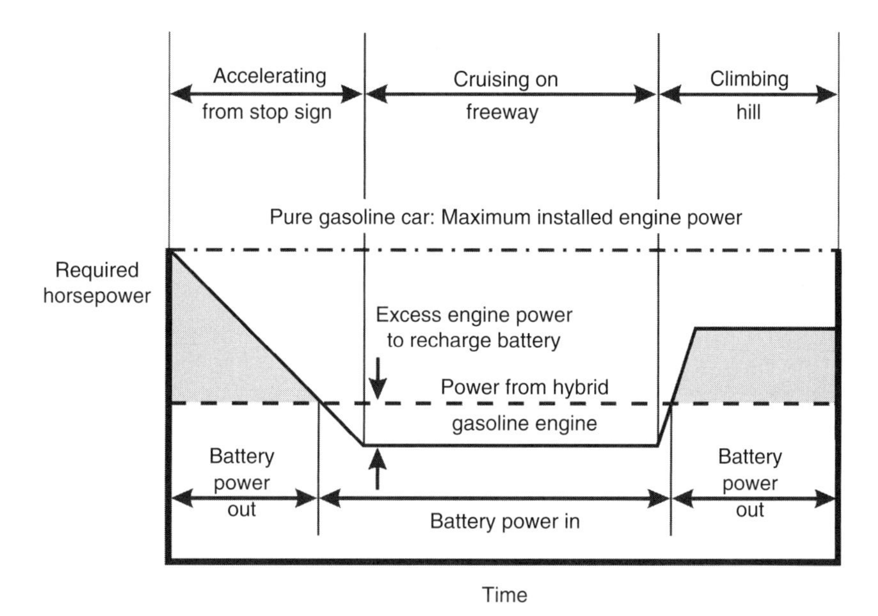

FIGURE 4.2 Why hybrids get good gas mileage.

The complex power distribution system is separated into two parts at vastly different power levels and voltages. One part is the "smarts" section or information section which is mainly computers and microprocessors (low power, low voltage). Another part is the "muscle" section or the power electronics that manages the power to/from the M/G and to/from the battery (high power, high voltage).

INHERENTLY MORE EXPENSIVE

HEVs have a price premium or price penalty. They are inherently more expensive because of the extra components compared to a conventional vehicle (CV). The M/G and the large, high-voltage battery add to the cost. In contrast to a CV, the HEV needs complex power management involving the HEV information (computers, software, algorithms, etc.) and HEV power (power transistors, cooling system, etc.). The control system is extremely complicated and represents a tremendous challenge; it is not easy.

HYBRIDS: MAINSTREAM OR ON THE FRINGE? SUCCESS OR FAILURE?

Chapter 18 addresses the question as to whether hybrids will capture a significant fraction of the new car market or whether hybrids will remain at a small percentage of the overall market. As an introduction to Chapter 18, several factors that affect the answer are now discussed. Hybrids will succeed or fail based on how well the control system integrates, distributes, and manages the power from two different sources. The complex interaction of the power flow between motor and engine must be controlled so as to be transparent and *safe*. In addition to these factors, the HEV must offer some aspect of performance that is superior to CVs, to justify its high price. The biggest concern is getting enough overall vehicle efficiency from the smallest and cheapest package.

Once again, the HEV must offer some aspect of performance that is superior to CVs to justify its high price. Performance enhancement in terms of powerful acceleration might be one advantage of a hybrid. According to *AutoWeek*, Toyota has tried to use hybrid as a performance-enhancing device; the idea does not work because of required electronic (control) sophistication. GM feels that existing hybrid components are too heavy and compromise weight distribution too much to be considered as a boost to vehicle performance. Also, GM feels a breakthrough in batteries is needed to make current hybrid technology attractive for performance enhancement (see Ref. [50] in Chapter 12).

HEV PROPULSION DESIGN OPTIONS

There are three common design options: series, parallel, and mixed. We will now discuss series hybrids followed by parallel hybrids. The hybrids with mixed design are discussed in Chapter 11.

HEV PROPULSION DESIGN OPTIONS: SERIES

Consider now some of the advantages of the series architecture. Figure 4.3 shows that the series HEV has only two draft shafts that are not connected. Hence, the engine can run at optimum rpm and throttle setting to give minimum fuel consumption. Further, control of engine operating point provides easier control to minimize emissions. The engine and generator are a unit, but this unit is not connected to the drive shaft. The engine and generator can be located anywhere ("in the glove box?"). Search Figure 4.3 for a transmission; none is to be found, a transmission is not needed.

Naturally, the series configuration does have some disadvantages. The generator, which is needed, is a heavy, extra component. The generator is sometimes called an alternator/rectifier. The capacity of the generator plus the battery maximum power must equal the total power of the HEV.

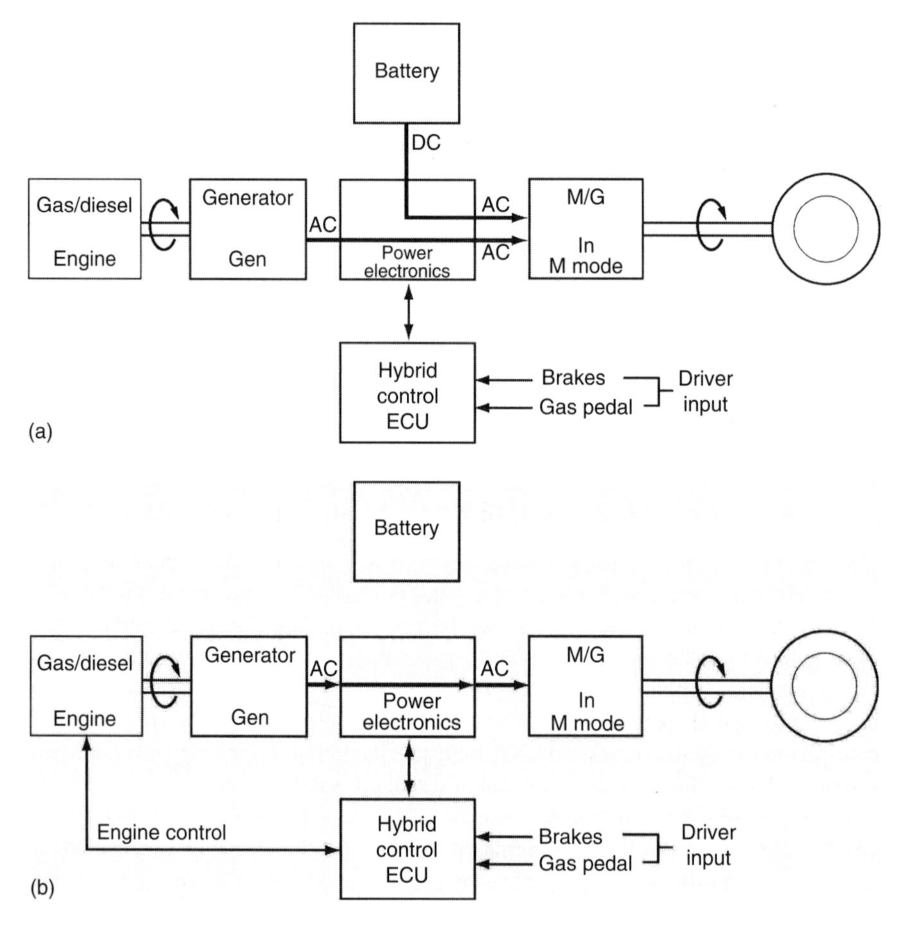

FIGURE 4.3 Components and geometry for series HEV. (a) Acceleration, going up hill, or operating in deep snow; (b) normal cruise operation.

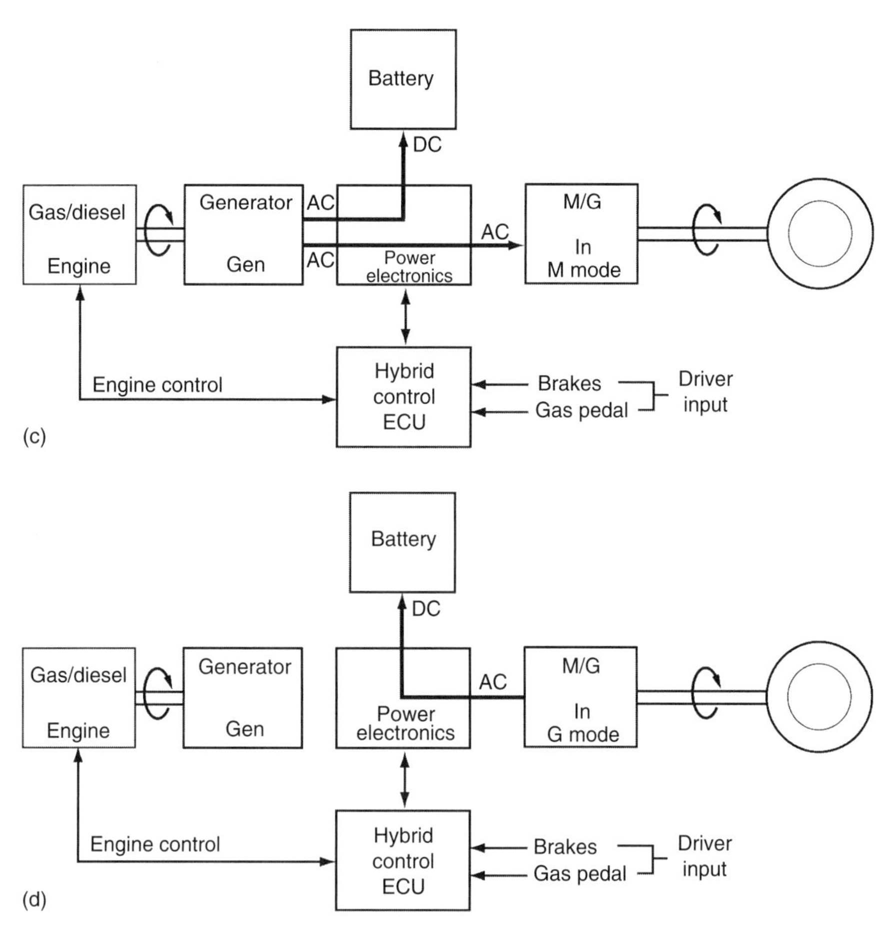

FIGURE 4.3 (continued) Components and geometry for series HEV. (c) Battery charging mode; and (d) regenerative braking.

Likewise, the traction motor, which is shown as M/G in M-mode in Figure 4.3a, must have power equal to total vehicle power for propulsion. The series HEV has double conversion of energy:

$$\text{Mechanical} \rightarrow \text{Electrical} \rightarrow \text{Mechanical}$$

In terms of components for this example, the energy flow is as shown.

$$\text{Gas engine} \rightarrow \text{Electrical generator} \rightarrow \text{Electrical motor} \rightarrow \text{Differential gears}$$

All mechanical power is converted to electric and then back to mechanical. Each conversion has associated losses.

Although a disadvantage relative to weight, the high power traction M/G does yield an advantage precisely because of its high power. During regenerative braking, large braking power may be available. If the M/G in G-mode does not have enough power rating to absorb the braking power, friction brakes are used. Regenerative braking efficiency suffers.

Designers and builders of an HEV must keep in mind that the hybrid driving experience cannot be too different from that of a CV. When stuck in a traffic jam, running the engine may be necessary to drive hotel loads such as air-conditioning; the car is stopped yet the engine seems to be racing. How strange! An alternate scenario may occur; the engine is decoupled from the drive shaft and required battery charging may have the engine running at 2500 rpm when the car is stopped.

The unexpected engine noise at constant rpm may be disconcerting to the driver and the passengers especially since the driver does not have his foot on the accelerator pedal. The engine noise is not necessarily loud, just unexpected.

To understand some of the complexity of the power electronics, note the type of electrical current at various locations and operating modes shown in Figure 4.3a through d. The battery is, of course, DC while the generators and M/G are usually AC. The power electronics converts back and forth between DC and AC. In addition, the direction of the current flow changes as operating mode changes.

DISCUSSION OF VARIOUS OPERATING MODES: SERIES

Figure 4.3a is the appropriate mode for acceleration, going uphill, or operating in deep snow. The heavy, solid, black lines represent flow of electrical power. Both the battery and generator are tapped to supply power.

Figure 4.3b shows the arrangement for normal cruise using the power from engine only. The battery is essentially bypassed; this assumes the battery has a satisfactory SOC. Because the generator can deliver the required power at different rpm, the engine can operate on its ideal operating line for minimum fuel consumption.

Figure 4.3c illustrates the division of generator power for charging the battery. Part of the generator output is routed by the power electronics to the battery. The remainder of the power goes to the M/G in M-mode. The M/G is the traction motor that supplies the torque to drive the vehicle. Power available at the wheels is obviously reduced.

One feature of hybrids is the ability to recapture some of the energy used to accelerate the vehicle. This is known as regenerative braking (Figure 4.3d). For regenerative braking, the M/G is in the G-mode. The engine can be turned off if desired. The torque slowing the wheels opposes shaft motion of the M/G which removes energy of vehicle motion and generates electrical power.

The starting of the engine by the generator is not shown. The generator connected to the engine shaft is also an M/G. For starting the engine, the M/G is in M-mode. The battery delivers the power. The electrical connection to the traction M/G is open.

HEV PROPULSION DESIGN OPTIONS: PARALLEL

Examine Figure 4.4 for the components of a parallel HEV. With parallel layout, a direct mechanical connection can be traced from the M/G to the drive wheels. Likewise, a direct mechanical connection

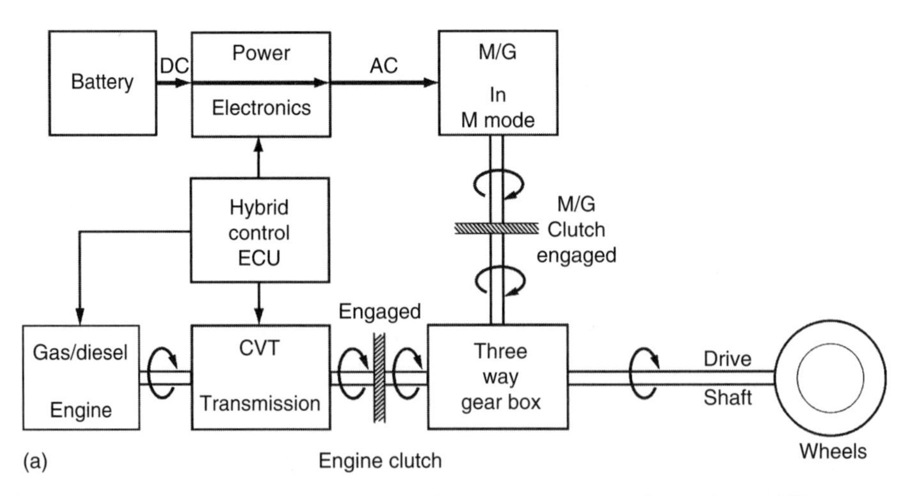

FIGURE 4.4 Components and geometry for parallel HEV. (a) Acceleration, going up hill, or operating in deep snow.

(continued)

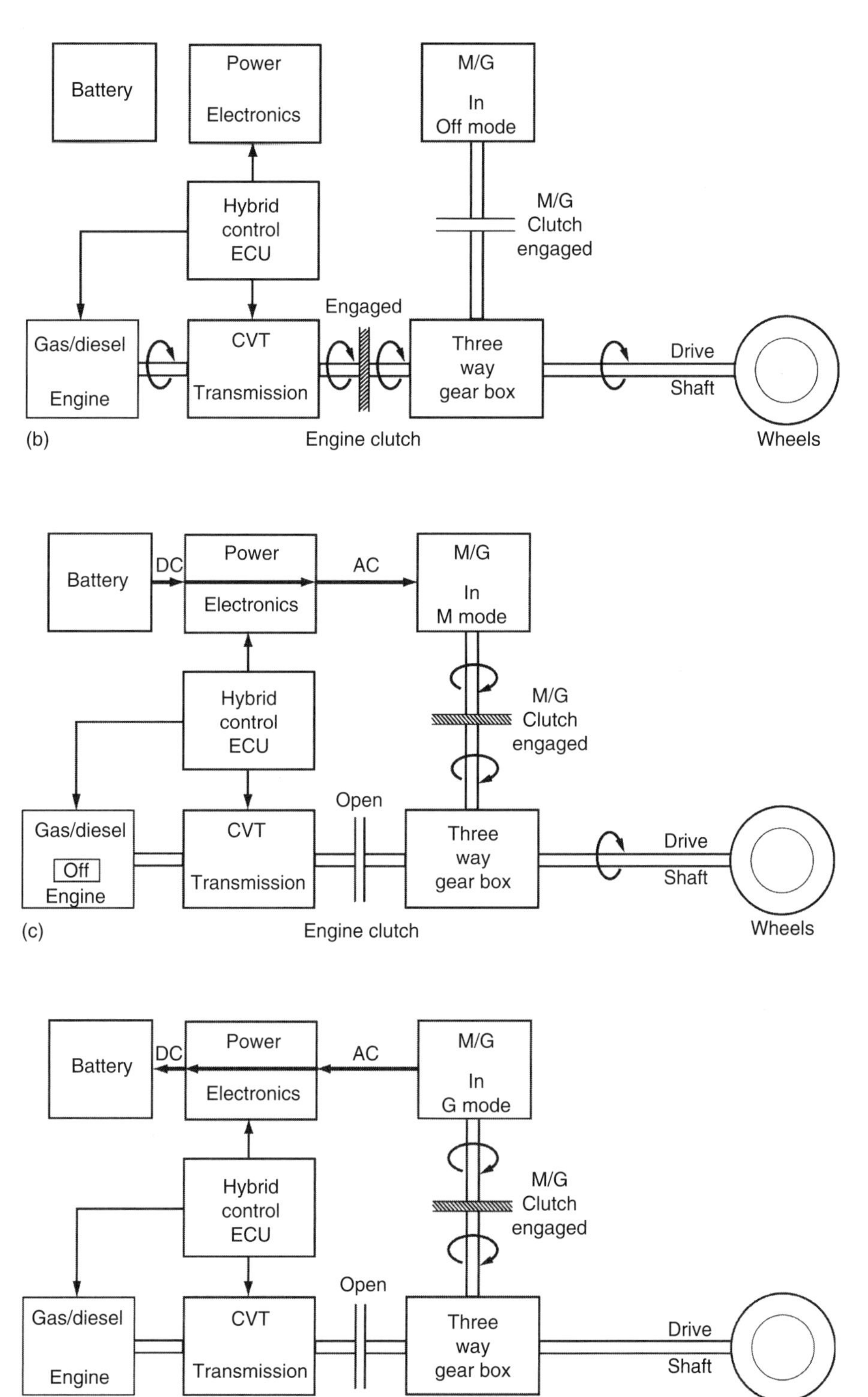

FIGURE 4.4 (Continued) Components and geometry for parallel HEV. (b) normal cruise operation; (c) electric-only mode; and (d) regenerative braking.

can be made from the engine to the drive wheels. Because of this direct mechanical connection a transmission is needed to match engine speed with drive shaft speed. In this application, a continuously variable transmission (CVT) is superior to a transmission with a limited number of fixed gear ratios.

Among the advantages of a parallel hybrid is the single energy conversion for both electrical and mechanical. This single energy conversion is shown diagrammatically as

Engine mechanical → Drive shaft mechanical

Motor electrical → Drive shaft mechanical

Except at low battery SOC, the engine runs only when the car is moving. Parallel hybrid does not need the heavy component, which is the alternator/rectifier or generator of Figure 4.3.

Considerable design flexibility exists in selecting the size of the M/G. The M/G can be sized to meet the required function at somewhat less than full vehicle power.

Some features of the parallel hybrid are the ability to operate with engine alone, electric motor only, or with both motor and engine supplying torque. The electric-only mode is very important for improved mpg during stop-and-go operations. The parallel hybrid can respond to the demand for large, near instantaneous changes in either torque or power. In contrast, the series hybrid is slower. The fast response is an advantage in traffic.

Among the disadvantages of parallel arrangement are the various added powertrain parts such as added clutches and transmissions. Typically a three-shaft transmission is needed; two input shafts and one output shaft. See Figure 4.4 which shows a three-way gearbox. Further, the engine throttle setting varies causing difficulty in controlling emissions. Varying engine throttle setting causes engine operation at higher specific fuel consumption. Because of the need to mechanically connect the engine with the drive shaft, choices for location of the engine are limited.

This is a repeat of an earlier discussion aimed at series hybrid. The same comments apply to power electronics for parallel hybrid. Note the type of current at various locations and operating modes shown in Figure 4.4a through d.

DISCUSSION OF VARIOUS OPERATING MODES: PARALLEL

Figure 4.4a shows the power flow for acceleration, climbing a hill, operating in deep snow, or gooey mud. This is the operating mode providing maximum power. From the electrical side, the battery is feeding power into the M/G set in M-mode. The M/G clutch is engaged and torque is fed into the three-way gearbox. On the engine side, engine torque is routed through and adjusted by the transmission. The engine clutch is engaged with engine torque being combined in the three-way gearbox. As a result, the HEV moves briskly down the road with a happy driver.

Normal cruise operation using the engine only is illustrated in Figure 4.4b. The M/G is turned off and power flow out of battery is zero. The M/G clutch is open. All power is supplied by the engine. The engine clutch is engaged and torque is applied to the drive shaft. The hybrid control determines the best gear ratio for the CVT and issues a command for that ratio. The gear ratio for the CVT is set to yield minimum fuel consumption.

An important mode of operation for achieving good mpg is electric-only. This mode is shown in Figure 4.4c. Frequently in this mode, the engine is not running. The engine clutch is open while the M/G clutch is engaged. The battery provides the power as shown by the heavy black line going through the power electronics box. The M/G is in M-mode creating torque that is transmitted by the shaft to the three-way gearbox.

In a sense, regenerative braking is the reverse of the electric-only mode. Compare Figure 4.4c and d. Shaft motion is in the same direction; however, the torques are reversed. The settings of the clutches are the same for each mode. The flow of power is reversed because of reversal of torques.

COMMENT ON TOYOTA PRIUS

Through clever use of a planetary gear set, several components shown in Figure 4.4 can be eliminated. *The planetary gear set replaces both the three-way gearbox and the CVT*. The clutches are also eliminated. Refer to Chapters 8, 11, and 16 for additional information on the Toyota transmission.

CHOICE OF SERIES, PARALLEL, OR MIXED

Series only or parallel only designs often do not meet performance requirements. An example is the Toyota Prius hybrid that has a mixed series/parallel design. The choice of series, parallel, or mixed hybrid depends on the driving cycle, for example, highway, freeway, or urban. The choice also depends on the vehicle function, for example, passenger car, taxi, truck, or bus.

As hybrid technology developed, the utility of series or parallel design became less significant. Mixed designs, rather than series or parallel designs, offer more flexibility. Also, in the early days of hybrids, power level was considered to be a defining factor for selection of series or parallel. Parallel was deemed appropriate up to 150 kW with application to the usual passenger car. Series design was considered appropriate above 150 kW with application in heavy duty vehicles such as trucks. The selection based on power level may not apply; the selection rule based on 150 kW is by no means rigid.

TECHNOLOGIES FOR A SUCCESSFUL HYBRID

Hybrids are complex and involve many different technologies. Here is a sampling of technologies along with a sampling of comments.

- *Electronics:* Improve efficiency of power transistors and operation safely at higher temperature.
- *Electronics:* Improve computers and microcontrollers especially with regard to cost.
- *Electromechanics:* Improve efficiency of motors and generators.
- *Electrochemistry:* The heart of batteries, ultracapacitors, and fuel cells.
- *Software:* Developing complex algorithms and strategies for control.
- *Materials:* Develop low cost, lightweight structures.
- *Simulation tools:* Essential for timely and economical hybrid design.
- *Basic data:* The different driving cycles, e.g., number of miles stop and go, etc.
- *Testing facilities:* Cold and hot weather testing (batteries and fuel cells are temperature sensitive).
- *Testing facilities:* Hardware in loop that allows testing of a component or components with the overall hybrid control system. Part of the test consists of hardware; the remainder consists of computer-generated inputs and outputs.
- *Advanced concepts:* For gasoline engines.
- *Advanced concepts:* For diesel engines.
- *Biology:* Development and usage of biofuels.

OPERATION WITH A "DEAD" BATTERY

To achieve long battery life, batteries are not usually discharged below SOC = 50% or so. As a result, "dead" is not SOC = 0% but SOC approximately 50%. When the battery is dead, power is still supplied by the gasoline or diesel engine. For example, if the HEV has a motor with a power of 30 kW and engine of 70 kW, then when the battery is dead, the engine can still supply the 70 kW power. Obviously, performance is degraded by a dead battery. Usually battery recharge is delayed so that all engine power is available for propulsion.

Hybrid functions are tested in deep snow, on muddy roads, or climbing a long, steep hill. By initially providing motor assist, the battery goes dead. Vehicle performance falls back on the engine only. Stop-and-go, creep-and-crawl city traffic taxes the battery and the control system. The dead battery must be avoided otherwise the benefits to improved fuel economy are lost.

SUMMARY

A hybrid vehicle combines any two power sources for propulsion of the vehicle. A special case within that broad definition is the HEV. Merging the components from pure electric vehicle (EV) and the conventional, pure gasoline vehicle yields an HEV.

Series, parallel, and mixed hybrids are described. Operating modes for both series and parallel designs are illustrated. The modes show the flow of power through various parts.

Hybrid operation with a dead battery falls back on the power of the gasoline engine only. A sampling of hybrid technologies is given.

First, is this hybrid vehicle classified as a series or parallel design? The hybrid bike is a parallel design. Second, what would need to be added to the kit to allow regenerative braking by the bicycle? Lots of expensive parts. Add to the kit an M/G and a battery. This makes the bicycle a three-way hybrid (*tribrid?*) having propulsion from the gasoline engine, the bike rider, and the M/G in M-mode. Adding the parts for regenerative braking very likely make the bike too heavy. Oh well!

5 Why Hybrids Get Good Mileage

Regenerative braking is almost perpetual motion. Note the word "almost."

INTRODUCTION

A hybrid vehicle has several design features and technologies that help it achieve increased mileage. Some of these features are hybrid specific and are not found in conventional vehicles (CVs). Among the hybrid-specific items, the major contributors to improved miles per gallon (mpg) are known as the heavy hitters. The devices that offer smaller, yet important, gains are called nickel and dime. All the minor gains add up and prove useful in increasing the mileage. Another class of techniques for better mpg is not unique to hybrids and is shared with CVs. These constitute the class of technologies that are common with other vehicles. CVs and hybrids vie for the consumer's money.

HYBRID SPECIFIC

Heavy Hitters Lineup

Major mpg gains are because of unique hybrid features, such as regenerative braking, electric-only propulsion, motor assist, and downsized engine. When combined, they yield 30%–40% gain in mpg.

Regenerative Braking

Regenerative braking recovers the energy of motion (kinetic energy) by electrical braking during stops. Electrical braking is in contrast to the more common friction braking. The stopping torques developed by the generator can also recover kinetic energy while slowing down. Once the driver takes his/her foot off the gas pedal, regenerative recovery of energy is activated. Also of significance is the recovery of energy (potential energy) while coasting downhill. See Chapter 9 for a full discussion.

An example of electrical braking is the exercise bike in the gym. Exercise bikes do not need to be plugged in; the rider generates electricity to power the display. The torque created by the generator opposes the motion of the pedals. When resistance level is increased, the added electrical power being generated is dissipated in a resistor buried somewhere in the bike.

Regenerative braking may yield 10%–15% fuel savings. A dual-purpose motor/generator (M/G) is required. When braking electrically, the M/G is in the generator mode (G-mode) as shown in Figure 5.1b. Arrows are shown for direction of shaft and wheel rotation as well as direction of the torque on the shafts. In the M-mode, the torques and wheel rotations are in the same direction; see the left-hand side of Figure 5.1. For opposing torques and rotations, which occurs in G-mode, electrical power flows into the battery. In G-mode, the hybrid is like a giant gym bike. The efficiency of regenerative braking is discussed in Chapter 9.

Regenerative braking may or may not be accurately described in the marketplace. Some car salespersons are more knowledgeable than others. Those who do not understand regenerative braking simply heap praise on it. Some descriptions border on perpetual motion machines. The consumer is normally aware only of overall mpg, and does not bother to check out the contributions of regenerative braking to the resultant mpg.

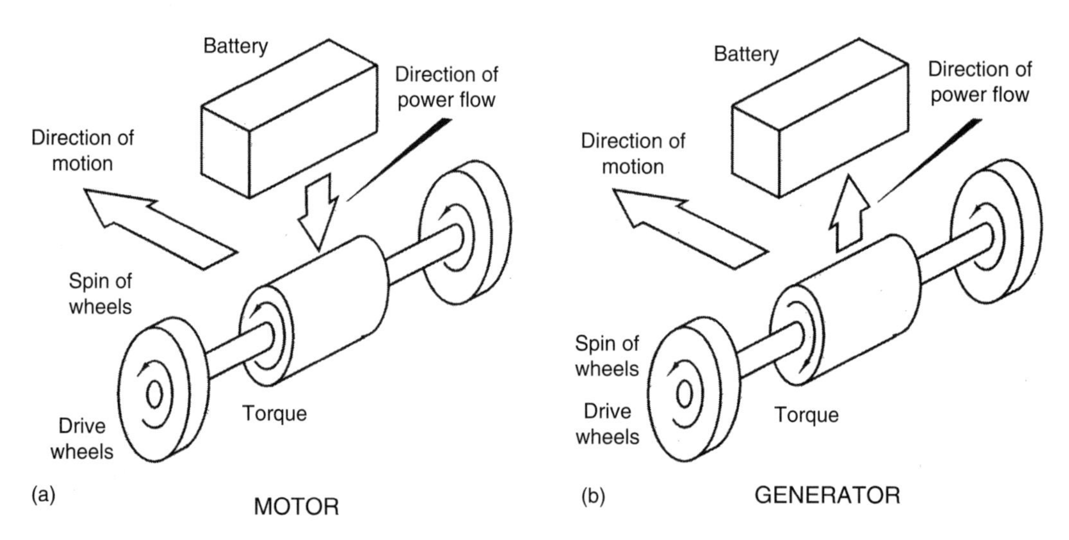

FIGURE 5.1 Regenerative braking. Note the direction of wheel spin and torque for (a) motor (motor assist) and (b) generator (regenerative braking).

Engine Downsizing

There are several reasons for engine downsizing. For instance, a loafing engine has poor fuel economy whereas a small, hardworking engine has better fuel economy. A smaller engine is possible because of the added power from the electrical motor; when power required exceeds that of the downsized engine, the electrical motor cuts in (see Figure 4.2).

Figure 4.2 is a graph of horsepower (hp) required to accomplish one of the three driving events: accelerating from a stop sign, cruising on the freeway, and climbing a hill. These events are arrayed across the top of the figure. The dash-dot-dash line is the installed horsepower for a pure gasoline car also known as CV. Whatever be the maximum horsepower for any part of the driving cycle, the installed power must be of that magnitude. For this figure, maximum power occurs at the start of acceleration. The heavy black line initially falls as vehicle gains speed. This is followed by a flat section at constant horsepower cruising on the freeway. The heavy black line climbs upward as the hybrid climbs a hill. The dashed line, which shows the power from the downsized engine, remains constant in this example.

Note that the power from the hybrid gasoline engine exceeds the horsepower required to cruise on the freeway. The excess power from the engine can be used to charge the battery. The battery power out (discharge) and battery power in (charge) are shown along the bottom of the graph. The gray-shaded areas are regions where the motor provides motor assist. Of course, this occurs only when required power exceeds engine power.

The charge and discharge of the battery is analogous to your checking account. Charge is the same as a deposit to checking; discharge is the same as cashing a check. Just as for a checking account, if check withdrawals exceed deposits, the checks bounce. If more power is withdrawn from the battery than put back in, the battery is an account overdrawn or the same as a dead battery.

A small engine lessens pumping loss, which is explained in more detail in Chapter 17. Pumping loss is due mainly to low pressure in the cylinder during the intake stroke. Suction lowers pressure acting on the piston. Variable valve timing has a positive effect on pumping loss as well on the torque and specific fuel consumption (SFC) curves. As discussed in subsequent paragraphs, the shift of belt, gear, or chain-driven auxiliary components to electrical motors effectively downsizes the engine by the amount of auxiliary power.

Figure 5.2 has a graph showing fuel consumption on the right-hand ordinate (vertical axis) as a function of engine load on the abscissa (horizontal axis). On the left-hand ordinate is the ratio

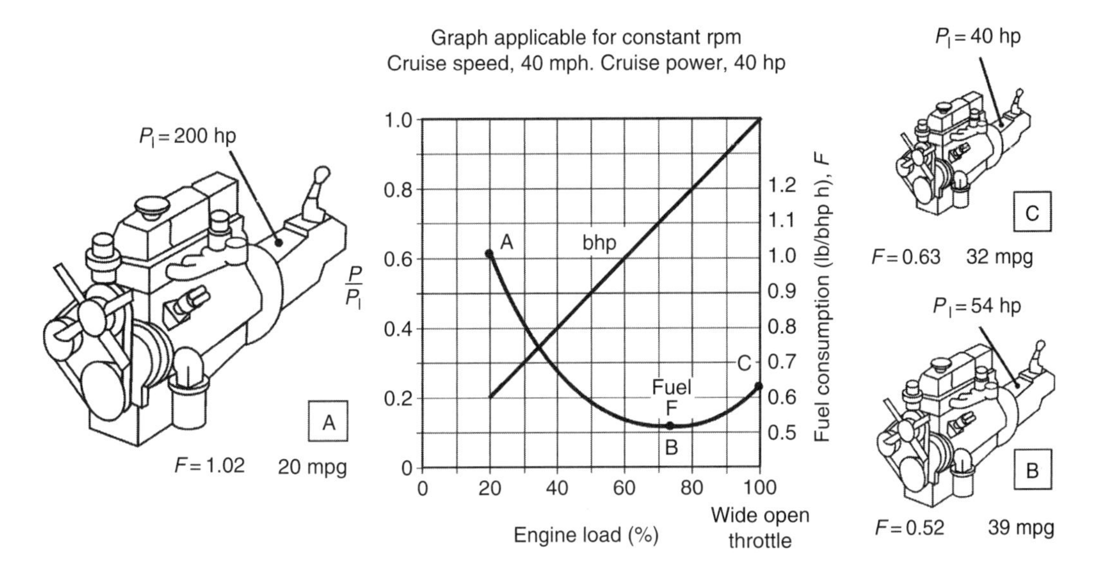

FIGURE 5.2 Benefits of downsizing the engine.

of engine power to maximum installed engine power, P/P_I. The curves are valid only for constant engine revolutions per minute (rpm). The engine load is expressed in terms of percentage of engine brake horsepower (bhp). The linear curve is bhp. Fuel consumption, which is in units of pounds of gasoline per bhp hour, is the curve with a bucket. The bottom of the bucket is located at minimum fuel consumption. When the fuel consumption curve goes upward, that is bad for mpg. Letters A, B, and C appear along the curve and correspond to the three engines with different horsepowers.

Figure 5.2 has three engines at different geometric scales to indicate the different installed maximum engine power. Subscript I is for installed. Each of the three engines has three numbers: F, which is fuel consumption; P_I, which is the installed maximum power at 100% wide open throttle (WOT); and mileage (mpg) to be expected from the particular engine.

The vehicle of Figure 5.2 is operating in cruise with a sp eed of 40 mph and a cruise required horsepower, P = 40 hp. The ratio, P/P_I, varies from engine to engine and is equal to the throttle setting.

Think of the large engine illustrated in Figure 4.2. That is engine A in Figure 5.2 and is intended for use in a conventional, pure gasoline vehicle. For this engine

$$\frac{P}{P_I} = \frac{40 \text{ hp}}{200 \text{ hp}} = 0.2 = 20\%$$

Point A on the graph is at 20% throttle setting or 20% engine load. Point A is at F = 1.02. Engine A is loafing along and, as a result, gives only 20 mpg.

Now focus on engine B which has been downsized so that

$$\frac{P}{P_I} = \frac{40 \text{ hp}}{54 \text{ hp}} = 0.74 = 74\%$$

At cruise, engine B is operating at minimum fuel consumption as shown by point B on the graph. From the graph, F = 0.52. The mpg is

$$\text{mpg} = \frac{0.52}{1.02}(20\ \text{mpg}) = 39\ \text{mpg}$$

Mileage increases considerably by downsizing. If small is better, perhaps engine C would be even better. However, fuel consumption has gone up from $F = 0.52$ to $F = 0.63$. Engine C provides only 32 mpg.

Based on this discussion, engine B is the ideal, downsized, engine for a hybrid electric vehicle (HEV). In Figure 5.2, the curve with a bucket has a narrow rpm range for minimum fuel consumption, F. F is near the minimum value only for throttle settings between 65% and 85%. The engine illustrated in Figure 5.2 does not have variable valve timing. With variable valve timing, the range of rpm for minimum F is much broader (see Chapter 10).

Electrical-Only: Motor Assist

When stopped at a traffic light, the engine can be shut off saving fuel. The use of electric-only mode to launch the vehicle from zero speed and for low speeds does delay starting the engine. For low acceleration, electric-only is feasible. For high acceleration, both motor and engine are necessary. Motor assist occurs when both the motor and the engine are operating.

The high torque electric motor allows the engine to operate within a narrower rpm band. A lower redline for rpm is feasible, and a high revving engine not necessary. Further, careful integration of the electric motor with the transmission allows the engine to operate at its best load points. The engine can operate at the best rpm and throttle setting.

Electrical-Only: Plug-In Hybrid Fuel Economy

How many mpg do you want? 100 mpg? 200 mpg? 1000 mpg? With suitable calculations, take your pick. To derive the equation for plug-in hybrid fuel economy, define some symbols:

D_T = Distance, total trip, miles
D_0 = Distance, electric-only, miles
D_E = Distance, engine operating, miles
G = Gasoline consumed, gallons
F_E = Fuel economy with engine operating, mpg
$F_{=D}$ = Effective fuel economy of plug-in hybrid, mpg
f = Ratio D_0/D_T

Some creative symbol selection has been used. The subscript = D in $F_{=D}$ is suggestive of the two prongs on the electrical plug which is plugged into the wall socket. The symbol f is the fraction of the total trip that is electric-only. Large f enhances $F_{=D}$. Three conditions exist:

$$D_0 > D_T,\ \text{then}\ F_{=D} \to \infty$$
$$D_0 = D_T,\ \text{then}\ F_{=D} \to \infty$$
$$D_0 < D_T,\ \text{then}\ F_{=D}\ \text{is equal to Equation 5.2}$$

The gasoline consumed during the trip is

$$G = \frac{D_E}{F_E} = \frac{D_T - D_0}{F_E} \tag{5.1}$$

The effective fuel economy of the plug-in hybrid, $F_{=D}$, becomes

$$F_{=D} = \left(\frac{D_T}{D_E}\right) F_E = \frac{D_T F_E}{\left(D_T - D_0\right)} = \frac{F_E}{(1 \quad f)} \tag{5.2}$$

For sample calculations, assume $D_0 = 610\,\text{mi}$ and $F_E = 40\,\text{mpg}$. For a trip of 70 mi, that is

$$D_T = 61 \text{ mi}, \; f = \frac{60}{61} = 0.983, \;\; F_{=D} = \frac{40 \text{ mpg}}{1 - 0.983} = 2440 \text{ mpg}$$

Now suppose the trip is

$$D_T = 70 \text{ mi}, \; f = \frac{60}{70} = 0.857, \;\; F_{=D} = \frac{40 \text{ mpg}}{1 - 0.857} = 280 \text{ mpg}$$

Finally assume

$$D_T = 300 \text{ mi}, \; f = \frac{60}{300} = 0.20, \;\; F_{=D} = \frac{40 \text{ mpg}}{1 - 0.20} = 50 \text{ mpg}$$

The increasing values of D_T reveal a trend. As D_T increases, $F_{=D}$ decreases, which agrees perfectly with your intuition. Greater use of the engine decreases effective mpg. For a quoted mpg for a plug-in hybrid, the essential data to accompany the claim must be: D_T, D_0, D_E, and F_E. Without these four values, the claim is unsubstantiated. Do not call the claim a scam but be skeptical.

Nickel and Dime Improvements

Many of the technologies offering lower percentage improvements in mpg are discussed in Chapter 11. A number of these nickel and dime technologies do offer the least cost and least modification of a production vehicle to create a hybrid. It is a case of small gains in mpg but accomplished economically.

Engine Idle-Off: Start–Stop

Driving cycles are important for optimizing hybrid performance. Cars in an urban driving cycle idle typically 20% of the time. By engine idle-off, mpg can be improved by 5%–8%. For frequent engine idle-off and restart, high-rpm spin at restart is necessary to spare the driver any discomfort. Start–stop creates temperature variations in the engine, which adversely affect emissions. To lessen start-up emissions, fuel injection can be delayed until the engine is spinning near the idle rpm.

Automobile Auxiliary Components

Currently, most of the items listed below are driven by the engine. How many can be electrically driven? Are gains in mpg possible by using a motor to drive the auxiliaries? Consider first, those auxiliaries that have as the source of power the engine belt driven alternator/rectifier. A typical luxury car has alternator power of 2–4 kW.

Auxiliaries in this category include

Entertainment systems	Active suspension
Window defrost	Active torque applied to roll bars
Ignition system	Ferromagnetic shocks
Electrically heated seats and mirrors	Electric windows
Power door locks	Headlights and running lights
Engine cooling fan; typically electrically driven	Electrically heated catalyst
Electrically heated windshield wiper fluid	Navigation systems
Radar and sonar warning systems	Sunroof
	Cabin ventilation

Electromagnetic valve actuation is discussed in Ref. [2]. Other auxiliary power absorbers are belt or gear driven by engine directly or indirectly. These include

Heating	Air-conditioning
Power steering	Antilock brakes and stability control
Water pump	Oil pump
Fuel pump and fuel injection	

Special Focus on Air-Conditioning

For a small car, the air-conditioning (A/C) requires 30%–50% of cruise power. A/C is not a trivial problem. An approach to reducing the A/C load is to use infrared absorbing windows and thermal insulation throughout the passenger compartment. When parked in the hot sun, the use of solar-powered cooling fans is proposed. Certain hybrids offer a control button to decrease the level of A/C to gain better mpg.

The National Renewable Energy Laboratory (NREL), which is part of the Department of Energy (DOE), conducted tests to determine improved fuel economy using means to reduce heat load in cabin [1]. Techniques used to reduce heat load in the cabin were

- Photovoltaic sunroof vent fan system (to reduce heat when parked in sun)
- Headliner insulation
- Infrared interior window reflective film especially on windshield
- Infrared reflective film on roof of vehicle
- Special glass in windows

Temperatures in the cabin after a heat soak on a 95°F day were reduced by 4°F at foot-level and 14°F at dashboard.

Tripled Savings by Shifting the Auxiliary Load

Improvements to lower the power consumed by accessories is a cost-effective way to increase mpg.

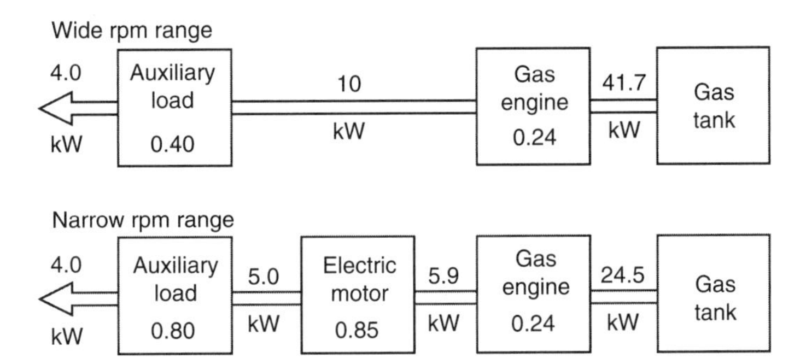

FIGURE 5.3 Shifting the auxiliary load from the engine to an electrical motor.

First, the low efficiencies of wide-rpm band, engine-driven components are eliminated. Obviously, the rpm of belt, gear, or chain-driven components vary with engine speed. At low speed, the auxiliaries must deliver adequate output. At high rpm, the efficiency drops off.

Second, the very low efficiency of engine (15%–25%) is replaced by the high efficiency of an electrical motor (80%–95%).

Third, the hybrid acquires the high efficiencies of narrow-rpm band, electrically driven devices. Efficiency is greatly improved because of the inherent electrical efficiency. Also auxiliary efficiency is greatly improved because of near fixed rpm operation. Besides the high efficiency advantage, other advantages accrue. When electrically driven, power is available regardless of engine status. Auxiliaries can be operated independent of the engine.

Figure 5.3 shows the gains by shifting the auxiliary load. For engine-driven auxiliaries operating over a broad rpm range, gasoline sufficient to provide 41.7 kW is withdrawn from the tank. For electrically driven auxiliaries, the power of the gasoline is only 24.5 kW. This is a very significant savings.

Damping Driveline Oscillations and Shudders

Abrupt fuel shut off often causes driveline oscillations and shudders. To use less fuel, the flow to the injectors should be turned off whenever the driver removes pressure on the gas pedal. As a result, the engine may shake and send shudders throughout the vehicle. The torques from the M/G can be used to neutralize these disturbing vibrations. Timely shutoff of fuel does yield a small percentage increase in mpg.

Another example applies only to those hybrids using a so-called slush-box transmission, that is, an automatic transmission coupled to the engine by a torque converter. Early torque converter lockup reduces transmission losses and saves fuel; however, the lockup sends shudders and other disturbing motions down the driveline. Once again, the torques from the M/G can be used to neutralize the disturbing vibrations. Timely, that is, prompt, lockup does yield a small percentage increase in mpg.

Vehicle Launch

High acceleration launch from zero speed is difficult for a conventional, gasoline vehicle. The engine torque is low and the torque requirements are high. The usual solution is a very low first gear. For a hybrid, even a small electric motor offers considerable torque and can help attain the first forward motion. High acceleration vehicle launch is part of motor assist occupying the low mph region starting from zero speed. Motor assist acts from 0 to 20 mph upward to 40 mph.

Common with CVs

Also-Ran

The words “also-ran” are intended to stress the commonality of a large pool of technology used by both hybrids and CVs. Because of the enormous scope of the common pool of technology, one viewpoint is that hybrids are but a minor variation of the CV. The variation may be minor but it is truly complex.

Electrical Power Steering

Hydraulic power steering uses shaft power from the engine to drive a hydraulic pump that in turn supplies the muscle (pressurized hydraulic fluid) to steer the car. Electrically actuated steering is more efficient and saves gasoline. For passenger cars, the 12V system is adequate for electrical power steering; however, for heavier vehicles (some light trucks and SUVs) the new 42V system is necessary.

Fast Warm-Up

To lessen cold-start pollution, methods for fast engine warm-up are necessary. Fortunately, fuel consumption is also lessened.

Aerodynamics

At even modest speeds, the major force retarding the motion of a car or truck is aerodynamic drag. In the 1930s, the first attempts were made to reduce aerodynamic drag. Art deco and streamlined shapes went hand in hand. The aerodynamic design was based less on science than on "what looks good." In this century, aerodynamic design has a firm foundation in theory, tests in wind tunnels, and computational fluid dynamics (CFD). Surprisingly, some of the 1930s aerodynamic designs had drag coefficients as low as modern cars. The Lincoln Zephyr and the Chrysler Airflow of 1930s had favorable drag coefficients. Many of the box-like sedans of the late 1920s had lower aerodynamic drag going backward than going forward!
Aerodynamic drag, D, is given by

$$D = \text{Drag} = \tfrac{1}{2}\rho V^2 C_D A \tag{5.3}$$

where
- D is the aerodynamic drag, N
- ρ is the air density, kg/m^3
- V is the vehicle velocity, m/s
- C_D is the drag coefficient, dimensionless
- A is the frontal area of vehicle, m^2

Power to overcome aerodynamic drag, P_A, in watts is

$$P_A = DV \tag{5.4}$$

Hence, combining equations gives

$$P_A = \left(\tfrac{1}{2}\rho V^2 C_D A\right)(V) = \tfrac{1}{2}\rho V^3 C_D A \tag{5.5}$$

The power to overcome drag varies as cube of the vehicle velocity. In the automotive press, the statement is made occasionally that P_A varies as the square of V; however, it varies as the cube of V. This statement is true for high speeds and upward; other retarding forces become less important. Chapter 12 introduces the concept of a crossover velocity, V_{CO}. When V exceeds V_{CO}, aerodynamic drag becomes the dominant retarding force.

Reduction of aerodynamic drag usually requires the vehicle body to be redesigned. Hence, the reduction of drag can be seriously considered only when the vehicle is being redesigned as a new model.

Although not aerodynamic drag, a discussion of hill climbing is included here. The equations will later be integrated into the discussion of hill climbing with a dead battery. The power to climb a hill is given by

$$P_{\mathrm{H}} = mgV\sin\theta \tag{5.6}$$

where

m is the mass of vehicle, kg
g is the acceleration of gravity, 9.8 m/s^2
θ is the steepness of the hill, angle between horizontal and the road surface

Consider the HEV discussed in "Operation with a 'Dead' Battery" in Chapter 4. In that discussion, the battery and M/G had a power of 30 kW while the gasoline engine had a power of 70 kW. To determine the loss of performance because of dead battery, assume the HEV is traveling at 80 mph (133 km/h). Assume that drag is due solely to aerodynamic drag; at 80 mph this is a reasonable assumption. What is the vehicle velocity when the battery is dead? As the discussion of Equation 5.3 shows, the power varies as cube of velocity. Therefore

$$V_2 = V_1\left(\frac{P_2}{P_1}\right)^{1/3} \tag{5.7}$$

Inserting numerical values

$$V_2 = 80\text{ mph}\left(\frac{70\text{ kW}}{100\text{ kW}}\right)^{1/3} = (0.89)(80\text{ mph}) = 71\text{ mph.}$$

The decrease because of a dead battery is not that significant because of the cube dependence.

For hill climbing, assume a velocity of 30 mph. Also assume that the aerodynamic and rolling friction drags are much smaller than the forces of hill climbing. Equation 5.4 shows that power varies linearly with velocity. What is the reduced velocity going up the hill with a dead battery?

$$V_2 = V_1\left(\frac{P_2}{P_1}\right) = 30\text{ mph}\left(\frac{70\text{ kW}}{100\text{ kW}}\right) = 21\text{mph} \tag{5.8}$$

A dead battery has a greater influence for hill climbing than for aerodynamic drag on level road.

Rolling Resistance of Tires

The rolling resistance of tires varies over a broad range. For the origin of rolling resistance see Figure 9.4b. The magnitude of resistance depends on the tire design. Some tires are designed for the fastest time going around a curve; these tires are selected for their handling capability. Other tires provide a comfortable ride with less handling capability. Other tires are designed for low wear and long life. The first assumption is that a hybrid designed for maximum mpg should have the absolute lowest rolling resistance; however, the marketplace often dictates otherwise. In a switch from one model year to another, Toyota changed the standard Prius tires from minimum resistance to a more comfortable ride.

Vehicles intended for significant off-road operation or for carrying large loads may not be able to use the low-rolling resistant tires.

Vehicle Weight

Vehicle weight is so important to good mpg that an entire chapter (Chapter 7) is devoted to that topic. Two approaches reduce obesity. First, weight can be reduced by using lightweight materials. Second, weight can be minimized by better design methods using the finite element method (FEM) and computational methods for stress and strength analyses. The aerospace industry adopted FEM about a decade before the automobile industry.

Reduced vehicle weight must be achieved within constraints. Other factors to be considered include lower cost, retention of strength, structural safety, unchanged vehicle performance because of weight reduction, and the use of recyclable materials. To provide advanced technology by 2010, the DOE, weight-reduction R&D goals include a 50% reduction of vehicle weight for structure and subsystems, use of affordable materials, increased use of recyclable and renewable materials, and materials suitable for high-volume manufacturing.

Engine Efficiency

CV and HEV mileage depends on efficiency of the gasoline or diesel engine. Several topics related to engine efficiency are discussed in Chapter 17. One crucial topic, important to HEV but less so to CV, is the Atkinson cycle. Another topic receiving considerable attention is heat recovery from the engine. The goal is to recover a portion of the heat dumped into the exhaust and radiator. The typical gas engine efficiency is 25%; the other 75% of energy from the fuel goes out the exhaust pipe or the radiator.

Several new engine technologies are receiving attention such as the stoichiometric, spark ignition, and direct injection engine. Another is the Mercedes Benz DiesOtto engine also known as the homogeneous charge compression ignition (HCCI) engine. Chapter 17 discusses these and other technologies.

A clever combination to improve power, reduce emissions, and reduce fuel consumption is dual injection used by Lexus V-6 in both CV and HEV applications such as hybrid Lexus GS450h. Injection can be into the manifold or directly into the cylinder. Manifold injection is called port injection. Fuel injected into the manifold passes through the intake valves enroute to the cylinder. With dual injection, both port and direct injections are used. At certain engine rpm and loads, the injection is mixed. For instance, at 2300 rpm at half throttle, the injection may be 30% port and 70% direct.

Cylinder deactivation, also known as displacement on demand (DoD), is compatible with existing engines. This technology results in a 10%–15% improvement in fuel efficiency.

Use of dual overhead cam (DOHC) allows variable valve timing on both the intake and exhaust valves. More freedom in design of cylinder heads (location for valves, spark plugs, etc.) is possible with DOHC. Some automakers offer both the single overhead cam (SOHC) and DOHC engines in the same power range. The advantages of the more expensive DOHC translates into better mpg.

Variable valve timing for both the intake and exhaust valves expands the engine operating map. High engine torque can be obtained over a broad rpm range. The extent of the favorable SFC contours in the engine map can be increased.

Powertrain

The powertrain extends from the engine to the wheels on the road. Improvements in transmissions, differential gearing, drive shafts, and shaft bearings all contribute to vehicle efficiency and improved mpg. These topics are discussed in Chapter 17. Multiple speed automatic transmissions begin to rival the manual shift transmission in efficiency. Automatics with five-speeds, six-speeds and even eight-speeds are beginning to appear in new cars. The Lexus 460/460L are equiped with the world's first eight-speed transmission.

The continuously variable transmission (CVT) offers the ability to match more precisely the actual engine operating point (torque and rpm) with the optimum engine operating line as seen in the engine map.

As discussed in Chapter 12, four-wheel drive (4WD) or all-wheel drive (AWD) adds weight and decreases powertrain efficiency. This can be avoided by disconnecting the front axle. The fuel efficiency gains are minimal.

Improved Lubricants: Friction and Wear

Lubrication, which is absolutely essential to a long happy component life, does reduce friction and improve efficiency. But even with lubrication, some friction remains. The friction may be in the boundary layer of oil on the cylinder wall. Better lubricants can make small incremental increases in engine efficiency. This technology does not require engineering changes to engines.

Improved Fuels

The consumer thinks of gasoline as a commodity which is the same from any gas station. Gasoline must meet certain standards for energy content, volatility, impurities, etc. All gasolines are not the same; however, all gasolines meet the standards. Extensive standards exist for gasoline and diesel fuels. With the advent of ethanol, new bounds in the form of standards are needed. The standards define the boundary between the acceptable and the unacceptable. A recent change in the specifications for the allowable sulfur content in diesel fuels will enable diesels to meet stringent new pollution requirements. This change in diesel fuel composition makes the competition between HEVs and CVs more exciting.

Range and Miles per Gallon: Cruise Mode

Estimates are made for the vehicle in cruise mode. Both pure gasoline and a series hybrid are used as examples. Define symbols and state the assumed values for the specific example of cruise:

E is the energy, kWh
P is the power, 9 kW for cruise
N is the number of gallons of gas in tank less reserves, 14 gal
V is the speed (velocity), 45 mph = 75 km/h = 20.8 m/s
E_G is the energy of a full tank of gasoline less reserves, kWh
e_G is the energy density of gasoline, 33.7 kWh/gal
η is the overall vehicle efficiency
t is the time, s

Equations for Range and Miles per Gallon

Range for the vehicle is

$$R = Vt \tag{5.9}$$

Time to depletion of energy at cruise assuming 100% efficiency

$$t = \frac{E}{P} \tag{5.10}$$

Accounting for the inescapable inefficiencies or, stated another way, the efficiency

$$t = \frac{\eta E}{P} \tag{5.11}$$

By combining the preceding equations, range becomes

$$R = \frac{\eta V E}{P} \tag{5.12}$$

Equations 5.9 through 5.12 apply equally well to pure gasoline and hybrid vehicles.

Pure Gasoline Car

The efficiencies for each of the components as well as the overall efficiency are shown in Figure 5.4 Replace E in Equation 5.12 by

$$E_G = e_G N \tag{5.13}$$

Using the assumed numerical values for the specific case of cruise

$$E_G = (14 \text{ gal})(33.7 \text{ kWh/gal}) = 472 \text{ kWh}$$

From Figure 5.4,

$$\eta = 0.13$$

The time to use all of gasoline, less reserve,

$$t = \frac{(0.13)(472 \text{ kWh})}{9 \text{ kW}} = 6.8 \text{ h} = 409 \text{ min}$$

Range, R, is

$$R = \frac{(0.13)(45 \text{ mph})(472 \text{ kWh})}{9 \text{ kW}} = 307 \text{ mi} = 512 \text{ km}$$

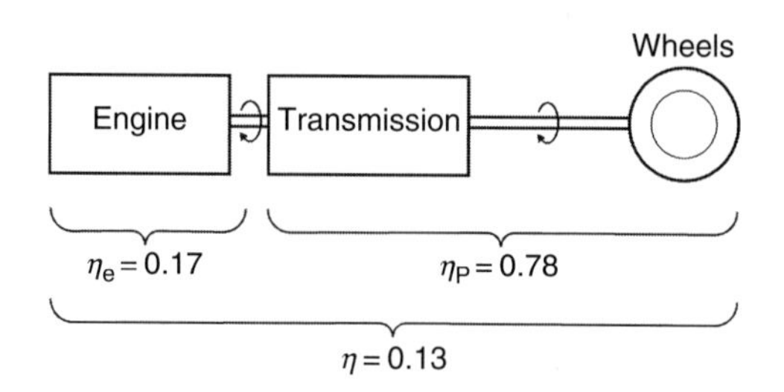

FIGURE 5.4 Components and efficiencies for a pure gasoline car. The powertrain efficiency, η_P, includes an automatic transmission with torque converter, drive shaft, differential, and effects of tire losses. Note that $\eta = (0.17)(0.78) = 0.13$.

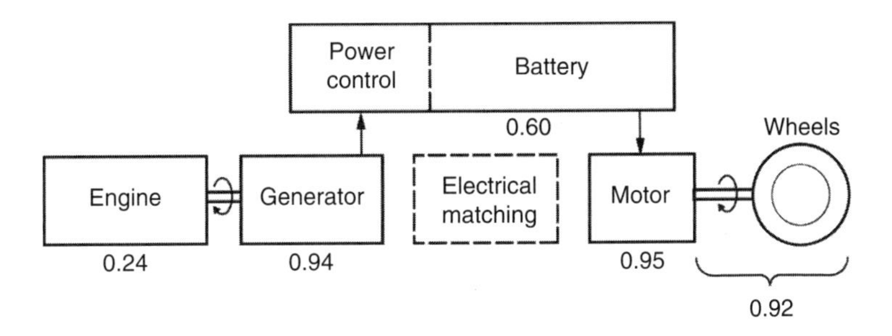

FIGURE 5.5 Components and efficiencies for a series hybrid car. The powertrain efficiency, η_P, includes only the drive shaft, differential, and effects of tire losses. Note that $\eta = (0.24)(0.94)(0.60)(0.95)(0.92) = 0.12$.

Miles per gallon is simply range, R, divided by number of gallons of gasoline, N,

$$\text{mpg} = \frac{307 \text{ mi}}{14 \text{ gal}} = 22 \text{ mpg}$$

Hybrid: Pure Electric Mode

The time and range to deplete a fraction of the energy stored in the battery is calculated. The efficiencies of Figure 5.5 are used. In addition, the fraction of battery energy that can be used is assumed to be

$$\Delta \text{SOC} = 25\% = 0.25$$

This value is based on an assumed initial state of charge (SOC) of 75% with a restriction that

$$\text{SOC} > 50\%$$

Further, the battery energy is assumed to be

$$E_G = 2 \text{ kWh}$$

The efficiency is based on the assumed values:

$\eta = 0.88$ = Efficiency taking power out of battery
$\eta = 0.92$ = Powertrain efficiency
$\eta = (0.88)(0.92) = 0.81$ = Overall efficiency in pure electric mode

Time to consume energy in battery is

$$t = \frac{\eta E_B \Delta \text{SOC}}{P} = \frac{(0.81)(2 \text{ kWh})(0.25)}{9 \text{ kW}} = 0.05 \text{ h} = 2.7 \text{ min}$$

Range in pure electric mode is

$$R = Vt = (45 \text{ mph})(0.05 \text{ h}) = 2 \text{ mi}$$

Hybrid: All Power through the Battery

Equation 5.8 applies with the energy being equal to the 14 gal of gasoline. The SOC of the battery is the same at the start and the end of the cruise period. For this reason, only gasoline energy contributes to range and mpg. Overall efficiency is shown in Figure 5.5.

$$\eta = 0.12 = \text{Overall efficiency}$$

Time to consume energy in gasoline tank is

$$t = \frac{\eta E_G}{P} = \frac{(0.12)(472\ \text{kWh})}{9\ \text{kW}} = 6.3\ \text{h} = 378\ \text{min}$$

Range is obtained using Equation 5.1

$$R = (45\ \text{mph})(6.3\ \text{h}) = 283\ \text{mi} = 472\ \text{km}$$

Miles per gallon is simply range, R, divided by number of gallons of gasoline, N:

$$\text{mpg} = \frac{283\ \text{mi}}{14\ \text{gal}} = 20\ \text{mpg}$$

Hybrid: Bypass the Battery

In cruise, if the battery has the design value of SOC = 75%, the battery can be ignored. All power generated goes directly to the motor. It is as though the battery was removed from the car. Status and efficiencies are shown in Figure 5.6.

Once again, Equation 5.8 applies with the energy equal to the 14 gal of gasoline. The SOC of the battery is the same at the start and the end of the cruise period. For this reason, only gasoline energy contributes to range and mpg. The greatly improved overall efficiency, which is shown in Figure 5.6, is $\eta = 0.19$ = overall efficiency.

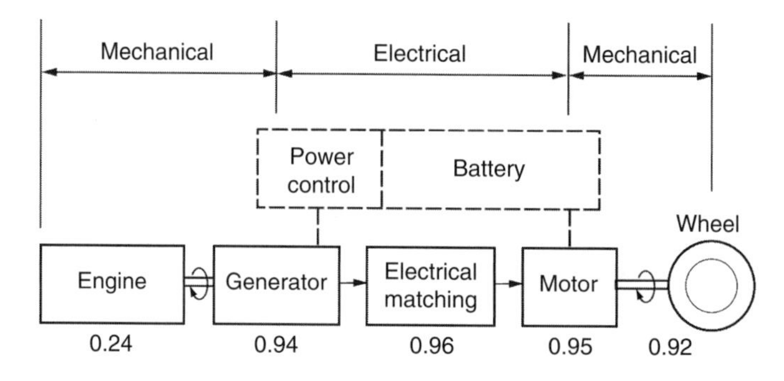

FIGURE 5.6 Components and efficiencies for a series hybrid car with the battery being bypassed. The powertrain efficiency, η_P, includes only the drive shaft, differential, and effects of tire losses. Observe that $\eta = (0.24)(0.94)(0.96)(0.95)(0.92) = 0.19$.

The time to consume the energy in gasoline tank is

$$t = \frac{\eta E_G}{P} = \frac{(0.19)(472\ \text{kWh})}{9\ \text{kW}} = 9.96\ \text{h} = 598\ \text{min}$$

Range is obtained using Equation 5.1

$$R = (45\ \text{mph})(9.96\ \text{h}) = 448\ \text{mi} = 747\ \text{km}$$

Miles per gallon is simply range, R, divided by number of gallons of gasoline, N:

$$\text{mpg} = \frac{448\ \text{mi}}{14\ \text{gal}} = 32\ \text{mpg}$$

Conclusions from the Three Different System Designs

The main gain is because of the much higher efficiency of a downsized engine operating at its optimum rpm and torque. The pure gas case had $\eta = 0.17$ while the hybrid had $\eta = 0.24$, which is a major improvement and accounts for most of the gain in mpg. A slight gain in powertrain efficiency occurred with changes $\eta_P = 0.78$ for the gas only car to $\eta_P = 0.79$ for the series hybrid.

The calculations are only as valid as the inputs for efficiency. An attempt was made to use real-life values for efficiencies.

SUMMARY

The main contributors to improved hybrid mileage are regenerative braking, electric-only propulsion, motor assist, and downsized engine that increase mpg by 30%–40% when used in combination.

Other lesser contributors are engine idle-off, electrically driven automobile auxiliaries, special focus on A/C, damping driveline oscillations and shudders because of fuel shutoff and/or early torque converter lockup, and vehicle launch.

Many techniques for achieving better mpg exist in an enormous technology pool that is common for both conventional and hybrid vehicles. Several of these common technologies were discussed.

Sample calculations of range and mpg were made for two different vehicle categories: CVs and HEVs (Table 5.1). For a series HEV, three different operating modes were examined. Range and

TABLE 5.1
Comparison of the Efficiency and mpg for HEV and CV

System Design	Efficiency, η	mpg	% Change
Pure gasoline	0.13	22	0
Hybrid through battery	0.13	20	−8
Hybrid bypass the battery	0.19	32	+48

Note: Two different operating modes are included.

mpg were calculated for electric-only mode, series hybrid with all power passing through the battery, and all power bypassing the battery. Of special interest is the range of a hybrid in electric-only mode. How far can the hybrid go before the battery goes dead? This question has been answered.

REFERENCES

1. P. Weissler, NREL looks at reduced A/C loading, *Automotive Engineering International, AIE*, October 2006, p. 22.
2. J. Kendall and R. Gehm, New powertrain enhancements from Valeo, *Automotive Engineering International, AIE*, December 2007, p. 36.

6 Multifaceted Complexity of Batteries

Battery technology will make or break hybrids and especially plug-in hybrids as well as electric cars.

INTRODUCTION

Three devices are important for hybrid vehicles: battery, fuel cell (FC), and capacitor. All three are electrochemical devices. Batteries and capacitors store energy. Storage of energy is essential for regenerative braking. In addition, batteries and capacitors provide energy for electric-only propulsion and electric assist. An FC converts a fuel to electrical energy.

The electrochemical devices discussed here are not heat engines. For operation, heat engines extract heat from a high-temperature heat source and dump energy at a low-temperature heat sink. The distinction is important since the efficiency of heat engines is restricted by thermodynamics. The limitations of the Carnot cycle apply to heat engines but have no influence on the electrochemical devices discussed herein. The internal combustion engine (ICE) is a heat engine.

BRIEF OVERVIEW OF THE THREE ELECTROCHEMICAL DEVICES

Table 6.1 shows that one (and only one) of two functions is performed.

TABLE 6.1
Functions of Electrochemical Devices

Electrochemical Device	Conversion of Fuel	Energy Storage
Battery	No	Yes
Capacitor	No	Yes
Fuel cell	Yes	No

Perhaps Table 6.2 is premature. It explains the differences between the three electrochemical devices. After consulting Table 6.2, you may have questions. This chapter provides answers to those questions.

Only the FC has a flow of chemicals in/out of the device. Because of this flow, an FC cannot be sealed. A battery may or may not be sealed. All three devices are electrochemical in nature and require an electrolyte to function. This statement applies to the supercapacitors (SCs) discussed below. Table 6.2 summarizes some salient features of the three major electrical energy sources for hybrids. (Naturally, electrical power is available from generators, but that is not the thrust of this chapter.) Both the battery and FC rely on chemical reactions for energy. The capacitor merely stores and dispenses electrical charges on command. Only for the battery do the electrodes undergo chemical change each charge/discharge (C/D) cycle. As an overstatement, but effective description, the electrolyte chews up the electrodes and then spits them back. This fact adversely affects battery life. The battery will not operate through as many C/D cycles as a capacitor because of electrode damage to each cycle. Since both the battery and capacitor do not require a flow of chemicals, each can be sealed. For short periods

TABLE 6.2
Distinctive Features of Electrochemical Devices

Feature	Battery	Fuel Cell	Capacitor
Chemical change of electrodes	Yes	No	No
Flow of chemicals	No	Yes	No
Energy storage	Yes	No	Yes
Sealed	Yes	No	Yes
Electrolyte required	Yes	Yes	Yes
Current flow between electrodes	Yes	Yes	No
Spontaneous[a]	Yes (when charged)	Yes (only during flow of fuel and oxidizer)	Yes (when charged)
Method of energy storage	Chemical change of electrodes and electrolyte	Not applicable; fuel cells do not store energy	Electrostatic fields

[a] Spontaneous in the sense, close the switch and an electrical current flows.

of time compared to the timescale for complete discharge, the battery supplies steady power. The FC has a steady power output. In an application in an electric vehicle (EV) or hybrid electric vehicle (HEV), the capacitor behaves as an unsteady, transient, power source. A characteristic of all three electrochemical devices is spontaneous behavior. The devices are spontaneous in the sense, close the switch and an electrical current flows.

Function of the Battery or Capacitor

For hybrid vehicles, the battery provides storage of energy that is essential for regenerative braking. In addition, the battery is a source of energy that is necessary for electric-only propulsion. Capacitors, which provide the similar capability, are high-power and low-stored-energy devices. Capacitors are suited to applications involving high peak power for short duration.

What can batteries do? Batteries are heavy and repeated deep discharge adversely affects life. Batteries have much better performance as a provider of peak power for HEV and/or fuel cell vehicle (FCV).

Function of an FC

An FC is similar to a battery in that it is an electrochemical device. An FC differs in that it cannot store energy within the FC. (Conceivably, the FC could be reversed for the electrolysis of water; the H_2 and O_2 represent stored energy.) The FC system has a fuel tank and stores energy in the form of hydrogen. An FC cannot be used for regenerative braking. An FC can be used for electric-only propulsion. The FC is more like an ICE than a battery. So long as the fuel flows, electricity is generated.

ELECTRICAL DEFINITIONS FOR HYBRID VEHICLES

Battery and Electrical Definitions

Hybrid vehicles are part electrical and part mechanical. The definitions to be discussed are essential for understanding the various components applied to hybrids. In addition, the electrical definitions apply to the electrical part of the hybrid vehicle.

Batteries are classified as primary batteries or secondary batteries. A primary battery cannot be recharged, and, therefore, is of little use in a hybrid. The chemical reactions providing the electrical energy are irreversible. A secondary battery, which has reversible reactions, can be charged and discharged. The basic unit of a battery is the voltaic cell, or just cell. Appendix 6.1 discusses the cell and gives the component parts, anode, cathode, and electrolyte, that form the basic unit. Batteries are composed of collections of cells. Typically, cells have a voltage of less than 4 V. Each cell has a voltage depending on the electrochemical potential of the chemicals. Nickel metal hydride (NiMH) has cell potential of 1.2 V. Lead acid has a cell potential of 2 V. To increase voltage, cells are placed in series end-to-end. A 12 V lead acid battery has six cells in series. Nominally the lithium-ion (Li-ion) battery has a cell voltage of 3.6–4.3 V.

Four features are used quantitatively to describe the battery: current, voltage, energy, and power. These features can be described by analogy or defined by precise scientific definitions that can be found in any physics textbook. Concise definitions are given in Ref. [17]. A water hose is useful as an analogy. The amount of water flowing through the hose is analogous to electrical current. The unit for electrical current is ampere. The water hose analogy for voltage is pressure. High pressure water in the hose has lots of voltage. An appropriate analogy for a storage battery is a water dam. When the water depth behind the dam is zero, the battery is dead. When the dam is full of water, the battery is charged.

Energy is the ability to do work. This definition of energy is meaningless without the technical definition of work. Work is the displacement of a force. Visualize a cart being pulled uphill. The force is that necessary to go uphill. For every foot of displacement (motion) up the hill, the greater the work done. Double the distance, and the work is doubled. Double the weight of the cart, and the work is doubled. Looking again at the hose analogy, turn the hose so that the water goes vertically upward thereby creating a fountain. The height that the water rises depends on the energy of the stream. Power is the rate of doing work; power is energy per unit time. Catch the water from the hose into a bucket. Using a stopwatch determine the amount of water captured in 1 min. The larger the amount of water, the greater the power of the flowing water.

Returning to the storage dam analogy, energy and power can be interpreted again. The water behind the dam is a source of energy. The water can do work when released. Hence, the stored water has energy. The total energy due to the water is the same whether the water is released slowly or rapidly. However, the power depends on the flow rate. A small stream has small power; a large stream has large power. Think of a water wheel that spins because of water motion. Spin rate depends on the rate of water flow. The water wheel is a source of power. In fact, dams are called hydroelectric power stations.

The ordinary flashlight is another example to illustrate the meaning of energy and power. Brightness of the light is indicative of power. For the same bulb and same brightness (power), changing the batteries for the flashlight does not change power. The common battery sizes are AAA (0.5 oz), AA (1.0 oz), C (2.5 oz), and D (5.0 oz). Since battery energy increases as the weight increases, the burn times for the series of batteries are AAA (1 h), AA (2 h), C (5 h), and D (10 h). The D battery supplies 10 times the energy as the AAA battery because the D battery weighs 10 times as much. This assumes constant specific energy among all four (A, B, C, and D). Energy and power are related by (energy) = (power)(time). For the example, the desired information was the burn time of the light with batteries of different weight. For this information, the desired form of the equation is (time) = (energy)/(power).

Another useful analogy concerns your vehicle. The gasoline in the gas tank or the energy stored in the battery represents energy and not power. The distance that the vehicle can travel before refueling depends on energy in the tank or, for a pure electrical vehicle, in the battery. The acceleration of the vehicle or the hill-climbing-speed depends on power.

As stated above, energy is the ability to do work. Displacement (movement) of a force of 1 N through a distance of 1 m does 1 J of work. The units for energy are joules or kilowatt hours (kWh). A watt is a joule/second. Power is the rate of doing work; power is energy per unit time or work per unit time. Units for power are watt (W). The prefix kilo means 1000; hence a 1000 W is 1 kW.

Battery Capacity

One factor rating a battery is capacity, C, with units of ampere hour (Ah). A rechargeable NiMH battery in the AA size has capacity equal to 2.5 Ah. For discussion, assume the battery current, I, is 0.5 A. The time, t, that the battery can operate before it goes dead is

$$t = \frac{C}{I} = \frac{2.5\ \text{Ah}}{0.5\ \text{A}} = 5\ \text{h} = 300\ \text{min} \tag{6.1}$$

The preceding calculation applies to a perfect battery and does not take into account the decline of terminal voltage.

Cell capacity, C, is determined partially by the mass of available reactants. The capacity of a perfect battery in either series or parallel assembly of cells is derived below. Let C_0 be the capacity of a cell. Let V_0 be the cell voltage. The energy of a cell, E_0, is the product.

$$E_0 = C_0 V_0 \tag{6.2}$$

Consider a battery with N cells. The energy of the battery, E, is

$$E = N C_0 V_0 \tag{6.3}$$

If connected in series, that is, end to-end, thc battery voltage is

$$V = N V_0 \tag{6.4}$$

The energy of the battery connected in series is

$$E = (NV_0)C_0 = VC_0 \tag{6.5}$$

When connected in series, the capacity of a battery is the same as the capacity of a cell. If connected in parallel, the battery voltage equals that of a cell, that is, $V = V_0$.

The energy of the battery connected in parallel is

$$E = (NC_0)V_0 = CV_0 \tag{6.6}$$

For a parallel connection, the battery capacity, C, is equal to NC_0. Compare battery energy for series and parallel connections; the equations are almost the same.

Battery Properties Used to Assess Performance

To help answer the questions as to how big is a battery, or how much does a battery weigh, certain terms are defined. Two words, specific and density, are used. The word specific refers to per unit mass. Thus, specific power has the unit kilowatt per kilogram where kilogram is a unit of mass. Suppose the specific power of a battery is 40 kW/kg. A battery of mass 2 kg would have power equal to 80 kW. Specific energy, which has the unit kilowatt hour per kilogram, is interpreted in a similar manner. Specific energy and specific power are used to answer the question, "How much does a battery weigh?"

The word density refers to a unit volume. Power density, which has the unit kilowatt per liter, and energy density, which has the unit kilowatt hour per liter, are used to answer the question

TABLE 6.3
Summary of Electrical Terms and Concepts Necessary for Understanding Hybrid Vehicles

Quantity	Typical Symbol	Units
Electrical potential	V	Volts (V)
Voltage	V	Volts (V)
Electrical current	I	Amperes (A)
Energy	E	Joules (J) or kilowatt hour (kWh)
Power	P	Watt (W); kilowatt (kW)
Mass	M	Kilogram (kg)
Resistance	R	Ohms (Ω)
Impedance	R	Ω
Specific energy	E_S	Kilowatt hour per kilogram ([kWh]/kg)
Specific power	P_S	Kilowatt per kilogram (kW/kg)
Energy density	E_D	Kilowatt hour per liter ([kWh]/L)
Power density	P_D	Kilowatt per liter (kW/L)
Capacitance	C	Farad (F)
Capacity	C	Ampere hour (Ah)
Distance	X	Kilometers (km)
Volume	V	Liter (L), cubic meter (m^3)
Time	T	Seconds (s), minutes (min), hours (h)

"How big is a battery?" Liter is a measure of volume. Additionally, the four defined quantities, specific energy, specific power, energy density, and power density are used to compare batteries. These numbers assist in the selection process. The various definitions are summarized in Table 6.3.

Importance of Energy Density and Specific Energy

Figure 6.1 highlights the significance of the battery energy density and specific energy. The data for Figure 6.1 are typical of NiMH batteries. First, note that as range is extended by adding more batteries, the weight becomes excessive. The tires go nearly flat, and the nose rises upward. For this application, the specific energy ([kWh]/kg) of the battery is too low. If the specific energy were doubled, the weight would be cut in half thereby helping to solve the sagging-springs problem.

Second, the storage area is overflowing with batteries as electric-only range is increased. This is not acceptable in the marketplace. For this application, the energy density ([kWh]/L) of the battery is too low. If the energy density were doubled, the available storage volume would be doubled.

Energy Density or Energy Density?

The words specific and density are carefully defined and widely used in the scientific and engineering disciplines (see Ref. [17] for definitions). This book uses specific to mean per unit mass and density to mean per unit volume. In battery world, the definitions for specific energy and energy density vary from book to book and even sometimes within the same book. Some authors use energy density for both specific energy ([kWh]/kg) and energy density ([kWh]/L). One battery reference book has a graph of energy density versus energy density. (Actually the plot was specific energy as a function of energy density.) These comments are to forewarn the reader that definitions must be carefully

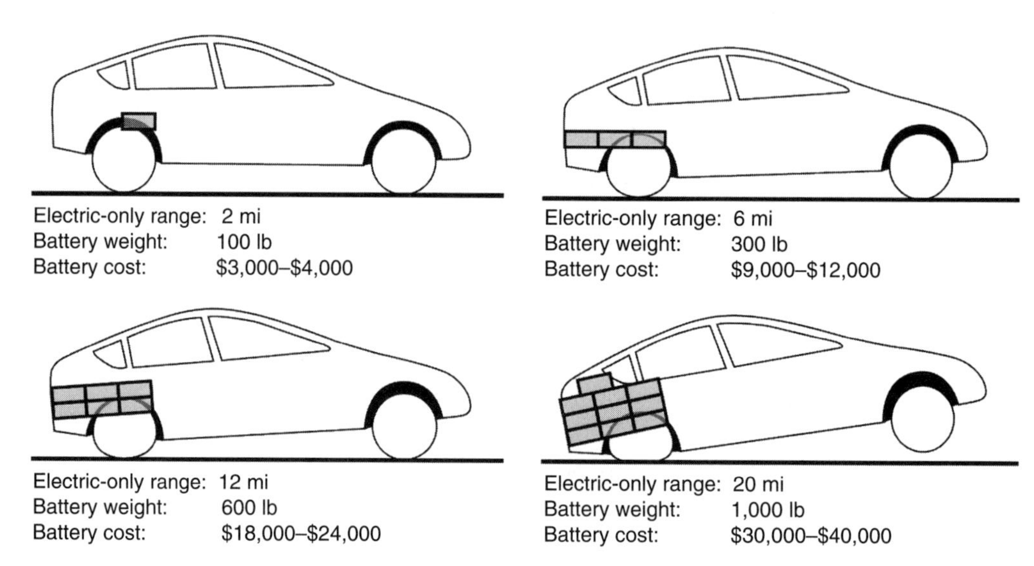

FIGURE 6.1 Effects of adding batteries for PHEV.

checked against usage. The best clues are the units used. That is, (kWh)/kg is always specific energy, kW/L is always energy density, etc. In the battery literature, the word density does mean per something. The task of the reader is to define the something.

BATTERY ELECTROCHEMISTRY

Electrochemistry of Cells

Chemical energy is converted to electrical energy by techniques of electrochemistry. A comparison with ordinary chemical reactions provides insight into the key features of electrochemistry. In ordinary reactions, the molecules are thoroughly mixed and must be in intimate contact. The molecules must collide. The chemical energy is released as heat. Combustion, as in the case of ICEs, is an example. To burn a hydrocarbon (HC) with air the molecules must be in intimate contact. The oxygen from air collides with the HC, a reaction occurs, new molecules are formed, and heat is released.

A cell is the device that converts chemical energy to electrical energy. To convert chemical energy to electrical energy, the reactants can be located far apart. Compared to distances (nanometers) between reactants in combustion, the reactants in a cell can be millimeters apart or a 1,000,000 times greater distance. Conversion is by means of oxidation/reduction reactions. These reactions involve a transfer of electrons. Removal of electrons, at the anode of the cell, is oxidation. Addition of electrons, at the cathode of the cell, is reduction. Transfer of electrons from anode to cathode is by means of wires external to the cell. The flow of electrons represents electrical energy.

Appendix 6.1 is devoted to the electrochemistry of voltaic cells. A cell consists of two electrodes: the anode and the cathode, and an electrolyte. The electrolyte may be liquid, solid, gels, or gaseous. The solid electrolyte is usually a plastic that must be an ionic conductor. As shown in Figure A.6.1, electrons flow from anode to cathode through a load to the cathode. The anode is negative and the cathode is positive. The circuit is completed within the cell by means of migration of ions within the electrolyte. The negative ions diffuse (move) toward the anode, and the positive ions move toward the cathode. The negative charge carriers complete a path in the clockwise direction in Figure A.6.1.

Electrochemistry of Eels

The electric eel can generate voltages up to 600V and can produce 1 A current. This is power of 600W. The voltage and power are sufficiently large as to be dangerous to humans. The skin of an eel is an excellent electrical insulator. Eels with damaged skin can electrocute themselves.

The electrical organ comprises 80% of the eel's length. The high voltage is generated by a stack of electrochemical cells placed in series. A hybrid vehicle battery may have a few hundred cells in series. Use of seawater as an electrolyte seems reasonable but such is not the case. With regard to electrical features, is an eel more like a battery or an FC? The answer is neither. After a strong discharge, the eel is a dead battery. The eel must recharge itself. The energy to recharge the cells comes from metabolism. The metabolism generates electrochemicals that recharge the individual cells. A battery requires an external power source for recharging. The eel recharges itself. An FC has electrochemicals that flow through the cell. An FC cannot store electrical energy while an eel can do so.

Computational Electrochemistry

Modern computational capabilities have advanced so that almost every device or technology has developed an ability to find details using the computer. For computational electrochemistry two thrusts have emerged. On the microscopic scale, the details of the electrodes are found. This includes the plasma sheath, ionic concentrations, local electrical potential, and surface reactions on the electrodes. At the macroscopic scale, models for simulation of batteries are developed for use in overall EV or HEV simulations. Such data are also available from real test data.

Discharge Characteristics

The perfect battery during discharge has constant voltage at the terminals. When depth of discharge (DOD) is 100%, the voltage drops precipitously to zero. Figure 6.2 has a perfect battery curve. The curves A, B, and C give terminal voltage as the battery is being discharged. Curve A has a modest current with curve B at higher value. Curve C is for a very large current and very rapid discharge. As the voltage curves droop in Figure 6.2, the useful state of charge (SOC) is decreased. The curves decline mainly because of battery internal impedance.

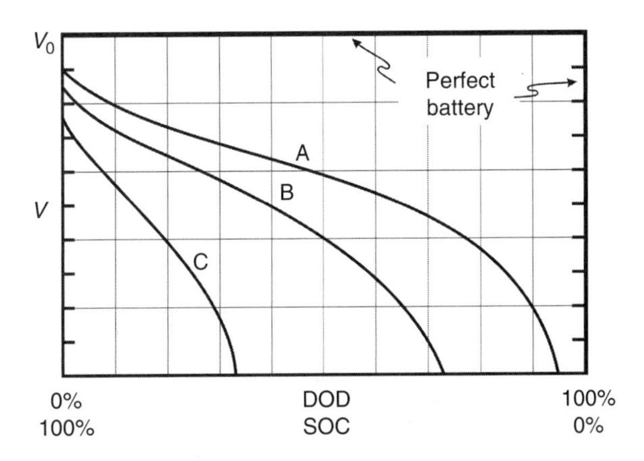

FIGURE 6.2 Voltage at the battery terminals depending on the SOC. Curves A, B, and C are for different discharge electrical currents.

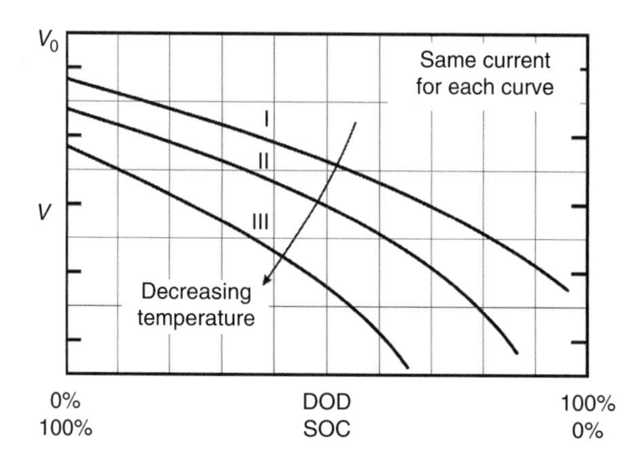

FIGURE 6.3 Temperature effect on voltage at the battery terminals.

Temperature Sensitivity

Batteries have a favorable temperature range. Some batteries operate at higher temperatures. For example, the lead acid (PbA) battery operates best at 104°F (40°C). An environment that is too cold may cause freezing of the electrolyte. Both FCs with water in the cell and lead acid battery with dilute sulfuric acid are susceptible to freezing. Batteries and FCs are chemical devices. Most chemical reactions are temperature sensitive becoming very sluggish as temperature decreases. Consequently, battery output decays as temperature falls.

An environment that is too hot may cause rapid deterioration of the electrodes. Conditions that are too hot can shorten battery life. High temperature also adversely affects the electrolyte. Some of the chemicals in the electrolyte may decompose at high temperature. The process of charging and discharging generates heat within the battery. As a result, during charging and discharging, battery temperature must be carefully monitored.

Battery voltage for different temperatures is shown in Figure 6.3. Each curve is for the same discharge current. As the temperature of a metal conductor, for example, copper wire, increases, the electrical resistance increases. In contrast, most electrolytes have decreasing resistance with increasing temperature. Curve I has a higher voltage than curve III. For curve I, the ohmic loss (current–resistance, IR, loss) is less since resistance is less at elevated temperature.

Self-Discharge

The lead acid battery has poor shelf life due to self-discharge. What causes loss of energy? With an open circuit, the potential difference between the electrodes builds sufficiently to cut off the ion migration. In this condition, zero current should flow, and zero self-discharge should not occur. However, ion motion because of diffusion may occur. Chemical instability in either the electrolyte or electrodes may destroy or transform energetic chemicals.

Self-discharge, which also happens during discharge through an external circuit, decreases terminal voltage. Instead of appearing as output power in form of the product (amperes)(voltage), chemical energy is dissipated as heat.

Performance Boundaries in a Hybrid or Electrical Vehicle

One boundary, which is illustrated in Figure 6.4, is the maximum allowable DOD. This boundary is required to prevent loss of battery life. The vertical line in Figure 6.4 represents the allowable DOD. Another performance boundary is due to the minimum allowable inverter voltage. As illustrated in

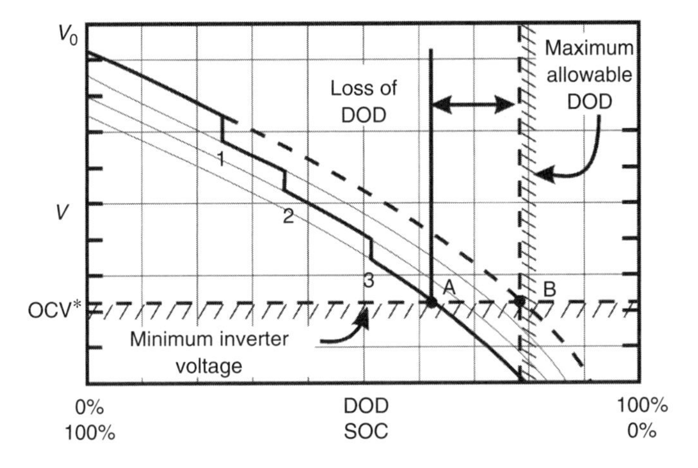

FIGURE 6.4 Memory effect and influence on useable energy from NiMH battery.

Figure 4.3, the battery is connected to the power electronics, which is usually the bidirectional inverter. The inverter tolerates swings in voltage; however, voltage below a certain value cannot be accepted.

Figure 6.4 shows step decreases in open circuit voltage (OCV) because of voltage depression. Without voltage depression, the upper curve (partially dashed) is not affected by voltage depression and turns off the battery at point B when OCV = OCV*. At point B, all available battery energy, which is consistent with the two constraints, has been used. With voltage depression, when OCV = OCV*, the battery is turned off at point A, and the actual battery SOC is much higher than the allowable value. Consequently, all available battery energy has not been used.

Battery Peculiarities

Two peculiar behaviors of batteries, NiMH battery voltage depression with memory effect and voltage recovery, are discussed below. The discussion confirms the fact that batteries are indeed complex.

Battery Peculiarities: NiMH Battery Voltage Depression with Memory Effect

The voltage depression is a memory effect in the battery chemistry. During discharge, the current is interrupted at three values of DOD denoted by points 1, 2, and 3 in Figure 6.4. Certain load patterns (history of power removal and charging) cause the memory effect. The load patterns involve repeated, narrow bands of change in SOC, symbolized as ΔSOC. A voltage depression results. The memory effect is created by increased concentrations of certain chemicals. A few deep discharges will eliminate the voltage depression; however, the need for a few deep discharges is viewed with disdain for an application.

Memory effect is less for NiMH than for Ni-Cd. Li-ion batteries do not exhibit memory effect.

Battery Peculiarities: Voltage Recovery

When battery power is extracted continuously as in Figure 6.2, the curves are smooth. However, as shown in Figure 6.5, when the current is interrupted as at points A, B, and C, partial recovery of voltage may occur in some batteries. Voltage depression, which is a memory effect, and voltage recovery demonstrate the complexity of electrochemical devices.

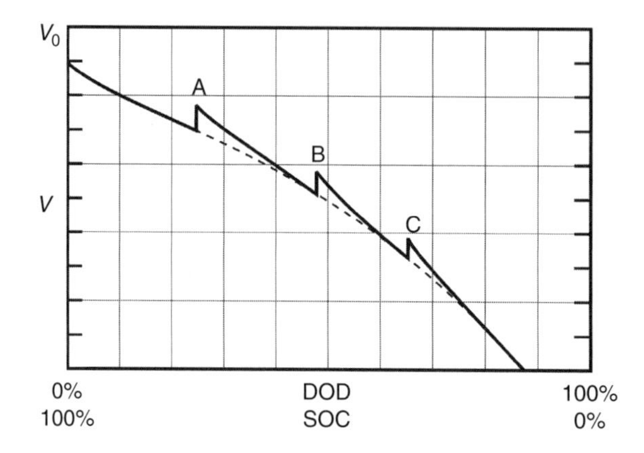

FIGURE 6.5 Partial recovery of terminal voltage after current has been interrupted.

Polarization

Refer to Figures 6.20 and A.6.4a that show the charge accumulation on the two electrodes. As you might expect, excess positive charges in the electrolyte occur at the negative electrode. Of course, excess negative charges in the electrolyte occur at the positive electrode. The electrodes act as the plates in a capacitor. The net charge accumulation at electrodes causes one of the capacitances in a battery model. Polarization at the electrolyte/electrode interface is important for electrochemical devices such as electrolytic capacitors.

Number of Different Possible Cells: Different Cell Chemistries

As discussed in Appendix 6.1 and Table A.6.2 many combinations of electrode material and electrolyte yield a cell. A few hundred possible cells can be formed. Many of the possible cells have been studied. In spite of this, only a few combinations of chemicals yield useful batteries. At one time, 14 different NiCad batteries were available for use in hybrids.

Battery Design

Specific power, specific energy, power density, and energy density of a battery provide insight to the mass (weight) and volume of a battery. Two components of a battery contribute to these quantities. First, the mass and volume of the chemicals determine the unattainable minimum mass sometimes called theoretical values. Second, the battery must be packaged; the mass and size of the packaging strongly affects values. In summary, for battery mass and volume

$$(\text{Total mass or volume}) = (\text{Chemicals}) + (\text{Structure})$$

This rather obvious relationship accounts for the variability between batteries of the same electrochemistry.

Practical aspects of cell design include the material to separate the electrodes. The separator material adds mechanical strength and maintains the desired separation distance. A high conductivity grid may be imbedded in an electrode to reduce internal resistance.

As stated in Appendix 6.1, the cell voltage is determined by the pair of electrodes that are selected along with the electrolyte. Cell energy can be determined from the overall cell reaction. Appendix 6.1 presents an example for the lead acid battery.

TABLE 6.4
Sample Values of the Theoretical Energy Density and Specific Energy for Two Common Batteries

Cell	Anode	Cathode	Cell Voltage (V)	Energy Density (W h/L)	Specific Energy (W h/kg)
PbA	Pb	PbO_2	2.0	37	70
NiCd	Cd	Ni oxides	1.2	33	60

$$2PbSO_4 + 2H_2O \rightarrow PbO_2 + 2HSO_4^- + 2H^+ + Pb$$

The electrochemist uses knowledge of the specific energy of each chemical species to ascertain overall cell energy. Recall specific means per unit mass. Multiplying by the mass of reactants, the cell capacity and energy are determined. Note that these values, which are called theoretical values, are for the mass of reactants and not the complete battery with its packaging (Table 6.4).

SOC and its companion quantity, DOD, are based on battery energy.

$$\text{SOC} = (\text{Actual energy})/(\text{Fully charged energy})$$

Note that

$$\text{SOC} = 100\% - \text{DOD}$$

PbA foil battery retains more charge in regenerative braking. Desirable features for a battery are quick (rapid) charge and discharge.

Current density, J = current density (A/area), affects battery efficiency through ohmic loss in the battery. Use the symbols A = cell plate area, I = cell current = JA, E = electric field V/m = ρJ, and ρ = resistance (Ω m)

$$V = \text{voltage drop} = \frac{I\rho L}{A} = \rho JL \tag{6.7}$$

The distance between electrodes is L. One can merely state that the voltage drop in a cell is due to current density. Larger electrodes reduce J, and the smaller separation of electrodes reduces L and battery volume.

The positive and negative electrodes must be separated by a fixed spacing. Two methods can be used to separate the electrodes. Make a box. One electrode is one wall of the box. The facing wall is the other electrode. Mechanical separation and spacing is by means of the box. The box can be filled with liquid or gel electrolyte. Two functions of the electrolyte are to provide conduction path for ions, which for Li-ion battery is Li^+, and to impose infinite (high) resistance to electrons. The electrons are forced to flow in the external load circuit.

An alternate method is to use a separator material and press the two electrodes against the separator. Use of liquid electrolyte is cumbersome. Two functions of the separator are to maintain spacing and to prevent short circuits. A solid separator that serves the dual function as a mechanical spacer and ion-conducting electrolyte is needed and is possible.

For an Li-ion polymer battery, the electrodes serve as housing and interstices (small spaces or holes) for lithium atoms and ions. See also intercalated discussed in "Charge/Discharge Cycle." Added weight reduces specific energy (W h/kg), which is a vital performance parameter. Added weight comes from

- Electrodes
- Separator
- Ionic conducting chemicals (Li^+ for Li-ion battery)
- Casing
- Pressure release
- Electrolyte
- Current collectors
- Electronics (for safety)
- Conductivity additives

BATTERY LIFE

WHAT IS BATTERY FAILURE?

What is battery failure? Gradual decline in energy and power? Battery power does decline with time. At some point, the performance of the hybrid is sufficiently impaired as to require a new battery. Gradual decline does not leave the driver stranded.

What is battery failure? Sudden loss of power? Sudden loss of power is a disaster with regard to customer satisfaction. A database is needed for occurrence of sudden loss of power. The battery must be carefully monitored and C/D cycles counted.

What causes failure? Any of the battery parts can fail. The electrodes are stressed each C/D cycle. The electrolyte may decompose or alter chemical properties. Mundane items such as seals may be a weak point.

STATE OF CHARGE: RESTRICTIONS AND BATTERY UTILIZATION

Experience has demonstrated that battery life can be extended by restricting deep discharge. For the battery of Figure 6.6, 50% of the battery energy is excluded due to battery life. The control system

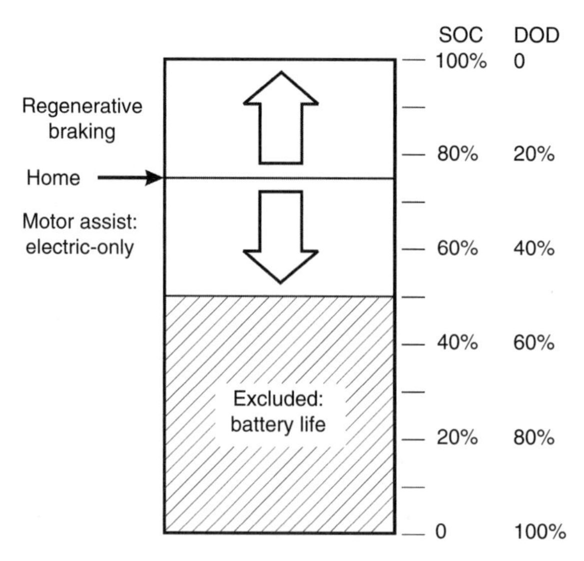

FIGURE 6.6 Battery utilization.

tries to maintain an SOC of 75% (DOD, 25%); this is home. From home value, the battery can provide 25% for motor assist or for electric-only operation. Also from the home value, 25% of battery is available to store energy from regenerative braking.

Determining Actual SOC and Battery Aging

For some batteries, measurement of OCV provides information concerning the life of the battery. Capacity reduction is the main determinant for defining battery end-of-life. This translates into measuring OCV. Determining actual SOC is not an easy problem. The example for NiMH battery gives some insight to problem of managing a battery both in power extraction and battery life.

Determining Actual SOC: Lead Acid Battery

Measure the specific gravity of electrolyte; it is a good indicator of SOC.

Determining Actual SOC: NiMH Battery

The logic for determining actual SOC hinges on a correlation between SOC and OCV [24]. That is, if you know OCV, you know SOC. Finding the correlation is necessary, of course. Determine a critical value of OCV* corresponding to the desired SOC. The critical value of OCV* depends on value of SOC that is desired, the age of the battery, and the effects of voltage depression (Figure 6.3).

Figure 6.7 gives OCV as a function of DOD for both a new and an old battery. At low DOD, aging increases OCV. At DOD near 40%, aging has no influence on OCV. At 75%, aging decreases OCV.

There is a connection between battery age and the critical value of OCV*. See Figure 6.8 which shows OCV as a function of DOD. Figure 6.8 is based on Figure 6.7. If the desired DOD is 23%, then OCV* critical value is 13.08 V for the new battery. As the battery ages, the value of OCV* increases to 13.2 V. If desired DOD is 50%, then OCV* critical value is 12.83 V for new battery.

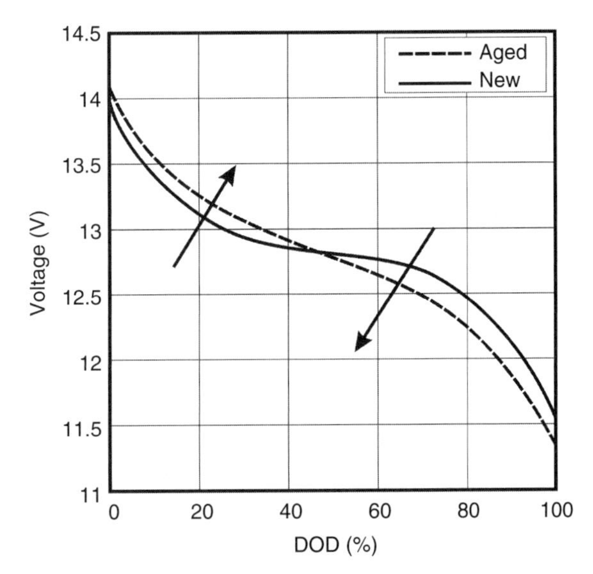

FIGURE 6.7 OCV, as a function of the SOC for new and aged batteries. (Reprinted with permission from SAE Paper *Advanced Hybrid Vehicle Powertrains 2006*, SP-2008, 2006 SAE International.)

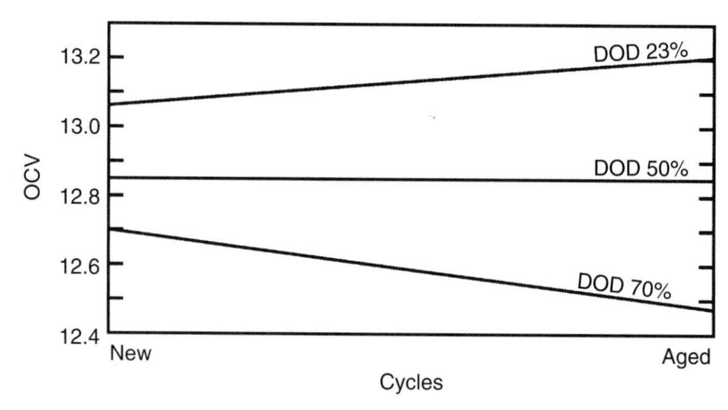

FIGURE 6.8 OCV, as a function of number of battery cycles with DOD as a parameter. Curves are for an NiMH battery and show the effects of aging. Values of OCV are shown for three different DOD: 23%, 50%, and 70%.

As the battery ages, the value of OCV* remains unchanged. If desired DOD is 70%, then OCV* critical value is 12.7V for a new battery and 12.47V for an aged battery. As the battery ages, the value of OCV* decreases.

When OCV is used to determine battery life, voltage depression confuses the battery management system (BMS). Using OCV for estimating battery life, two effects are superimposed. One is the aging and the other is voltage depression. The BMS predicts a battery life that is too short and that is false. The BMS needs to remove the effects of voltage depression before battery life is assessed. Sufficient data concerning NiMH batteries are available to permit removal of the effects of voltage depression [10].

BATTERY CHARGING AND DISCHARGING

Charging Rate

Long battery life depends on an optimum C/D protocol. What limits the charging rate? The charging rate of a lead acid battery is limited by excessive gassing and excessive battery temperature. There is a connection between charge rate and battery life as dependent on battery temperature.

$$\text{Battery heating} = I^2 R \tag{6.8}$$

The energy being transferred to the battery by charging is

$$E = Pt = VIt \tag{6.9}$$

Solving for current I gives

$$I = \frac{E}{Vt} \tag{6.10}$$

Combine Equations 6.8 and 6.10 to give

$$\text{Battery heating} = \left(\frac{E}{Vt}\right)^2 R \tag{6.11}$$

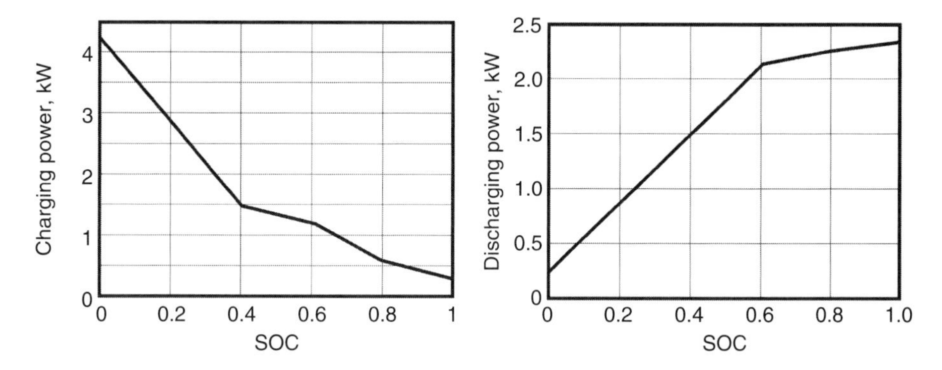

FIGURE 6.9 Charging and discharging limits for valve-regulated lead acid battery based on Advisor simulation program.

Heating varies inversely as charging time, *t*, squared.

Capacity of a battery is given in ampere hours. Sometimes charge and discharge rates are stated in terms of capacity. Overcharging causes battery damage.

The LHS graph in Figure 6.9 shows the permissible charging power. As SOC increases the charging power falls from 4.4 to 0.3 kW. The RHS graph in Figure 6.9 shows the permissible power during discharge. As SOC increases, the discharging power increases from 0.3 to 2.3 kW. The curves are based on the Advisor simulation program developed by the National Laboratory for Renewable Energy [4].

Discharging Rate

What limits discharging rate? There is a connection between charging rate and battery life. Damage to electrodes may be induced by high currents. Chemicals may be decomposed. A high discharge rate causes droop of terminal voltage (see Figures 6.2 and 6.4). When terminal voltage droops, efficiency also falls.

There are two types of lead acid batteries: the flooded or wet cell type and the recombinant or valve regulated type (valve-regulated lead acid). The absorbed glass mat and gel cell batteries are mixtures of the two types.

Thermal Runaway

Thermal runaway occurs when an increase in temperature changes conditions such that a further increase in temperature occurs. Ultimately this leads to a destructive result. Thermal runaway becomes thermal *ruin*away.

Batteries may be driven into thermal runaway if severely overcharged for a prolonged period at high temperature. Some batteries fail gracefully as a result of thermal runaway. Other batteries may ignite and catch fire, explode, or create external arcs that endanger the vehicle.

BATTERY DATA

Three batteries are discussed in this section: PbA, NiMH, and Li-ion.

Lead Acid Battery

For 100 years, the PbA has been popular for automobile use because of its low cost. The PbA has a high rate of discharge and is capable of high current. The high current needs to be limited for long life. At low current, deep discharge does not decrease life. At low temperature, the discharge of the battery is satisfactory while charging is poor. Accurate determination of SOC can be based on the specific gravity of the electrolyte. Lead acid batteries are used in submarines. Even the nuclear submarines have batteries for propulsion. The mixture of battery/electrical motors and steam turbines/generators/electrical motors make the submarine a hybrid vehicle.

The battery has short shelf life; self-discharge slowly causes the battery to go "dead." For lead acid battery, self-discharge varies from 3% to 20% per month (see Table 6.7). The charge rate is limited by excessive outgassing and excessive battery temperature. Typical safe charge rate is $I = C/5$ for 5 h where I is current and C is capacity. The charge rate varies according to

- Start $C/5$ charge at SOC near zero
- End $C/5$ at SOC 80%–85%
- Last 20% SOC use $C/10$ rate

Specific energy for the lead acid battery is in the range of 25–35 W h/kg. Specific power is near 150 W/kg.

Different plate design is used depending on the DOD allowed. If the lead acid battery is used for starting (automobile engines) with limited DOD, thin plate construction is used. If the battery is used for propulsion as for an HEV or an EV, the deep discharge cycle requires thick plates for long life.

The lead acid battery provides only poor cold weather starting at 10°C (50°F). Cold weather starting is a problem for diesel-powered construction equipment that is cold soaked on the job site in winter. Maybe an SC would be a big help. Life of a lead acid battery is approximately 1000 cycles at 80% DOD, which is about 3 years service in an EV or HEV. Life can be extended to 5 years by precision controlled C/D. This emphasizes the importance of proper controlled C/D.

Overcharging of a lead acid battery may cause electrolysis of the electrolyte. The H_2 and O_2 can explode. For the same reason, maintenance-free lead acid batteries can also explode in spite of being sealed.

A battery that is little used and self-discharges tends to build up lead sulfate on the plates. Attempts to control the sulfation of the battery by additives are rarely effective.

R&D on lead acid batteries leverages existing networks of manufacturing, distribution, experience, recycling, etc. An advance in lead acid technology is graphite foam. Weight and size of the lead acid battery is trimmed. Fast recharging is possible. Memory effects, which plague some batteries due to quick recharges, are avoided.

Nickel Metal Hydride Battery

Some technical details concerning the NiMH battery are presented. The anode complex is a metal hydride. The cathode is nickel hydroxide. Memory effect is less than for nickel-cadmium (Ni-Cd) batteries. The electrolyte is potassium hydroxide (KOH). Safety aspects of KOH are discussed in Chapter 21.

NiMH batteries provide reasonably good power and energy for HEV applications. Specific energy density for the NiMH battery is near 70 W h/kg. Power density is near 200 W/kg. Battery life is approximately 600 cycles with DOD 80%. Recharge time is 35 min from DOD 80%. Cooling helps fast recharge. Cooling is essential for HEV operation. For almost any application, a charger specifically designed for the NiMH battery should be used.

Improvements in NiMH batteries leading to greater specific power or power density mean smaller weight and volume. In turn, this means increased cooling requirements. The necessity for increased cooling capability increases volume, cost, and noise from cooling fans (see Ref. [13] of Chapter 3). The comments above apply equally well to other battery types.

LITHIUM-ION BATTERY

Li-ion batteries provide both high power and high energy for HEV and EV applications. Two main types of Li-ion batteries exist. One type uses a liquid or gel for the electrolyte. Throughout this book, Li-ion means the battery using liquid or gel electrolyte. The other type of battery uses a solid polymer in the dual role as separator and electrolyte. These batteries are always denoted as Li-ion polymer. Some Li-ion polymer batteries have a porous separator, which has the pores filled with gel to improve ionic conductivity.

Lithium-Ion Battery for HEV

Propulsion (traction) Li-ion batteries will have little in common with Li-ion batteries for cell phones, laptops, etc. The major challenges for the Li-ion battery used for propulsion are

- Safety
- Cost
- Calendar life

Will Li-ion batteries prove too costly and too dangerous for applications in HEV or EV?

SAFETY

Chapter 21 has a more extensive discussion of Li-ion battery safety.

Mandatory Safety Devices

For service outside the laboratory, safety devices lessen the possibility of harmful failure. These devices include

- Shut-down separator (overtemperature condition)
- Tear-away-tab (excess internal pressure)
- Vent (excess internal pressure)
- Thermal interrupt (overcurrent/overcharging)

Typically, these safety devices permanently and irreversibly disable the cell. The goal is to develop an Li-ion battery that does not require safety devices for safe operation, or at least, requires fewer safety devices.

"Smart" Li-Ion Batteries

Smart Li-ion batteries have the mandatory safety devices that draw a small current continuously. The small current has the effect of being a self-discharge. When considering self-discharge, ascertain whether or not the figure includes the small current drain.

COST

Cost is a major factor for any battery used in a hybrid. The electrochemical technology (selection of anode and cathode materials, etc.) for Li-ion battery application in hybrids is evolving. Further, manufacturing methods are also being developed. Hence, current costs are likely to be reduced in the future.

Li-ion batteries have a 50% greater cost than NiMH and cost 40% more than Ni-Cd. However, a lower cost of ownership is anticipated for Li-ion compared to NiMH. The expected life of NiMH is 2 years. For the same application, the expected life of Li-ion is 4 years. For the same battery energy, the cost of NiMH is $1560 and Li-ion is $2000.

NiMH: cost/year = $780/year
Li-ion: cost/year = $500/year

A higher cost per energy, $/W h, prevails for Li-ion polymers than for the Li-ion. Advances in battery technology will likely reverse this fact. Li-ion has potential for low-cost batteries even though today $/W h is greater.

CALENDAR LIFE

Battery aging occurs due solely to its age (zero age at manufacturing). Battery life does not depend just on the number of C/D cycles or is not due to the date of first charging. The clock starts at manufacturing. Permanent capacity loss occurs due to aging. The aging is temperature dependent.

GUIDE TO LONG BATTERY LIFE

Battery life depends on calendar age, details of C/D cycling, exposure to high temperature, and other factors. Suggestions are made below to extend the life of Li-ion and Li-ion polymer batteries. Some suggestions may be incompatible with operation in a hybrid. Improved batteries will make these restrictions less important.

- Charge early and often.
- When out of service for weeks, reduce charge to 40% and store in cool place.
- Never deep cycle; never SOC near 0%.
- Store dormant battery in refrigerator (never below −40°C [−40°F]).
- Exposure to high temperature in hybrids may cause rapid degradation of battery.
- Buy only when needed (shelf life starts at manufacture).

ELECTROCHEMISTRY

LITHIUM-ION BATTERY

Recall that electrons are always removed from an anode. The polarity of a battery normally does not change; the positive terminal remains positive during charge or discharge. The anode switches as charge (positive terminal) changes to discharge (negative terminal). Usually when the word anode is used for batteries, discharge is being considered so that the anode is negative. The Li-ion battery is named for the lithium ions that flow between the electrodes. During charging, the Li-ions move from the positive electrode (anode) to the negative electrode (cathode) as shown in Figure A.6.3. During discharging, the Li-ions move from the negative electrode (anode) to the positive electrode (cathode) as shown in Figure A.6.2. The ions are moving up an electrical potential hill.

Some technical details concerning the present-day Li-ion battery are presented. Typically, the anode is lithium within an atomic lattice of graphite or tin oxide. For many Li-ion batteries, the cathode is manganese, cobalt, or nickel oxide. The electrolyte may be aqueous which poses a safety hazard. Alternatively, the electrolyte may be a gel, which is somewhat safer.

LIQUID/GEL ELECTROLYTE

Liquid electrolyte consists of an Li-salt in organic solvent. The gel has an Li-salt and is used as a replacement for the liquid electrolyte or for use with a porous polymer. When using liquid or gel, a mechanical problem exists with electrode spacing and separation.

Solid Electrolyte Interphase

The liquid electrolyte reacts with the carbon negative electrode and creates a thin layer known as the solid electrolyte interphase (SEI) which is transparent to Li^+ and opaque to electrons. This layer forms on first charging. The conductivity of SEI is important with regard to the cell internal impedance.

Temperature Limits

For the Li-ion battery, typical safe limits for temperature during operation are from −20°C (−4°F) to 60°C (140°F).

Charge/Discharge Cycle

During charge, the material of the positive electrode is oxidized and the negative electrode is reduced. This is the familiar redox reaction discussed in Appendix 6.1. During charge, the lithium atoms are de-intercalated from the positive electrode and intercalated into the negative electrode. The word intercalated means a reaction where the lithium atoms are inserted into the host without significant structural change to the host. The word de-intercalated means a reaction where the lithium atoms are reversibly removed from a host once again without significant structural change to the host. The reverse process occurs during discharge.

Depending on the Li-ion chemistry, the battery can be fully charged in 45 min. Other batteries can be 90% charged in 10 min. For full (or near full) DOD, the lifetime C/D cycles are nominally 200. Some claims are made for 300–500 cycles.

The charger for Li-ion must be tailored to the specific battery design. Overcharging damages active materials in the battery. Only a narrow band of voltage exists between fully charged at 4.1 V and overcharge at 4.3 V. That is why a carefully designed BMS is necessary to prevent overcharging. Controlling the voltage of every cell in a stack is a big problem.

LITHIUM-ION ALKALINE BATTERY

The information presented here on the Li-ion alkaline battery is mainly from Refs. [20,21]. Also data from Ref. [13] of Chapter 3 is presented here.

Advantages

The battery can be made into a wide variety of sizes and shapes. However, the Li-ion battery has less flexibility than the Li-ion polymer. Li-ion batteries do not experience the memory effect that is found in NiMH and Ni-Cd. The high OCV of about 3.0 V (NiMH OCV = 1.2 V) is a big advantage. Due to mandatory safety circuits, the Li-ion battery has apparent self-discharge. Otherwise, the "dumb" battery has low or zero self-discharge. To understand dumb batteries, see "'Smart' Li-Ion Batteries."

The Li-ion battery offers high specific energy, E_S: $90 < E_S < 180$ W h/kg and high energy density, E_D: $200 < E_D < 400$ W h/L. Batteries have been made with energy density of 120 W h/kg with total stored energy equal to 35 kWh. This is about 15 times more energy than the Prius II battery. The advantages with regard to specific energy and energy density can be stated in various ways. For the same space (volume), the Li-ion has less weight compared to NiMH. For the same space, an Li-ion battery has twice the capacity (C, Ah) of an NiMH battery. For the same weight, the Li-ion battery has 15%–20% more energy than an NiMH battery. For the same battery energy, an Li-ion battery has lower weight with less volume than an NiMH battery. Li-ion has higher specific energy (W h/kg) than lead acid with the ratio being Li-ion/lead acid = 4. Also, the same ratio for specific energy (W h/kg) for NiMH is Li-ion/NiMH = 2–3. In summary, the Li-ion offers vehicle designers (1) lower weight, (2) smaller size, (3) potential for lower cost, and (4) greater electric-only range. Safety remains an issue.

Other advantages include low loss of capacity per cycle:

ΔC = loss of capacity
ΔN = number of cycles

Hence, $\Delta C/\Delta N$ is the loss per cycle and is less for Li-ion than for NiMH. Some sources state the transient current can be several times rated current, other sources state that the transient current cannot handle high discharge rates. Compared to other batteries, Li-ion offers rapid recharge. The current (A) for the C/D cycle is limited to $1C$ to $2C$ (C = capacity, Ah). Li-ion batteries can be recycled. A battery life of 1000 cycles with deep discharge is anticipated.

For application as propulsion battery, the flat discharge curve (discharge curve more nearly approaches that of a perfect battery) is particularly attractive. More energy can be extracted from the battery with less help from the power electronics control.

Disadvantages

The main obstacles for the Li-ion battery are cost, durability, safety, and (at least initially) low specific energy ([kWh]/kg). Recall that specific means per unit mass. The word durable means highly resistant to decay or wear; long lasting and enduring. According to some reports, Li-ion batteries are less durable than Ni-Cd or NiMH. As discussed above, battery aging occurs due solely to its calendar age (zero at manufacturing). Aging is noticeable at 1 year. The battery frequently fails at 2–3 years; life up to 5 years has been observed. The Li-ion has transportation restrictions. Some Li-ion cannot be shipped on commercial aircraft.

Deep discharge state (DDS) is unique to Li-ion. Without DDS, a few hours are required to fully recharge battery; with DDS, 40–50h are necessary to recharge or may never recharge. Li-ion may abruptly fail (without warning) with high discharge currents. The battery has SOC restrictions: 0%–30% excluded, 30%–85% available, 85%–100% excluded.

Li-ion battery is highly susceptible to reverse polarity. As with all batteries, under certain conditions of charging and temperature, thermal runaway may occur. For small mobile devices, the technology of Li-ion battery is well developed, but for HEV applications the technology is still nascent. Protection circuits are required.

LITHIUM-ION POLYMER BATTERY

Advantages

The Li-ion polymer has most of the advantages of the Li-ion batteries discussed above but with additional advantages enumerated here. The solid polymer is transparent to Li^+ and opaque to electrons. This improves safety by giving a battery more resistant to overcharge. Also safety is improved; less chance (zero chance) to spill the electrolyte. Specific energy (W h/kg) is 20% higher than for Li-ion. Thin cells, as thin as credit cards, are possible. Thin strips or sheets of electrode/separator/electrode allow great freedom and flexibility in the size and shape of a cell. The result is a lightweight cell. The metal shell for each cell is eliminated. Electrode sheets and separator sheets can be laminated.

Safety is improved with the solid polymer. A wider margin of safety for overvoltage results from the polymer. Likewise, a wider margin of safety for high temperature exists. However, protective circuitry is still needed. The dry, solid, polymer serves as both separator and electrolyte. Also, the dry, solid, polymer simplifies fabrication; the three components (+electrode, −electrode, and separator) can be laminated. This leads to rugged and low maintenance cells. Recycling is relatively easy.

Disadvantages

The Li-ion polymer has disadvantages. The dry, solid polymer has low ionic conductivity. To improve performance, a gelled electrolyte can be added by means of a porous polymer. Gel fills the pores. For early designs, the following statements are true. The battery has slightly lower capacity, C (Ah), compared to liquid/gel Li-ion battery. Also, compared to Li-ion, the polymer has lower energy density. Compared to the Li-ion, the polymer has decreased C/D cycle lifetime (500 for Li-ion polymer). Early batteries were expensive to manufacture. With the lack of high-volume applications, a range of standard sizes had not been developed.

ADVANCED TECHNOLOGY

Nanostructure

Nano is the prefix denoting 10^{-9}. For nanostructure in batteries, the characteristic size is 10^{-9} m. Nanopores in a polymer would have a size near 10^{-9} m. Use of nanostructures greatly increases surface area and effective electrode area, A. Internal resistance is lower. Battery current, I(A), on both C/D can be increased. Consider the Equation 6.12

$$I = JA \tag{6.12}$$

where
J is the current density (A/m^2)
A is the area (m^2)

Current density, J, is almost fixed by cell voltage, electrolyte conductivity, electrode spacing, etc. Hence, more surface area, A, increases cell current. More current means more power as shown by Equation 6.13.

$$P = IV = \text{power (W)} \tag{6.13}$$

where V is the cell voltage. However, nanostructure gives little change in energy content of the battery.

Application of Nanotechnology

Toshiba has applied nanotechnology to Li-ion polymer batteries. Some performance gains include faster charge time; greater capacity, C (Ah); and longer cycle life.

A123 Systems has also applied nanotechnology leading to nanostructured phosphate chemistry for the cathode. For an application in an HEV, the typical $LiCoO_2$ will likely be replaced with lithiated metal phosphate. Batteries are currently available for conversion of Prius, Ford Escape, Honda Civic hybrid, etc. to plug-in hybrids.

Advanced Anodes (Negative Electrode upon Discharge)

The anode for many Li-ion batteries is some form of carbon. Carbon-based rounded edge morphology is a newer version of a carbon material for an anode. Another new anode material is stable intermetallic compounds.

The 3M Corporation is studying use of tin or silicon metalloids for the anode instead of carbon. An alternate anode material uses amorphous silicon. Cold weather performance is improved; at −40°C the usual Li-ion battery has capacity reduced by 80%–90%. The new 3M battery has capacity reduced by 40% at −40°C. At the same temperature for both the 3M and a conventional Li-ion battery, capacity is improved by 30%.

Advanced Cathodes (Positive Electrode upon Discharge)

One cathode being researched uses multi-doped $LiNiO_2$; the dopants help stabilize the Ni. Another cathode uses layered Li_2MnO_2. Nanosized titanate for the positive electrode may offer several advantages including: three times the power of existing Li-ion, fully charged in 6 min, 20,000 C/D cycles, and operation 50°C–75°C. The battery will not explode and does not have thermal runaway. The battery does not contain a graphite coated metal electrode.

Advanced Electrolytes

Some of the difficulties leading to unsafe batteries are due to the decomposition of the electrolyte. More stable salts, for example, propylene carbonate based, are an improvement. Electrolyte additives help to form more stable passivation films and retard flammability of electrolyte solvents. Passivation is the treatment of a surface to give greater protection from corrosion. Passivation, which helps to maintain desirable surface properties, employs surface coatings or thin films.

Battery Management System

The BMS monitors each cell for voltage, current in/out, and temperature. The BMS avoids the stress of heat and overtemperature. By corrective action of the BMS, the effects of excessive charging or discharging are eliminated or lessened. The BMS is essential for long battery life and optimum mpg. The BMS works with the hybrid control system.

For the Li-ion cell, OCV is 3.6 V while end-of-charge voltage ranges from 4.1 to 4.3 V. The high cell voltage means fewer cells for $V_{BATTERY}$. With fewer cells, monitoring each cell is easier. An accurate assessment of SOC can be based on $V_{BATTERY}$. Two benefits are easier: battery pack management and longer battery life.

APPLICATION OF LITHIUM-ION BATTERY IN HYBRIDS

Application of Li-Ion Battery

The information presented here on the Li-ion polymer battery is partially from Ref. [20]. The liquid electrolyte is replaced with a polymer–gel electrolyte or with a solid electrolyte. The polymer increases performance and significantly reduces hazards in case of a collision involving a hybrid. The safety aspects of Li-ion batteries are discussed in Chapter 21. The Li-ion polymer battery may be the best hope for an economically viable EV and HEV during the next two decades. The Li-ion battery is now compared in Figure 6.10 for power density and energy density with various storage devices and energy conversion methods.

Note the axes for Figure 6.10; the axes are interchanged compared to Figures 6.14 through 6.16. The energy storage devices in Figure 6.10 are all electrochemical except for the flywheel. The flywheel is discussed in Chapter 16. Electrochemical storage or fuel conversion falls into three categories: ultracapacitor (UC) (double layer capacitor [DLC]), batteries, and FC. The DLC yields very high power density but abysmal energy density. Batteries are centered near 100 W h/kg and 100 W/kg. Battery energy density increases with advancing technology going from lead acid to Li-ion polymer. The symbol "✯" stands for an HEV and includes both the engine and battery densities. FCs provide large energy densities at a power density near 100 W/kg. Gasoline, which has dominated vehicle propulsion for a century, provides both high power and energy densities.

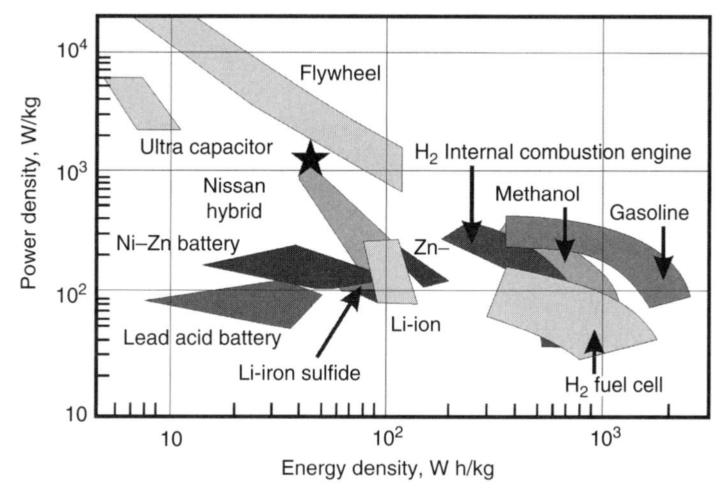

FIGURE 6.10 Specific power (shown as power density) and specific energy (shown as energy density) for various storage devices and energy conversion methods. (Reprinted with permission from SAE Paper *Electric and Hybrid-Electric Vehicles*, PT-85, 2002 SAE International.)

The hydrogen ICE, hydrogen and methanol FCs, and gasoline engine are shown. The power out and the mass define power density. For energy density (W h/kg), the reciprocal of the fuel consumption is used. See Figure 10.3, which has minimum fuel consumption of 230 g/kWh for an economical gasoline engine. The reciprocal is 4300 W h/kg, which is properly located on the graph.

Table 6.5 provides comparisons of different energy storage devices. SC has a large value for *P*/*E* that is consistent with Figure 6.10. Except for the Li-ion tailored for an HEV, all batteries have *P*/*E* near unity. Once again this agrees with Figure 6.10.

Three important batteries for EV and HEV are lead acid (less so with the advance of technology); NiMH; and Li-ion:alkaline and Li-ion:polymer. Energy and power values are given in Table 6.6. A range of values appear in Table 6.6 due to changes in cell design, packaging, and electrochemistry.

TABLE 6.5
Comparisons for Different Energy Storage Devices

Storage Technology	*P*/*E*	*E*/*P*	
Batteries		min	h
Lead acid	6.0	9.0	0.16
NiMH	2.7	22.2	0.37
Li-ion EV	7.0	8.4	0.14
Li-ion HEV	36.0	1.8	0.03
Capacitor			
Electrolytic	10^9	μs	—
Electronic double layer	800	3 s	—

Note: *P* is the power density (kW/kg) and *E* is the energy density (kWh/kg). The values for *P* and *E* are maximum.

TABLE 6.6
Summary of Battery Properties for Three Important Battery Types

Battery	Specific Energy (W h/kg)	Specific Power (W/kg)	Energy Density Power (W h/L)	Density (W/L)
Lead acid	30–40	180	60–75	—
NiMH	30–80	250–1000	140–300	—
Li-ion: alkaline	150–200	300–1500	250–350	—
Li-ion: polymer	130–200	up to 2800	300–640[a]	—

[a] Due to better packaging, 20% better than Li-ion alkaline.

Three important battery characteristics are shown in Table 6.7. The self-discharge is given as percent per month. Price of a battery in terms of energy per U.S. dollars is listed. Obviously, W h/US$ is a volatile number and changes due to changing technology as well as currency fluctuations. In Table 6.7, C/D represents the charge/discharge cycle for a battery.

APPLICATION OF BATTERIES IN HYBRID AND ELECTRIC VEHICLES

Batteries in Spacecraft

Satellites experience night and day as they move in their orbit. The amount of time in darkness and sunshine is more pronounced in low earth orbit. Even in geostationary orbit, the satellite will be eclipsed by earth. Typically, batteries provide the power when the satellite is in darkness. In sunshine, solar cells provide the power.

Many aspects of satellite batteries are similar to hybrid vehicle batteries. Since the satellite cannot pull into the service station if a dead battery occurs, reliability is vital. Both satellites and hybrids require long life. Both vehicles restrict DOD of the battery. One big difference is cost. Cost of batteries is important for hybrids; however, for satellites cost is subordinate to long life.

TABLE 6.7
Summary of Battery Characteristics for Three Important Battery Types

Battery	Self-Discharge[a] (%/Month)	Consumer Price (W h/$)	C/D Efficiency (%)
Lead acid	3–20	—	70–92
NiMH	30	1.37	66
Li-ion: alkaline	5–10	2.80	99
Li-ion: polymer	—	—	—

[a] Temperature dependent.

Prius I Battery

A brief summary of technical facts follows:

- NiMH
- Energy: 1.8 kWh
- Peak power: 21 kW for 10 s

- Regenerative braking: 20 kW for 10 s
- Cold capacity: 5.0 Ah
- Needs advanced thermal management
- Nominal cell voltage: 1.2 V
- Capacity 6.5 Ah at 20 h discharge
- 288 V

Nissan Battery

- Li-ion
- "D" cells
- Peak specific power: 1.7 kW/kg
- Capacity 18 Ah at 20 h discharge
- Cold capacity: 6.0 Ah
- Nominal cell voltage: 3.6 V
- High cost, expensive battery

The Prius and Nissan batteries show that a matrix of rechargeable cells can be made to give large peak powers on a repeated basis with excellent life performance [12].

Application of Li-Ion Battery: Two Approaches

One approach is to use off-the-shelf, high production, commercially available Li-ion batteries. The Tesla sports car, at least initially, took this approach. The advantage is lower cost. The disadvantage is that several 1000 cells are needed to obtain the required battery energy (kWh). Each cell must be monitored for long life and safety. Another approach is to use high cost (for now), limited production, large cells. The advantages include lower cost and less critical monitoring. The disadvantage is cost.

Application of Li-Ion Battery: Mitsubishi Pure EV

This is a lightweight EV intended for production; for this reason, it may not properly be called a concept vehicle. Mitsubishi intends to launch the EV in 2010. There are two paths for the transition to the FCV. The first path is HEV $\rightarrow$ FCV. The second path is EV $\rightarrow$ FCV, which is the Mitsubishi path. The EV will use the Li-ion battery and rotor-in-wheel as was done in the concept motor in wheel electric vehicle hybrid above. According to Mitsubishi the Li-ion battery has six times the energy density ([kWh]/kg) as lead acid, and the Li-ion has two times the energy density ([kWh]/kg) as NiMH. The range using a 150 kg Li-ion battery was 150 km in 2006 with a projected range of 250 km by launch time in 2010. In 2006, the fleet tests using 150 kg lithium battery were giving 150 km range (for more details see Ref. [22]).

Application of Li-Ion Battery: Nissan and NEC

The long-term commitment of Nissan and NEC to Li-ion batteries is indicated by the creation of a separate independent company to manufacture the batteries for any customer. Various Li-ion battery designs are intended for a range of vehicles including HEV, EV, plug-in HEV (PHEV), and fuel cell hybrids (FCXs) [30].

Interchangeable Power Packs

Some novel ideas have been stated for solving the problem of a dead battery. One suggested approach is to remove the dead battery and install a fully charged battery. This is done with flashlights. Because of high voltage, safety is an issue. Finally obtaining reliable connections every time must be demonstrated (see Chapter 16).

Another novel suggestion is to have a small trailer to be towed behind the vehicle. The trailer has a battery and may have a small engine to help charge the battery. For a trip that is longer than the normal range of the vehicle, the trailer would be towed. This suggestion is applicable to EV, PHEV, and HEV (see Chapter 16).

Comparisons for Different Energy Storage Devices

The two major categories for energy storage using electrochemistry are batteries and capacitors. Electrolytic capacitors are not suitable for transportation. The electronic double layer (EDL) has capability to be used in EV and HEV.

Table 6.5 compares different energy storage devices and gives the ratios of P/E and E/P. From E/P, a characteristic time is obtained. For the lead acid battery, this time is 9 min. Assume an application where the run time is 9 min. For this application, the lead acid battery will have neither excess power nor excess energy. If the application becomes 20 min, the battery must be sized to fit the energy requirement. Note in passing that for longer application times, more energy is required. The battery will then have excess power. Of course, the battery can be used at lesser power.

BATTERY SELECTION

Direct Effect on Miles per Gallon

Low internal impedance is very important due to influence on efficiency and terminal voltage. Terminal voltage droops as current is removed from the battery due to internal impedance. The losses heat the battery internally making cooling essential. High cycling efficiency is desired and implies high efficiency charging and discharging.

The four primary characteristics of a battery are high energy density ([kWh]/L), high power density (kW/L), high specific energy ([kWh]/kg), and high specific power (kW/kg). These four quantities determine how big and how heavy the battery will be in the hybrid. Compact battery construction affects the density values. Cell geometry has a larger effect on internal impedance than electrochemistry. The battery should have significant surge capability. The hybrid typically needs 10 s or so spurt of power.

Customer Satisfaction

Battery life and battery replacement costs are a dominant concern of hybrid purchasers and owners. Long life is essential. Reasonable life can be attained with controlled (limited) discharge. In contrast to sudden failure, the battery should age gracefully. Add high reliability to the list of features.

Although not so important for HEVs, which charge the battery on-the-run, the battery is key to the PHEV. The hybrid should provide easy and safe charging management (avoid overcharging with attendant damage). Easily maintained SOC is necessary. Quick recharge is a winner. Leakage current must be small; self-discharge of the battery is not acceptable.

Safety

Safety is discussed in Chapter 21. A totally sealed battery may avoid spillage of toxic chemicals. A physically robust battery is likely to be safe in collision. Explosions or fires are, of course, not acceptable (see Figure 6.11 and the section on Battery Safety).

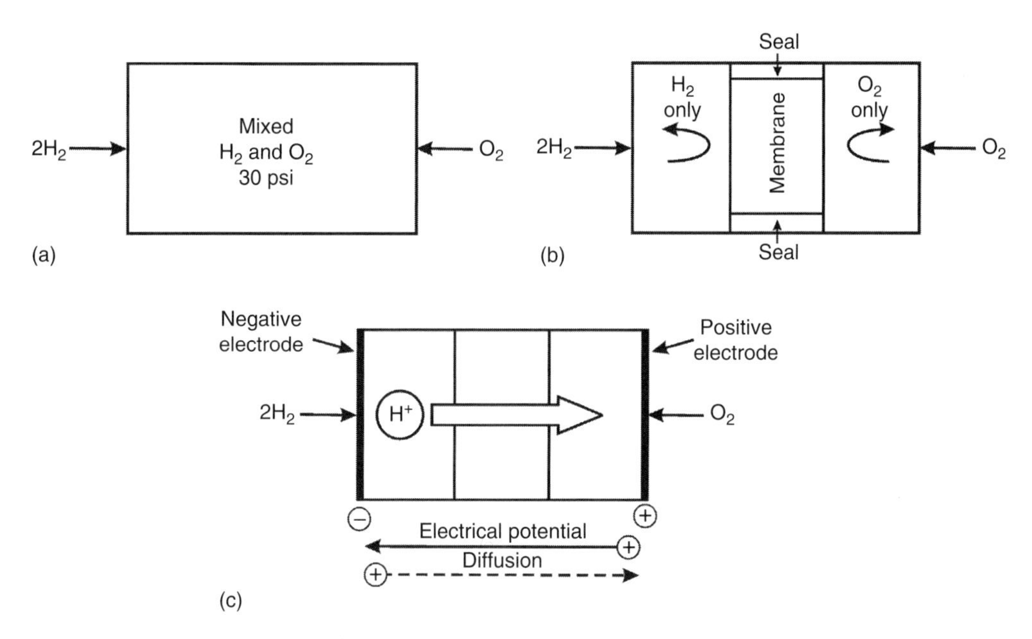

FIGURE 6.11 Step-by-step introduction to battery safety.

Hybrid Cost on the Showroom Floor

A significant fraction of HEV cost is the battery pack. Focus on cost makes for a more marketable HEV.

Permissible Operating Environment

The automobile operates in a severe environment and, therefore, the battery also operates in the same environment. The hybrid must operate over a broad temperature range. Stable performance at low temperature is expected. The battery needs to provide lively power with excellent cold temperature performance at –49°F (–45°C). Special need for preheating or molten electrolytes will likely eliminate a battery from serious consideration.

Avoiding Unanticipated Recalls

Avoiding unanticipated recalls is achieved by avoiding unsuspected battery peculiarities. Examples of two peculiar behaviors are memory effect (Figure 6.4) and voltage recovery (Figure 6.5), which occur for some batteries. These abnormalities confirm the fact that batteries are indeed complex. Some batteries generate gas during recharging. Excellent gas recombination (in those batteries that generate gas) avoids potential problems.

Recycling Batteries

As discussed in Chapter 3, one motivation of the hybrid owner is to drive green. For this reason, recycling of batteries is uppermost in their minds. Of course, environmental regulatory agencies require recycling. Because of the use in CV, the lead acid battery has a vast infrastructure for recycling. This experience and know-how can be extended to the so-called exotic batteries.

Cost of Battery

As with most of the electrical components in EV or HEV, major cost reductions are anticipated due to employment of tools of mass production. Such tools include robots, automated design, and dedicated inspection. Further, shortcuts and innovative modifications accrue with experience. Hunt for and eliminate, if possible, the costly parts of the battery. Use low cost base materials. Adopt modularity of design.

Materials in batteries are largely irritable. Examples of cantankerous characteristics include toxicity, scarcity of raw materials, highly dense (heavy) materials, and expensive catalysts (platinum catalyst in FCs).

BATTERY PROPERTIES

Battery Safety

Batteries inherently have dangerous chemicals. For an application in hybrids, batteries have high voltage. Consequently, battery safety must be a primary focus. One desirable feature, but not necessarily a firm constraint, is an electrolyte that is not aqueous (for safety).

What do you do to make a bomb? Combine the most powerful chemicals available. What do you do to make a battery? Combine the most powerful chemicals available. The three diagrams in Figure 6.11 clarify one aspect of battery and FC safety. To provide the most power and energy, the most energetic chemicals are selected. In Figure 6.11a, hydrogen and oxygen are being added to an empty box. The $2H_2$ and O_2 are mixed at high pressure forming a very dangerous bomb. Separation of the reactants is a basic safety requirement. A membrane has been added in Figure 6.11b. The membrane is opaque to both the $2H_2$ and O_2. Further, the $2H_2$ and O_2 should not leak around the membrane. For this reason, seals become an important safety item. With the additions to Figure 6.11b, the FC is safe but will not function. The ions in the electrolyte must move between the electrodes. Just any membrane will not work. The special proton exchange membrane (PEM) allows the protons, which are H^+, to pass through thereby completing the internal electrical circuit.

In Figure 6.11c, arrows are shown for the direction of motion due to electric potential and diffusion. The H^+ move uphill against the electrical potential. The motion of H^+ is due to diffusion.

Battery Efficiency and Internal Impedance of Battery

Internal impedance of a battery affects the battery efficiency; high impedance means low efficiency. Also, high impedance means high thermal loads. Hence, one of the most important battery specifications is the internal resistance. This statement is true for all electrochemical storage devices, not just batteries. The droop of voltage is strongly affected by internal impedance. Battery models have been developed. One such model that gives voltage droop as a function of SOC is the Advisor program and software [4].

Internal impedance affects fuel economy because of the losses within the battery. Lost energy in the battery becomes heat. Battery cooling is necessary. Equation 6.14 is the ratio of power into the load, P_L (kW) to the power from the battery is P_B (kW). Other symbol definitions are the internal resistance of cell, R_C; cell OCV, V_C; and the electrical current, I.

$$\frac{P_L}{P_B} = 1 - \frac{IR_C}{V_C} \tag{6.14}$$

The importance of R_C is evident. The larger the value of R_C, the larger the loss of energy stored in the battery.

Monitoring Individual Cells

In a stack, what happens when one cell goes bad during charging or discharging? What happens if charging or discharging continues? A single high resistance cell in the stack can disable the whole stack; the defective cell must be identified and bypassed. The SOC in each cell should be uniform. There are obvious costs for monitoring every cell in the stack. Also losses occur due to monitoring every cell in the stack.

Cells in middle of the stack tend to have less heat transfer and hence get hotter than those on the end. Liquid cooling may be necessary; usually air cooling is adequate. Certain designs create special, battery-unique problems. For example, NiCad cells cannot be connected in parallel; one cell transfers power to the other. An unstable condition occurs.

When connected in series, each cell of the NiMH batteries must be monitored. One cell may become fully discharged and go into polarity reversal.

Thermal Management

During charging, some of the energy from the generator is converted to heat. The result is that the energy stored is less than the charging energy (Figure 6.12). When an attempt is made to remove energy from the battery during discharge, once again, some of the energy becomes heat. On both charge and discharge, heat is generated inside the battery and must be removed to prevent excessively high temperatures. High temperature causes decreased life or even sudden battery failure.

Usually fan-driven air cooling suffices for batteries in HEV. However, some applications require liquid cooling, which significantly adds to cost of the vehicle.

HYBRIDS USING ELECTROCHEMICAL STORAGE

Three selected hybrids using electrochemical storage are presented next. Those hybrids mated to a gasoline or diesel engine are the most numerous and are discussed in almost all chapters. Various hybrids are of interest to the military as well [27]. The two choices for electrochemical storage are batteries and the electrochemical double-layer capacitor (EDLC). A choice for conversion of fuel to electrical energy is the FC. The hydrogen-fueled ICE is another candidate for this chapter. Possible hybrid combinations are

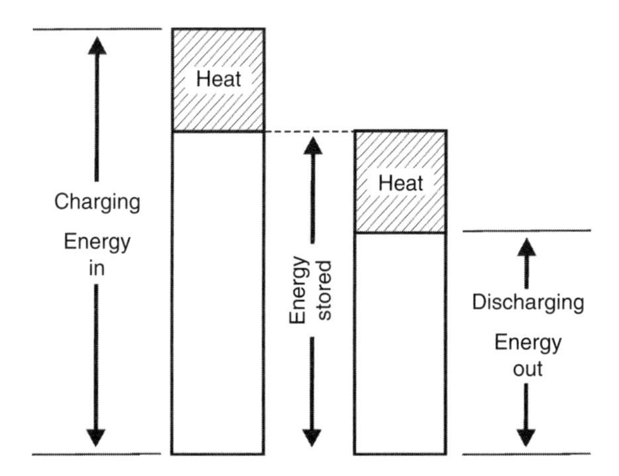

FIGURE 6.12 Thermal management of battery.

- FC/battery
- FC/electrochemical capacitor
- Hydrogen ICE/battery
- Hydrogen ICE/electrochemical capacitor
- Tribrid, three-way hybrid
- Battery/electrochemical capacitor

Since these combinations are exploratory, the studies have been by means of simulations. The combinations of FC/electrochemical capacitor hybrid, battery/electochemical capacitor hybrid, and tribrid are discussed below.

FC/Electrochemical Capacitor Hybrid

In what way does an electrochemical or UC complement an FC, beyond storage? The FC may have time lags between command for power and supplying power. The UC fills in the gap in power as FC begins to produce power. For regenerative braking, a battery may not be able to absorb the high power in a short time whereas the UC can do so. Note that an FC cannot absorb power from regenerative braking.

The Honda FCX also discussed in Chapter 3 has combined power of 80 kW (110 hp). The UC and FC are combined giving higher initial acceleration and overtaking acceleration. UC and FC combined to give higher initial acceleration and overtaking acceleration. The FCX vehicle has a higher efficiency, η, FC. The higher overall efficiency is 55% which is twice that of a hybrid and three times that of a CV.

Cold weather operation has been a problem for FCs. Start-up and power generation at −20°C (−4°F) have been achieved. Driving range is a credible 430 km (258 mi).

Battery Electrochemical Capacitor Hybrid

In what way does an electrochemical capacitor or UC complement a battery, or vice versa? In this example, the battery is a lead acid that has some deficiencies to be discussed. The UC is cost competitive with NiMH. However, the UC cannot replace the NiMH. The combination of the UC with the lead acid battery can replace NiMH with cost saving. The lead acid could replace NiMH in spite of lower W h/kg. The main hurdle and reason to reject lead acid is sulfation. Sulfation is a deposit on an electrode that affects battery performance (Figure A.6.7). Sulfation occurs because of two reasons. First, frequent high ampere pulses and, second, operation at partial SOC.

The following example is based on Ref. [15]. The scheme to combine the UC with the lead acid battery divides the storage devices according to voltage. The operating voltages are as shown in Table 6.8.

TABLE 6.8
Different Combinations of Power Sources for the UC and Battery Hybrid Depending on Voltage Range

Energy Source	Voltage Range	Mode: Power In/Out
UC only	28 < volts < –45 V	Regenerative braking; modest motor assist
UC + lead acid	24 < volts < –28 V	Heavy current draw
UC + lead acid	volts < 24 V	No power out

Tribrid, the Three-Way Hybrid: Fuel Cell, Hydrogen ICE, and Battery Hybrid

The three-way hybrid, tribrid, has three components to size. With the complexity of three major components, can improvements be made in mpg and cost?

The strategic motivations for the study concern less oil import, energy security, and reduced CO_2. An examination of technology with a new combination of parts is the tactical incentive to undertake the study. This study not only gives some insight into a new hybrid concept but also information on electrochemical energy storage.

Two major barriers to fuel cells are expense and the number (very few) of hydrogen refueling stations. The characteristics of the hydrogen ICE include small efficiency, η, limited range (due to fuel storage), but with low cost. Briefly the characteristics of the hydrogen fuel cell are bigger η, extended range (for the same fuel storage), but with high cost. The efficiency of the crop of 2006–2007 demonstrators are FCV 40%–50% and CV 10%–16%.

Goals for future technology are performance oriented. Acceleration: 0–60 mph in 10 s; and hill climbing: continuous grade 5.7% at 55 mph. Other goals are cost oriented. Future cost for the hydrogen ICE (\$35/kW); hydrogen fuel cell (\$45/kW); and storage goal for hydrogen (0.06 kg/tank kg). Other cost and performance goals on the electrical side are motor (5 kW/kg); controller (1 kW/kg); motor + controller (\$12/kW); and Li-ion (\$30/kW).

What is the role of each power source for a tribrid, three-way hybrid? The role varies with the hybrid concept. The fuel cell is the primary source of power and is used over the entire power range. The hydrogen ICE is used only during high power demands. The battery has three roles: store energy from regenerative braking, return energy to vehicle when operating at low power, and provide transient fill-in for the fuel cell lag of few seconds. In contrast to the usual HEV, the battery controller seeks to maintain SOC near a fixed value and does not attempt to store extra energy in battery.

To provide the scope and flavor of the study, Figure 6.13 shows component sizes and power for different values of hybridization, H, and fuel cell fraction, f. Hybridization determines the split in power between the engine and motor. Hybridization, H, is defined in Chapter 11. The same definition is used here. Fuel cell fraction, f, which is defined special for this discussion, is

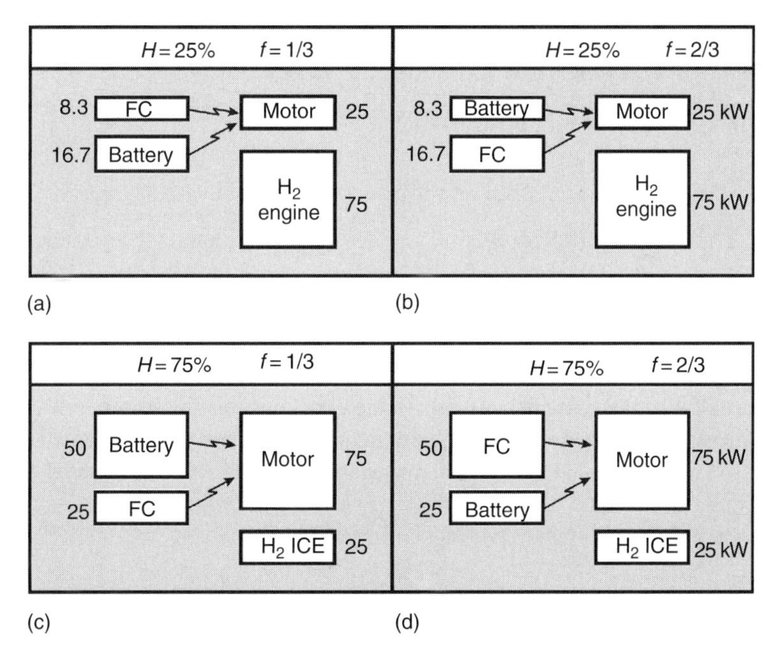

FIGURE 6.13 Sketches of a tribrid, three-way hybrid, for different values of hybridization, H, and fuel cell fraction, f. The numbers adjacent to each component are power (kW). (From Kwon, J., Sharer, P., and Rousseau, A. *Applications of Fuel Cells in Vehicles 2006*, SP-2006, Society of Automotive Engineers, p. 13.)

$$f = \text{(Fuel cell power)/(Motor power)}$$

with H and f in mind (Figure 6.13).

When H is small (H = 25%), the engine is emphasized (Figure 6.13a and b). For large H (H = 75%), the motor is emphasized (Figure 6.13c and d). For Figure 6.13, the total vehicle power is 100 kW. When f is small (f = 1/3), the battery is emphasized (Figure 6.13a and c). For large f (f = 2/3), the FC is emphasized (Figure 6.13b and d). The aim of the study is to vary H and f while in quest of maximum mpg or minimum cost.

Kwon et al. completed a study of the tribrid, three-way hybrid [23]. Their study differs slightly from the preceding discussion. Total power is adjusted to maintain constant performance [23]. The results can be summarized for fuel economy and costs relative to the baseline vehicle.

In Table 6.9, since the baseline vehicle is a hybrid, it has a good 46 mpg. Adjusting f, the fuel economy is increased by 48% from 46 to 68 mpg. Compared to the baseline vehicle, which is a CV, the cost for the least expensive vehicle is 11% less. The cost of the most expensive vehicle is 7% more.

TABLE 6.9
Fuel Economy and Cost Gains for the Tribrid, Three-Way Hybrid

		Fuel Economy	Cost Ratio
Miles per gallon baseline vehicle	$H = 56\%;\ f = 0$	46 mpg	—
Best mpg	$H = 56\%;\ f = 0.35$	68 mpg	—
Cost baseline vehicle	$H = 0;\ f = 0$	—	1.00
Least cost vehicle	$H = 56\%;\ f = 0$	—	0.89
Most cost vehicle	$H = 56\%;\ f = 0.35$	—	1.07

MAPS BASED ON BATTERY ENERGY AND POWER

RELATION OF BATTERY TO OVERALL HYBRID

As a result of the equation

$$\text{(Battery energy)} = \text{(Battery power)(Time)}$$

a plot of energy as a function of battery power has contours of constant time as straight lines. These straight line contours are shown in Figure 6.14. Two types of hybrids are illustrated: PHEV and HEV.

The box for possible PHEV designs is based on constant cruise using a power of 20 kW. Also, the hybrid is a series arrangement. See Chapter 11 for the equations relating battery power to hybridness, H. For H = 1, the design is a pure EV. For H less than unity, battery requirements are reduced. The installed battery energy is E_I, and useable battery energy is E_B. The two are related by

$$E_B = (\text{DOD})(E_I)$$

where DOD is the allowable depth of discharge. The box for the PHEV is tall in energy but narrow in power.

To cruise for 4 h, the PHEV with $H = 0.8$ requires a battery that yields power equal to 15 kW and installed energy equal to 120 kWh.

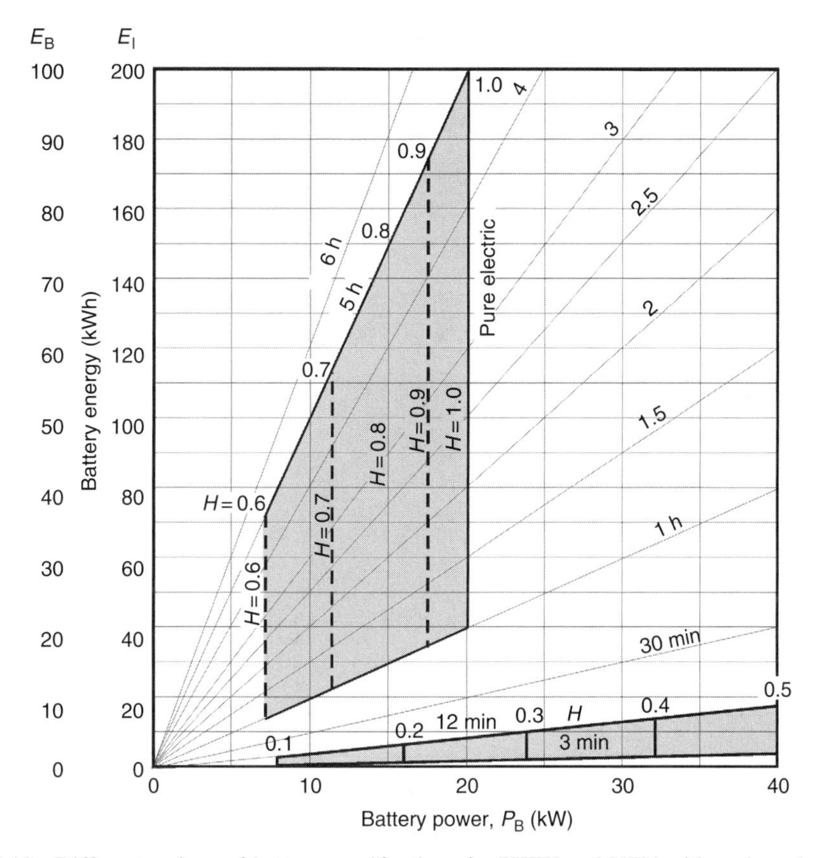

FIGURE 6.14 Different regions of battery specifications for PHEV and HEV with regions defined using battery energy (kWh) as function of battery power (kW).

The box for possible HEV designs is based on a constant maximum power driving the wheels of 80 kW. For H = 0.4, the battery power is equal to (0.4)(80 kW) = 32 kW. The maximum power is required for short duration only. The box for the HEV is narrow in energy but broad in power. Time durations for maximum battery drain are short; the figure shows 3–12 min. As H decreases, the battery requirements also decrease. The run times, 3–12 min, apply to electric-mode for HEV.

The straight lines are also distance driven at a constant speed. Assume a speed of 60 km/h or 1 km/min. Then the constant 3 min line is also 3 km distance. The 1 h line becomes 60 km; 2 h, 120 km, etc.

Plug-In EV or EV: Adequate Power, Large Energy

Figure 6.15 can be used to determine PHEV using different batteries. For the PHEV, assume a battery mass equal to 500 kg, which is a heavy battery. Further assume an allowable DOD equal to 0.50. On the left-hand side are horizontal lines for lead acid, NiMH, Li-ion, and Li-polymer batteries. Select the battery, for example, Li-ion. To help determine the performance, put a sheet of paper parallel to, and above, the horizontal line for the Li-ion battery. All points not covered by the paper are accessible. Read off some points. For H = 0.6, a run time of 3 h is possible. For H = 1.0, that is, an EV, a run time of only 1 h is possible.

Suppose the DOD allowed is increased to 100%. Repeat the same performance analysis using the horizontal lines for lead acid, NiMH, Li-ion, and Li-polymer batteries on the right-hand side of the graph. Use Li-ion again for comparison. As a pure EV with H = 1.0, 2 h cruising is possible.

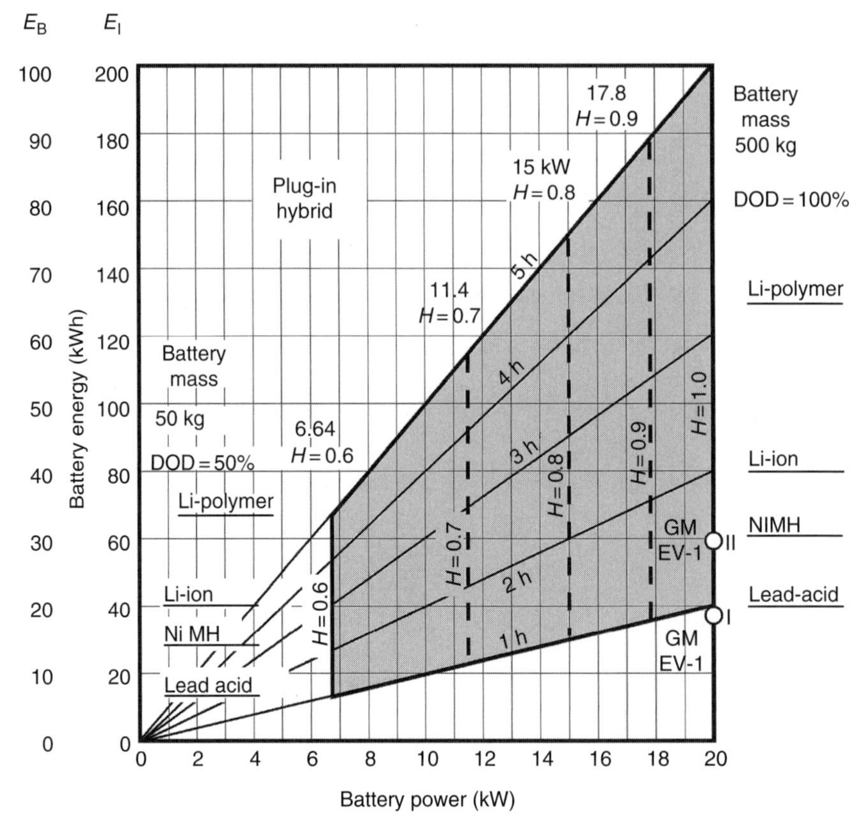

FIGURE 6.15 Box for adequate power but large energy; the expanded box for the PHEV of Figure 6.14.

For $H = 0.7$, cruising of 3.5 h becomes possible. Assuming a speed of 100 km/h (62 mph), the PHEV range is 350 km (210 mi).

General Motors produced one electric vehicle (EV-1) with two different batteries. GM EV-1 I had lead acid batteries while GM EV-1 II had NiMH batteries. The two GM EV-1 I and II are plotted in Figure 6.15.

HEV: High Peak Power, Small Energy

On the left-hand side of Figure 6.16 are horizontal lines for lead acid, NiMH, Li-ion, and Li-polymer batteries. As before, select the battery, for example, Li-ion. To help determine the performance, put a sheet of paper parallel to, and above, the horizontal line for the Li-ion battery. For $H = 0.5$, only 3 min run time at maximum power is possible. As H is decreased, the run time is increased. For this graph to apply, the gasoline engine must operate—the graphs are not for electric-only operation.

FUEL CELLS

Fuel Cell: Electrochemistry

An FC is the reverse of electrolysis (Figure A.6.5). For the apparatus shown in Figure A.6.5, the process is chemically reversible. An FC is more complicated than electrolysis; however, the overall reaction is the reverse of electrolysis, that is,

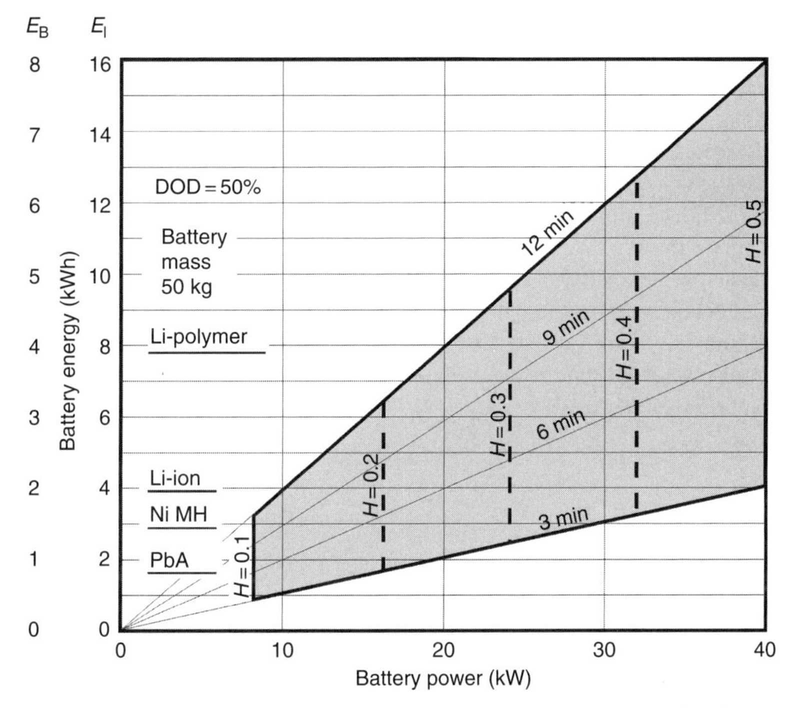

FIGURE 6.16 Box for high peak power and small energy; the expanded box for the parallel HEV of Figure 6.14.

$$2H_2 + O_2 \rightarrow 2H_2O$$

In an FC, hydrogen and oxygen produce energy and water. Some distinctions between reversible cells and FCs are highlighted in Figure 6.17. For electrolysis and the FC, chemical symbols are not shown on anode and cathode. This signifies that the electrodes do not undergo chemical change, at

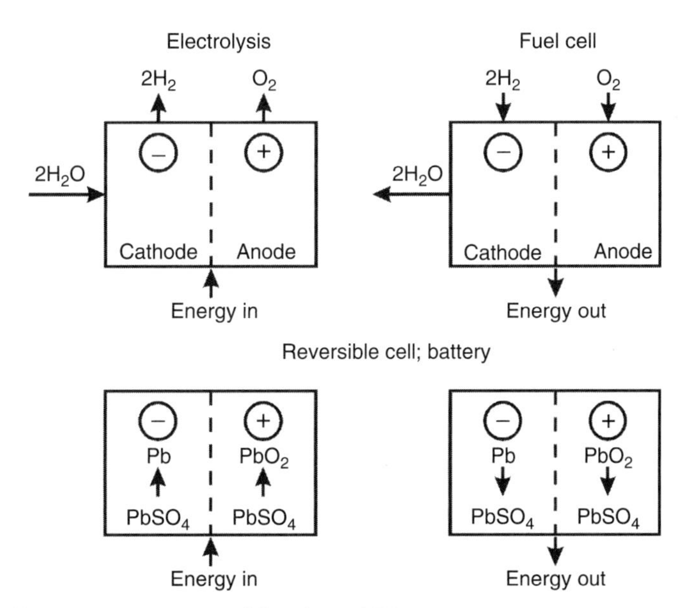

FIGURE 6.17 Differences between reversible cells and FCs.

least not intentionally. Further, for an FC, a flow of chemicals in and out occurs in addition to energy in or out.

For the reversible cell, which is combined to form a battery, each electrode undergoes chemical change. The change is reversed from discharge to charge. Only energy flows in or out of the cell. The cell can be fully sealed. Both the reversible cell (when in a charged condition) and the FC are spontaneous in the sense that if an electrical circuit is completed external to the device, power will flow out. Of course, the FC requires the input of the chemicals.

Figure 6.2 shows the sag in terminal voltage as a battery is discharged. FCs also have a droop in output voltage as current density, J (A/cm^2), within the cell increases. At low J, the sinking voltage is due to lack of adequate electrocatalysis. For a broad range of J, the voltage decreases linearly due to ohmic losses, that is, (current-resistance [IR] loss). For very high J, the voltage drops rapidly due to decreasing transport of ions in the electrolyte. This droop in voltage has an adverse effect on FC efficiency (Equations 6.7 and 6.14).

Fuel Cell: Types

The words fuel and cell go together like ham and eggs. As is the case with batteries, cells are placed in series to obtain higher voltages. Putting FCs in series creates a fuel battery. Instead of FC battery, FC stack is the conventional term. Simply stack is frequently used.

Many different FCs can be derived from the array of components shown in Figure 6.18. With regard to fuel for the FC, three major factors are use of liquid fuels, the off-board production of the fuel, or the on-board generation of hydrogen from hydrogen-bearing compounds. On-board means the FCV has equipment on the vehicle to process the chemicals. Off-board means the processing by central facilities. Gasoline refining is an example of off-board. Hydrogen and oxygen are not the only possible chemicals to power an FC. On the upper left, both liquid hydrazine and methanol are shown. Both chemicals are favorable from electrochemical potential. Since methanol has carbon, some CO_2 will be formed. Hydrazine, which does not have carbon, is a rocket propellant.

In the mid-left side, off-board options for utilizing hydrogen are shown. The hydrogen is available in a hydrogen service station. The only remaining problem is to store the hydrogen in the vehicle. Four different avenues are shown including the potential use of the marvels of nanotechnology. High pressure storage (6000 psi) requires considerable energy to compress the hydrogen. Cryogenic storage can be done at different pressures and temperatures. At 600 psi in

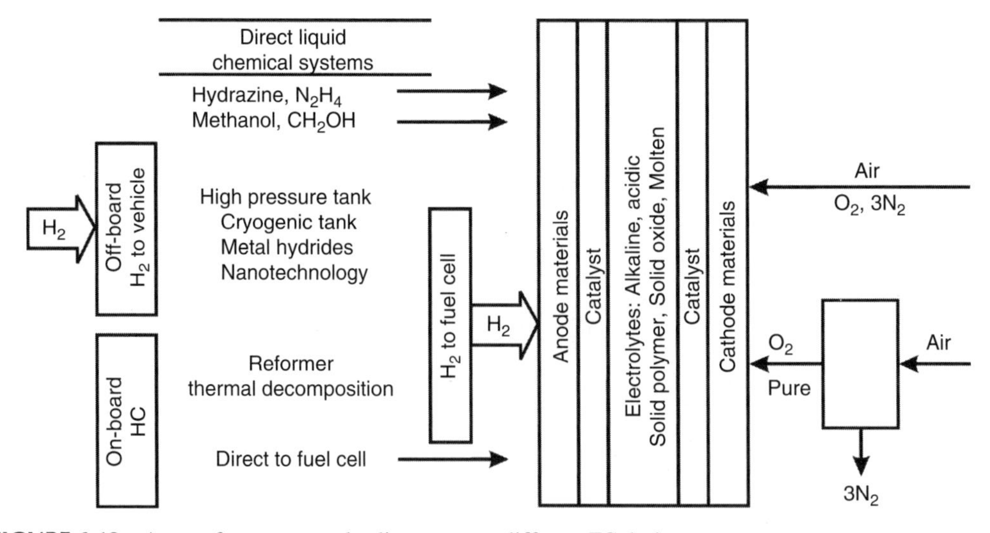

FIGURE 6.18 Array of components leading to many different FC designs.

the hydrogen tank, a temperature of −180°C (−292°F) is adequate for low boil-off. With 14.7 psi in the hydrogen tank, a much lower temperature of −270°C (−454°F) is necessary. Experience has been gained with cryogenic storage by the storage of liquid natural gas. Nanotechnology may provide a mesh that stores hydrogen in the pores.

In the lower left corner, on-board options using HC fuels that yield hydrogen by reformers or by thermal decomposition are shown. Ford is working on an FCV, the Ford Focus FC5, using a methanol reformer. Of course, another entirely different FC design might use the HC directly.

Focus now on the FC itself. The fuel, typically but not exclusively hydrogen, is fed to the anode. Different materials can be used for the anode and the cathode. Choice of materials is dictated more by durability and reliability. A variety of electrolytes are possible, although the PEM seems to be the winning choice. The catalyst must be matched to the electrochemistry occurring at the electrodes. A major aspect of the catalyst is the availability and cost.

On the right-hand side the source of oxygen is shown, which is air for most FCs. The nitrogen flows through the FC. A hydrogen and oxygen FC has superior performance using pure oxygen. However, the better performance using pure oxygen does not justify the increase in cost and complexity.

Comparing PEM and alkaline, KOH, FCs, the gas purity requirements differ. PEM can tolerate carbon monoxide (CO) while alkaline cannot accept CO. CO is formed when gasoline or methanol are reformed. About 40% of fuel energy becomes heat in the FC. Cooling needs also differ. Some PEM cells require liquid cooling. The water in the PEM FC can freeze whereas the alkaline cell does not freeze at 0°C. For insight into the design and production of FCs see Refs. [31,32].

Fuel Cell: Characteristics

Power regulation of an FC is by the flow rate and partial pressure of hydrogen fed to anode. Air pressure of air fed to the cathode is related to electrical current demand on FC; 10 s are needed to increase pressure. Pressure can be decreased faster.

Operating pressure leads to two main designs: low pressure and high pressure FCs. The low pressure cells have slow response time of about 10 s. The low pressure cells cannot follow the demands of a typical driving cycle. Power is needed for the hydrogen and oxygen supply system. Typically, 10% of the FC power is consumed by auxiliary equipment. The high pressure FC has fast response and can follow the demands of a typical driving cycle. However, the fast response absorbs 25% of the power due to the necessary auxiliary equipment.

Fuel Cell: Efficiency

One characteristic of FCs is that highest efficiency is attained at a small fraction of maximum output power. This is illustrated in Figure 6.19. The region for high fuel efficiency is narrow extending from 10 to 40 kW. The span of high efficiency for the H_2 ICE is greater going from 10 to 60 kW.

FCs have little sensitivity to scale. A tiny FC is as efficient as a big FC. From the efficiency viewpoint, FCs are ideal for cruise (Table 6.10).

Fuel Cell: Applications

The transition to FCs may involve a series of steps. Development of the hydrogen ICE, which is a smaller challenge than the FC, will create demand for hydrogen. The vehicle with a hydrogen ICE can be dual-fuel so that failure to find a hydrogen station is not a disaster; just fill up with gasoline. The early FCV can be sold or leased to fleet operators, both commercial and government, who will have the incentive and can build their own hydrogen refueling station. FCV will unlikely be dual-fuel. So, when the hydrogen is near empty, a hydrogen refueling station must be found. GM is leasing FC-powered minivans to the U.S. Postal Service. The motor in the van has 60 kW (80 hp) [29].

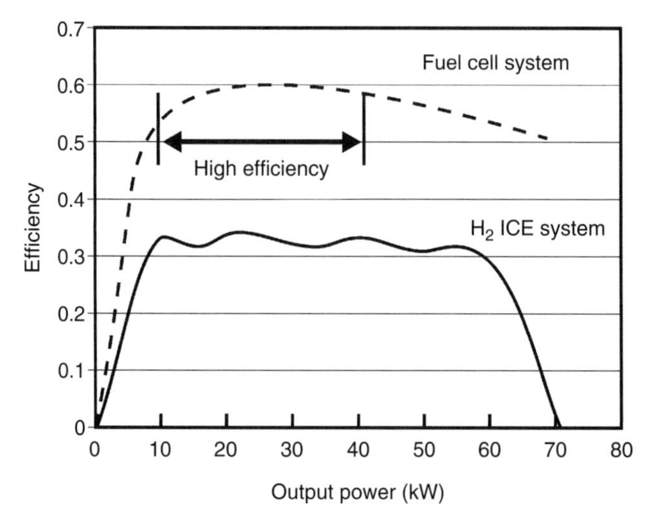

FIGURE 6.19 FC and H_2 ICE efficiencies as a function of output power. (Reprinted with permission from SAE Paper *Applications of Fuel Cells in Vehicles 2006*, SP-2006, 2006 SAE International.)

The automobile operates in a severe environment and, therefore, the FC also operates in the same environment. In addition to operation over wide temperature ranges, the FC must be reliable, durable, and efficient. Numerous requirements are dictated by the application. Cost is dominant, and a goal of $30/kW by 2010 is the target.

Design trade-offs include cell pressure, stoichiometry (ratio of hydrogen to oxygen), and humidification. FC operates best with humid air and hydrogen. The Andromeda stack of Ref. [18] provides the following performance:

10 ± 5 kW
240–384 V ± 15 V
Transient: 10%–90% in 2 s
Weight: 115 ± 5 kg
Power density: 1.3 kW/L
Specific power: 0.7 kW/kg

Power density and specific power are at peak power. The H_2O from the cell is used to provide the humidity.

FC costs are no longer dominated by the stack. The added equipments that feed and control the fuel have become the major expensive items. Cost reduction is possible once mass production is needed to meet demand. With experience, innovative design modifications can lead to sizeable cost reductions.

Transition to FCs has two aspects: emissions and efficiency, η. Even if the first FCs on the market have moderate efficiency, η, the emissions will be very low. With either gas or diesel

TABLE 6.10
Fuel Cell Efficiency for Peak and Cruise Levels of Load

Mode	Power Out	Efficiency	Duration
Peak	100%	$\eta = 40\%$	Limited
Cruise	20%–30%	$\eta = 60\%$	Most of time

TABLE 6.11
Comparison of AFC and PEM FCs

Feature	Alkaline Fuel Cell	Proton Exchange Membrane
Electrolyte	KOH	Polymer
Operating temperature	80°C	85°C
Catalyst: cathode	Silver	Platinum
Catalyst: anode	Platinum	Platinum
Typical maximum power	100 kW	250 kW
Efficiency[a]	50%	45%–55%

[a] Includes lost power to drive FC auxiliaries.

Note: The AFC stack needs to warm up to 70°C (158°F) for full rated power. Warm-up time is approximately 15 min.

engines, exhaust gas aftertreatment is not cheap. This expense will be saved in FCV. With regard to CO_2 and global warming, FCV gets rid of CO_2. FCs can solve the zero emissions vehicle (ZEV) standard that was at one time required by California Air Resources Board (CARB). ZEV remains a goal of CARB.

Tables 6.11 and 6.12 compare the advantages of the alkaline and PEM FCs.

TABLE 6.12
Advantages and Disadvantages of the Alkaline and PEM FCs

Alkaline FC

Advantages
Quick to start
Has high power density, kW/kg
Economical materials

Disadvantages
Corrosive electrolyte
Cannot tolerate carbon dioxide, CO_2 (maximum 50 ppm)

PEM FC

Advantages
Proven in space applications
Quick start
Long life (life limited by life of PEM)
Ease of volume manufacturing
Solid, noncorrosive electrolyte
Produces potable water

Disadvantages
Expensive metal catalyst
Cannot tolerate sulfur
Cannot tolerate CO
Low operating temperature; waste heat cannot be recovered; waste heat is wasted.

Fuel Cell: Hybridness

The definition of hybridness, which is used in Chapter 11 needs a slight modification for FCV. Logically, replace the engine with the FC giving

$$H = \frac{\text{Battery power}}{\text{Fuel cell power} + \text{Battery power}} \tag{6.15}$$

Many FCVs have high values for H (80%–95%). As FCVs develop, what will be the best balance between battery power and FC power?

Hydrogen Production

The possible methods for production of hydrogen are as follows:

1. Steam and methane; water gas reaction; good for on-board production. Two chemical steps produce the hydrogen.

$$CH_4 + H_2O \rightarrow CO + 3H_2$$
$$CO + H_2O \rightarrow CO_2 + 2H_2$$

2. Partial oxidation.
3. Autothermal: Uses methods of 1 and 2 above. The steam and methane absorb energy. The partial oxidation releases energy. The combination requires less energy to produce hydrogen.
4. Electrolysis: One big advantage of electrolysis is the use of power from "green" sources. Wind power, wave power, nuclear fusion, nuclear fission, and solar power can be used.
5. Methanol reforming: Conditions are favorable for on-board reforming. Also direct use of methanol as the fuel in FC may be possible. Methanol reforming generates pollutants due to the carbon in the methanol.
6. Autothermal reforming of gasoline and diesel fuels: Existing service stations can be used for off-board reforming. Likewise on-board reforming is an alternative. Reforming of gasoline and diesel fuels generates many pollutants that must be eliminated. The process is subject to whims of prices for gasoline and diesel fuels. As time passes, gasoline and diesel fuels will become expensive and scarce. This is a stopgap measure. On a mass basis of the energy content of gasoline, slightly more than one-half is due to the carbon with, of course, slightly less than one-half due to the hydrogen. If the energy from carbon is not used in the system, then more than one-half of the energy of the gasoline is thrown away.
7. Photo-electrolysis of water: Ultraviolet (UV), light illuminating titanium oxide electrodes immersed in water splits the water in H_2 and O_2. Unfortunately sunlight has very little UV light. Most UV is absorbed by ozone O_3 in the atmosphere. Research is being conducted on other metals that might permit photo-electrolysis of water using the broad spectrum sunlight.

Reforming: Hydrogen Production

Reforming on board the FCV may be undesirable. This conclusion is based on cost, complexity, and inefficiencies compared to a central facility. Refer to Table 6.13 for a summary.

TABLE 6.13
Factors Affecting Design the Decision for On-Board or Off-Board Fuel Supply

Item	On-Board	Off-Board
Fuel tank	Almost same as gasoline	Heavy, high pressure or cryogenic
Refill tank	Normal, same as gasoline	High pressure at pump. Need compressor on-board?
On-board equipment	Likely heavy; must haul everywhere; fewer mpg	None (compressor ?)
Infrastructure	Methanol similar to gas	Chicken-and-egg problem; about $20 billion infrastructure to justify mass production of FCV
Cost	?	?
Fuel purity	?	Electrolysis very good
Reliability	?	?
Safety	Safety narrowed to FCV	Whole system hydrogen safe
Environmental issues	Pollutants made in vehicle How to dump safely?	Pollutants mainly in central station; easier to dump safely

Note: A methanol reformer is used as an example.

CAPACITORS

ELECTROLYTIC CAPACITORS (SCS)

Capacitors have long life, are very efficient, and tolerate abuse such as high temperatures. Capacitors have very low specific energy ([kWh]/kg) compared to batteries. This fact is discussed more below. Capacitors offer rapid charge, which is an advantage for regenerative braking. Capacitors offer rapid discharge, which is an advantage for bursts of high power. The cost of capacitors, $/kW, is more than the PbA battery but less than NiMH or Li-ion batteries.

An electrolytic capacitor is a cell in which a very thin layer of dielectric (nonconducting material) is deposited on an electrode by an electrical current. Because of the thinness of the material, large capacitance is achieved in a smaller volume than the usual capacitor. An SC has an electrode and an electrolyte that forms a sheath (boundary layer) on the electrode surface. The electrode/electrolyte interface is a dielectric with an effective thickness of a few angstroms. When combined with a material of high surface area, such as carbon black, a high-capacitance, low-voltage energy storage capacitor is formed. In theory, a few milligrams of material with an area of 1 cm^2 can yield 1 F of capacitance. The EDLC is formed by using both a positive and negative set of electrodes.

What is learned from the data of Table 6.14? With regard to specific power, the PbA battery and the SC possess nearly equal power densities. This means that for the same level of power, both

TABLE 6.14
Comparison of PbA and SC

Energy Storage	Specific Power W/kg	Specific Energy W h/kg
Lead acid (PbA)	50	400
Supercapacitor (SC)	40	2
Ratio (SC)/(PbA)	0.80	0.005
Ratio (PbA)/(SC)	1.25	200

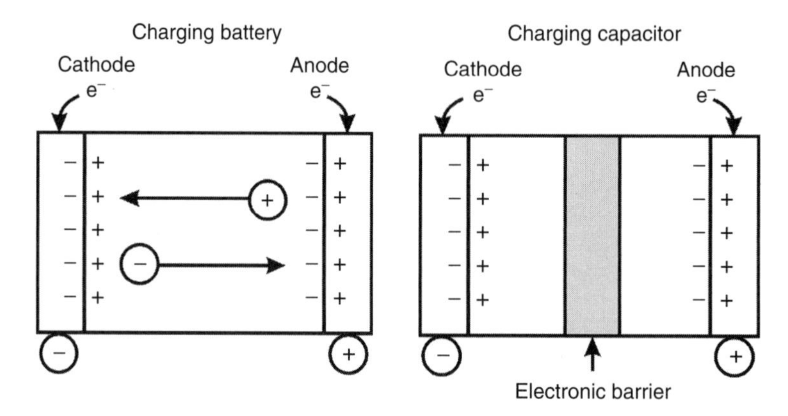

FIGURE 6.20 Comparison of battery and SC.

will weigh nearly the same. However, with regard to energy, the PbA battery provides 200 times as much energy as the SC for the same mass. The SC requires considerable volume compared to the PbA battery.

Features of SCs

When charging, the positive electrode is the anode; removal of electrons causes the electrode to become positive (see the right-hand sketch in Figure 6.20). Only low voltage operation is possible otherwise the electrolyte is changed chemically. Only a few volts can be used for SCs and electrolytic capacitors; higher voltages cause electrolysis of electrolyte.

As shown in Appendix 6.2, the energy stored depends on the capacitance, C. Energy stored is directly proportional to C. Double C, and the stored energy is doubled. The unit for capacitance is farad (F). SCs provide very high capacitance, $C = 1.0\,\mathrm{F}$, in a small volume and small electrode area equal to $1.0\,\mathrm{cm}^2$ with 1 mg of electrolyte. Porous carbon electrodes are used, which have a surface area $2000\,\mathrm{m}^2/\mathrm{g}$. Dilute sulfuric acid electrolyte gives big capacitance and good specific power (kW/kg). Organic or salt electrolyte is favorable for specific energy (W h/kg).

Very rapid charging is possible for SCs and they require no maintenance. SCs do not deteriorate with use; the electrodes do not undergo chemical reactions every C/D cycle. Compared to other storage devices, SCs are relatively inexpensive.

A battery being charged completes the current loop by means of ion migration between electrodes. This motion of electrodes is shown in the LHS figure. A battery has reduction/oxidation reactions, called redox reactions for short. In the capacitor, redox reactions do not occur. Completion of the current loop is not needed for a capacitor. Consequently, an electronic barrier or separator is inserted to block migration of ions in the electrolyte. A dipole sheet is formed at each electrode. This layer stores energy. During charging, most of the ions may be depleted from the electrolyte and appear in the layer.

The two layers of (−)(+) at the cathode and (+)(−) at the anode give the name DLC. The voltage applied to the layer is limited by onset of electrolysis. The limit for a single layer is less than 3 V for organic and 1.25 V for aqueous electrolytes. For DLC, the two-layer capacitor limits are 6 and 2.5 V.

Charge/Discharge Efficiency

Capacitors have higher C/D efficiency than NiMH batteries. This means less heat is generated inside the capacitor that in turn requires less cooling. Under discharge, capacitors have less voltage droop than an NiMH battery as high power flows out of the device (Table 6.15).

TABLE 6.15
Typical Specifications for a Super Capacitor (SC)

Specification	Value
Discharge voltage[a] (V)	120
Capacitance (F)	20
Maximum current (A)	200
Mass (kg)	24
Volume (L)	17

[a] Voltage at the terminals when operating under load.

Characteristic Time Constant

Chapter 8 discusses the enormous challenges of integration and control software. Components must engage and disengage smoothly; the change between hybrid modes should be imperceptible (almost) to the driver. To sell the hybrid, the driving experience must be pleasant without annoying jerks, shudders, and shakes. Further, perceived mismatches between driver commands and car responses are forbidden.

The capacitors, in combination with the resistance, R, in the hybrid circuit, along with the resistance, R, in the battery itself have a time constant, τ, which is simply

$$\tau = RC \tag{6.16}$$

At what value of the time constant, τ, does the driving experience become unpleasant with annoying jerks, shudders, and shakes? Delays in commanded action can cause harshness, vibration, and noise (HVN). The delays may be traced to the time constant, τ. Adding and removing energy is delayed. The hybrid engineer has one more checkpoint to investigate during the design and testing. Model battery circuit diagrams provide the resistances and capacitance for analysis. The Advisor battery simulation program is discussed in connection with Figure 6.9.

An example of a time constant and annoyance to humans is the flicker of a light. This example is not from hybrids. Standard frequency for power in the United States is 60 Hz. Flicker of light is not noticeable at 60 Hz. At 50 Hz and below, the flicker becomes very annoying. The threshold time constant for flicker is near $\tau = 1/50\,\text{Hz} = 0.020\,\text{s}$. The bleed down of the voltage of an inverter capacitor is controlled by the RC time constant. This is a safety issue discussed in Chapter 21.

Energy Stored

Appendix 6.2, which discusses capacitors, gives more information on the energy stored in a capacitor. The equation for storage is

$$E = \frac{1}{2} CV^2 \tag{6.17}$$

The symbol C is capacitance of the capacitor. An important fact is that when voltage is reduced to one-half, the energy is reduced only to one-quarter. This fact is favorable for the application of SC in EV and HEV.

Comparison of Lifetimes or Cycles of Operation

Lifetimes or cycles of operation are important factors in selecting a battery or SC. More information is given in Chapter 3.

Attractiveness of SC for Trucks and Buses

An extensive discussion on the attractiveness of the SC, compared to a battery, for trucks and buses is given in Chapter 3. Table 6.16 summarizes information that is relevant to this short section as well as the preceding sections.

Some Analogies

To this point, three sources of electricity for hybrids have been defined: batteries, FCs, and capacitors. Think of an interstate, say I-40. The steady flow of eastbound traffic is analogous to the flow of electrons within the external circuit of either a battery or an FC (Figure A.6.1). The steady flow of westbound traffic is analogous to the flow of ions in the internal circuit of either a battery or an FC. Once again correlate with Figure A.6.1. The flow of traffic and electrical current are steady.

Using a similar analogy, the unsteady nature of the capacitor can be illustrated. Think of traffic driving into a parking lot. The cars in the parking lot are analogous to electrical charges in a capacitor. The parking lot eventually fills. Likewise, the capacitor eventually fills as determined by maximum voltage. For both electrical current and traffic into the parking lot, the flow is unsteady. Compare with the sketch at the right-hand side of Figure 6.20. The number of cars (and electrical charges) is analogous to the amount of energy stored within the capacitor.

TABLE 6.16
Comparison of Ultracapacitors (UCs) and Batteries as Electrical Storage Devices

	Batteries	UCs
Lifetime without maintenance	1–5 years	10+ years
Number in lifetime of high rate discharge/charge cycles	1,000–10,000[a]	1,000,000[b]
C/D efficiency	40%–80%	90%–98%
Charge time	1–5 h	0.3–30 s
Discharge time	0.3–3 h	0.3–30 s
Specific energy (W h/kg)	10–100	1–10
Specific power (kW/kg)	~1	<10
Adequate energy to meet peak power duration	Yes	Yes
Limitation on SOC	Low SOC limits life	No effect SOC on life
Cost $/kWh	Lead acid least; other than types three to ten times more	Slightly more lead acid[c]
Working temperatures	−20°C to +65°C (−4°F to +150°F)	−40°C to +65°C (−40°F to +150°F)

[a] Depends on the application and BMS.
[b] Less dependent on application and monitoring.
[c] Significantly cheaper than all batteries except lead acid.

SUMMARY

HEVs are a combination of electrical and mechanical components. This chapter focuses on the electrical side. Several definitions essential for the understanding of electricity are given. Also terms to quantify batteries are presented.

Three main sources of electricity for hybrids are batteries, FCs, and capacitors. Each device has a low cell voltage, and, hence, requires many cells in series to obtain the voltage demanded by an HEV. Broadly speaking, the FC provides high energy but low power, the battery supplies both modest power and energy, and the capacitor supplies very large power but low energy. Many cells in series require monitoring of the stack.

The components of an electrochemical cell include anode, cathode, and electrolyte. The current flow both internal and external to the cell is used to describe the current loop.

A critical issue for both battery life and safety is the precision control of the C/D cycle. Precision control requires accurate determination of the SOC. Overcharging can be traced as a cause of fire and failure.

The droop of voltage at the battery terminals hinges to a large extent on internal impedance. Increasing current increases voltage droop. The sag is related to efficiency. Big sag equals big losses. The wilting of terminal voltage is also correlated with the battery cooling requirements. Of course, the losses ultimately affect fuel economy.

The temperature sensitivity of batteries is discussed along with each battery's best operating temperature.

Applications impose two boundaries or limitations on batteries. The first limit, which is dictated by battery life, is the minimum allowed SOC. As a result, not all the installed battery energy can be used. The battery feeds energy to other electrical equipment, which is usually the inverter. This equipment can use a broad range of input voltage, but cannot accept a low voltage. The second limit is the minimum voltage allowed from the battery.

An FC is the reverse of electrolysis. For the hydrogen and oxygen FC, besides power out, pure water is produced. The FCV will emit zero CO_2, which is important for global warming. Further, other emissions will be low. FCs signal the independence from petroleum as an energy source. The FCV will likely be a hybrid with a fusion of an FC and a battery. In addition to low emissions, the FC offers good efficiency. Efficiency is poor at high power; 40% or so. Efficiency is good at partial power; somewhere near 60%.

FCs cannot be mentioned without the caveat of the hydrogen infrastructure, or lack thereof. As presented in this chapter, the rungs in the hydrogen ladder start with fleets of demonstrators to establish early demand. Development and production of the hydrogen ICE expands demand. Finally, FCV enter mass production, and the hydrogen infrastructure grows hand in hand. Favorable government policy is essential for success.

Perhaps, or perhaps not, capacitors are a viable option for hybrids. Hybrid trucks, buses, and high performance sports cars are likely to be the pioneer applications. A capacitor stores electrical charges that, in turn, can be released as energy. Capacitors do not have chemical reactions that account for their long life. The three main types of capacitors are discussed. Capacitors have very rapid charge or discharge. The word rapid implies high power. For regenerative braking, absorbing all the energy is a problem, and capacitors offer a solution. A weakness of capacitors is low energy density.

Battery safety pivots on ample control of highly energetic chemicals. Highly energetic translates into potentially or inherently dangerous. Complete control would be preferred to ample or sufficient control. Only after extensive testing is confidence high enough to consider a battery for an application in a hybrid. Due to danger from fire, certain types of Li-ion batteries cannot be legally hauled in the cargo bay of commercial aircraft.

Battery selection is a major decision of the hybrid engineers. The factors affecting the decision are enumerated above and include performance, direct influence on fuel economy, customer satisfaction (long life and reduced threat of replacement costs), safety, cost, permissible operating environment (winter and summer, etc.), and the avoidance of unanticipated recalls by the manufacturer.

Some hybrid designs are presented as a capstone for this chapter's earlier discussion. These include FC/battery, FC/electrochemical capacitor, and the tribrid, which is a cute name for a three-way hybrid. Since these combinations are exploratory, the studies have been by means of simulations. Using the graph of energy as a function of power, two hybrid examples are highlighted; these are PHEV in cruise and a parallel HEV while using maximum power during acceleration or hill climbing.

APPENDIX 6.1: ELECTROCHEMISTRY OF A VOLTAIC CELL AND FUEL CELL

INTRODUCTION

By electrochemistry, chemical energy is transformed into electrical energy. With electrical power, the motor/generator (M/G) may be directly used. In contrast, by combustion, chemical energy is transformed into heat. A heat engine, such as the ICE, must be used.

COMPONENTS OF A CELL

The basic element of a battery is a cell. Batteries are formed by combining cells. The components of a cell are the anode, cathode, and electrolyte. The three components are illustrated in Figure A.6.1. Electrons always flow out of the anode; removal of electrons corresponds to the chemical reaction known as oxidation. Electrons flow into the cathode. At the cathode, the chemical reaction is reduction. The overall reaction is known as redox for reduction/oxidation. The third component is the electrolyte, which separates the anode and cathode.

In Figure A.6.1, notice the closed loop for the electrical current. The current of electrons circulates clockwise. The current in the external circuit is carried by electrons typically by copper wires through a load. The minus and plus electrodes are immersed in the electrolyte or are separated by a solid electrolyte. To complete the clockwise current loop within the electrolyte, the electrical conduction is by means of ions. An ion is a molecule with extra electrons giving a negatively charged ion. Alternatively, an ion that has a deficiency of electrons is a positively charged ion. Each ion is represented by a circle with either (−) or (+).

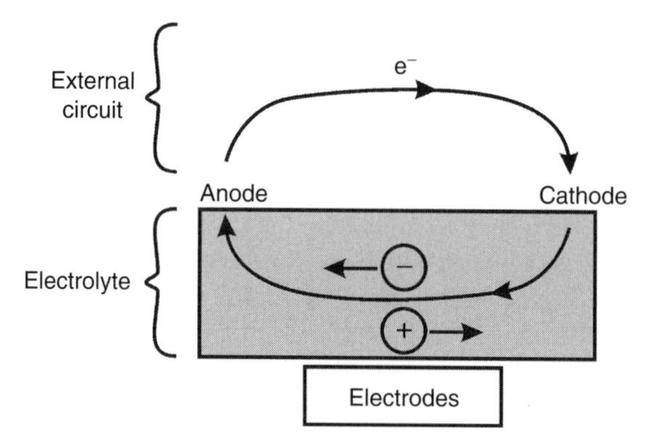

FIGURE A.6.1 Nomenclature for the electrodes. This is a circuit for a cell which is converting chemical energy to electrical energy. The motion of negative charges is clockwise and forms a closed loop through external wires and load and the electrolyte in the cell.

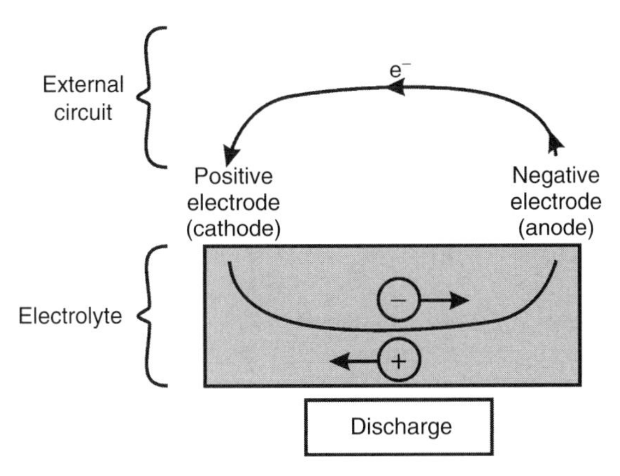

FIGURE A.6.2 Direction of current flow during discharge of a cell. Instead of anode and cathode to identify the electrodes, positive and negative are used.

Resistance to the electrical conduction through the electrolyte is a major source of battery internal impedance. In turn, the internal impedance has a major influence on battery efficiency during both charging and discharging. Ultimately, battery efficiency affects the mpg of a hybrid.

Refer to Figure A.6.2 (discharge) for more information on cell properties. During discharge, the positive electrode is the cathode. This means chemical reduction is taking place at the cathode. In the external circuit, the electrons are repelled by the negative electrode and attracted by the positive terminal. The electrons are moving downhill in the electrostatic field, so to speak. In the electrolyte, the motion of the negative ion is against the electrostatic field. Does the uphill motion of the ions violate the laws of physics? Uphill motion of the negative ion is due to diffusion as well as local electric fields.

For the case of discharge, normally a load for the cell would be shown. The load is omitted here.

Refer to Figure A.6.3 (charge) for more information on cell properties. During charge, the positive electrode is the anode; electrons always are removed by the external circuit from the anode. Note the reversal of current. For charging the cell, the electron current is clockwise, while for discharge the electron current is counterclockwise. Hence, the name assigned to the positive electrode switches from anode (charge) to cathode (discharge). Further, the chemical process at the positive electrode switches from oxidation (charge) to reduction (discharge). Table A.6.1 summarizes

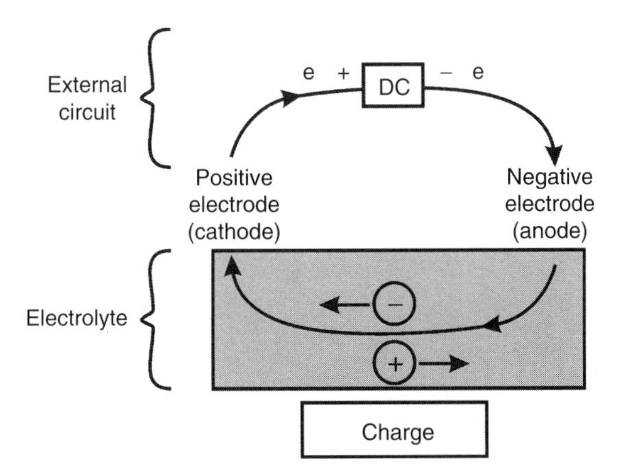

FIGURE A.6.3 Direction of current flow during charging a cell. Instead of anode and cathode to identify the electrodes, positive and negative are used.

TABLE A.6.1
Nomenclature for Electrochemistry

	Charging Electrolysis			Discharging Electromotive Force (EMF), Cell		
Electrode	**Electron Motion**	**Chemical Reaction**	**Electrode Polarity**	**Electron Motion**	**Chemical Reaction**	**Electrode Polarity**
Anode	Out	Oxidation	Plus	Out	Oxidation	Minus
Cathode	In	Reduction	Minus	In	Reduction	Plus

the nomenclature used for electrochemistry. Note the DC power supply that elevates the voltage of the positive electrode.

Voltage Distribution in Cell and External to Cell

Three cases are shown in Figure A.6.4a through c. Figure A.6.4a is an open circuit; the cell is neither being charged nor discharged. The voltage at the terminals, points "a" and "d" is known as the OCV. Clusters of positive charged ions are at the negative electrode while excess positive charges are at the positive electrode. The voltage in the external circuit spans from "a" to "d." At "b" the switch is open, and the OCV appears across the open switch. Within the cell, the voltage is shown as linear with distance. In reality, the local voltage within the cell is more complicated.

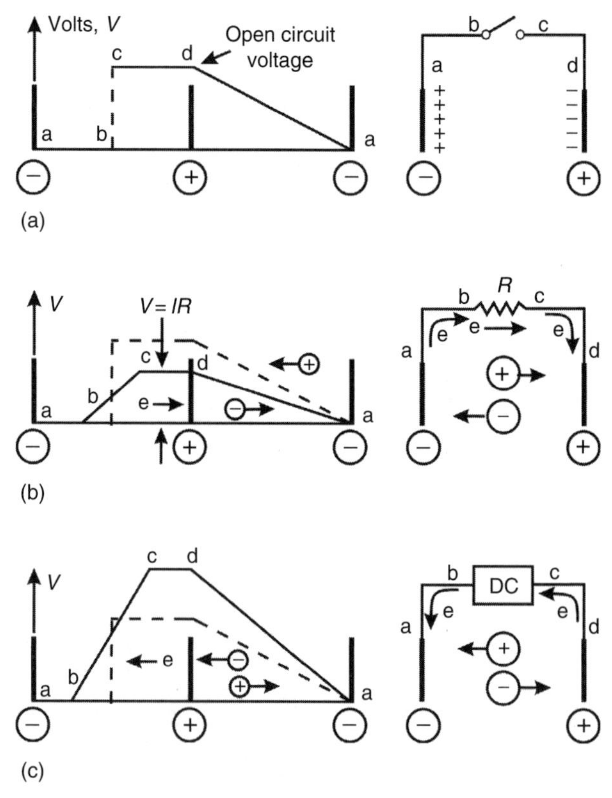

FIGURE A.6.4 Voltage distribution within a cell "d" to "a" and in the external circuit "a," "b," "c," to "d." The vertical heavy black bars represent the positive and negative electrodes of the cell shown in previous figures.

Figure A.6.4b is the case where the cell is being discharged through a resistance R (Ω). The current is I (A), and the voltage drop across the load is

$$V = IR$$

In the diagram, the voltage V is shown between "b" and "c." The voltage across the cell equals V and is shown from "d" to "a." The dashed curves repeat the voltage distribution for the open circuit case. Note the voltage is less when power is being taken from the cell. The drop in terminal voltage is due to cell internal resistance.

The diagram in Figure A.6.4c is for charging. The external DC power supply increases the voltage above that for open circuit. Current is reversed compared to discharge.

When an electron or negative ion moves from negative toward positive, the charge gains energy. This occurs in the external circuit for discharge and within the cell for charge. When an electron or negative ion moves from positive toward negative, the charge loses energy. This occurs in the external circuit for charge and within the cell for discharge. Combining these observations, a statement can be made about transfer of energy or power. For discharge, the external circuit gains energy while the cell loses energy. Conversely, for charge, the external circuit provides energy while the cell gains energy.

ELECTROLYSIS

Electrolysis is presented here. Two objectives are met with one discussion. Electrolysis provides a tutorial in one aspect of electrochemistry. Further, hydrogen production, which is of great interest, can be by means of the electrolysis of water. Figure A.6.5 shows the anode and cathode immersed in an electrolyte, which is not pure water. In practical application, the electrolyte is either diluted acidic or basic so that ions are numerous. The DC power supply provides the energy to dissociate the water into hydrogen and oxygen. The electrolysis equipment can be located at a hydrogen refueling station where hydrogen is made to sell to the customers. Gobs of electricity are needed. Copious amounts of water are also needed.

Both the anode and cathode are enclosed in a closed, dashed-line, approximate circle. The circle assists with writing the chemical equations. Focus on the cathode, which is negative. Crossing the

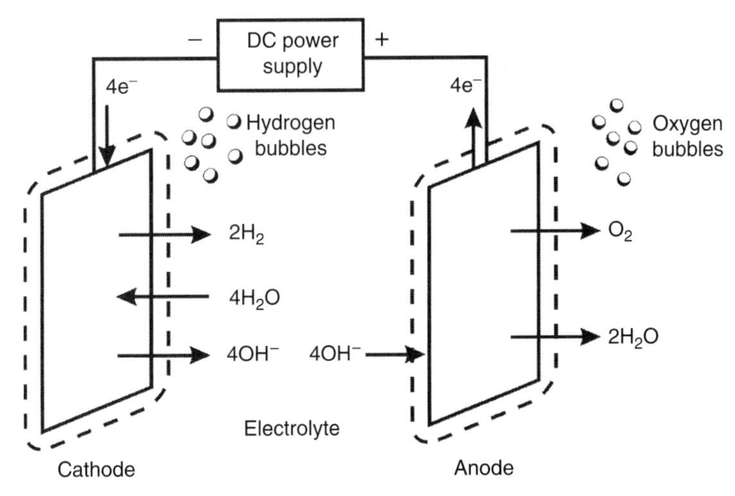

FIGURE A.6.5 Electrolysis of water. The electrodes are immersed in an electrolyte, and the hydrogen and oxygen gases which are created bubble upward in the water.

circle and going in are $4e^-$ and $4H_2O$. Coming out are $2H_2$ and $4OH^-$. For electrolysis of water, the chemical equation for the cathode is

$$\text{In} \rightarrow \text{Out}$$

$$4e^- + 4H_2O \rightarrow 2H_2 + 4OH^-$$

Note that the number of atoms balance as well as the number of electrical charges. The left-hand side has four negative charges as does the right-hand side.

Focus on the anode, which is positive. Crossing the circle and going out are $4e^-$, O_2, and $2H_2O$. Going in is $4OH^-$. For electrolysis, the chemical equation for the anode is

$$4OH^- \rightarrow 4e^- + O_2 + 2H_2O$$

The overall cell reaction is obtained by adding the anode and cathode equations. This is

$$4e^- + 4H_2O + 4OH^- \rightarrow 4e^- + O_2 + 2H_2O + 2H_2 + 4OH^-$$

The overall reaction equation can be simplified to

$$2H_2O \rightarrow 2H_2 + O_2$$

The DC power supply provides the energy to break the chemical bond of $2H_2O$ molecules.

ELECTROCHEMICAL POTENTIAL

The electrochemical potential series is a classification of chemicals according to reduction/oxidation half-reactions in order of reducing or oxidation strengths. The combination of any half reaction with the reverse of one further down the series will give a spontaneous chemical reaction. The difference in electrochemical potential provides insight to the cell OCV for the combination of chemicals. The half reaction (oxidation)

$$H_2 \rightarrow 2H^+ + 2e$$

is the reference reaction and is assigned the value of zero.

A question concerning a cell is "What combinations of chemicals will give a spontaneous reaction?" The process of discharge requires a spontaneous reaction. Another question concerning a cell is "What is the OCV?" Answers to the questions can be found using the electrochemical potentials (V). Through the hard work of electrochemists, extensive tables of electrochemical potentials (V) exist. A few examples are given in Table A.6.2.

The chemicals used in a battery can be classified according to their electrochemical potential (V). Combining two chemicals with widely separated potential provides information about the OCV for the cell. The values are combined algebraically using

$$V_C = V_B - V_A$$

where

V_C is cell voltage

V_B and V_A are the electrode potentials from Table A.6.2

TABLE A.6.2
Selected Values of Electrode Potential for an Oxidation Series

Electrode Reaction	Standard Electrode Potential Volts
$Li \rightarrow Li^{+} + e$	3.045
$Al \rightarrow Al^{3+} + 3e$	1.660
$\frac{1}{2}Pb \rightarrow \frac{1}{2}Pb^{2+} + 2e$	0.126
$H_2 \rightarrow 2H^{+} + 2e$	0.000
$Cu \rightarrow Cu^{2+} + 2e$	−0.337
$Cl^{-} \rightarrow \frac{1}{2}Cl_2 + e$	−1.360
$\frac{1}{2}Pb + H_2O \rightarrow \frac{1}{2}PbO_2 + 2H^{+} + e$	−1.455
$2F^{-} \rightarrow F_2 + 2e$	−2.650

As the first example, use Li and F for the proposed cell. The numbers are

$$V_C = 3.045 - (-2.650) = 5.695 \text{ V}$$

This is a great battery with a very high cell voltage. However, the combination of chemicals makes a great bomb also. Likely to be very dangerous. As a second example, consider the lead acid battery.

$$V_C = 0.126 - (-1.455) = 1.581 \text{ V}$$

Since the cell voltage for PbA battery is 2.0 V, the electrode potential does not predict cell voltage without modification. Accurate prediction of cell voltage goes beyond this discussion. Nonetheless, the first step is to use the electrode potentials as illustrated above.

ELECTROCHEMISTRY OF CELL (BATTERY)

Lead Acid Battery

Even after more than 100 years of service to the automotive world, the PbA battery remains an important energy storage device. Although not on the leading edge of battery technology, the lead acid battery is for some applications the battery of choice. Frequently the lead acid battery is denoted by PbA. The Pb comes from the chemical symbol for lead.

During charging, electrons are removed from the positive electrode; therefore, the plus terminal is the anode. The DC power supply, in the same manner as electrolysis of water in Figure A.6.5, provides the energy. As before, the approximate circle with the dashed line is a tool to assist with the accounting of electrical charges and atoms. Focus on the anode. The anode reaction is given by

$$PbSO_4 + 2H_2O \rightarrow PbO_2 + HSO_4^{-} + 3H^{+} + 2e^{-}$$

The arrows for the chemicals $PbSO_4$ and PbO_2 do not cross the dotted line. The reason to portray these chemicals on the surface of the electrode is to emphasize the actual location. The horizontal arrows show chemicals entering or leaving the electrolyte.

Focus on the negative electrode, which, during charging, is the cathode. When electrons enter an electrode, it is always the cathode. During charging, the cathode reaction is given by

$$H^+ + 2e^- + PbSO_4 \rightarrow HSO_4^- + Pb$$

$PbSO_4$ is removed from both electrodes during charging.

The overall cell reaction is obtained by adding the anode and cathode equations. This is

$$2PbSO_4 + 2H_2O + H^+ + 2e^- \rightarrow PbO_2 + 2HSO_4^- + 3H^+ + 2e^- + Pb$$

The overall reaction equation can be simplified to

$$2PbSO_4 + 2H_2O \rightarrow PbO_2 + 2HSO_4^- + 2H^+ + Pb$$

The energy that is stored in a cell of the battery can be determined from the overall reaction. The energy stored in the cell is not the same as the energy supplied by the DC source. The difference is the losses.

The overall reaction for the PbA cell shows two significant features of a PbA cell. First, water is removed from the electrolyte to be replaced by hydrogen sulfate. Since the specific gravity of the electrolyte increases as H_2SO_4 replaces H_2O, the specific gravity becomes a measure of the SOC of a battery. Fully charged, the specific gravity is 1.215. Second, $PbSO_4$, which is formed on both electrodes during discharging of the cell, is removed by charging.

All batteries have certain electrochemical features that are part of folklore. Some batteries have a memory effect that depends on the anode–cathode reactions. Two items of folklore for PbA battery follow. If the discharged battery is allowed to stand for a time, the lead sulfate ($PbSO_4$) becomes less soluble and crystallizes out on the surfaces of both electrodes. This partially covers the surface of the electrode and reduces the current available from the battery. In cold weather, the dilute sulfuric acid may freeze, which damages the electrodes and their retaining structures.

Figure A.6.5 shows the electrolysis of water. When the DC power supply is removed and the electrodes are connected via a load, zero current will flow. Electrodes dipped in water do not make a battery. The chemicals in PbA do make a battery. A cell can be charged and discharged.

During discharge, electrons enter the positive electrode making it the cathode. As before, the cathode reaction is determined by the arrows crossing the dashed circle. Comparing Figure A.6.5 for charge and Figure A.6.6 for discharge, every arrow is reversed. Consequently, the cathode reaction is obtained by reversing the arrow in the chemical equation.

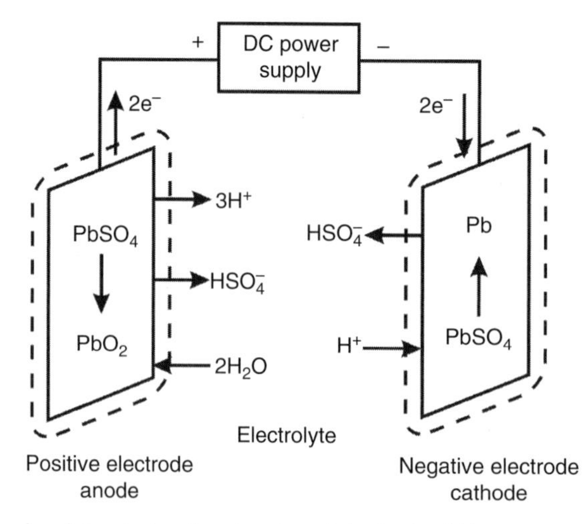

FIGURE A.6.6 Electrochemistry during the charging of the lead acid battery.

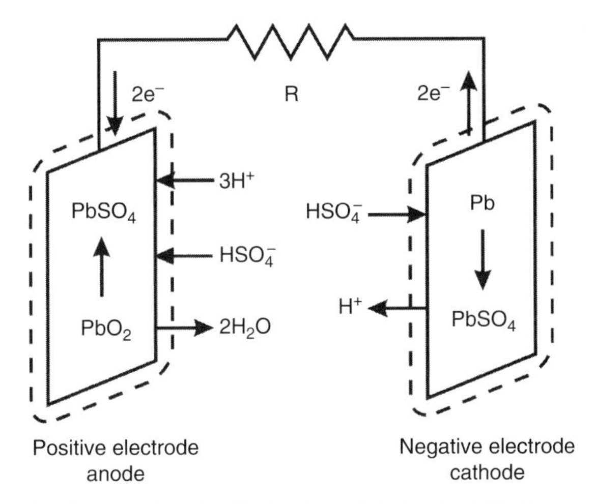

FIGURE A.6.7 Electrochemistry during the discharging of the lead acid battery.

$$H^+ + 2e^- + PbSO_4 \leftarrow HSO_4^- + Pb$$

Compare with the anode reaction for charging. The overall reaction is

$$2PbSO_4 + 2H_2O \leftarrow PbO_2 + 2HSO_4^- + 2H^+ + Pb$$

The current in the external circuit flows through the resistance, R. The power dissipated in the resistance comes from the chemical changes shown in the overall reaction above (Figure A.6.7).

Voltage and Power from a Cell

Figure A.6.8a shows cell voltage as a function of current. Positive current, which corresponds to discharging, causes the voltage to droop. A major part of the voltage droop is due to battery internal impedance. At zero current, the battery voltage is known as the OCV and is shown in Figure A.6.8.

For negative current, which corresponds to charging, the voltage across the battery terminal must be greater than the OCV. This is shown in Figure A.6.8b where voltage exceeds the open circuit value. For positive power, positive current is necessary. Positive means power is being extracted from the battery, that is, the battery is being discharged. Obviously, negative current and negative power go hand in hand, and the battery is being charged.

ELECTROCHEMISTRY OF FC

Hydrogen/Oxygen FC

The components of the hydrogen and oxygen FC are illustrated in Figures A.6.9 and A.6.10. The electrodes are porous and are transparent to oxygen, hydrogen, and water vapor, but are opaque to the electrolyte. The electrolyte separates the hydrogen from the oxygen; this separation prevents explosions or fires from the highly reactive oxygen and hydrogen mixture that might otherwise form. Seals are also necessary to prevent accidental mixing of the high-energy chemicals. The anode is fed hydrogen while the cathode receives oxygen. Each electrode has a catalyst to accelerate the reactions. The catalyst varies; compare the PEM and alkaline fuel cells (AFC). The AFC feeds water into the anode channel. The cathode receives air and uses the oxygen. The nitrogen from the air passes unaltered through the ducting.

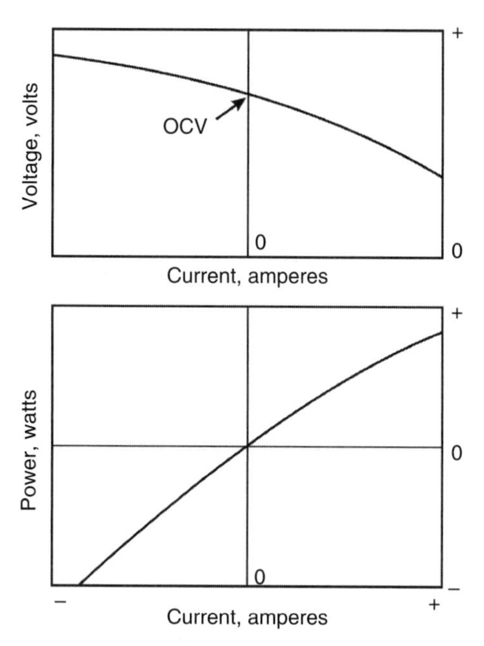

FIGURE A.6.8 Voltage and power during the discharging and the charging of a cell.

The potential difference (voltage) that develops between the two electrodes is small and decreases with output current through the load. The load is shown as a resistance, *R*. Terminal voltage is less than 1 V when drawing current.

Electrolysis, which is the inverse of the hydrogen/oxygen FC, has been discussed above. Figure 6.5 compares the FC with the reversible cell and demonstrates the inverse relationship between electrolysis and the FC. Figure A.6.9 shows the reactions for the hydrogen/oxygen FC with an alkaline electrolyte, KOH. The anode reaction is

$$2H_2 + 4OH^- \rightarrow 4H_2O + 4e^-$$

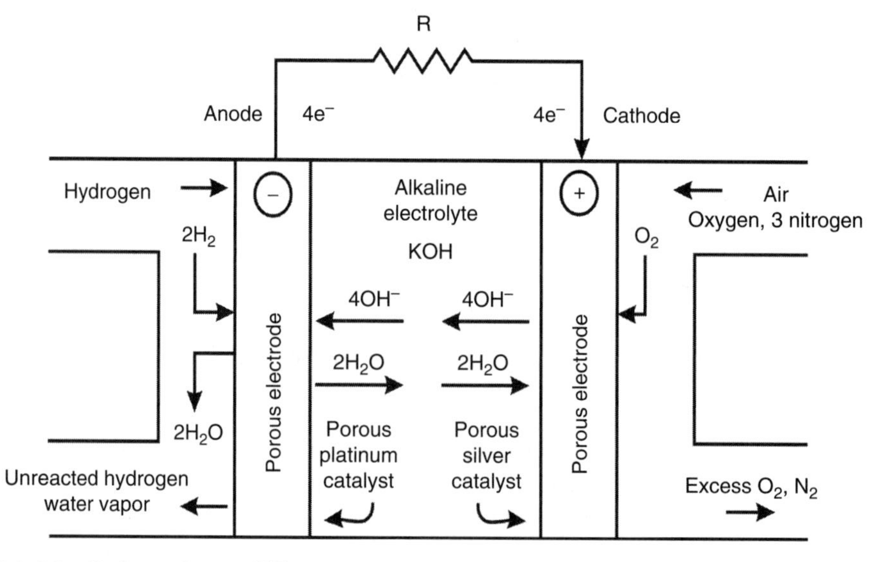

FIGURE A.6.9 Hydrogen/oxygen FC with an alkaline electrolyte, KOH.

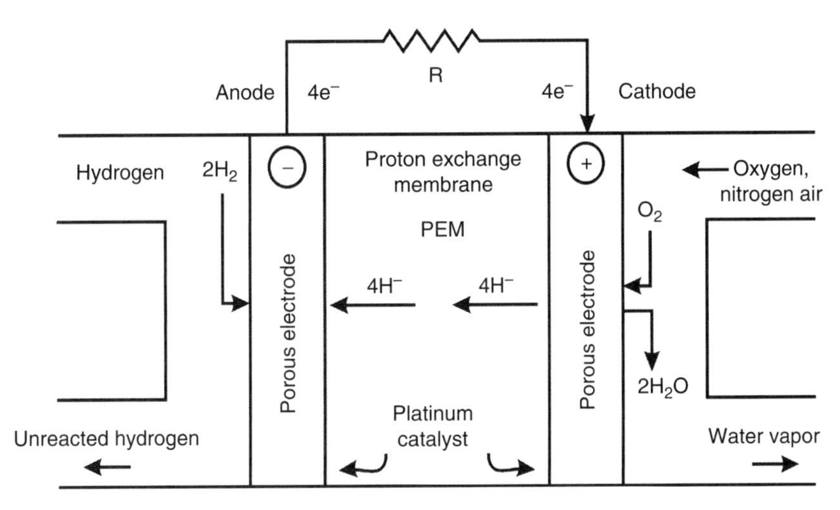

FIGURE A.6.10 Hydrogen/oxygen FC with PEM solid electrolyte.

The cathode reaction is

$$O_2 + 4e^- + 2H_2O \rightarrow 4OH^-$$

Adding the reactions for the electrodes gives the overall FC reaction.

$$2H_2 + 4OH^- + O_2 + 4e^- + 2H_2O \rightarrow 4H_2O + 4e^- + 4OH^-$$

which simplifies to

$$2H_2 + O_2 \rightarrow 2H_2O$$

Compare with the overall equation for electrolysis.

Much of the FC R&D is focused on the hydrogen/oxygen FC with PEM solid electrolyte as illustrated in Figure A.6.10. Recall the definition of a proton, which is the hydrogen atom ion, H^+. The anode reaction is

$$2H_2 \rightarrow 4H^+ + 4e^-$$

The cathode reaction is

$$O_2 + 4e^- + 4H^+ \rightarrow 2H_2O$$

Adding the reactions for the electrodes gives the overall FC reaction.

$$2H_2 + O_2 + 4e^- + 4H^+ \rightarrow 2H_2O + 4H^+ + 4e^-$$

which simplifies to

$$2H_2 + O_2 \rightarrow 2H_2O$$

The overall reaction is once again the reverse of the electrolysis reaction.

APPENDIX 6.2: CAPACITORS

The voltage for electrolytic capacitors is limited to a few volts; however, for energy storage in hybrids, the voltage must be a few hundreds of volts. The capacitors must be placed in series. Capacitors in series have reduced capacitance, C. The number of farads is reduced. The equation for capacitors in series is

$$C = \frac{C_0}{N} \tag{A.6.1}$$

The capacitance of a single capacitor is C_0. The number of capacitors in series is N. For N capacitors in series, the capacitance is C (F).

The operating voltage of a single capacitor is V_0. The voltage of the N capacitors is V. The required number of capacitors is

$$N = \frac{V}{V_0} \tag{A.6.2}$$

A single capacitor may have very large farads; however, with N in series, the farads are reduced by a factor of N.

The energy stored in a capacitor, E, depends on the voltage squared as shown.

$$E = \frac{1}{2} C V^2 \text{ J} \tag{A.6.3}$$

The voltage squared dependence is fortunate. Suppose the minimum voltage allowed, due to the inverter, is half the voltage of the fully charged capacitor. In this case 75% of the capacitor energy is used in spite of the one-half voltage restriction.

REFERENCES

1. L. Pauling, *General Chemistry*, W.H. Freeman, San Francisco, CA, 1947. (Nobel Prize Winner; text used for freshman chemistry at CalTech).
2. S. Glasstone, *Textbook of Physical Chemistry*, 2nd ed., D. Van Nostrand, New York, 1946.
3. I. M. Gottlieb, *Electric Motors and Control Techniques*, 2nd ed., TAB Books, McGraw-Hill, New York, 1994.
4. A. Malikopoulos, Z. Filipi, and D. Assanis, Simulation of an integrated starter alternator (ISA) system of HMMWV, *Advanced Hybrid Vehicle Powertrains 2006*, SP-2008, Society of Automotive Engineers, Organized by M. A. Theobald, J. Moore, M. E. Fleming, and M. Duoaba, 1994, p. 11.
5. A. Landgrebe, I. Weinstock, K. Heitner, T. Duong, R. Sutula, and P. H. Maupin, Overview of technologies for batteries for electric and hybrid vehicles, *Hybrid Gasoline-Electric Vehicle Development*, J. M. German (Ed.), SAE PT-117, 2005.
6. B. H. Mahan, *University Chemistry*, 3rd ed., Addison-Wesley, Reading, MA, 1975.
7. R. H. Cole and J. S. Coles, *Physical Principles of Chemistry*, W.H. Freeman, San Francisco, CA, 1964.
8. F. Daniels and R. A. Alberty, *Physical Chemistry*, John Wiley & Sons, New York, 1957.
9. D. Linden, Batteries and fuel cells, *Electronic Engineers Handbook*, 2nd ed., D. G. Fink and D. Christiansen (Eds.), McGraw-Hill, New York, 1982, pp. 7.50–7.65.
10. F. V. Conte and F. Pirker, Nickel metal hydride batteries' voltage depression detection in vehicle application, *Advanced Hybrid Vehicle Powertrains 2006*, SP-2008, Society of Automotive Engineers, Organized by M. A. Theobald, J. Moore, M. E. Fleming, and M. Duoaba, 2006, p. 150.
11. M. H. Westbrook, *The Electric and Hybrid Electric Car*, Society of Automotive Engineers, Warrendale, PA, 2001, p. 198.

12. R. Hodkinson and J. Fenton, *Lightweight Electric/Hybrid Vehicle Design*, Society of Automotive Engineers, Warrendale, PA, 2001, p. 253.
13. H. P. Neff, Jr., *Basic Electromagnetic Fields*, 2nd ed., Harper & Rowe, Cambridge, 1987, p. 616.
14. Editorial staff, *Reference Data for Radio Engineers*, 6th ed., Howard W. Sams, Indianapolis, IN, 1982 (Chapters 5 and 6).
15. A. W. Stienecker, M. A. Flute, and T. A. Stuart, Improved battery charging in an ultracapacitor-lead acid battery hybrid energy storage system for mild hybrid electric vehicles, *Advanced Hybrid Vehicle Powertrains 2006*, SP-2008, Society of Automotive Engineers, Organized by M. A. Theobald, J. Moore, M. E. Fleming, and M. Duoaba, 2006, p. 153.
16. Y. Murakami and K. Uchibori, Development of fuel cell vehicle with next-generation fuel cell stack, *Applications of Fuel Cells in Vehicles 2006*, SP-2006, Society of Automotive Engineers, Organized by R. K. Stobart, and S. Kuman, 2006, p. 1.
17. P. M. B. Walker, General Editor, *Cambridge Dictionary of Science and Technology*, Cambridge University Press, Cambridge, 1990 (45,000 entries).
18. W. L. Mitchell and A. Toro, Advanced fuel cell development for automotive operation, *Applications of Fuel Cells in Vehicles 2006*, SP-2006, Society of Automotive Engineers, Organized by R. K. Stobart, and S. Kuman, 2006, p. 7.
19. N. Amada, M. Fukiro, M. Tahara, and T. Liyama, Development of a performance prediction program for EV's powered by lithium-ion batteries, *Electric and Hybrid-Electric Vehicles*, PT-85, Society of Automotive Engineers, R. K. Jurgen (Ed.), 2002, p. 17.
20. M. Origuchi, T. Miyamoto, H. Hoeika, and K. Katayama, Development of a lithium-ion battery system for EV's, *Electric and Hybrid-Electric Vehicles*, PT-85, Society of Automotive Engineers, R. K. Jurgen (Ed.), 2002, p. 3.
21. S. W. Moore and P. J. Schneider, A review of cell equalization for lithium ion and lithium polymer batteries, *Advanced Hybrid Vehicle Powertrains*, SP-1607, Society of Automotive Engineers, Organized by D. N. Assanis, and F. Stodolsky, 2001, p. 87.
22. J. Yamaguchi, Hybridization gathers momentum, *Automotive Engineering International*, Society of Automotive Engineers, 2006, p. 40.
23. J. Kwon, P. Sharer, and A. Rousseau, Impacts of combining hydrogen ICE with fuel cell system using PSAT, Society of Automotive Engineers, *Applications of Fuel Cells in Vehicles 2006*, SP-2006, Organized by R. K. Stobart, and S. Kuman, p. 13.
24. A. Malikopoulos, Z. Filipi, and D. Assanis, Simulation of an integrated starter alternator (ISA) for HMMWV, *Advanced Hybrid Vehicle Powertrains 2006*, SP-2008, Society of Automotive Engineers, Organized by M. A. Theobald, J. Moore, M. E. Fleming, and M. Duoaba, 2006, p. 11.
25. I. Matsuo, T. Miyamoto, and H. Maeda, The Nissan Hybrid Vehicle, *Electric and Hybrid-Electric Vehicles*, PT-85, Society of Automotive Engineers, R. K. Jurgen (Ed.), 2002, p. 477.
26. D. Schneider, Who's resuscitating the electric car? *American Scientist*, September–October, 2007, pp. 403–404.
27. G. Jean, Powering up, *National Defense*, September 2006, pp. 34–35.
28. R. Smith, Feeding power-hungry cars, *Automotive Engineering International*, Society of Automotive Engineers, July 2005, pp. 42–43.
29. K. Jost, GM fuel cell for US postal service, *Automotive Engineering International*, Society of Automotive Engineers, September 2004, pp. 10–13.
30. S. Birch, Battery breakthrough on the horizon? *Automotive Engineering International*, Society of Automotive Engineers, July 2007, pp. 32–33.
31. S. Kumar and S. Radhakrishnan, Flow simulation improves robustness of fuel-cell design, *Automotive Engineering International*, Society of Automotive Engineers, September 2007, pp. 52–53.
32. D. Carney, Fuel cells power up; as the alternate-propulsion technology moves from lab to limited production, car makers are looking for new design solutions and materials to reduce cost, *Automotive Engineering International*, Society of Automotive Engineers, September 2007, pp. 54–56.

7 Obesity: Bad in Humans, Bad in Cars

Fuel efficiency standards are killing people.

INTRODUCTION

The two major topics discussed in this chapter are the effect of vehicle weight on fuel economy and the effect of vehicle weight on the safety of the driver and passengers. Also, CO_2 emissions versus weight are discussed toward the end of this chapter.

Throughout this chapter, the words mass and weight are used interchangeably even though each has a distinct definition. The other two important words in this chapter are injury and damage. Injury refers to the effects an accident has on the occupants of a vehicle, such as physical impairment, wounds, or distress. Damage refers to the effect of a collision on a vehicle such as bent metal, broken glass, torn upholstery, and wrinkled frame.

The focus of this chapter is weight. Weight reduction is as big a factor as powertrain efficiency in the field of fuel economy. Car manufacturers try their best to reduce weight. For example, a steel hood may be replaced with an aluminum one saving 80 lb. The normally aspirated engine may be replaced, which is a complicated task, by a turbocharged version saving 110 lb. Even use of premium fuel saves weight; a higher compression ratio can be used giving the same horsepower in a smaller engine. This book is also about hybrids. A normal hybrid battery adds 100 lb, which can go up to 500 lb in the case of plug-in hybrids.

See Ref. [18] for an overview of vehicle weight. The move to global architecture requires design to the most stringent requirements. For example, the toughest crash safety standards might be in England. Then the cars sold in the United States and the Orient are too heavy for local requirements. A global standard might be the answer. Weight adds up; weight audits are needed. No part is too small or insignificant to be ignored. Weight reduction is a balancing act. Cost, size, and content must be balanced. Within a class segment, weight reduction should not affect interior volume, feature content, or safety. A low price on the showroom floor makes introduction of higher cost, lighter weight materials, a difficult decision.

Fuel economy gains can be made by changes in the powertrain (e.g., more efficient engines) or weight reduction. What is the best place to invest development funding? Car makers must answer this question, and cannot choose both. Lower vehicle weight means the same performance can be obtained with smaller engines. In aviation, the aircraft companies must guarantee weight and fuel economy of a new aircraft. Purchase contracts for either military or commercial aircraft have penalties (fines) if performance specifications are not met. The automobile industry is different, although the customer can inflict the ultimate penalty by buying a competitor's car.

VEHICLE WEIGHT AND FUEL ECONOMY

FIRST PASSING COMMENT

For an aircraft, most people intuitively feel that weight is bad. This comment is to convince you that weight in hybrids (or conventional cars) is equally bad.

Think of a battery-powered model airplane. The modeler enthusiast wants as much flight time as possible. Long flight (fun) time means a large battery. A large battery means the model airplane might not get off the ground—and even if it did, it would be a dud at acrobatics. So, a trade-off exists between performance and battery weight. A heavy battery (vehicle weight) is not conducive to high performance.

Desired performance for model aircraft: long flight time and agile flight

Desired performance for hybrid: long electric-only range and agile driving

Let us now consider a hybrid. With the added weight of the battery, the hybrid starts with a negative balance in its checking account. For example, extra weight increases rolling resistance that in turn reduces miles per gallon (mpg). Before the hybrid can deliver more mpg it has to get out of the "extra weight pothole," which is a negative mpg place. For the new GM Chevrolet Silverado full dual mode hybrid pickup and GMC Sierra full dual-mode hybrid pickup, the hybrid extra weight pothole was avoided. The hybrid components added 132 lb (60 kg). The substitution of aluminum for hood, liftgate, etc. reduced the weight by 150 lb (68 kg).

Nickel metal hydride batteries are used in model airplanes just as in current hybrids. Li-ion batteries are used in model airplanes and are coming in hybrids. The same trade-off for model aircraft exists for hybrid cars, though in a little less obvious form. For model aircraft, the trade-off may be more compelling.

Second Passing Comment

Is all the extra weight necessary? The typical driver of a typical car weighs 200 lb. A small car weighs 3000 lb. The percentage of useful load, that is, the driver to vehicle weight, is a mere (200 lb)/(3000 lb) = 7%. The typical bicyclist weighs 200 lb, and the typical bicycle weighs 60 lb. The percentage is an astonishing (200 lb)/(60 lb) = 333%. Compare 7% and 333%.

How to Improve Fuel Economy?

Three main methods can improve fuel economy. These are

- Improve the efficiency, η, of the power source, e.g., the gasoline engine.
- Improve the efficiency, η, of the powertrain, e.g., the transmission.
- Reduce the resistances to vehicle motion.

Weight affects several of the resistances to vehicle motion. A reduction of weight decreases fuel consumption. One big resistance at higher speeds is aerodynamic drag that is discussed in Chapters 2 and 12. Aerodynamic drag is not discussed here.

How to Decrease Vehicle Weight?

One approach is better engineering design. The use of computer codes, which predict stresses and strains, gives the required bending and torsional stiffness using less material. One such popular computer code is finite element analysis. It was developed by NASA for designing weight-sensitive rockets and spacecraft.

Another approach is material replacement. Frequently, weight can be saved by replacing steel with composite materials, aluminum, magnesium, or plastics. For example, plastic fenders may become common. Lightweight engines offer reduced weight for the same power. Turbocharging or supercharging is commonly used to reduce engine weight. Reference [16] describes a combination of aluminum and magnesium engine block developed by BMW. Reducing vehicle weight was one motivation but maintaining an even weight balance (50/50 F/R) was another challenge.

Composite materials have been and are now being widely used in aircraft and spacecraft structures. These materials offer great strength and rigidity with low weight. In the aircraft industry, production volumes are low being less than 100 units per year and occasionally 1000 units per year. Manufacturing methods suitable for aircraft are not cost effective for automobiles. Cars with composite structures have not been manufactured on a large scale (e.g., 100,000 units per year). To keep costs reasonable, additional R&D focused on manufacturing is needed. Misconceptions exist concerning the possible improvements in fuel economy because of weight reduction.

> "… Americans could still drive cars and fly on planes, but they would be made out of lightweight steel with a goal of doubling the fuel efficiency, …" [12,13].

Consider Equation 7.8 that relates fuel efficiency to weight. Also consider the comments on weight from Chapter 1 in conjunction with Figure 1.6. Refer to Figure 7.3, which gives the best-effort weight reduction (20%) after the 1970s oil crisis. Does the doubling of fuel efficiency stated in the quote seem reasonable in view of these insights? Does lightweight steel offer the big gains necessary to double fuel efficiency? In the area of materials, Freedom Car R&D efforts include aluminum, magnesium, and composites. Lightweight steel is an also-ran material. Doubling of fuel economy using only lightweight steel seems unlikely.

Interplay: Production Lines and Vehicle Design

A hybrid vehicle designed from the road up as a hybrid will have less weight than a conventional vehicle (CV) modified to be a hybrid. The hybrid-only design can be optimized with regard to placement of components, structural design, and other factors. Car makers must live in the real world of existing factories and assembly lines. Even if the hybrid were designed solely as a hybrid but were produced on a dual-car (CV and hybrid electric vehicle [HEV]) assembly line, some compromises are necessary. Compromises translate into weight. In order of decreasing weight, the vehicles are

1. Hybrid derived from a modified CV
2. Clean-sheet design hybrid assembled on a dual-car line
3. Clean-sheet design hybrid assembled on a hybrid-specific line

As hybrids make further inroads into the new car market, the third scenario seems the most likely.

Weight Factors Affecting Fuel Economy

Three important factors that adversely affect fuel economy as weight increases are

- Rolling friction due mainly to tires
- Power to climb grades
- Power for acceleration

Each factor will be discussed.

Power and Fuel Consumption

Appendix A has the definition for brake-specific fuel consumption (BSFC), which is defined as the pounds of fuel required for each brake horsepower hour. For our purposes, think of BSFC as being a constant equal to 0.48. BSFC is constant for a fixed point on the engine operating map. For 100 hp, 48 lb of fuel are used each hour. If power is doubled to 200 hp, the fuel consumed is also doubled to 96 lb. Increasing installed power will increase fuel consumption.

Rolling Resistance Due Mainly to Tires

The retarding resistance, F, because of tire hysteresis is given by the equation

$$F = f W \tag{7.1}$$

where f is the rolling resistance coefficient with value between 0.007 and 0.040. The value depends on tire construction, tire pressure, road surface condition (sand has larger value than asphalt), and weakly on speed. The resistance increases linearly with weight, W. The power, P, required to overcome rolling resistance is

$$P = F V = f W V \tag{7.2}$$

where V is vehicle velocity. Power for V=60 mph (26.8 m/s), f=0.01, and W=3000 lb (m=1362 kg) is P=3.6 kW (4.8 hp). Note that the power scales directly as vehicle mass (weight).

Power to Climb Grades

Performance up a grade is determined by two quantities. One is the velocity, V, going uphill. The other factor, of course, is the steepness of the hill or percent grade. A road that climbs 6 ft while going 100 ft along the road has a grade of 6%. The grade can also be defined as the angle of the road relative to the horizontal, θ. The power, P, is

$$P = m g V \sin\theta \tag{7.3}$$

For a modestly steep 6% grade (θ = 3.4°), V = 60 mph (26.8 m/s), g = 9.81 m/s^2, and W = 3000 lb (m = 1362 kg), the added power to climb the hill is 21 kW (29 hp). This is in addition to power required to overcome all other resistances. Note that power and fuel consumption scale directly as vehicle mass (weight).

Power for Acceleration

The instantaneous power, P, required for acceleration at any velocity is

$$P = F V = m a V \tag{7.4}$$

F is the retarding (inertial) force because of acceleration. An estimate for constant, average acceleration, a, is

$$a = \frac{V_{60}}{t_{60}} \tag{7.5}$$

Combining Equations 7.4 and 7.5 produces the instantaneous power

$$P = \frac{m V_{60} V}{t_{60}} \tag{7.6}$$

Frequently, the maximum installed power is that required to meet acceleration requirements. The installed power, P_I, required for acceleration is

$$P_I = \frac{m V_{60}^2}{t_{60}} \tag{7.7}$$

The symbol, t_{60}, is the time 0–60 s. A value of 8 s is assumed for an example. Using the same numerical values as used above, the power at V_{60} = 26.8 km/h (60 mph) is

$$P_I = \frac{(1362\,\text{kg})(26.8\,\text{m/s})^2}{8\,\text{s}} = 122\,\text{kW (163 hp)}$$

To repeat, this power is in addition to that required to overcome all other resistances. In fractional form, Equation 7.7 is

$$\frac{\Delta P_I}{P_I} = \frac{\Delta P_C}{P_C} = \frac{\Delta W}{W} - \frac{\Delta t_{60}}{t_{60}} \tag{7.8}$$

Note that the power scales directly as vehicle mass (weight).

Correlation between Acceleration and Fuel Economy

Since fuel economy changes rapidly during acceleration, the total weight (or gallons) of fuel used to increase velocity is more meaningful. This is discussed at length in Chapter 12. As a preview of Chapter 12, due solely to the inertia forces ($F = ma$), the weight of fuel used does not depend on acceleration. Loss of fuel economy during heavy acceleration is only part of the story. See "Aggressive Driving; Rapid Acceleration" in Chapter 12. Also study Equation 12.16.

Correlation between Weight and Fuel Consumption at Cruise

Two quantities are used to specify usage of fuel. One is fuel volume used to travel a certain distance. This is frequently gallons per mile or liters per 100 km, and is known as fuel consumption (F_C). The other is distance traveled for a unit volume of fuel. Units are mpg, or km/L, and is known as fuel economy (F_E). Note that economy and consumption are reciprocals; $F_E = 1/F_C$. In Figure 7.1, F_C, which is plotted on left-hand vertical axis (the ordinate), increases upward. F_E decreases upward, which is because of the reciprocal relationship. The conversion factor between mass (kg) and weight is mass (kg) = 0.454 lb (weight). Figure 7.1 shows data on fuel consumption as a function of vehicle weight. The data are for two classes of vehicles: gasoline and diesel vehicles. Each data point represents a car of that weight and fuel consumption. The higher the point in the graph, the greater the fuel consumption, F_C. Weight and fuel consumption usually correlate as straight lines whereas weight and fuel economy do not. The data have considerable scatter; however, a statistically best fit has been made using straight lines.

Equations for the lines are

$$F_C = 0.00700\,m + 0.025 \quad \text{for gasoline}$$
$$F_C = 0.00411\,m + 0.414 \quad \text{for diesel}$$

where F_C is L/100 km, and m is the vehicle weight (mass). Because of the linear relationship, the constant, C, equals 1.0. The fractional change of fuel consumption because of weight change is

$$\frac{\Delta F_C}{F_C} = +C\left(\frac{\Delta W}{W}\right) = +C\left(\frac{\Delta m}{m}\right) = -C\left(\frac{\Delta F_E}{F_E}\right) \tag{7.9}$$

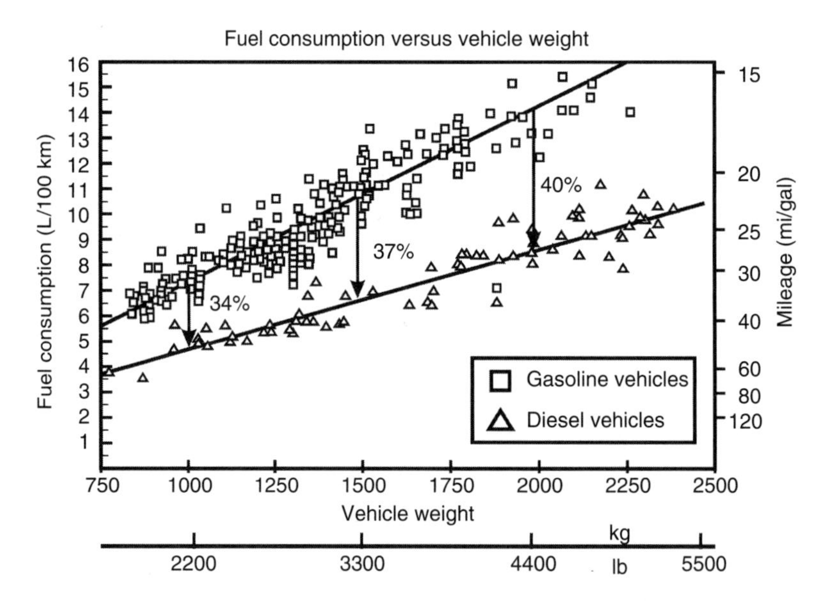

FIGURE 7.1 Correlation between fuel consumption and vehicle weight for gasoline and diesel-powered vehicles. (From Fuel Economy in Automobiles, Wikipedia, the Free Encyclopedia, September 2006.)

A numerical example, using values from Figure 7.1, shows that Equation 7.9 correctly applies to the data. Use the data

$m_1 = 1500\,\text{kg}$	$F_{C1} = 10.7\,\text{L/100 km}$
$m_2 = 2000\,\text{kg}$	$F_{C2} = 14.0\,\text{L/100 km}$

The results are $\Delta F_C/F_C = 0.267$, and $\Delta m/m = 0.286$. The fact that the two are not exactly equal is because of reading values for F_C from the graph of Figure 7.1. The conclusion is that an $X\%$ increase in vehicle weight will give an $X\%$ increase in fuel consumption and an $X\%$ decrease in fuel economy.

Note in passing that for a lightweight vehicle, the diesel provides 34% better fuel consumption than a gasoline vehicle. For a heavyweight vehicle, the diesel provides 40% better fuel consumption than a gasoline vehicle.

Fuel Economy for Cars, Trucks, and Both

Several interesting points can be made using Figure 7.2. In 1976, cars had a low mileage of 15 mpg, trucks had 12 mpg, and both combined had a fuel economy of 14 mpg. Since the fraction of trucks in the fleet was low, the mpg for both was close to that for a car. In contrast, in 2005, cars had a significantly higher mileage of 25 mpg, trucks had 18 mpg, and both combined had a fuel economy of 21 mpg. Since the fraction of trucks in the fleet was about one half, the mpg for both is about halfway in between the cars and trucks.

From Figure 7.2, the change in fuel economy can be determined using the fractional equations for both 1976 and 2005. Why calculate the fractional change? The results based on actual data validate the fractional equations. For 1976, the values used are

W_T = truck weight = 4600 lb	F_T = truck fuel economy = 12 mpg
W_C = car weight = 3600 lb	F_C = car fuel economy = 15 mpg

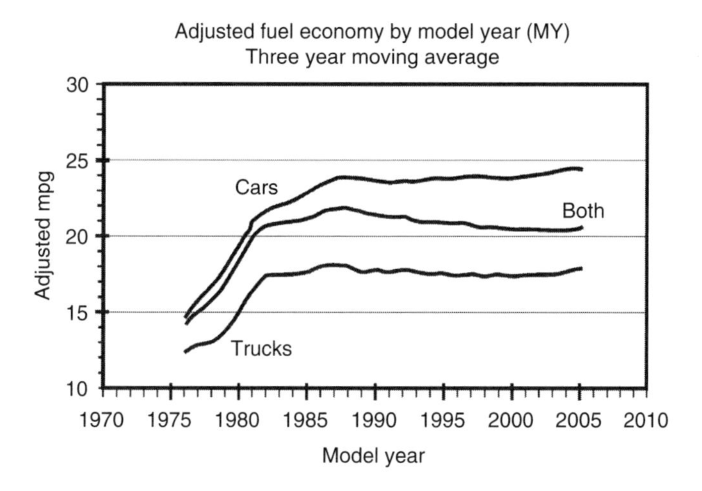

FIGURE 7.2 Adjusted fuel economy for cars, trucks, and both a mixed fleet of cars and trucks. (Reproduced from Environmental Protection Agency, Light-duty automotive technology and fuel economy trends: 1975 through 2006, executive summary, EPA42-S-06-003, July 2006.)

Using these values and the equation

$$F_T = F_C\left(1 - \frac{\Delta W}{W}\right) \tag{7.10}$$

truck fuel economy was predicted to be

$$F_T = 15(1 - 0.28) = 12 \text{ mpg}$$

For 2005, the values are

W_T = truck weight = 4800 lb	F_T = truck fuel economy = 18 mpg
W_C = car weight = 3600 lb	F_C = car fuel economy = 25 mpg

The prediction using the fractional equation is

$$F_T = 25(1 - 0.29) = 18 \text{ mpg}$$

The equations yield valid answers.

Weight and Performance

A new point can be added to Figure 7.3. The average weight of a car in 2006 was 4142 lb (1879 kg). How fast can automobile industry change the character of their cars? Looking at Figure 7.3, in the span of 5 years immediately following the gas crisis, dramatic changes were made in weight, W; power, P; and engine efficiency, η. Using values from Figure 7.3, estimates can be made of the fractional changes. From Equation 7.6, a fractional relation can be found involving installed power, P_I; vehicle weight, W; and t_{60}, which is (note that $\Delta W/W = \Delta m/m$)

$$\frac{\Delta P_I}{P_I} = +\frac{\Delta W}{W} - \frac{\Delta t_{60}}{t_{60}} \tag{7.11}$$

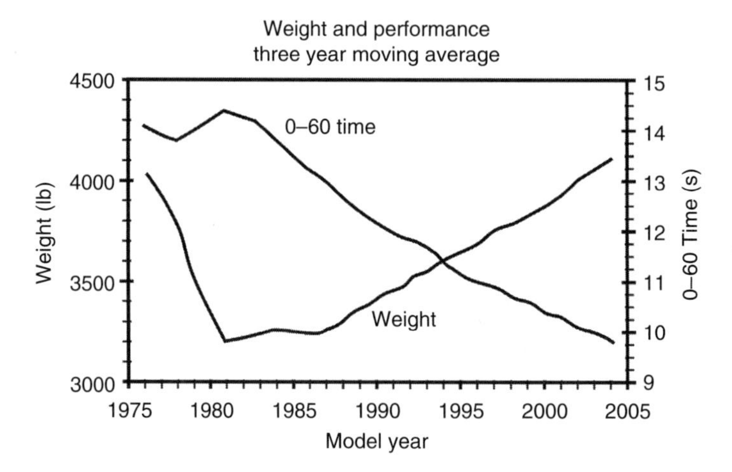

FIGURE 7.3 Weight and performance for years 1975–2005. Three year moving average. (Reproduced from Environmental Protection Agency, Light-duty automotive technology and fuel economy trends: 1975 through 2006, executive summary, EPA42-S-06-003, July 2006.)

Appropriate values are obtained from Figure 7.3 for the years 1976 and 1981:

1976	1981
$W = 4050\,\text{lb}$	$W = 3200\,\text{lb}$
$t_{60} = 14.2\,\text{s}$	$t_{60} = 14.5\,\text{s}$

Equation 7.11 yields for the change in installed power.

$$\frac{\Delta P_\text{I}}{P_\text{I}} = \frac{(3200 - 4050)}{3625} - \frac{(14.5 - 14.2)}{14.35} = -0.234 - 0.021 = -0.26 \tag{7.12}$$

Appropriate values are obtained from Figure 7.2 for the years 1976 and 1981:

1976	1981
$F_\text{E} = 14.5\,\text{mpg}$	$F_\text{E} = 21\,\text{mpg}$

Using these values, the improvement in fuel economy is 0.37. The power $\Delta P_\text{I}/P_\text{I}$ accounts for 0.26 of that amount. Assume engine efficiency provided the amount of 0.37 less 0.26 or 0.11. From Appendix 7.2,

$$\frac{\Delta \eta}{\eta} = \frac{\Delta F_\text{E}}{F_\text{E}} + \frac{\Delta P_\text{I}}{P_\text{I}} \tag{7.13}$$

The gain in engine efficiency is

$$\frac{\Delta \eta}{\eta} = 0.37 - 0.26 = 0.11 \tag{7.14}$$

In summary, during the 5 year span:

$$\frac{\Delta P_{\mathrm{I}}}{P_{\mathrm{I}}} = -26\%; \quad \frac{\Delta W}{W} = -23.4\%; \quad \frac{\Delta \eta}{\eta} = +0.11\%$$

Power decreased by 26%, weight was trimmed by 23.4%, and engine efficiency improved by 11%.

During the years 1976–1990, carburetors were replaced by fuel injection. Fuel injection increases engine efficiency and lowers emissions. The improvement in engine efficiency by 11% was perhaps because of fuel injection.

Let us review the three figures introduced at this point. Figure 7.1 has the correlation between weight and mpg, which is basis for Equation 7.8. Figure 7.2 provides data to calculate $\Delta F_{\mathrm{E}}/F_{\mathrm{E}}$. See the calculations following Equation 7.11. With certain assumptions, Figure 7.3 gives weight and acceleration allowing an estimate of the installed power. Equation 7.10 is the appropriate equation. Figures 7.2 and 7.3 use the same database; information can be traded between figures. Combining knowledge from both figures, Equation 7.12 can be derived. Refer to Appendix 7.2.

Engine Efficiency Improvement to Match Weight Gain

How large of an improvement in engine efficiency can compensate for a weight gain? The engine efficiency offsets the gain in weight maintaining equal fuel economy. Insert Equation 7.10 into Equation 7.12 giving

$$\frac{\Delta \eta}{\eta} = \frac{\Delta F_{\mathrm{E}}}{F_{\mathrm{E}}} + \frac{\Delta W}{W} - \frac{\Delta t_{60}}{t_{60}} \tag{7.15}$$

For equal fuel economy, $\Delta F_{\mathrm{E}}/F_{\mathrm{E}} = 0$. With fixed, maximum acceleration $\Delta t_{60}/t_{60} = 0$. The result is

$$\frac{\Delta \eta}{\eta} = \frac{\Delta W}{W} \tag{7.16}$$

If weight increases by 10%, then efficiency must also increase by 10%. In the years 1975–2006, improved component efficiencies such as η_{E} were used to compensate for increases in weight (see Table 12.2).

VEHICLE WEIGHT AND VEHICLE SAFETY

Weight Reduction and Vehicle Safety

A correlation exists between fuel economy and vehicle mass (weight). Is there a correlation between vehicle fuel economy and safety? A "yes" answer implies that safety depends on vehicle mass. To quote from Ref. [5], "But one risk of downsizing is that smaller cars involved in crashes with larger vehicles tend to have higher numbers of fatalities." The answer is yes.

The quote at the beginning of the chapter "Fuel efficiency standards are killing people" was intended to grab your attention. We now need to qualify that statement. Fuel efficiency and possible increased traffic injuries is a controversial subject. The problem is the decrease in weight as a means to increase fuel economy. With the information contained in this chapter, the reader can assess the issues involved. The reader can form his/her own opinion. References [4–9] provide additional background information on weight and safety. Reference [8], which has a particularly optimistic viewpoint, states that raising fuel economy standards will keep drivers safe. Additional discussion of safety and vehicle size or weight appears in Ref. [14].

TABLE 7.1
Velocity Change during Collision

Vehicle	Collision Damage	Initial Velocity	Final Velocity	Change of Velocity	Final System Momentum
Heavy	None	−30	10	40	−30
Light		30	−50	−80	
Heavy	Modest	−30	3.3	33.3	−30
Light		30	−36.7	−66.7	
Heavy	Major	−30	−3.3	26.7	−30
Light		30	−3.3	−53.3	
Heavy	Severe	−30	−20	10	−30
Light		30	−20	−50	

Note: All velocities are miles per hour. A plus sign means the vehicle is moving toward the right side and minus is toward the left side.

The possibility of injury during a collision depends on, among many other factors, the acceleration experienced during the crash.

$$\text{Acceleration} = \frac{\text{Change in velocity}}{\text{Time interval}} = \frac{\Delta V}{\Delta t} \tag{7.17}$$

The time interval, Δt, is the time period for which the vehicles are in direct contact, and the velocity changes by an amount ΔV. The equations to obtain the results shown in Table 7.1 are from homework problem 4.22 found in Ref. [1]. Similar equations can be found in almost any book on dynamics. Table 7.1 provides information on the numerator of Equation 7.17, that is, the change in velocity. This is only part of the information needed to determine acceleration; at this point, time interval is missing.

Equation 7.17 gives acceleration averaged over the entire collision event. It cannot predict acceleration since Δt cannot be found. However, Equation 7.17 is nonetheless valuable. It does give correct relative accelerations of the heavy and light vehicles. The ratio of accelerations of the light to heavy vehicles is proportional to the weight disparity, Q. Of significance is the fact that the influence of the crumple energy can be determined. The decrease in acceleration with increasing crumple energy is found using Equation 7.16. Reference [10] discusses crumple zones and the use of simulation in the development of crash systems.

Weight disparity, Q, is defined as ratio of the weight of the heavy vehicle divided by the weight of the light vehicle.

$$Q = \frac{\text{Weight of the heavy vehicle}}{\text{Weight of the light vehicle}} \tag{7.18}$$

For the example of Table 7.1 and Figure 7.4, Q has a value of 2.0. The two vehicles in Table 7.1 are the heavy and the light vehicle, the former being twice as heavy as the latter, $M = 2m$. The mass of the heavy vehicle is M, while the mass of the light vehicle is m. Also, in general, $M = Qm$. Initially both are moving at 30 mph toward a head-on collision. The heavy vehicle is moving toward the left side (west) at 30 mph and reverses direction rebounding at 10 mph. The change in velocity is 40 mph. The light vehicle is moving toward the right side (east) at 30 mph and reverses direction rebounding at 50 mph. The change in velocity is 80 mph. Figure 7.4 is a series of sketches that illustrate the data of Table 7.1. Figure 7.4a has the two vehicles moving toward each other. At the collision point, the velocity arrows reverse with the light vehicle headed west.

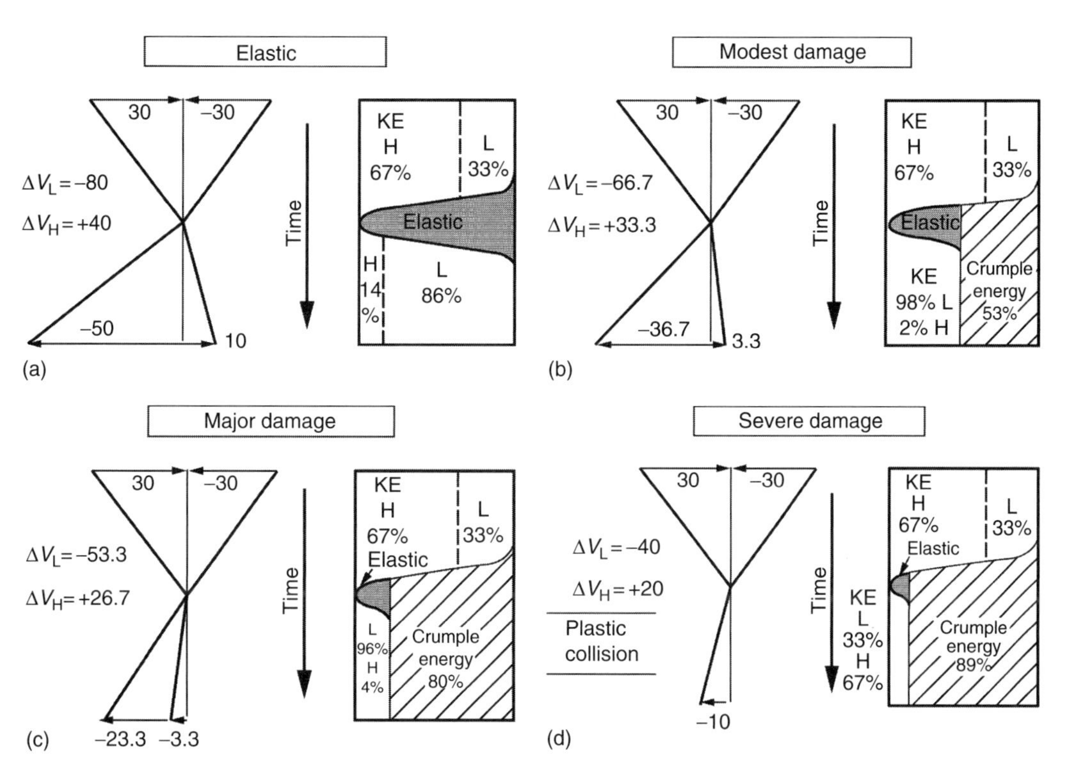

FIGURE 7.4 (a) *Zero damage*, elastic collision with all KE stored in elastic energy at moment of collision; (b) *modest damage*, portion of KE, 57%, absorbed by the crumple zone showing the decrease in change of velocity for each vehicle, $\Delta V_H = +33.3$ and $\Delta V_L = -66.7$; (c) *major damage*, additional energy absorbed by the crumple zone; and (d) *severe damage*, the plastic collision results in the two vehicles sticking together and sliding along road toward the west. Eighty-nine percent of KE absorbed by crumple zone. Sketches for approaching and rebounding velocities for a head-on collision between two unequal-weight vehicles with the heavy being twice that of the light. Variation of the rebound velocities as the energy absorbed by crumpling of structure increases.

An energy diagram appears on the right side. Increasing time runs downward. Before collision, the kinetic energy (KE) is divided with the heavy having two third. Upon first contact, the structures begin to compress like a spring. When both vehicles have zero velocity, all KE has been transformed into elastic energy. During reversal of motion, each vehicle is shoved by the spring. This shove accelerates the vehicles on their new rebound path. Energy once in the spring is now again KE; however, a transfer of KE between vehicles has occurred. Light has 86% and heavy has only 14%.

Figure 7.4b shows that with energy absorbed by the crumpling of structure, the rebound velocities are reduced. In Equation 7.17, ΔV is reduced. At the collision point, some of the initial KE has bent the structure and some has been stored elastically, that is, the structure will exert an elastic force on the rebounding bodies. For this collision, 53% of the initial KE is absorbed by the buckling of the vehicle bodies. Of the remaining 47% in the form of KE, the light vehicle carries away 98%.

Figure 7.4c has 80% of the KE converted to crumple energy. Note that the heavy does not reverse direction but continues at a reduced velocity. With greater crumple energy, ΔV for both heavy and light is once again reduced. The light vehicle carries away the lion's share, that is, 96% of the remaining KE.

Finally, in Figure 7.4d, 89% of the KE has been converted to crumple energy. At this point, the two vehicles stick together and slide along the roadway in the westward direction at 10 mph. The fraction of KE apportioned to light and heavy is unchanged before and after the collision and remains at 33% and 67%, respectively.

The change in velocity creates the safety problems. Four levels of damage to the vehicles are assumed. The damage in the elastic collision does not even scratch the paint. This is not realistic but does illustrate the change in velocity. The case is known as an elastic collision. The last case is for a plastic collision in which the two vehicles stick together and go sliding along the road in the left-hand direction.

Some of the vehicle's initial KE crumples the bodies of both cars. For the plastic collision, the residual KE energy is only 11% of the initial value for both cars. The energy expended bending metal is 89%. As more energy is absorbed in bent metal, the change in velocity decreases from 40 mph down to 20 mph for the heavy car and from 80 mph down to 40 mph for the light car. In every case, for $Q = 2$, the light car has twice the change in velocity compared to the heavy car.

Seeming Paradox

On one hand, the increasing energy in crumpling indicates that the structures have been more severely damaged. On the other hand, for increasing energy in crumpling, the deceleration has decreased. Note the decreasing ΔV. Deceleration levels can be correlated with injury to the people in the vehicles. The vehicles suffer more, and the people suffer less.

Added Comments on Acceleration

The preceding analysis, in connection with Table 7.1 and Figure 7.4, provides insight into the numerator, ΔV, of Equation 7.17. The denominator, Δt, is not available from this analysis. Light versus heavy ΔV is twice as big, which suggests acceleration of the light is twice that of heavy. This is indeed true since the forces on each vehicle, F_L and F_H, are equal during the collision (subscript L is for light, and H is for heavy).

$$F_L = a_L m = a_H M = F_H \tag{7.19}$$

Since the mass of the heavy, M, is twice that of the light, m, $M = 2m$, the accelerations are related by $a_L = 2a_H$. The conclusions are consistent. Since the light and heavy are partners in the collision, one would expect that Δt is the same for both; that is exactly what it is.

Acceleration Based on Crumple Zones and Driver Motion

An alternate equation for acceleration involves the distance traveled when the vehicle has constant acceleration, a. It is

$$a = \text{acceleration} = \frac{(\text{velocity})^2}{2(\text{stopping distance})} \tag{7.20}$$

The questions are: What velocity to use in Equation 7.20? How should the velocity be measured or defined for this case? If hitting a rigid, brick wall, the velocity is that of the vehicle relative to the wall. For any two vehicles in collision, the velocity to use for either vehicle is that relative to the center of gravity (CG) of the combined two vehicles. The CG effectively acts like a rigid, brick wall. The velocity of the CG is

$$V_{CG} = \frac{MV + mv}{M + m} \tag{7.21}$$

The velocity of the light vehicle relative to the CG, u, is

$$u = v - V_{CG} \tag{7.22}$$

and the velocity of the heavy vehicle relative to the CG, U, is

$$U = V - V_{CG} \tag{7.23}$$

Hence, the acceleration for the heavy vehicle is

$$a = \frac{U^2}{2S} = \frac{U^2}{2(S_C + S_P)} \tag{7.24}$$

For the acceleration of the light vehicle, replace U by u in Equation 7.24. For deceleration, S equals the distance to stop and is the sum of the crumple zone plus the driver motion relative to vehicle. The sum is

$$S = S_C + S_P \tag{7.25}$$

The symbol, S_C, equals the length of the crumple zone exterior to the passenger cabin. A typical value for S_C is 1.0 m (3 ft). The symbol, S_P, is the driver motion relative to vehicle because of telescoping of the steering column, collapse of the air bags after deployment, and any motion of the seat. A typical value for S_P is 0.3 m (1 ft).

Since the sum, S_C+S_P, appears in the denominator of Equation 7.24, and since small acceleration is desired to lessen or avoid injury, a large value for the sum, S_C+S_P, is sought. Figure 7.5 illustrates driver motion relative to the firewall of the vehicle. The uppermost sketch shows the driver just at moment of impact. The air bag is deployed. Now focus on the middle sketch. As time elapses, the diver moves forward, and the steering column telescopes while the driver surges ahead. Note that seat belts are elastic and stretch when loaded [11]. Seat belts behave similar to a rubber band although requiring greater force to stretch. In the bottom sketch, the dashboard is buckled (crumpled) giving an added distance, S'_P. Actually, very little empty space exists under the dash so S'_P may be zero or very small.

A sample calculation using Equation 7.24 is

$$a = \text{Acceleration} = \frac{(25\ \text{m/s})^2}{2(0.6\ \text{m} + 1.5\ \text{m})} = 156\ \text{m/s}^2 \tag{7.26}$$

An acceleration of 156 m/s^2 is about 16 G's, which is detrimental to the human body.

Equation 7.24 is complete in the sense that acceleration can be evaluated. Equation 7.24 gives the average acceleration during the time interval starting at first contact until each vehicle is at rest just before commencing rebound. Equation 7.24 forms the basis for determining crumple lengths, S, to compensate for weight disparity. With longer crumple length, the acceleration of the driver or passenger in a light vehicle can be made equal to that in the heavy vehicle. The total crumple length scales as Q^2. This is important for mitigating weight disparity. Figures 7.7 and 7.8 illustrate different values of Q and the effect on vehicle design.

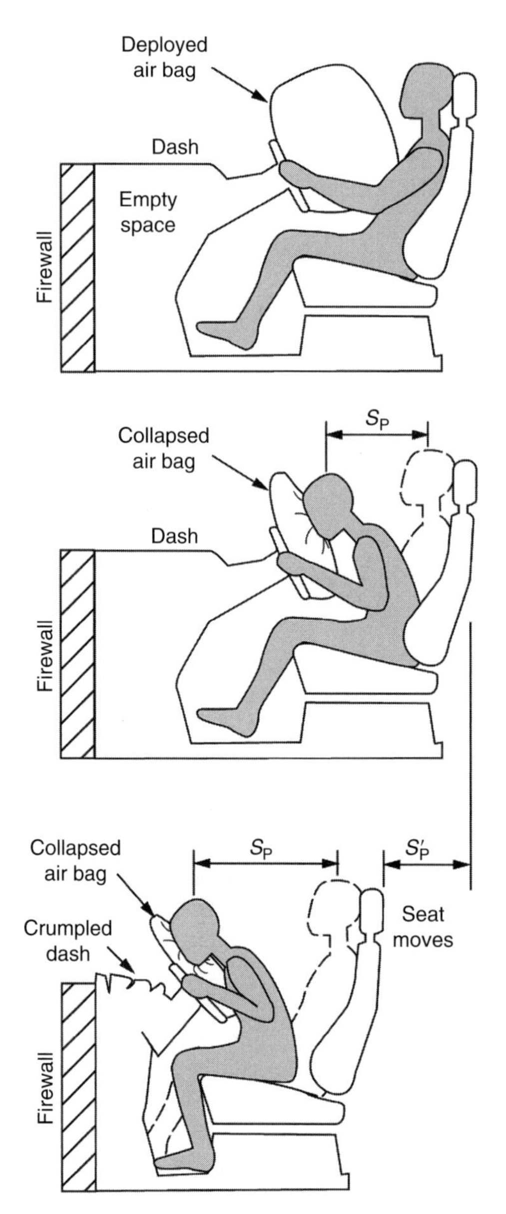

FIGURE 7.5 Illustration of driver motion relative to vehicle.

Fractional Form for Acceleration Based on Crumple Zones and Driver Motion

The fractional form for the Equation 7.24 is

$$\frac{\Delta a}{a} = 2\left(\frac{\Delta U}{U}\right) - \frac{\Delta S}{S} \tag{7.27}$$

Since acceleration, a, depends on U^2, a factor of 2 appears before $\Delta U/U$. A 10% increase in U causes a 20% increase in acceleration, a. Also, since S appears in the denominator of Equation 7.24, a 10% increase in S causes a 10% decrease in acceleration, a.

Velocities Relative to the CG

As discussed with regard to Equations 7.19 through 7.24, the formula of Equation 7.24 requires the velocity relative to CG. Using the numerical values for the sketches of Figure 7.4, the drawings have been converted to the CG as reference. For the examples, $M = 2m$, and $V_{CG} = -10\,\text{mph}$, see Figure 7.6. The path of the CG is denoted by a dashed line.

Several interesting observations can be made. Refer to Figure 7.6a. Note the symmetry of the rebound velocities with incoming velocities. Also, note the change in velocity is given by the differences

$$\Delta v_L = u - u_0 = -40 - 40 = -80$$

and

$$\Delta V_H = U - U_0 = +20 - (-20) = +40$$

For $M = 2m$, the CG is at $X/3$ from the heavy vehicle, where X is the separation of the two vehicles. This separation remains true at any time before or after the collision. At the end of Figure 7.6b, note that

$$\frac{13.3}{13.3 + 26.7} = \frac{1}{3}$$

The same result applies to Figure 7.6c as well. With regard to $X/3$, Figure 7.6d is indeterminate.

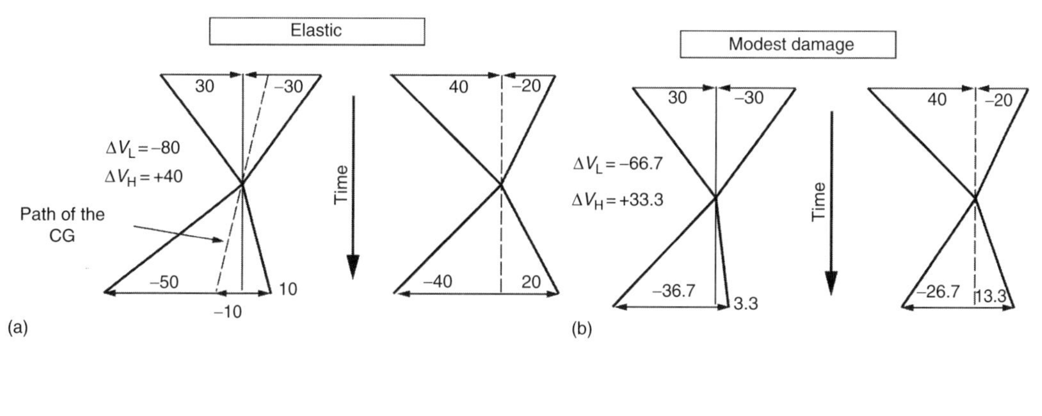

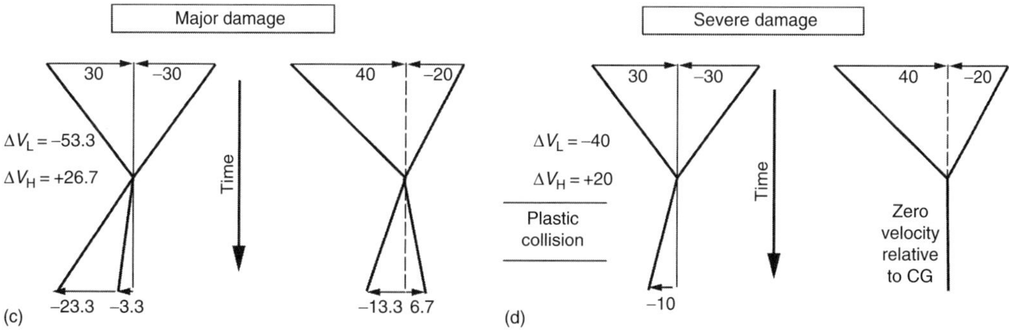

FIGURE 7.6 (a) Velocities relative to the CG, elastic collision, and zero damage with maximum deceleration; (b) velocities relative to the CG, modest damage; (c) velocities relative to the CG, major damage; and (d) velocities relative to the CG, plastic collision, and severe damage with least deceleration.

Continuing on Figure 7.6d, note the two vehicles are stuck together and are sliding along the road at −10 mph. Since the vehicles are stuck together, the CG is traveling with the vehicles. Therefore, the velocity of either vehicle relative to the CG is zero as shown.

Compensation for the Double Acceleration

As stated above, since the mass of the heavy vehicle, M, is twice that of the light vehicle, m, $M = 2m$, the accelerations are related by $a_L = 2a_H$. Can the crumple zone be used to make equal accelerations for both the heavy and light vehicles, that is, $a_L = a_H$. If so, for what value of $S = S_C + S_P$ can compensation for double acceleration occur?

$$a_H = \frac{U^2}{2S_H} = \frac{u^2}{2S_L} = a_L \tag{7.28}$$

The crumple lengths are S_H for the heavy and S_L for the light. Solving for the acceleration ratio, $a_L/a_H = 1.0$, the result is

$$\frac{S_L}{S_H} = \frac{u^2}{U^2} = \frac{(v - V_{CG})^2}{(u - V_{CG})^2} \tag{7.29}$$

For an example, pick any two relative-to-CG values from Figure 7.6. From Figure 7.6c, the values are

$$\frac{S_L}{S_H} = \frac{-13.33^2}{+6.66^2} = 4.0$$

For the example of Figures 7.4 and 7.6, the crumple length for the light vehicle must be four times as long as for the heavy vehicle. The following two conclusions can be drawn: crumple length can compensate for weight disparity, but with demands on crumple length; and a guiding design principle is to not downsize light vehicles when decreasing weight. Reduce weight, but retain or enlarge the body size to incorporate long crumple lengths. This is a challenge for the structural engineer and the stylist.

Figure 7.7 is drawn for weight disparity $Q = 2.0$. Figure 7.8 uses $Q = 4/3$.

Compensation for Realistic Weight Disparity

The disparity of the previous discussion, that is heavy being twice the weight of the light, does not match the fleet of vehicles currently on the road. The weights of 3600 lb for the light vehicle and 4800 lb for the heavy vehicle more closely characterize the fleet. Weight disparity is $Q = (4800)/(3600) = 4/3$.

As before, use 30 mph for each vehicle with a head-on collision. The change in velocities, ΔV, are in the ratio $Q = (4800)/(3600) = 4/3 = 133\%$. The accelerations are in the same ratio. The length of the crumple zones are in the ratio:

$$\frac{S_L}{S_H} = \frac{4800^2}{3600^2} = Q^2 = \left(\frac{4}{3}\right)^2 = 1.78$$

Figure 7.8 shows the compensation for acceleration because of this realistic weight disparity. The resulting design seems feasible from the structural viewpoint. Even though the light vehicle receives

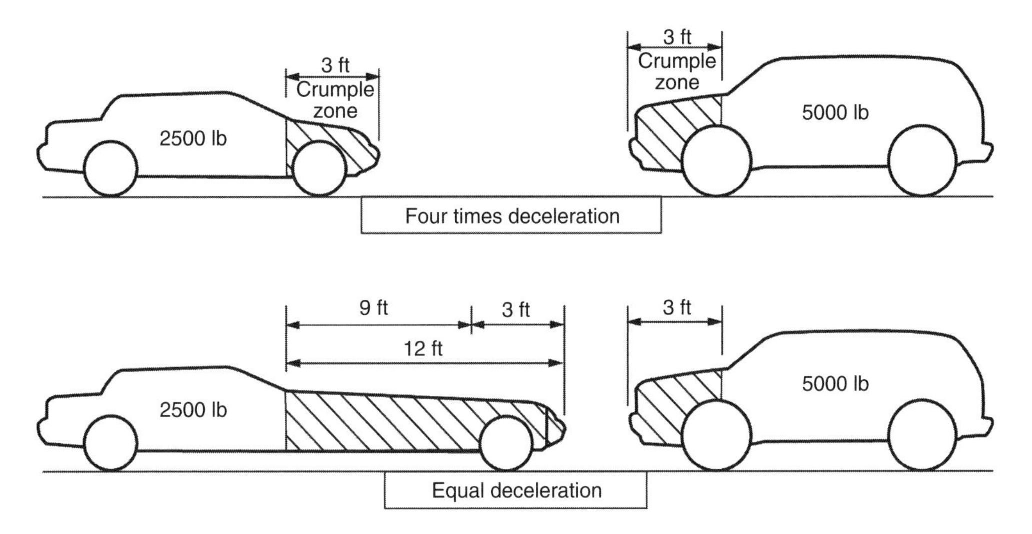

FIGURE 7.7 Compensation for double acceleration by use of crumple zone four times in length. Obviously, the design with the compensating crumple zone is unacceptable.

one third more deceleration (passenger injury) when crumple zones are an equal 3.0 ft, with the added crumple zone of 2.3 ft, the deceleration is equal.

With cross-hatching to highlight the crumple zones in Figure 7.8, the styling impressions are distorted. Figure 7.9 compares the 3600 lb car before and after the crumple zone is added. The long nose is acceptable and may even be attractive. The connection between style and safety is discussed in Ref. [19].

NHTSA Action

The prevailing opinion is that by aggressive action, National Highway Traffic Safety Administration (NHTSA) could reduce traffic injuries and fatalities while still allowing an increase in vehicle fuel

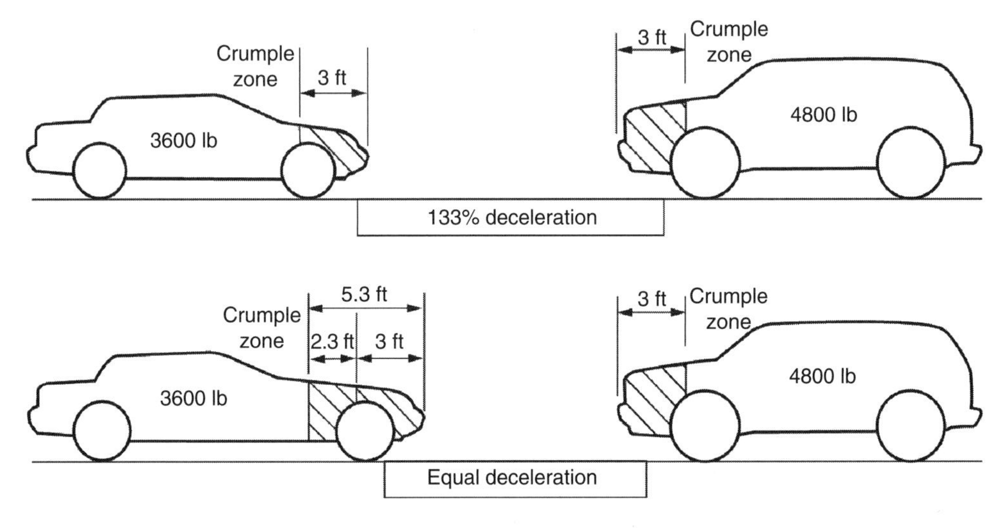

FIGURE 7.8 Compensation for acceleration because of realistic weight disparity. Crumple zone $Q^2 = (4/3)^2 = 1.78$ times in length gives equal acceleration in a feasible package.

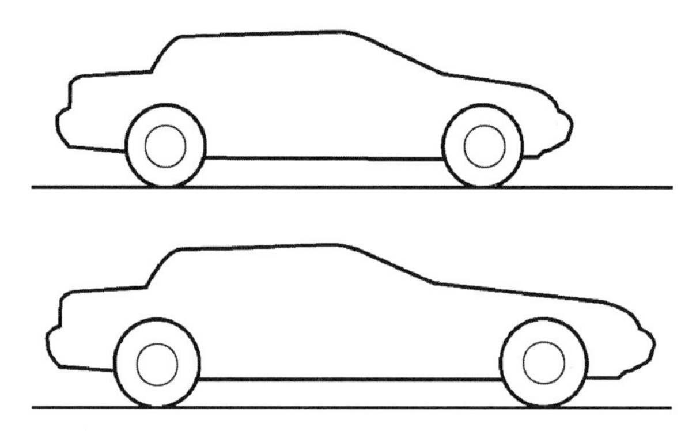

FIGURE 7.9 Styling comparison of original vehicle with vehicle having added crumple zone.

efficiency. Safety and fuel efficiency are not incompatible. The costs to simultaneously increase mpg and to preserve or enhance safety can be recouped in lower gas costs and lower insurance premiums.

Weight Disparity

Table 7.1 and Figure 7.4 provide insight to the problem of weight disparity. How can the problem be solved? The solution is rightly a function of government. Congressmen are sensitive to the feelings of the car-buying public. "Preservation of choice" is a term used frequently when some new bill is being considered. Even if the government decided to homogenize the vehicles on the road, it would take more than a decade to do so. Replacing old vehicles with new ones takes much longer than the average age of a vehicle on the highway. In 2006, United States had 220 million lightweight cars on the road. The annual sales of new vehicles is less than 20 million per year. Turnover of the fleet of cars is slow. Eliminating the weight disparity enhances safety, and could also decrease fuel consumption.

Four options are presented in Figure 7.10. Figure 7.10a aggravates the problem by widening the weight gap between sport utility vehicle (SUV) and small cars. One benefit is better mpg for small cars. Figure 7.10b suggests that the ideal solution is to shrink the SUV until its weight is equal to a small car. The problem of disparity is solved. The fuel economy of the SUV is greatly enhanced. However, a shrunken SUV cannot provide the same interior volume, trailer towing capability, and room for the same number of passengers.

Figure 7.10c expands the small car and shrinks the SUV until each is of equal weight. The weight disparity problem is solved. Improvement in small car mpg cannot be achieved by weight decrease, obviously (weight is being increased). The mpg for the SUV becomes more favorable. As shown by Figure 7.10d, shrinking both the SUV and the small car by the same percentage does offer the opportunity to improve mpg for both classes of vehicles. The disparity is not solved. At first thought, enhanced safety seems to have slipped away. However, the energy absorption by crumpling of structure is more effective because of the reduced vehicle weights.

Effect of Footprint on Safety

One focus above has been the relationship between vehicle weight and safety. Footprint, which is a geometrical factor, is used to define fuel efficiency for light trucks and SUVs. The discussion of Corporate Average Fuel Economy in Chapter 12 is related; see Figure 12.4 which is the target fuel economy, mpg, for light trucks as a function of footprint. Footprint is the product of wheelbase and average track. Vehicle track is a dominant factor for vehicle rollover. Determination of the fuel efficiency curve in Figure 12.4 required consideration of track and rollover safety [15].

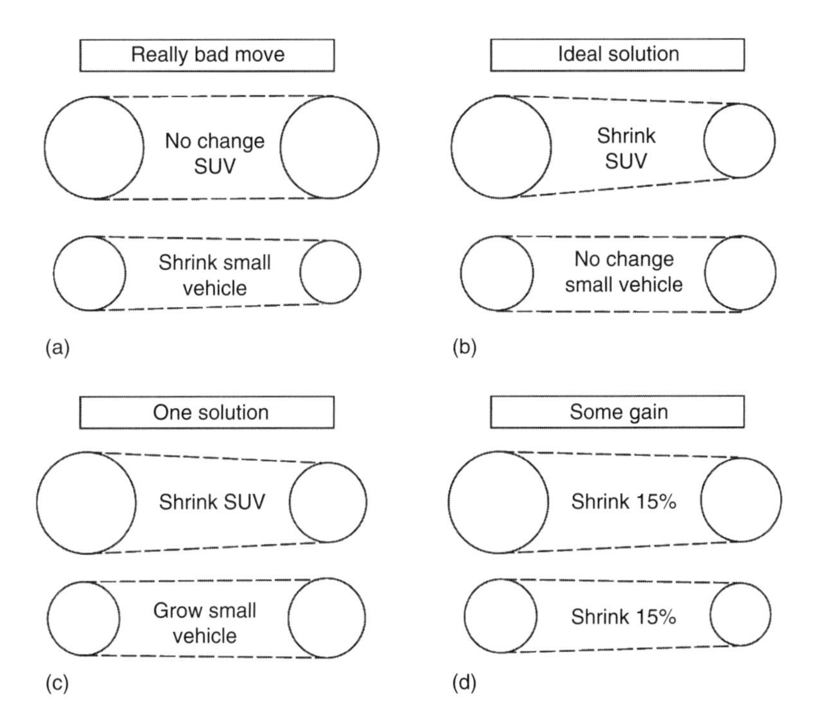

FIGURE 7.10 Possible approaches for solving weight disparity.

CO_2 Emissions versus Vehicle Weight

Figure 7.1 shows fuel consumption increases with weight. Since the combustion of hydrocarbons produces CO_2, an increase in CO_2 is expected with increasing weight. The more the fuel burnt, the more the CO_2 emission. Figure 7.11 substantiates that expectation. For both diesel and gasoline engines, the CO_2 emissions almost double when the vehicle weight increases from 1500 to 5500 lb. Introduction of a gasoline HEV reduces CO_2 emissions by 27% for a lightweight car (2000 lb) and 20% for a heavy vehicle (5500 lb). Introduction of a diesel HEV reduces CO_2 emissions by 24% for a lightweight car

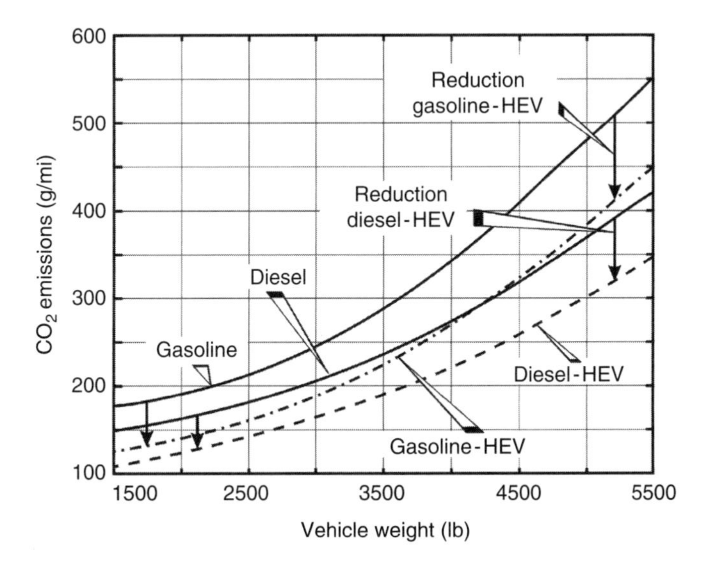

FIGURE 7.11 CO_2 emissions versus vehicle weight. Amount of reduction in CO_2 because of use of hybrids.

(2000 lb) and 18% for a heavy vehicle (5500 lb). CO_2 is a greenhouse gas (GHG). Hybrids can help reduce GHG. Hybrids can contribute to a reduction in GHG but are not a panacea [17].

Just as the points in Figure 7.1 have much scatter, the data to create Figure 7.11 also had scatter. The numerous points are not shown. The curve labeled gasoline is the lower bound of the field of scattered points for gasoline-powered cars. Likewise the curve labeled diesel is the lower bound for all diesel-powered cars. Hence, the reductions because of use of hybrid technology will almost certainly be larger than shown. The curves have been drawn to show the minimum of the reductions.

SUMMARY

This chapter covers the somewhat inflammatory and controversial statement "Fuel efficiency standards are killing people." Sufficient information is given to help the reader arrive at his/her own conclusion.

This chapter has two major topics: (1) weight versus fuel economy and (2) weight and safety. With regard to weight versus fuel economy, correlation data are presented for gasoline and diesel cars. The fractional form is derived (in Appendix 7.1) and then used throughout. The fractional form can be interpreted as a percentage change. Predictions of percentage changes are made for many examples.

The years following the gas crisis of the mid-1970s was a time of considerable changes in the automobile industry. Two graphs depict the changes from 1975 to 2005. From 1976 to 1981, automobile companies made considerable progress reacting to market demands. A big improvement in F_E occurred during these years with a flattening of the F_E curves for almost two decades. The fraction of light trucks and SUVs increased from a minor fraction to 50%. Advances in technology were neutralized by added weight and greater acceleration thereby keeping F_E almost constant.

The major issue with regard to weight and safety is whether or not drivers of small cars are at a serious disadvantage relative to the behemoths on the road. The collision diagrams (Figures 7.4 and 7.6) provide insight to that question. Acceleration (or in the case of collision, deceleration) of the human body inside the car causes injury. Equations 7.17 and 7.24 provide information about acceleration.

Equation 7.17 gives acceleration averaged over the entire collision event. It cannot predict acceleration since Δt cannot be found. However, Equation 7.17 is nonetheless valuable. It does give correct relative accelerations of the heavy and light vehicles. The ratio of accelerations of the light to heavy vehicles is proportional to the weight disparity, Q. Of significance is the fact that the influence of the crumple energy can be determined. A decrease in acceleration with increasing crumple energy is found using Equation 7.17.

Equation 7.24 is complete in the sense that acceleration can be evaluated. Equation 7.24 gives the average acceleration during the time interval starting at first contact until the vehicle is at rest just before commencing rebound. Equation 7.24 forms the basis for determining crumple lengths, S_c, to compensate for weight disparity. With longer crumple length, the acceleration of the driver or passenger in a light vehicle can be made equal to that in the heavy vehicle. The total crumple length scales as Q^2. This is important for mitigating weight disparity. Two drawings illustrate different values of Q and the effect on vehicle design. Different design approaches for eliminating weight disparity are illustrated by Figure 7.10.

The connection between CO_2 emissions and vehicle weight is displayed in Figure 7.11. Hybrids can help reduce GHG. Hybrids can contribute to a reduction in GHG but are not a panacea.

APPENDIX 7.1: MATHEMATICS OF FRACTIONAL CHANGE

The concept of fractional change is useful in two cases. First, an analytical representation of a correlation such as Figure 7.1 assists interpretation and application of data. Second, an analytical representation of a correlation assists with the rapid and easy application of data.

The analytical form

$$Z = a X^{\mathrm{n}} Y^{\mathrm{m}} \tag{A.7.1}$$

can usually reproduce the correlated data to the desired accuracy. Equation A.7.1 is used to curve fit the data. The logarithm of Equation A.7.1 is taken yielding

$$\ln Z = \ln a + n \ln X + m \ln Y \tag{A.7.2}$$

Differentiation of Equation A.7.2 gives

$$\frac{\Delta Z}{Z} = \frac{\Delta a}{a} + n\left(\frac{\Delta X}{X}\right) + m\left(\frac{\Delta Y}{Y}\right) \tag{A.7.3}$$

The fractional change is the form $\Delta X/X$ and can be interpreted as a percentage (when multiplied by 100). A certain percentage in Y can be used to determine the resultant percentage change in Z.

The exponents n and m multiply the fractional change. Either can be positive or negative. For a linear correlation, as shown in Figure 7.1, the exponents are unity ($n = 1.0$ or $m = 1.0$). For a square relationship, $n = 2$, and the fractional change, $\Delta X/X$, is doubled.

APPENDIX 7.2: DERIVATION OF EQUATION 7.12

The cruise fuel economy is given by

$$F_{\mathrm{EC}} = \frac{\rho_{\mathrm{G}} V_{\mathrm{C}}}{B_{\mathrm{C}} P_{\mathrm{C}}} \tag{A.7.4}$$

The symbols are defined as ρ_{G} = density of gasoline, kg/L; V_{C} = cruise velocity, m/s; B_{C} = specific fuel consumption (SFC), g/(kWh); and P_{C} = power, kW, required at cruise. Subscripts E and C are for economy and cruise, respectively. The fractional form of the preceding equation is

$$\frac{\Delta F_{\mathrm{EC}}}{F_{\mathrm{EC}}} = -\frac{\Delta B_{\mathrm{C}}}{B_{\mathrm{C}}} - \frac{\Delta P_{\mathrm{C}}}{P_{\mathrm{C}}} \tag{A.7.5}$$

Assume a linear relation between installed power and cruise power, which is

$$P_{\mathrm{C}} = f P_{\mathrm{I}} \tag{A.7.6}$$

where f is a fraction less than unity with a usual value between 0.1 and 0.2. Further assume that f is a constant for any specified cruise speed. Equation 7.7 provides the installed power. In fractional form

$$\frac{\Delta P_{\mathrm{I}}}{P_{\mathrm{I}}} = \frac{\Delta P_{\mathrm{C}}}{P_{\mathrm{C}}} = \frac{\Delta W}{W} - \frac{\Delta t_{60}}{t_{60}} \tag{A.7.7}$$

Equation A.7.7 assumes that f is constant. Combining Equations A.7.5 and A.7.7 yields

$$\frac{\Delta F_{\mathrm{EC}}}{F_{\mathrm{EC}}} = -\frac{\Delta B_{\mathrm{C}}}{B_{\mathrm{C}}} - \frac{\Delta W}{W} + \frac{\Delta t_{60}}{t_{60}} \tag{A.7.8}$$

The SFC, B (g/[kWh]), can be related to engine efficiency, η. The relation is

$$B\eta = C \tag{A.7.9}$$

The constant, C, has a value depending on the fuel. From Appendix 7.1, the specific energy for gasoline is equal to 11.77 (kWh)/kg or 0.01177 (kWh)/g. C is the reciprocal of this number, which is

$$C = \frac{\text{g}}{0.01177\ \text{kW}_\text{G}\ \text{h}} = (85)\frac{\text{g}}{\text{kW}_\text{G}\ \text{h}} \tag{A.7.10}$$

The kW_G h indicates that this is the energy contained in the gasoline and would be the energy from an engine at 100% efficiency. From Figure 10.2, which is the energy map for the Prius II engine, the minimum value for B = 230 g/(kWh). From Equations A.7.9 and A.7.10, the engine efficiency is η = 85/230 = 37%. The worst value from the engine map is B = 600 g/(kWh). The corresponding efficiency is η = 85/600 = 14%. From Equation A.7.9

$$\frac{\Delta B}{B} = -\frac{\Delta \eta}{\eta} \tag{A.7.11}$$

Substitute Equation A.7.11 into Equation A.7.5; then substitute Equation A.7.7 into the result. Your efforts give

$$\frac{\Delta \eta}{\eta} = \frac{\Delta F_{\text{EC}}}{F_{\text{EC}}} + \frac{\Delta P_{\text{I}}}{P_{\text{I}}} \tag{A.7.12}$$

Equation 7.12 has been derived.

REFERENCES

1. G. W. Housner and D. E. Hudson, *Dynamics: Applied Mechanics*, Vol. II, 2nd ed., D. Van Nostrand Company, Princeton, NJ, 1959, 392 pp.
2. Fuel Economy in Automobiles, Wikipedia, the Free Encyclopedia, September 2006.
3. Environmental Protection Agency, Light-duty automotive technology and fuel economy trends: 1975 through 2006, executive summary, EPA42-S-06-003, July 2006.
4. R. Bamberger, Automobile and light truck fuel economy: The CAFE Standards, Issue Brief for Congress, Library of Congress, March 12, 2003.
5. P. R. Portney, Chair, Federal fuel economy standards program should be retooled, *National Academies*, Washington, DC, Media Release, July 31, 2001.
6. M. Ross and T. Wenzel, Losing weight to save lives: A review of the role of automobile weight and size in traffic fatalities, *American Council for an Energy-Efficient Economy*, July. (Report submitted to National Research Council, March 2001).
7. Fuel efficiency standards and the laws of physics, *Source Watch*, February 2006. Available at www.sourcewatch.org.
8. The facts about raising auto fuel efficiency, *National Environmental Trust*, September 2006. Available at www.net.org.
9. S. Barlas, CAFE safety implications debated, *Automotive Engineering International*, Society of Automotive Engineers, Warrendale, PA, June 2006, p. 77.
10. M. Schrank, Advancing crashworthiness Simulation, *Automotive Engineering International*, Society of Automotive Engineers, Warrendale, PA, June 2006, p. 77.
11. T. Harris, How seatbealts work, http://auto.howstuffworks.com/seatbelt.htm.
12. A. Lovins, *Winning the Oil Endgame*, Lecture at Monterey Institute of International Studies, Monterey, CA, December 7, 2006.

13. A. B. Lovins, E. K. Datta, O. E. Bustnes, J. G. Koomen, and N. J. Glasgow, *Winning the Oil Endgame*, Rocky Mountain Institute, Snowmass, CO, 2004, 309 pp.
14. M. Alpert, Saving gas and lives, *Scientific American*, October 2007, pp. 20–21.
15. Average Fuel Economy Standards for Light Trucks, Model Years 2008–2011, National Highway Safety Administration, Department of Transportation, 49 CFR Parts 523, 533, and 537.
16. D. Carney, Material issue, *Automotive Engineering International,* Society of Automotive Engineers, Warrendale, PA, August 2005, pp. 55–58.
17. P. Moreton, Diesel goes hybrid? *Automotive Engineering International,* Society of Automotive Engineers, Warrendale, PA, April 2007, pp. 24–26. Also correction on p. 70 of May 2007 issue.
18. L. Brooke, Mass reduction—the next frontier; automakers turn their attention to reducing vehicle weight in quest for greater fuel economy, *Automotive Engineering International*, Society of Automotive Engineers, Warrendale, PA, July 2007, pp. 70–73.
19. T. Costlow, Style versus strength; software helps designers and body engineers get the most in looks and safety, *Automotive Engineering International*, Society of Automotive Engineers, Warrendale, PA, December 2007, pp. 56–58.

8 Vital Role of the Control System

All functions must engage and disengage smoothly without jerks, shudders, shakes, or clanks.

HISTORY OF AUTOMOBILE CONTROL

This abbreviated history starts with the Ford Model T; my grandfather had a 1922 open three-door model in which he used to go fishing. The 1920s Ford Model T had driver-controlled ignition timing: a manual lever-controlled advance/retard. Later, for other cars, manual control was replaced by centrifugal spark advance. The driver of a 1920s Ford Model T controlled the mixture composition lean/rich (fuel/air ratio). Advances in carburetors eliminated the chore of mixture control, and automatic chokes replaced the manual choke.

Control of engine temperature advanced through the 1930s. First, a blanket was placed to cover the radiator on cold blistery days. This was followed by adjustable radiator shutters, which were rotatable, narrow, slats in the grill. Finally, thermostats were placed in the coolant path.

Scientists taking air samples in the Los Angeles basin recognized in 1950 that hydrocarbons (HC) and NO_X caused the smog. Automobiles were considered to be one of the culprits. Exhaust emissions were added as new variables to be controlled by the automobile engineer. Slowly, smog introduced a new dimension to automobile control—emissions. Engine controls in the 1980s evolved to master fuel economy and the level of emissions.

EXAMPLE OF ENGINE CONTROLS FROM 1980s

Reduction of pollutants was mandated and by the 1980s had become quite complex. Yet, the 1980s controls were primitive compared to the hybrid control systems of today. The balance of the conflicting effects between fuel economy and pollution levels was accomplished by

1. Optimized control of ignition timing
2. Optimized control of fuel/air ratio (carburetors were used in 1980s)
3. Air injection into exhaust to reduce pollutants
4. Exhaust gas recirculation (EGR) to use fuel more efficiently
5. Prevention of engine knock or ping
6. Prevention of dieseling
7. Diagnostics of faults in engine and in various sensors

To get and maintain stoichiometric fuel/air ratio, 14.7 for gasoline, numerous sensors were required as follows:

1. Temperatures
2. Oxygen content
3. Gasoline flow rate
4. Pressures
5. Throttle position

FUNCTION OF CONTROL SYSTEM

Major functions of the control system are to maximize mpg and to minimize exhaust emissions. Minor functions of the control system include component monitoring and protection. Examples for hybrids are battery state of charge (SOC), battery temperature, electrical motor overheating, and gas engine overheating. The battery merits special attention to avoid failure and to assure long life. The control system generally provides a fail-safe mode in the event of failure. This gives a limp-home capability that helps the manufacturer to retain some credibility with the customer. Another control function, which is minor in cost but valuable in practice, is onboard diagnostics (OBD).

For a hybrid, integration and control software are as important as hardware, if not more so. The challenges of integration and control software are enormous. Components must engage and disengage smoothly; the change between hybrid modes should be imperceptible (almost) to the driver. The potential for optimized control is illustrated by the Honda FCX fuel cell vehicle, in which range was increased by 30% over prior versions. Of the 30% increase, 9% is attributable to increased fuel tank size. The rest is because of a better control system. See Ref. 13 in Chapter 3.

ANALOGY WITH SYMPHONY ORCHESTRA CONDUCTOR

A symphony orchestra conductor creates beautiful music by coordinating the strings, woodwinds, percussion instruments, and brass. The hybrid control system creates a beautiful driving experience by coordinating the flow of energy between the gasoline engine, battery, and electrical motors. The hybrid will sell if the driving experience is pleasant without annoying jerks, shudders, and shakes. Further, perceived mismatches between driver commands and car responses are forbidden.

A difficult area to achieve smooth operation is regenerative braking. Peculiar brake pedal sensations are hard to eliminate. The new electronic wedge brake may help ease this problem. See Chapter 12 for more details.

FRINGE BENEFITS OF CONTROL SYSTEM

Because of the massive computing capability of hybrid control systems, fringe benefits accrue. Two such benefits are diagnostics of the electronic control units (ECU) and the useful OBD. Numerous advances have been made in computer technology; modern hybrid controls would not have been possible without computers. These advances include speed, or operations/second; reliability; software complexity and scope; and the massive storage of data. Most importantly, price was decreased, which is a vital factor for automobiles.

The size of a software program can be expressed as the number of lines of code. The size of the program to control a modern engine is about the same as that required by the hybrid master controller. The cost of memory and CPU limits the size of acceptable programs.

SIMULATION AND MODELING

As applied to hybrid vehicles, simulation is an imitation of a real hybrid. Another description of simulation is that it is a low-cost, easy-to-manipulate copy of the expensive hybrid system being considered. Simulation requires representation of the characteristics or behavior of the system (hybrid vehicle). Modeling also involves the representation of the behavior or characteristics of the system. A damped spring-mass system can be modeled by a differential equation. The dynamics (bounce, pitch, cornering) of a complete vehicle can be modeled by combining many of the differential equations. However, the solution of all the interacting models requires the capability of a computer. Complex collections of interacting models become simulations. The level of complexity separates and defines the differences between simulation and modeling. In practice, the words simulation and modeling have been used indiscriminately so that the differences have been erased.

To focus the discussion, refer to Figure 4.1 once again. The system consists of the engine, motor, transmission, battery, control system, and wheels. Each component must be represented by a mathematical model (algebraic or differential equation) or a table of lookup values. The engine map of Figure 12.5 is one method of defining or representing the engine. The various components are connected and can interact. For stated inputs and initial conditions, the behavior of the system can be determined. Of course, the validity of the predicted behavior depends on the validity of the models for each component.

When the interactions between components are too complex for analytical solution, it is necessary to resort to computers. For stated inputs and initial conditions, computer simulations provide the outputs or system behavior of complex systems. Simulation shows the effects of alternate components, alternate input conditions, and alternate arrangements of components (series vs. parallel hybrids). Simulation is an economical method to sort out various designs and to screen various new concepts. As the development time for a new vehicle decreases, finding fast answers and narrowing options will require more dependence on simulation.

DESIGN SEQUENCE

Procedures and goals are evolving for software and control design. Car makers recognize that in this era of uncertainty, flexible control systems should also be adaptable to clean diesels, ethanol-fueled cars, fuel cells, and plug-in HEV. Current strategy is to replace point designs for a single vehicle by the plug-and-play approach; that is, the software can control a variety of engines, motors, batteries, etc. Automated software generation, so-called auto-coding, is gaining popularity and creates nearly error-free code. Proprietary software was common in the early days; however, open-source software has been or is developed by suppliers to the automotive manufacturers. Real-time operating systems (RTOS) are replacing time- and task-oriented software systems [1]. Two problems with the design and control of hybrids are the packaging and integration of components, and control and coordination.

Several approaches exist to reduce fuel consumption. First, reduce losses due to vehicle dynamics such as the mass (weight) of the car and the body shape as it affects aerodynamic drag by reducing AC_D where A is the frontal area of the car and C_D is the drag coefficient. Also reduce rolling friction of tires. The aim is to increase efficiency of each and every component.

The design engineers have different ways to increase conversion efficiency and design a hybrid. Incorporate the latest proven technology. Optimize existing powertrain components, for example, direct injection for gas engines. Use new powertrain components, for example, six- or seven-speed transmissions. Combine and integrate optimized powertrain components into the hybrid.

Combining electric motors and gasoline engines opens up many more degrees of freedom that must be controlled. Simulation is a powerful and essential tool which is required long before prototyping. Simulation yields accurate sizing of components and helps develop control algorithms.

Individual component hardware is selected. These pieces can be tested in facilities offering hardware in (control) loop (HIL). Having a collection of tested parts and pieces, attention can be focused on prototypes and concept vehicles. Finally, after comparison of the possibilities, production vehicles come into view. HIL is analogous to a flight simulator in which a pilot is inserted in the loop.

Hybrid design involves determination of the degree of hybridness, size of the gas engine, size of the battery, and size of electrical motors and generators. For a discussion on hybridness, see Chapter 11.

ALGORITHM DEVELOPMENT

What is an algorithm? An algorithm is a set of rules that specify a sequence of actions to be taken to solve a problem. Each rule is precisely and unambiguously defined so that in principle it can be carried out by a machine (computer). A sample control algorithm is

- Control electrical motor and gas engine torque to equal required drive torque
- Operate engine at most efficient points
- Maintain battery SOC

The time from powertrain definition to executable computer code is a matter of weeks by modern automobile manufactures. Algorithm development is part of the sequence above; hence, it must be done on a tight, compressed, schedule.

CONFLICT: FUEL ECONOMY AND EXHAUST EMISSIONS

Why are emissions and fuel economy in conflict? Three examples will be given; many more examples exist.

1. When more energy is extracted by the gas engine (thereby increasing engine efficiency) the exhaust temperature goes down. At lower temperatures, the chemical reactions associated with the combustion of the unburned HC may not occur.
2. Increase in compression ratio, which enhances fuel economy, also raises temperature in combustion chamber. Increased temperature increases both CO and oxides of nitrogen represented by NO_X.
3. Large percentage of EGR helps fuel economy; however, HC are increased.

EXHAUST EMISSIONS: ORIGIN OF POLLUTANTS

Here is a partial list of the origins of pollutants:

- Interruption of chemical reaction chains because of the short dwell time in the combustion chamber. Chemical equilibrium does not prevail.
- Spatially varying fuel/air ratios within combustion chamber. A local volume may be fuel rich while an adjacent volume may be fuel lean.
- Combustion chamber wall effects because of relatively cold wall that quenches reactions.
- Impurities in fuel.
- Additives to fuel.

VARIOUS OPERATIONAL MODES TO BE CONTROLLED

The complexity of the control problem can be appreciated by considering the various operational modes and the interplay between many components. The switching from one mode to another must be smooth without jerks, shudders, or clanks. Otherwise the car cannot be sold to the demanding public.

Operational Mode	Engine	Engine Clutch	M/G	M/G Clutch	Battery	Electrical Current	Vehicle Motion
Starting[a]	Off→on	On	Motor	On	Discharge	Small	At rest
All-electric	Off	Open	Motor	On	Discharge	Big	Moving
Starting[b]	Off→on	On	Motor	On	Discharge	Big	Moving
Engine alone	On	On	Off	Open	—	None	Moving
Engine[c]	On	On	Generator	On	Charge	Medium	Moving

Operational Mode	Engine	Engine Clutch	M/G	M/G Clutch	Battery	Electrical Current	Vehicle Motion
Acceleration[d]	On	On	Motor	On	Discharge	Big	Moving
Braking[e]	Off	Open	Generator	On	Charge	Very big	Moving

[a] Engine cold; car has been parked. Engine must be warmed rapidly for control of emissions.
[b] Car is in all-electric mode. Additional power is needed; hence start engine.
[c] Engine is both driving the wheels and charging the battery.
[d] Acceleration from traffic light, climbing a hill, or driving in snow or mud.
[e] Setting of components for regenerative braking.

The engine has several variables to be set or controlled as operating modes change in Figure 8.1. These include

- Ignition timing
- Tuned intake manifold
- Camshaft angle for exhaust valves
- Camshaft angle for intake valves
- Fuel injector settings: not a simple squirt of fuel
  - Timing of injection
  - Fuel flow rate
  - Percentage each of port and direct injection for dual injection systems
- EGR, valve setting
- Displacement on demand; cylinder deactivation (if so equipped)
- Waste gate for turbocharger

Each time the driver pushes or releases the throttle, the operational mode may or may not need changing. The hybrid control system makes the decision. If the operational mode does not need changing, as a minimum, the control system must reset each of the engine variables listed above. If the operational mode does need changing, in addition to the engine variables, all the other components must have their settings readjusted.

The potential customer demands seamless integration of all the components either during operational mode changes or during any inputs by the throttle or brake pedals. All functions must engage and disengage smoothly without jerks, shudders, shakes, or clanks. The unexpected is to be avoided; the customer is accustomed to hearing and feeling the engine revolutions per minute (rpm) decrease when lifting his/her foot off the gas pedal. While stopped, a series hybrid may have the engine running at 2500rpm to charge the battery. The customer expects an idling engine at stop. Another example involves the brakes. For a gentle push on the brake pedal, the customer expects minor deceleration. If the regenerative braking is a little overactive, the stopping sensation may exceed the customer's expectations. An uneasy feeling

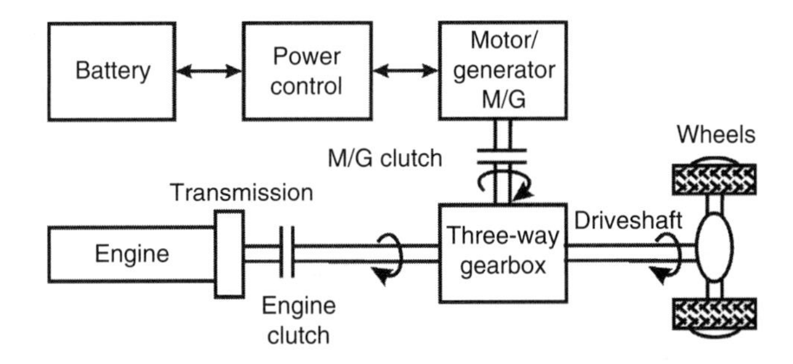

FIGURE 8.1 Control system orchestrates changes between the various operational modes.

of not being in control may follow. Reference [6] reports that at about 80 mph, the Lexus GS 450h Hybrid causes a mild surge/deceleration sensation. This is an unnatural and uncomfortable feeling. The same sensation was detected on the Prius II at 80 mph. Otherwise the feeling was "Love this car!"

Compare Figures 8.1 and 8.3, which are different versions of a parallel hybrid. Figure 8.3 shows a belt-driven, mild hybrid.

ELEMENTS OF CONTROL THEORY

One basic element of a control system is a feedback network. The essence of control theory is in feedback. The goal is to maintain a close relationship between desired and actual quantities. The desired quantity is the input, and the actual quantity is the output of the feedback network.

The basis for control is the feedback loop, one form of which is illustrated in Figure 8.2a. An input, which is the desired value of the controlled variable, is made. The circle represents a summation point where the input and actual values are compared to form the error signal. The error signal is fed into one or more control elements that move actuators at the plant. The word "plant" is the

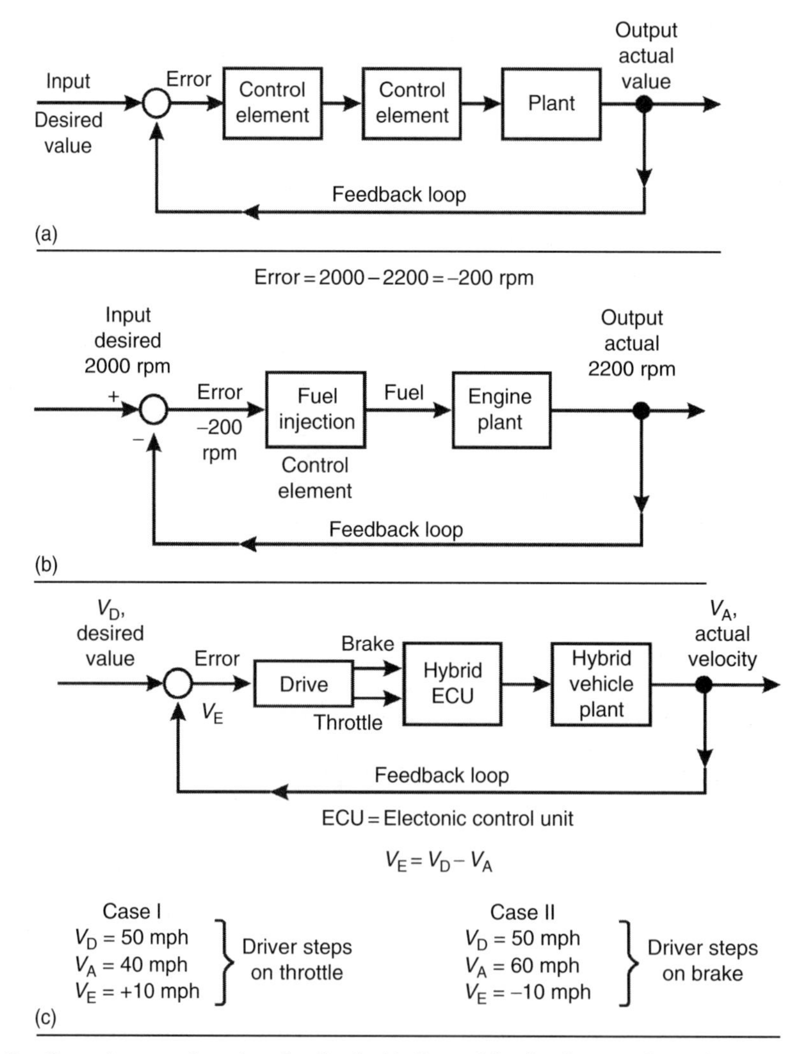

FIGURE 8.2 Control system based on feedback. (a) General feedback control may have one or more control elements in series and or in parallel, (b) simple example of control of engine speed, rpm, by feedback, and (c) control of a hybrid vehicle by feedback.

name for the object being controlled. The actual value of the controlled variable from the plant is transmitted via the feedback loop to the summation point.

Feedback networks can be unstable and oscillate wildly being of no use. Mathematical tools have been developed to detect and correct instability. Using these tools, stability can be designed into the control system. Figure 8.2b, which is a simple example, shows control element and plant. As shown by Figure 8.2a, a feedback network may have several control elements. Control element is a general term and identifies the element, singly or in combination with others, that receives the error signal and changes the plant. In control theory, plant is a general term that identifies the object being controlled. For the example of rpm control (Figure 8.2b), the fuel injector is the control element and the gasoline engine is the plant.

As an example, control of the rpm of a gasoline engine will now be discussed using the feedback network of Figure 8.2b. The circle is the summing point where the actual rpm (2200) is subtracted from the desired rpm (2000) giving an error of −200 rpm. The feedback loop extends from the output to the summing point. The input, which is 2000 rpm, enters the summing point. The fuel injector responds to the negative error, −200 rpm, by decreasing the flow of fuel to the engine. With less gasoline, engine speed will drop.

An important development in control theory is optimal control, which is discussed later in the chapter.

A modern control system for hybrids may use only a few, or even only one, feedback control loops similar to the one shown in Figure 8.2c [3]. The sensors from 1980s controls have also multiplied in number to meet today's control needs. Some of the features of the 1980s controls are found in modern cars, for example, EGR to use fuel more efficiently, prevention of engine knock or ping, prevention of dieseling, and diagnostics of faults in engine and in various sensors. The conflicting effects between fuel economy and pollution levels remain to be controlled.

One control scheme that uses a single feedback loop has desired velocity as the variable being controlled (Figure 8.2c). The driver of the hybrid is one of the control elements and acts on the error. If the hybrid is moving too slowly, the driver steps on the gas pedal. If the hybrid is moving too fast, the driver steps on the brake pedal.

OVERVIEW OF CONTROL SYSTEM: CARTOON VERSION

To provide insight to the function of the hybrid control system, a cartoon is used. The cartoon is reminiscent of your days in the third grade. However, in defense of the cartoon, you will learn two features of hybrid propulsion. First is introduction to the components within the control system and their interactions. Second, as a byproduct of the discussion, you will learn about brake by wire.

Brake by wire replaces a direct mechanical and hydraulic connection between the brake pedal and the disk brakes at the wheels. The direct connection is replaced by a wire from brake pedal running to a master control unit or hybrid control ECU. Braking is controlled by microprocessors. Two major issues with regard to brake by wire are complexity and reliability; and liability and lawsuits. In the late 1920s, mechanical brakes were being phased out by hydraulic brakes. Similar issues surfaced then. Who would want to replace good mechanical brakes with hydraulic brakes that are likely to leak and fail? In 1920s, the world was not quite so prone to sue. See Chapter 9 for more information on braking. Note that the continuously variable transmission (CVT) in a real hybrid, or any car for that matter, would not be connected directly to the drive wheels as shown.

Names have been assigned to the cartoon characters who control the hybrid (Figure 8.3). The various components have a person in charge. Debbie the driver provides the driver input, which in this case is braking. Debbie steps on the brake pedal with a slight force waking up the Virtual Brakeman known by his initials VB.

VB: "Hey Sebastian, Debbie wants to stop."
Sebastian: "How hard is she pushing on the pedal?"
VB: "Slightly."

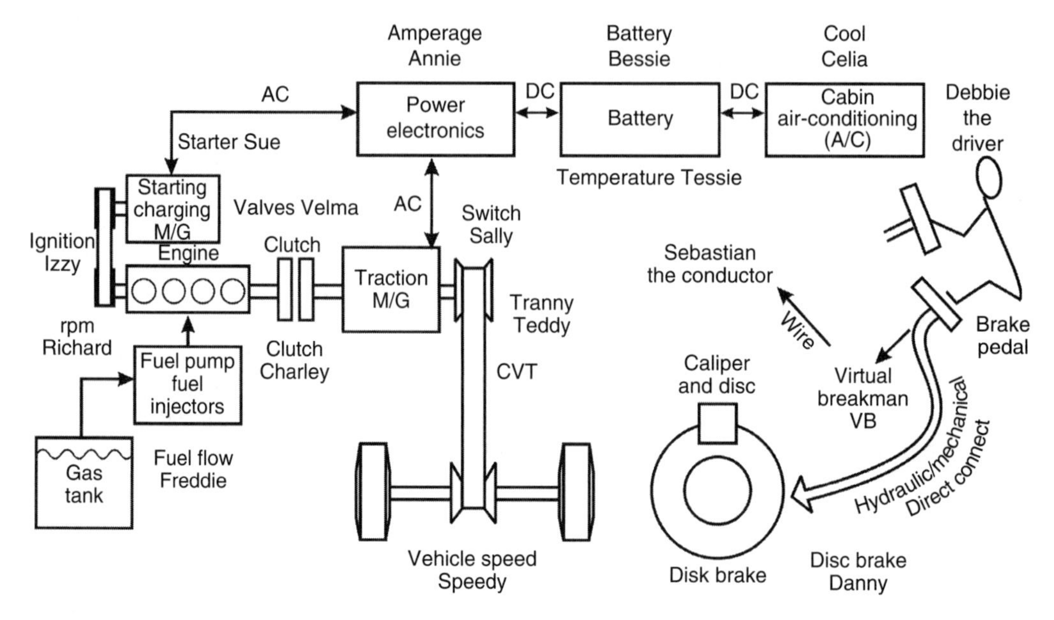

FIGURE 8.3 In a cartoon format, the components and control of a parallel hybrid and brake by wire.

Sebastian determines that 210 lb of brake force is just about right.

Sebastian: "Fuel Flow Freddie, Debbie wants to stop, taper off the fuel injectors."

Note: If fuel flow is abruptly stopped, the engine gives a shudder that shakes the car and everyone inside. Hence, fuel flow is gradually reduced or is damped by the M/G.

Sebastian: "Battery Bessie, what is the ess oh see?" (SOC of battery)
Battery Bessie: "Sebastian, it is 62%; we can take a few amperes."
Sebastian: "Clutch Charley, disengage the clutch, we are going to stop."
Clutch Charley: "As good as done."
Sebastian: "Temperature Tessie, is each cell in the battery cool?"
Temperature Tessie: "The complete battery is cool; we can take on some electrons."

Note: When Temperature Tessie says she can take on some electrons she is talking shop but means that regenerative braking is allowed. If the battery were hot, regenerative braking would be prohibited. All braking would be done by friction brakes.

Sebastian: "Speedy, how fast are we going?"
Speedy: "42 mph."
Sebastian: "Switch Sally, what rpm and torque do you want for your M/G to generate the best braking?"
Switch Sally: "Rpm of 2400 and a torque of 329 ft·lb_f."
Sebastian: "Tranny Teddy, set the CVT to give 2400 rpm at the traction M/G. Amperage Annie, how many amps can be given to the battery?"
Amperage Annie: "About 430 A."

Sebastian determines that the rpm and torque will yield 128 lb of regenerative braking force. Hence, Disc Brake Danny must provide 210 – 128 = 82 lb of frictional braking. The caliper squeezes on the disk to yield that braking force.

Sebastian: "Disc Brake Danny, give me 82 lb frictional braking force."
Disc Brake Danny: "Right on boss, here is exactly 82 lb force."

About a second later, Debbie changes her mind and pushes hard on the brake pedal for 480 lb braking force. Sebastian needs to do it all over again!

OVERVIEW OF CONTROL SYSTEM: ECU VERSION

Refer to Figure 8.4. In the real world of hardware and software, the cast of fictional cartoon characters are replaced by microprocessors in little black boxes distributed throughout the hybrid car. These black boxes are called ECUs.

In this transition from cartoon to reality, Sebastian becomes the hybrid controller, which is the master, overall, in-charge controller. The battery ECU supplants both Temperature Tessie and Battery Bessie. In addition, the battery ECU monitors the resistance and temperature of every cell in the battery. A battery may have 200 cells. Also, values for voltage and current are provided. The function of each ECU will now be discussed.

DESCRIPTIONS OF ECUs

HYBRID ECU

The hybrid ECU is in command of all other ECU and selects the operational mode based on the driver's input, for example, how hard the gas pedal is pushed. Other information for selecting an operational mode includes the signals from subordinate ECUs, which give the status of components. Of special importance is the battery SOC.

The hybrid ECU is responsible for system-wide energy management. Typically, the goal of control is to minimize fuel consumption. Other performance attributes may be the goal. For each gallon

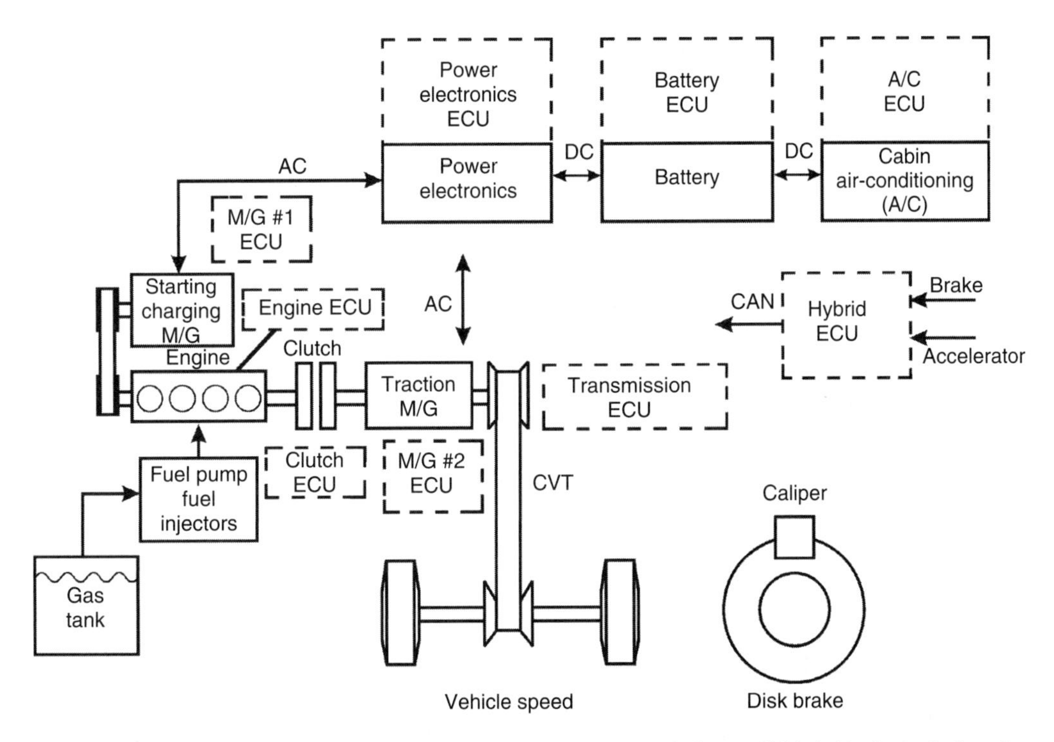

FIGURE 8.4 In serious engineering format the components and control of a parallel hybrid; also brake by wire.

of gas, the hybrid ECU tries to provide maximum mpg. To do this, the hybrid ECU allows or prohibits engine shutoff. The hybrid ECU commands the amount of torque and power from the motor and engine. Also, the amount and timing of power generation to charge battery is commanded.

Engine ECU

See the earlier discussion in connection with Figure 8.1, which presents the engine variables. Variables needing control included ignition timing, tuned intake manifold, etc.

M/G ECU

The M/G ECU provides for M/G safety and reliability based on M/G temperature and rpm. It responds to commands for switching from motor to generator and vice versa.

CVT ECU

The CVT provides the correct gear ratio to control the torques and angular speeds of the M/G and engine.

Power Electronics ECU

Having the power from a battery or a fuel cell is only the first step. The power must be delivered to the M/G (in motor mode) at the voltage and current needed. For regenerative braking, the power must be accepted from the M/G (in generator mode). The power electronics can be expensive, bulky, and hot. The lower the efficiency of the power electronics the greater the heating and the greater the cooling needs. Reliability is another problem to be solved. Cooling plates that use brazed aluminum construction and provide uniform temperature across the plate have been developed. Integral glycol cooling loops can be used.

The function of this ECU is to receive commands from hybrid ECU, to control inverter energy flow both ways, that is, charge and discharge, to control switching of M/G between motor and generator modes, and to control engine starting sequence. The engine must be started initially from cold-start to warm-up and after each idle-off event.

Even though having quite high efficiency, power electronics generate considerable heat making cooling essential. A function of the power electronics ECU is to monitor and protect the high-power transistors.

Battery ECU or Battery Management System

The battery ECU, also known as the battery banagement system (BMS), monitors and measures temperature and assures cooling is adequate. The BMS monitors each cell for voltage, current in/out, and temperature. The BMS avoids the stress of heat and overtemperature. By corrective action of the BMS, the effects of excessive charging or discharging are eliminated or lessened. The BMS is essential for long battery life and optimum mpg. The BMS works with the hybrid ECU.

Air-Conditioning ECU

For some hybrids the air-conditioning (A/C) compressor is driven by an electrical motor. In other hybrids, the A/C compressor is engine driven. In heavy stop-and-go traffic, the A/C ECU can request the hybrid ECU to start the engine to obtain power for driving the A/C compressor.

CONTROL AREA NETWORK

As shown in Figure 8.5, the communication connections between the ECUs are by means of the control area network (CAN). The CAN is a fast, high data rate network enabling communication between ECUs. Most data can be updated every 10 ms. Checking to assure data reliability is provided.

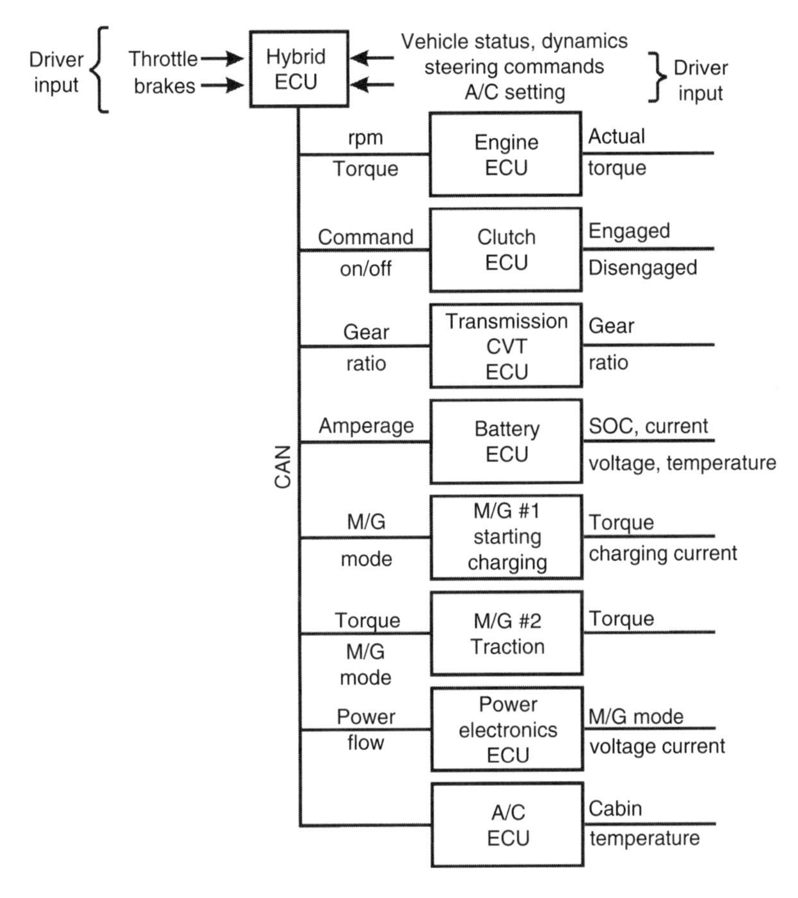

FIGURE 8.5 Connections between the various ECUs by means of the CAN.

CONTROL VARIABLES: VARIABLES CONNECTING THE ECUs

The quantity on the connecting wire is a digital string of ones and zeros. What does the string represent? What physical quantity is measured and acted upon? The control variables connecting the various ECUs fall into one of three categories: mechanical, electrical, and discrete. The control variables connecting various ECUs for mechanical components are torque, gear ratio for CVT, and the rpm of each component. The control variables connecting various ECUs for electrical components are torque, current in amperes, and voltage in volts.

Another more poetic way to inquire about the appropriate control variables is to ask what is the coin of the realm? Does the system use dollars, euros, francs, pesos, or yen to conduct business? By analogy, what coins (or variables) fit the realm (or control system)?

Discrete variables are like yes/no or on/off. Hybrids have a few such variables. These are

- M/G mode: motor or generator?
- Gear ratio: *n*-speed transmission—which of the *n* gears?
- Clutch: engaged or disengaged?

ENGINE MANAGEMENT

First identify the engine ECU in Figure 8.5. Now look at what goes on inside the engine ECU (Figure 8.6). The inputs and outputs are usually torque based. The scope of control is far beyond the control of just fuel injection and ignition timing. Many sensors are required; for example, the O_2

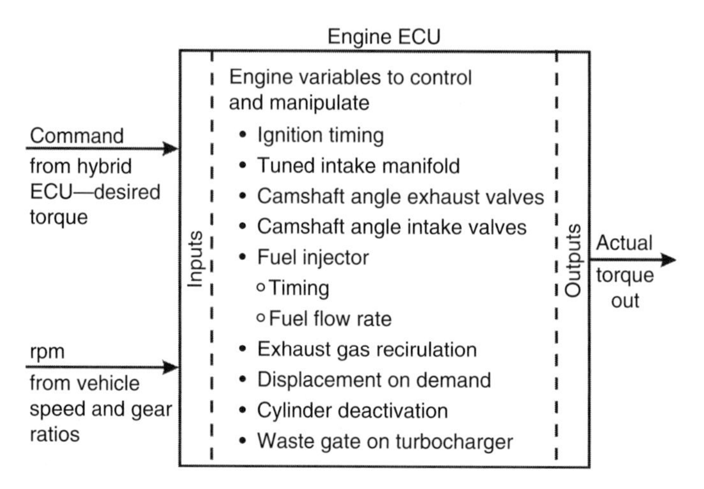

FIGURE 8.6 Engine management by the engine ECU.

Changes to Engine and Motor to Move to Optimum Point

Point	Engine Speed	Engine Torque	Change in Gear Ratio	Throttle Command	Engine Torque Change	Motor Torque Change
A	Low	Low	Lower	Push	Increase	Decrease
B	High	Low	Higher	Push	Increase	Decrease
C	High	High	Higher	Release	Decrease	Increase
D	Low	High	Lower	Release	Decrease	Increase

sensor provides data that permit precise control of the fuel/air ratio. EGR influences both emission levels and fuel economy (see the next section).

EXHAUST GAS RECIRCULATION

When combustion gases are mixed with the incoming air, the process is called EGR. Some features of EGR are

- Combustion gases have molecules with low values of dissociation energy; these molecules decrease combustion temperature.
- Decreased combustion temperature decreases NO.
- High EGR (15%) yields higher HC.
- High EGR may cause poor idling.
- EGR can be done internally by overlapping opening of the intake and exhaust valves.
- EGR can be done externally by connecting the intake and exhaust manifolds with a tube; an EGR valve is inserted to control percentage EGR.
- EGR increases the pressure in the cylinder during intake that decreases pumping losses.

ENGINE EFFICIENCY MAP

An engine efficiency map is frequently a plot of torque as function of engine speed, rpm. Note these are the same variables used to connect the various ECUs. Contours of brake-specific fuel consumption (BSFC), lb of fuel/brake horsepower (bhp) hour, are superimposed on the map. In the United States, BSFC has units of lb of fuel/bhp h. In Europe and Japan, fuel consumption is expressed as L/(kWh). Note the contours are fuel consumption so that a smaller number is better.

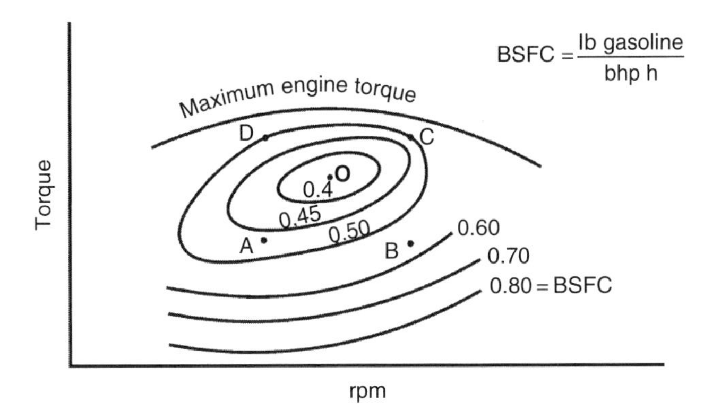

FIGURE 8.7 Engine efficiency map showing contours of BSFC. Optimum fuel economy occurs at point O.

Optimum fuel economy is at point O, which is at the center of BSFC = 0.4 contour. Points A to D are not located at the optimum location. Changes must be made to move engine operation from A, B, C, or D to point O as shown in Figure 8.7. The changes involve both engine and motor.

LOAD LEVELING

One simple approach to hybrid control is known as load leveling. Essentially load leveling is "keep the battery charged." The timing and amount of charging varies with the driving cycle. The concept is explained more by Figure 8.8.

In the crosshatched regions, the electric motor supplies power to meet instantaneous power demands of the driving cycle. The battery is being discharged. Areas below the average power are battery-charging events. At the end of the driving cycle, the battery SOC is the same as at the start. This is the goal of load leveling.

CONTROL COMPLEXITY AND DIFFICULTY

Control complexity and difficulty are somewhat analogous to the game of Sudoku. Sudoku puzzles range from very easy to very difficult (unsolvable?). Using a scale of Sudoku for reference, hybrid control complexity and difficulty are beyond very difficult. However, with enough resources, Sudoku and hybrid control can be solved.

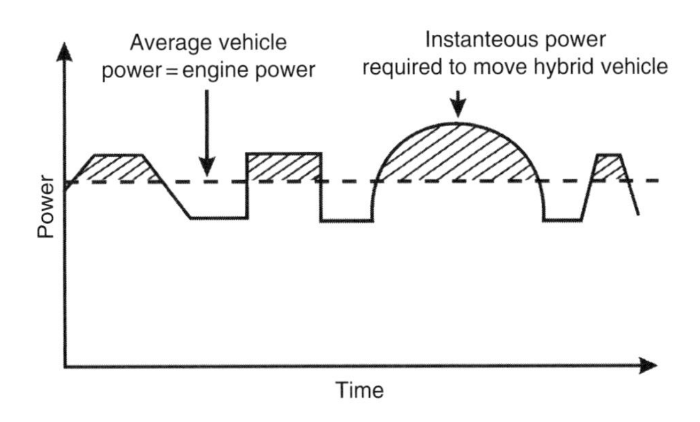

FIGURE 8.8 Concept of load leveling.

Why the difficulty? Hybrid control involves many variables; this is termed multivariable by the experts. Further, the problem is time-varying although an assumption can be made that time is not important. The words that make the analysts cringe are nonlinear. Many techniques have been developed for linear problems, but solution of nonlinear problems involves ad hoc methods at best. Even computer routines can be choked by nonlinearity.

Finally, multiple uncertainties exist. For example, the driving cycles are only models or representations of real stop and go. The effects of altitude and weather are other uncertainties. The engine efficiency map discussed above represents the average engine off the production line. Some engines are better, some are worse. It is difficult to find the optimum split between electric motors and gas engines.

ADAPTIVE CONTROL STRATEGY

Some control systems have the capability to adapt to the environment in which they operate. In the hybrid technical literature little mention, or none, is made of adaptive control. Hybrid control systems are in their infancy and have not reached maturity. Soon, however, the capability of hybrid control systems will be expanded to include adaptive control.

ROBUSTNESS OF CONTROL SYSTEM

Control systems have an attribute known as robustness. As the name suggests, inputs outside the normal design boundaries do not cause the control system to spin out of control. The system has reserve capability to handle the unexpected. A certain robustness of the controller allows tolerance to disturbed signals and the temporary malfunction of components.

FOUR CONTROL STRATEGIES

Of the four strategies, a rules-based control algorithm is the least complex. An example is discussed shortly. Optimization techniques as applied to hybrid control fall into two categories based on time. Static optimization is independent of time. Frequently, static optimization becomes map-based optimization; maps of component efficiency are used as a guide to share power between engine and motor. Dynamic optimization, which is time dependent, introduces a new variable, time. Introduction of time, along with all the other variables, significantly increases complexity.

To understand the importance of static and dynamic optimization, refer to Figure 4.2. Focus on the heavy curve that shows the required power for the three driving modes. Where the curve is horizontal, the situation is static, and time is not important. The curve is horizontal for cruising on the freeway and in the latter part of climbing a hill. Static optimization yields good results. Where the curve is sloping, that is, accelerating from a stop sign and early part of hill climbing, dynamic optimization is essential to get best performance. With static optimization, the best management of a hybrid does not occur. For the driving cycle of Figure 4.2, static optimization may yield 35 mpg while dynamic optimization yields 39 mpg. Any transition between driving modes is dynamic.

Time-dependent problems typically have a characteristic timescale. This timescale, which is an inherent part of the problem, varies from microseconds to eons. Based on typical 0–60 times, the timescale for acceleration is seconds. Battery performance decays over a span of years; hence, the characteristic time is measured in years. The strategy used for control must be able to discern and manipulate control variables with timescales from milliseconds to years.

The intelligent control strategies utilize techniques developed in other contexts and then adapted to hybrid control. These are fuzzy logic control (FLC), neural network (NN), and combined FLC and NN. Each technique is discussed.

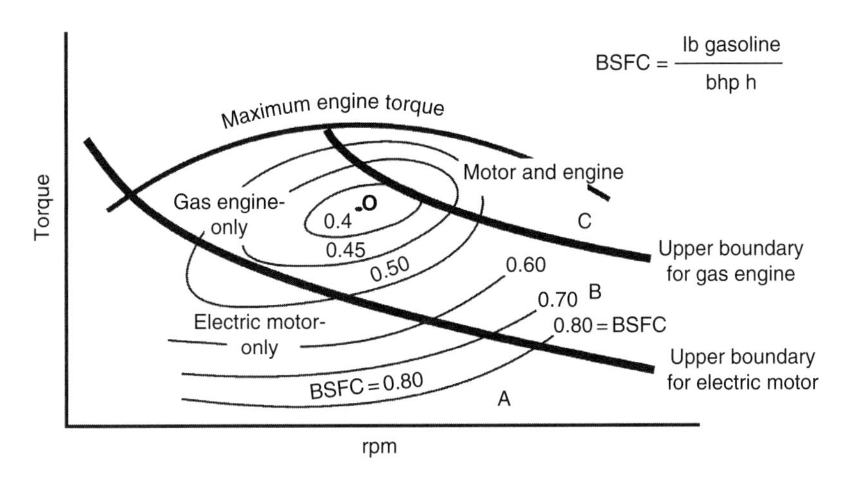

FIGURE 8.9 Rules-based control strategy using an engine efficiency map.

RULES-BASED CONTROL ALGORITHM

Rules-based control algorithms can cope with the various operating modes of a hybrid. Three typical modes are the normal mode, typically, cruising, acceleration, and hill climbing; charging mode; and regenerative braking mode.

The rules-based algorithms are developed using engineering insight and intuition, analysis of the engine efficiency charts (Figure 8.7), and the analysis of electrical component efficiency charts. Control using rules-based algorithms is simple but does not give as good results as intelligent control strategies or optimization.

The map of Figure 8.9 has two bold lines drawn across the map. The lines, which are drawn using engineering insight and intuition, divide the map into three regions: A, B, and C. In region A, only the electric motor is used. In region B, only the gas engine is used, while in region C, which requires high power, both the electric motor and the gas engine are used.

Contours for higher engine efficiency are located in region B between the bold lines. The bold lines are drawn to include as much of the low fuel consumption region as reasonable and exclude the areas of high fuel consumption.

OPTIMIZATION

Optimization is a formal mathematical procedure that involves

- Objective (cost) function
- Constraints
  - Inequality, e.g., maintain battery $SOC_{max} > SOC > SOC_{min}$
  - Equality, e.g., engine + motor torque = required torque to move vehicle

The goal of optimization is to minimize or maximize the objective (cost) function. For the case of hybrid, the usual goal is to minimize fuel consumption. The cost function is

$$\text{Cost function} = \text{fuel consumed} = \sum\left(\frac{\text{fuel}}{\text{time}}\right)(\text{time spent in each cell})$$

The symbol $\sum$ means to sum, or to add up, the contribution from each cell in Figure 8.10.

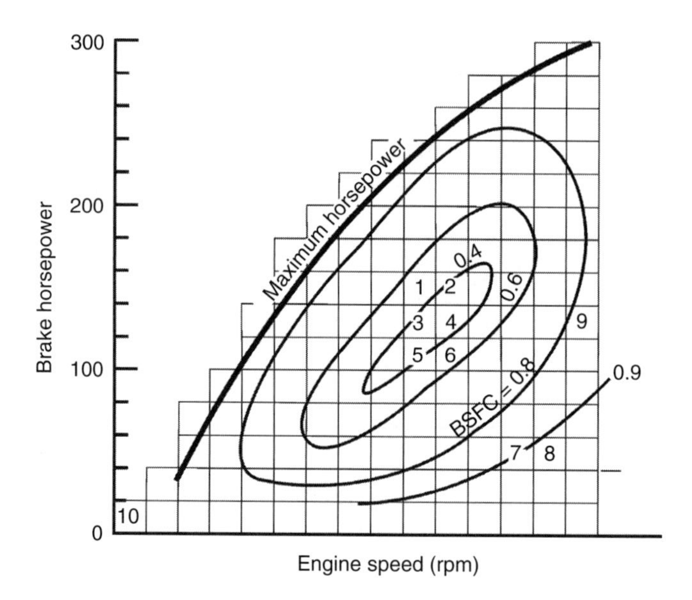

FIGURE 8.10 Engine efficiency map using bhp, and engine speed (rpm), with contours of BSFC. The grid overlay defines cells used for optimization. As discussed in the text, cell 4 is good and cell 9 is not so good.

Figure 8.10 is an engine efficiency map with a grid overlay; the overlay creates cells. Each cell has with it a bhp. For example, cell 4 has a power of 125 bhp. The fuel used/time is

$$\frac{\text{Fuel}}{\text{Time}} = \text{BSFC} \times \text{bhp}$$

For cell 4, this has a value

$$\frac{\text{Fuel}}{\text{Time}} = (0.4 \text{ lb/bhp h}) \times (125 \text{ bhp}) = 50 \text{ lb/h}$$

In contrast, for cell 9 this has a value

$$\frac{\text{Fuel}}{\text{Time}} = (0.81 \text{ lb/bhp h}) \times (125 \text{ bhp}) = 101 \text{ lb/h}$$

Consequently cell 9 is a very expensive cell compared to cell 4. Hopefully, cell 9 can be avoided and almost all time spent in cells 3 and 4. In fact, cell 9 is 101/50 = 2.0, or twice as expensive as cell 4.

The other item appearing in the cost function is "time spent in each cell." In fact, that is the dominant aspect that optimization determines. Stated another way, the secret of optimization is to spend most of the time in cell 4; however, because of demands of the driver, this may not be possible. What is the best way to divide engine and motor power to spend most of the time in or near cell 4?

Think of a driving cycle which encompasses stop–go, traffic jams, acceleration from traffic light, and an occasional panic stop. A constraint in this optimization was

$$\text{Engine torque} + \text{motor torque} = \text{required torque to move vehicle}$$

For this reason, the split between engine and motor torque varies from instance to instance. Because of the changing nature of the driving cycle, the time spent in each cell will be distributed all over the map. Obviously, at a traffic jam the engine should be turned off, and engine operation moves to cell 10. For modest acceleration, how much power from the engine and how much from the motor are needed so as to keep the engine operating in cells 3 and 4 or even 5 or 6? Optimization provides the answer.

HARDWARE-CONSTRAINED OPTIMIZATION

Honda IMA hybrids have the motor connected to the engine. Consequently, the motor and engine must turn at the same rpm. This fact simplifies, but also limits, the optimization of the hybrid control unit.

FUZZY LOGIC CONTROL: WHAT IS IT?

FLC is a method of computing precise results, called crisp results, from imprecise inputs, called fuzzy inputs.

- Crisp values are precise values.
- Fuzzy values are full or partial members of a range of numerical values.

FLC is a logical system that generalizes the classical two-valued logic for reasoning under uncertainty. FLC has already been applied successfully to control powertrains, engine, and transmissions. Currently FLC is being studied for hybrid control. FLC is so complicated that some researchers have devoted their career to studying FLC.

FLC can use logic rules such as

If X, then Z.

An example occurs when you are taking a shower.

If shower water is cold, then turn on more hot water.

Other logic rules are

If X and Y, then Z
If X or Y, then Z
If X not Y, then Z

An example concerns engine speed control.

If engine rpm is too low and rpm is getting lower, then inject more fuel.

The logic rules are not precise statements but are a verbal, fuzzy statement of desired action. FLC has the tools and procedures to convert imprecise fuzzy logic statements to quantitative results.

FUZZY LOGIC CONTROL APPLICATION TO HYBRID CONTROL

The energy management and control strategy using FLC should maximize fuel economy, minimize emissions, and distribute the driver's request for power between two sources: engine and motor. Further, FLC should maximize fuel economy at any point in operation, that is, provide dynamic or instantaneous optimization. If required, FLC should maximize some other attribute such as acceleration of vehicle.

Finding the optimum power split between engine and motor is difficult because of the reasons cited earlier, which are multivariable, time-varying, nonlinear systems with multiple uncertainties. FLC is well suited to do the job of defining the logic split [2]. The FLC consists of many rule-based logic statements in the form of "If X, then Z." The job of the controller is to calculate to what degree each rule is satisfied.

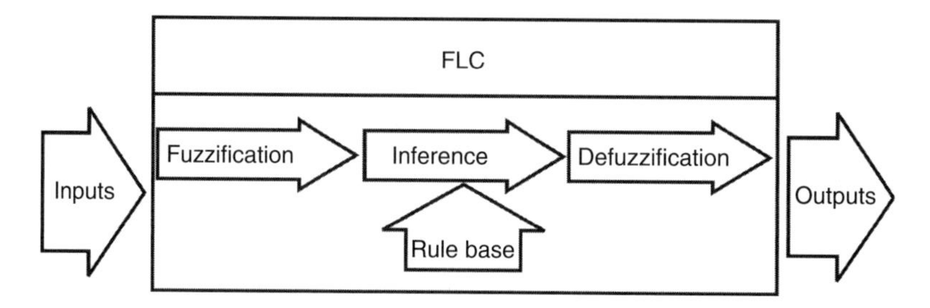

FIGURE 8.11 Structure of an FLC. (Reprinted with permission from SAE Paper *Electric and Hybrid-Electric Vehicles*, PT-85, 2002 SAE International.)

As shown in Figure 8.11, within the FLC there are four processes:

1. Fuzzification, which is the change from crisp values to fuzzy values. For engine speed, the crisp value might be 2000 rpm. An earlier example was

   If engine rpm is too low, then inject more fuel.

   The fuzzy value associated with the "If X, then Z" statement would be for X, rpm < 2000.
2. Rule base has a collection of rules; several hundred rules may be developed and applied. Continuing with the example, FLC has as one of the rules "If X and Y, then Z" stated above.
3. Inference applies the defined rules to the inputs. For the example, if the input rpm = 2200, then the rule does not apply; however, for an input of 1800 rpm, the rule is satisfied.
4. Defuzzification transforms the results of the inference process to crisp outputs.

NEURAL NETWORK CONTROL: WHAT IS IT?

Everyday (a single central processing unit, CPU, or rarely a few CPUs) computers are especially good at fast arithmetic and doing precisely what the programmer programs them to do. Everyday computers are not so good at interacting with noisy data and interacting with data from the environment. Everyday computers do not handle massive parallelism, which is many computers, each with a CPU, simultaneously solving the problem. The usual computer has difficulty with fault tolerant computing and adapting to circumstances.

In contrast, NN computing is useful where an algorithmic solution cannot be formulated and where lots of examples of the desired behavior are available. NN computing is especially valuable where a need exists to discern structure from existing data.

NN, which are a different paradigm for computing, are based on the parallel structure of animal brains. NN possesses the following properties:

- Simple processing elements
- High degree of interconnection
- Simple scalar messages
- Adaptive interaction between elements

This section is based on Ref. [4].

NN CONTROL: APPLICATION TO HYBRID CONTROL

When applied to the control of a hybrid vehicle, NN control allows both input/output representations of complex, nonlinear phenomena. NN computing weighs an input by some factor, W, which is a simple scaling number, for example, $W = 2.2$. If in parallel computing there are 10 inputs, then all 10 inputs interact with each other using 10 computers. Training data are input, and NN learns by adjusting the weighting factors, *W*, until the desired output is achieved. In other words, NN have the capability to learn. After training, real data are then inserted and output results are obtained.

One suitable application for NN computing is your income tax. The income tax return has Form 1040, Schedules A, B, C, and D. These items are linked; a change in one generates a change in another. Each form or schedule could be calculated in parallel, which is a feature of NN computing. For purposes of training NN, prior-year tax returns could be entered. From a pragmatic point of view, the customary single computer method is entirely adequate. NN is much too powerful even if it is appropriate.

COMPARISON OF FUZZY LOGIC AND NNs FOR CONTROL

Table 8.1 provides a comparison of FLC and NN control giving the advantages and disadvantages of each. See [5] for more details.

COMBINED FUZZY LOGIC AND NNs FOR CONTROL

A study of the Table 8.1 indicates that many disadvantages of either control system are eliminated by combining the two approaches to control. Five benefits accrue by the combination:

1. Model-free approach. An explicit model of the plant is not needed; however, qualitative knowledge about the overall system must be provided.
2. Adaptability. The system can modify itself in order to consider varying drivers and driving cycles as well as changing characteristics of the plant, for example, an aging battery.
3. Tolerances against faults and uncertainties. A certain robustness of the controller allows tolerance to disturbed signals and temporary malfunction of components.
4. Nonlinearity. Supervisory control requires taking the nonlinear behavior of overall system and its components into account.
5. Real-time operation. Systems that are mainly experiencing transient operations make real-time control unavoidable.

TABLE 8.1
Comparison of Features of Fuzzy Logic Control with NN Control

Advantages	**Neural Networks**	**Fuzzy Logic Control**
	No mathematical model of plant needed	No mathematical model of plant needed
	Learning ability	Knowledge representation
	Generalization and association ability	Uncertainty tolerance
	Uncertainty tolerance	Fault tolerance
	Various models and parametric approaches	Simple design of systems and rule bases
	Real-time operation	Expert knowledge
	Optimization ability	Good hardware implementation
	Nonlinearity	Real-time operation
	Fault tolerant	Nonlinearity
Disadvantages	**Neural Networks**	**Fuzzy Logic Control**
	No knowledge representation	No learning ability
	No expert knowledge	Problems with changing an existing system
	Convergence problems	No optimization methods
	Severe hardware and software requirements	

ELECTRICAL COMPONENTS: ELECTRICAL MOTORS AND GENERATORS

For electrical motors and generators, the following data are needed: efficiency maps, torque as function of input current and rpm, and heat dissipation rate. Efficiency maps usually have contours of efficiency superimposed on a graph using rpm as abscissa and torque as the ordinate.

For some efficiency maps, the part of the map with positive torque represents the motor while the negative torque region is the generator.

SUMMARY

Success of a hybrid in the marketplace depends strongly on smoothness of transitions between operational modes. That smoothness in turn hinges on the control system. To repeat the opening thought, all functions must engage and disengage smoothly without jerks, shudders, shakes, or clanks. The analogy of hybrid control being similar to the conductor of a symphony highlights the requirement for a pleasant driving experience. Discordant notes and unpleasant shudders repel the customer.

Control theory, that is, the basic mathematics, has evolved since World War II; the war gave tremendous impetus to control theory and practice. Parallel to advancements in control theory has been the capability of computers. In the automotive world, requirements for pollutants drove engine controls starting in the mid-1960s and continuing into the twenty-first century. As advances are made in other fields, these advances find a way into hybrid control. Examples are FLC, optimization, and NNs. As the title to this chapter states, hybrid control plays a vital role.

REFERENCES

1. T. Costlow, Taking Control of Hybrids. *Automotive Engineering International*, Society of Automotive Engineers, Warrendale, PA, November 2007, pp. 29–30.
2. B. Baumann, G. Rizzoni, and G. Washington, Intelligent control of hybrid vehicles using neural networks and fuzzy logic, *Electric and Hybrid-Electric Vehicles*, Society of Automotive Engineers, PT-85, Warrendale, PA, R. K. Jurgen (Ed.), 2002, p. 257.
3. B. Baumann, G. Rizzoni, and G. Washington, Intelligent control of hybrid vehicles using neural networks and fuzzy logic, *Electric and Hybrid-Electric Vehicles*, Society of Automotive Engineers, PT-85, Warrendale, PA, R. K. Jurgen (Ed.), 2002, pp. 261–263.
4. L. Smith. An Introduction to Neural Networks. Available at lss@cs.stir.ac.uk.
5. B. Baumann, G. Rizzoni, and G. Washington, Intelligent control of hybrid vehicles using neural networks and fuzzy logic, *Electric and Hybrid-Electric Vehicles*, Society of Automotive Engineers, PT-85, Warrendale, PA, R. K. Jurgen (Ed.), 2002, p. 262.
6. A. Bornhop, Long Term Test Lexus GS 450h. *Road & Track*, Newport Beach, CA, February 2008, pp. 84–85.

9 Regenerative Braking

With regenerative braking, your friction brakes should last the lifetime of your hybrid.

INTRODUCTION

Function of Regenerative Braking

The obvious function of braking is to stop or reduce vehicle speed. In addition, brakes must operate so as to maintain directional stability. The brakes must avoid locked wheels that have an adverse effect on steering and stability.

For regenerative braking, the goal is to recover as much vehicle kinetic energy (KE) as possible. Regenerative braking is not a new concept. The hybrid vehicles of the 1900s used regenerative braking. Regenerative braking has been used in trolleys and street cars for 100 years; the generated power goes back into the power line.

Integration of Functions and Components

Regenerative braking requires special sensors and other equipment. This equipment, with slight modification, can be used for the antilock braking system (ABS), traction control, stability control, and brake force distribution by the automatic control of braking forces at each of the four wheels.

Logic Relative to Size of Regenerative Brakes

Brakes are a critical component for safety. A fail-safe mode requires 100% braking capability; hence, friction brakes must be as large as for a conventional car without regenerative braking. Peek under a hybrid—the brakes are as big, if not bigger, than a conventional car. As subsequent discussion will show, mechanical, or friction, brakes do 90% of the stopping in a panic stop.

If regenerative braking were 100% efficient, friction brakes would be unnecessary. During normal driving, much of the braking load is carried by regenerative braking; hence, friction brakes should last very long.

Brake Cooling, Aerodynamic Drag, and Regenerative Braking

The design of an automobile is a very complex process involving many different skills and talents. Frequently the design process leads to compromises based on judgment and experience. Occasionally the demands combine favorably and reinforce each other. One such example concerns brake cooling, aerodynamic drag, and regenerative braking.

Fancy chrome wheels with shiny spokes do provide ample cooling for friction brakes. However, these wheels add to the overall drag of the vehicle. The flat plate hub caps, such as those shown in Figure 1.6, offer lower drag with increased mpg. With the flat hub caps, brake cooling may be a problem. Now enter regenerative braking, which markedly reduces the amount of energy dissipated as heat by friction in the brakes. Combining the three gives low drag, cool brakes, and increased mpg without compromise.

BRAKES AND TIRES PRIMER

WHY DISCUSS TIRES?

Almost all forces acting on the vehicle originate in, or are transferred to, the tire patch on the road. Forces between the tire and the road are essential for control of the car. Knowledge of tires is necessary to understand braking, for example, wheel lock. Regenerative braking adds new dimensions as the braking tasks are shared with the friction brakes. Regenerative braking must avoid locking any wheel, especially the rear wheels, and have a fail-safe mode of operation.

SLIP, SLIDE, AND SKID

The three words slip, slide, and skid describe how forces (relative to tires) are developed and how tires behave. Slip is depicted in Figure 9.1. Without slip, one revolution of the tire moves the wheel a distance equal to the circumference. However, with slip, the distance moved by one revolution is greater when braking (Figure 9.1a). Slip during braking is due to stretching of the tire. When braking, the desired result is to move the minimum distance; naturally, slip increases the distance.

Think of a rubber band. Make two marks on the unstretched rubber band. Now stretch the rubber band. The distance between the marks is obviously greater. The same concept applies to the slip of tires. Without slip, forces between the tire and the road cannot be developed. No slip, no force.

When the vehicle is moving forward with traction, the rubber in the tire patch is compressed. The compression causes slip. With slip, the distance moved by one revolution is less when tire provides traction (Figure 9.1b). Once again, the desired result is lessened due to slip. Traction means you want to move along the highway. With slip, the distance is decreased.

Return to your rubber band with two marks. Stretching is easy, but compression is harder. When rubber band is compressed, the distance between the two marks is less.

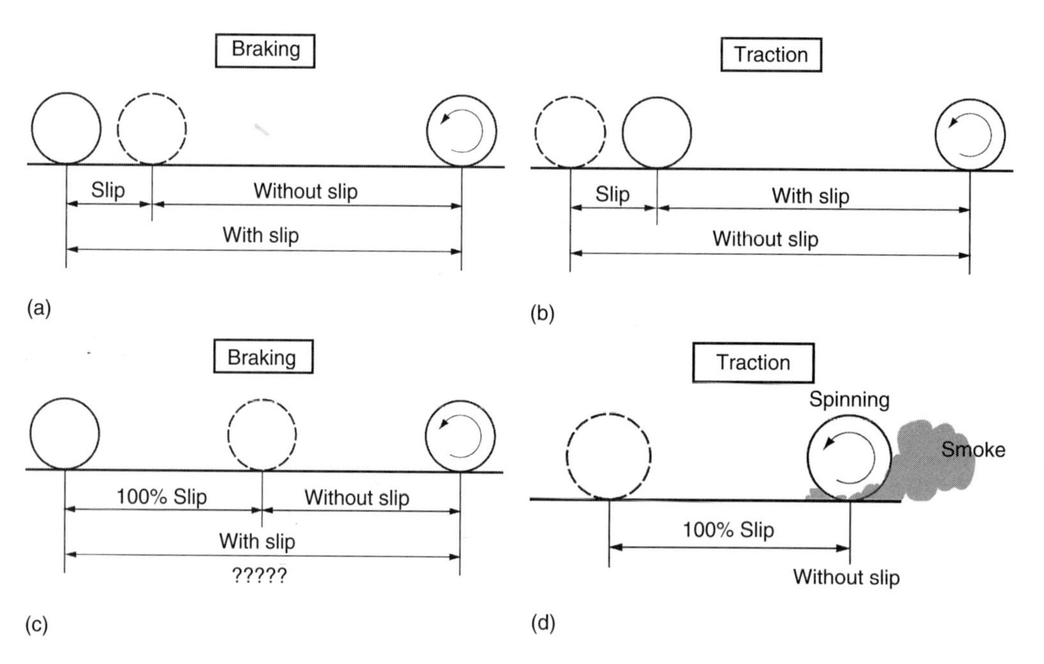

FIGURE 9.1 Slip affects both braking and traction. (a) Slip due to braking. Rubber in tire tread is stretched causing slip. (b) Slip due to traction. Rubber in tire tread is compressed causing slip. (c) Distances for 100% slip. For traction the wheel does not move. For braking, the car moves an unknown distance with the wheel locked. The series of ????? in the drawing suggest the unspecified distance. This distance can be calculated, of course. (d) For traction with 100% slip, the car does not move at all. The tire is being shredded as it generates copious smoke.

Slip spans from 0% to 100% as shown in Figure 9.1. Sliding implies relative motion between the road and the tire patch. For small slip, the tire sticks to the road without sliding. Slip includes both sticking and sliding. For slip between 0% and about 10%, which is on the linear part of the curve in Figure 9.5, the tire sticks to the road. For slip between 10% and 30%, portions of the tire patch toward the rear are sliding and other regions within the patch stick. At larger slip, the tire patch slides over the road. Skid applies to braking and is used in the same sense as slip.

Slip equal to 100% does occur either in traction or braking; conceptually the result of 100% slip is interesting. For braking, as shown in Figure 9.1c, the wheel is locked. The tire slides to a stop at an unspecified distance. When stopped, the location of the wheel is not the same each time. The wheel stops when all KE has been dissipated. For braking, this is not the desired trend.

For traction, 100% slip can occur. With a powerful car, at launch, the car can just sit there with the tire spinning. A smoky launch is a sign of masculine virility.

Relation between Tire Force and Weight on Tire

The force on the tire depends on the weight being carried by the tire.

$$F = \mu W \tag{9.1}$$

where

F is the force on the tire, pounds or newtons
μ is the friction coefficient, dimensionless
W is the weight on the tire, pounds or newtons

The force on the tire is also the force on the road surface. The two forces are equal and opposite. The value of the μ, the friction coefficient, changes for different road surfaces, such as dry, covered with ice or snow, etc. The friction coefficient depends on the tire design.

Weight on Each Wheel

The weight on each wheel is determined by total vehicle weight and the location of the center of gravity (CG). Figure 9.2 shows various factors affecting weight on front and rear tires. Equations are given to calculate the weight on front and rear tires. Available braking force takes into consideration the weight on each tire. Also vehicle stability and traction forces depend on the weight on each tire.

The total weight of the car is the sum of the weights on each wheel; thus

$$W = W_{F0} + W_{R0} \tag{9.2}$$

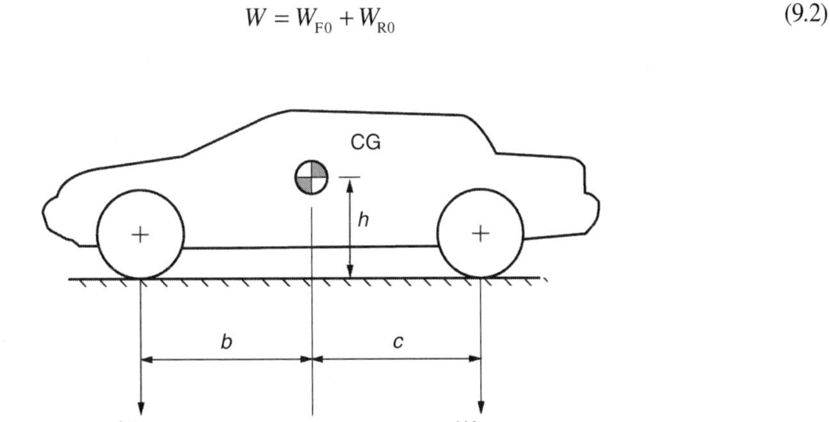

FIGURE 9.2 Weight on front and rear tires. Weight affects both braking and traction capability.

The subscripts F and R refer to front and rear while 0 refers to the car at rest; there are no dynamic loads due to braking or otherwise. The weight on the front is

$$W_{\mathrm{F0}} = \left(\frac{c}{b+c} \right) W \tag{9.3}$$

The weight on the rears tires is

$$W_{\mathrm{R0}} = \left(\frac{b}{b+c} \right) W \tag{9.4}$$

The equations above are identical to the balance of a teeter-totter board. Note two aspects. First, adding Equations 9.3 and 9.4 leads to Equation 9.2. Second, the distance c, which is defined in Figure 9.2, is associated with the front. So if c is large, then the weight on the front is also large.

Braking "Dive": Shift of Weight to Front Tires

Braking causes a shift of weight from the rear to the front tires. Because of this, the use of front wheels only may be adequate for a regenerative braking system. The shift of weight tends to unload the rear tires; unloading the rear tires is not desirable because of the loss of directional stability. See Figure 9.3 for identification of symbols for deriving the shift of weight.

The braking force, F_{B}, illustrated in Figure 9.3, is related to the acceleration (actually deceleration), a, and the mass of the car, m. The equation is

$$F_{\mathrm{B}} = am \tag{9.5}$$

An additional torque is created by the force, F_{B}.

$$T = F_{\mathrm{B}}\, h \tag{9.6}$$

The height of the CG above the road surface, h, is shown in Figure 9.3. The torque, T, is balanced by a change in weight on the tires.

$$T = bw_{\mathrm{F}} + cw_{\mathrm{R}} \tag{9.7}$$

The symbols are defined in Figure 9.3. As before

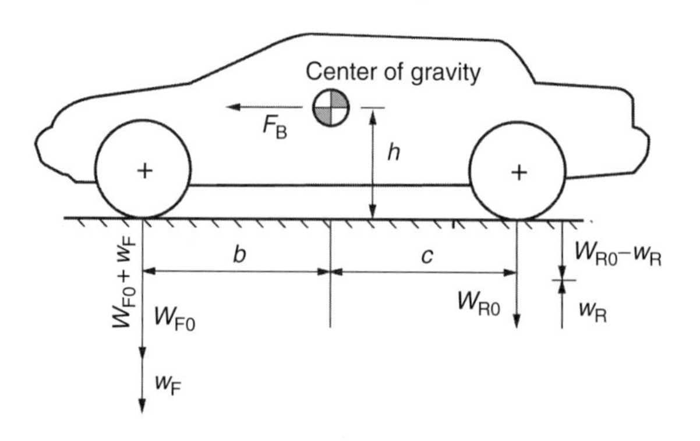

FIGURE 9.3 Change of weight on front and rear tires because of braking.

$$W = W_{F0} + w_F + W_{R0} - w_R \tag{9.8}$$

Subtracting Equation 9.2 from the previous equation yields

$$w_F = w_R \tag{9.9}$$

Combining Equations 9.7 and 9.9 leads to

$$T = (b + c) w_F \tag{9.10}$$

The change in front tire load can be found by combining Equations 9.6 and 9.10

$$w_F = \frac{F_B h}{b + c} \tag{9.11}$$

Equation 9.11 gives w_F in terms of known quantities. Using Equation 9.9, w_R can be found.

Tire Patch

The tire patch is the footprint on the road. It is the surface in contact with the road (Figure 9.4). Even though the tire patch is not much larger than the palm of your hand, every time you drive your car, your safety and well being depends critically on that surface of contact. Control forces, such as braking, depend on road–tire interactions in the tire patch.

Longitudinal Tire Forces

Tire forces are divided into longitudinal and lateral. Longitudinal tire forces are along the tire tread. Longitudinal forces are relevant for braking and traction (acceleration). Design of braking, stability, and traction control systems uses near maximum longitudinal friction forces.

Longitudinal Tire Forces: Traction

Longitudinal forces depend on whether the tire is being used in braking or traction. The value of friction coefficient, μ, depends on slip. For small values of slip, μ varies linearly with increasing slip

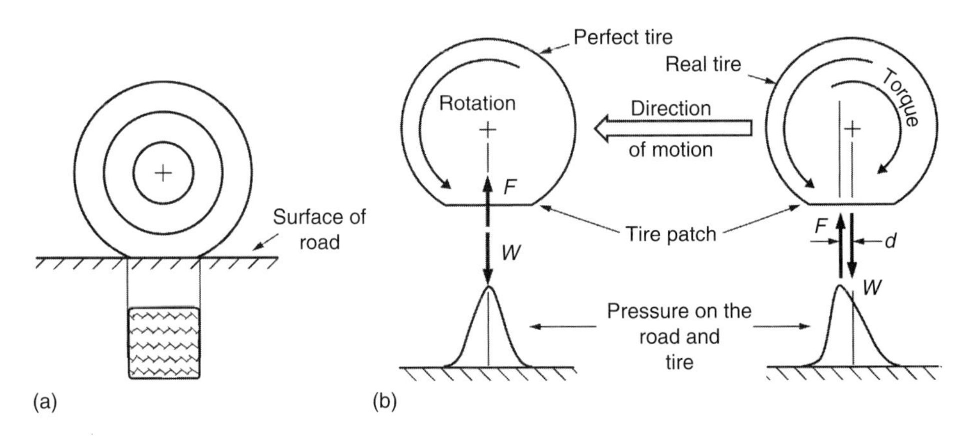

FIGURE 9.4 Tire patch. (a) Surface in direct contact with the road and (b) origin of rolling friction because of tire hysteresis.

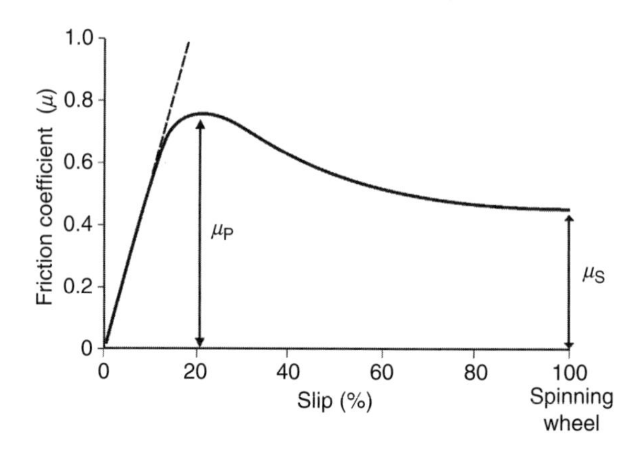

FIGURE 9.5 Traction friction coefficient as a function of longitudinal slip of the tire.

up to about 10%. Beyond that, the μ curve falls over reaching a peak value μ_P at about 20% slip. At 100% slip, which is a spinning wheel without car motion, the value of μ_S is about 0.4–0.5 for the tire and road surface (Figure 9.5).

Longitudinal Tire Forces: Braking

The variation of friction force, or the variation of μ, is shown in Figure 9.6 for three road surfaces. As the braking force increases, the friction force rises linearly with increasing slip. As with traction μ, the curves bend over reaching a maximum value. On ice, the tire breaks through maximum friction force very quickly making driving on ice very treacherous. For dry road, the loss of friction force from peak value to locked condition is only about 20%, which does not seem like much, but has pronounced effect on stability.

Both locked and rolling tires are shown. The slope of initial linear part of curve affects design of ABS.

Deceleration or acceleration can be measured in G, which is the ratio of actual acceleration, a, measured in m/s^2, to the acceleration of gravity, g, where $g = 9.81$ m/s^2. The friction coefficient, μ, is related to G in a very simple manner. This relationship is

$$F_B = \text{Braking force} = m\,a \tag{9.12}$$

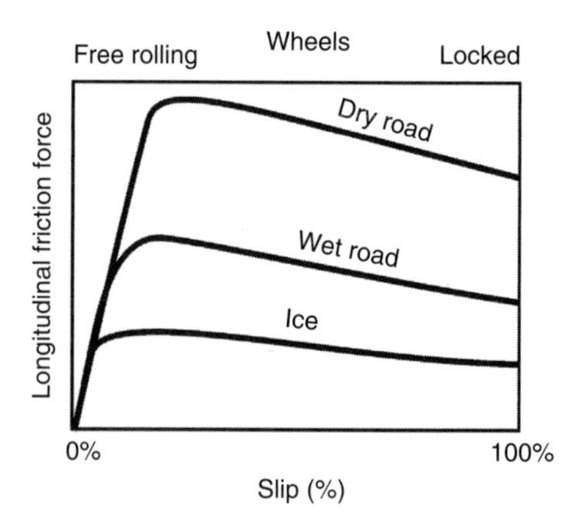

FIGURE 9.6 Friction force because of braking as a function of longitudinal slip for three different road surfaces: dry road, wet road, and ice.

In terms of μ, the braking force is

$$F = \mu W = \mu m a \tag{9.13}$$

The mass of the vehicle is m, kg, the weight of the vehicle is W, N or lb

Setting Equation 9.12 equal to Equation 9.13 gives

$$\mu = \frac{a}{g} = G \tag{9.14}$$

The maximum deceleration, measured in Gs, equals the maximum value of μ from Figure 9.6.

Locked Wheels

Locked wheels cause significant reduction in vehicle control. Locked wheels have lower friction forces. Locked rear wheels lead to loss of control, whereas locked front wheels result in significant loss of steering control. Figure 9.6 shows friction force reduction because of locked wheels during braking. Figure 9.7 has a curve of lateral force for locked wheels. Compare with the rolling tire.

Lateral Tire Forces

Lateral tire forces are important for both steering and cornering. In Figure 9.7, slip angle is defined by the sketch on the left-hand side. Slope of the curve at zero slip angle is C_α and is known as cornering stiffness. Both locked and rolling tires are shown. Typically, new tires are less stiff than old tires. Also tread design plays a role; lug tires are less stiff than rib tires.

Directional Stability

Figure 9.8 has two drawings. On the left-hand side, the torque about the CG is shown for rear tires only. For directional stability, torques that tend to decrease yaw angle are favorable. The rear wheels provide stabilizing torques. On the right-hand side, the torques due to the front wheels are shown. The front tires produce the destabilizing torques that tend to increase yaw angle.

A balance exists between the two torques from the front and rear wheels. If the rear wheel torque is reduced by lower tire forces, the vehicle may become unstable. As noted in the preceding discussion, locked wheels reduce tire forces.

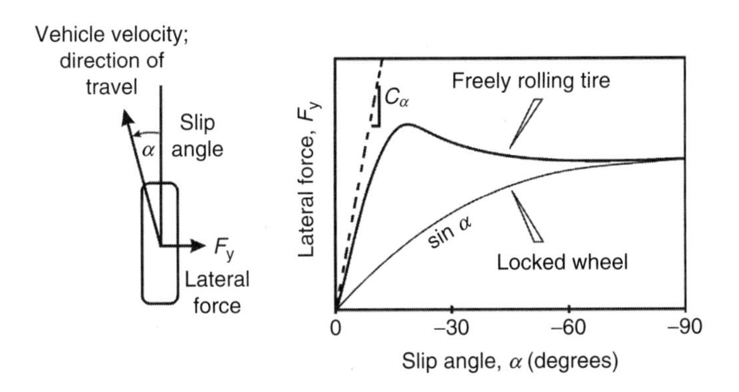

FIGURE 9.7 Lateral force for a tire as a function of slip angle. C_α is the initial slope for the freely rolling tire and is known as cornering stiffness. The lateral force of the locked wheel varies as sin α as shown.

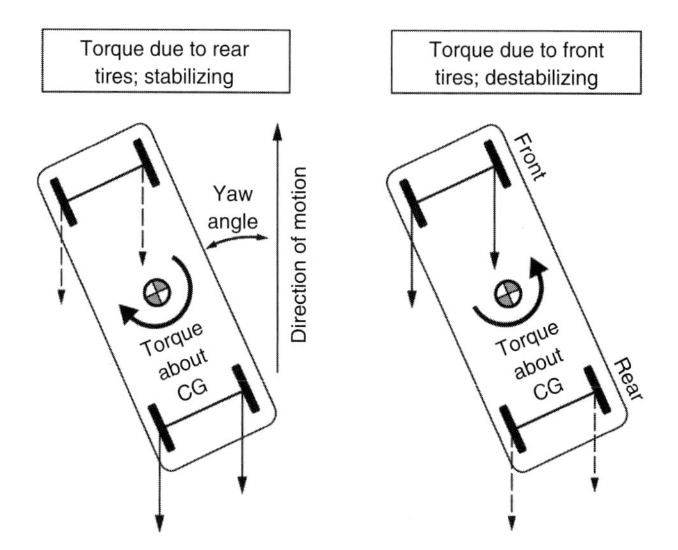

FIGURE 9.8 Effect of front and rear tires on directional stability. Front tires create a torque about the CG that tends to increase angle of yaw. Hence, front tires cause directional instability. Torque about the CG because of the rear tires tends to decrease yaw angle. As a result, forces on rear tires are essential to directional stability.

Antilock Brake System

Antilock brake system (ABS) can be easily integrated with regenerative braking. Look ahead to Figure 9.15, which has a wheel angular velocity sensor, called an ω-sensor, in the upper left-hand corner. This ω-sensor is needed for both regenerative braking and ABS. The aim of ABS is to prevent wheel locking due to over zealous brakes. Wheel angular velocity is measured for all four wheels: ω_{LF}, ω_{RF}, ω_{LR}, and ω_{RR}. The subscripts LF, RF, LR, and RR represent left front, right front, left rear, and right rear.

Each of the four angular velocities is compared with a reference angular velocity. The reference velocity can be the average of all four measured values. A reference angular velocity can be deduced from wheel radius, *r*, and vehicle velocity, *V*

$$\omega_R = \frac{V}{r} \tag{9.15}$$

Using ω_R has two potential difficulties. First, any change in wheel size upsets the calculations until a new radius is determined. Second, low-tire pressure effectively decreases radius, *r*, giving a false reference angular velocity and delays onset of ABS.

Regenerative "Coasting"

Regenerative braking returns a fraction of vehicle KE back to the battery. The term regenerative coasting is rarely used. However, slowing down and coasting down a hill may provide much more recharging of the battery than regenerative braking. Table 9.1 shows the relative amount of energy available for three conditions.

Energy to Be Recovered by Regenerative Braking

Table 9.2 lists the energy to be recovered, assuming 100% efficient regenerative braking, for three different vehicle velocities. Also shown is the fraction of battery energy that is restored. Since KE

TABLE 9.1
Regenerative Coasting Showing Relative Battery Recharge by Various Operational Conditions of Hybrid Vehicle

Recharge Battery by	Relative Amount of Charge
Slowing with regenerative brakes from 30 mph to 0 mph	1
Slowing without friction brakes 60 mph to 30 mph	3
Descending a hill; 1000 ft change in elevation	33

varies as the square of the velocity, 60 mph has four times as much energy as 30 mph, and 90 mph has nine times as much energy. Most regenerative braking is effective near the 30 mph region. Two factors limit recovery of vehicle KE. First, the battery may not be able to accept the energy because of very high state of charge (SOC) or excessive temperature. Second, as the vehicle slows, the generator cannot provide enough voltage to exceed battery voltage. After this point, KE cannot be stored in the battery. A new DC–DC converter is being developed that is positioned between M/G and energy storage. It provides a better matching between the M/G and the battery, greater efficiency, and better hybrid packaging. In addition to greater energy to be recovered from regenerative braking, the converter provides better acceleration in motor assist. Further a limp-home mode is incorporated [4].

Bands of SOC for Battery

Battery life imposes some strict reductions in available energy. When the battery has deep discharge, that is, SOC near zero, the number of charge/discharge cycles is severely reduced. Figure 9.9 illustrates four important levels of SOC. There are four design decisions: that is, values for SOC_{MAX}, SOC_1, SOC_2, and SOC_{MIN}. Usually SOC_{MAX} equals 100%. SOC_2 is selected to allow energy storage due to regenerative braking. The band between SOC_{MAX} and SOC_2 is for regenerative braking.

The band from SOC_1 to SOC_2 depends on capability of control system and driving cycle. SOC below SOC_1 triggers use of the gas engine to drive the generator to move into the desired band of SOC. The distance the hybrid can reliably travel on electricity alone depends on the band from SOC_{MIN} to SOC_1. If lucky enough, the battery may be at SOC_2, and then the energy from SOC_{MIN} to SOC_2 can be used.

TABLE 9.2
Energy to Be Recovered by Regenerative Braking

Hybrid Speed (mph)	Energy		E/E_{BAT} (%)
	Joule	kWh	
30	116,000	0.032	6
60	467,000	0.130	26
90	1,050,000	0.292	58

Data for calculating energy
m, mass of vehicle = 1300 kg; E_{BAT}, energy of battery = 2 kWh; SOC, allowable change in battery state of charge = 0.25 (see Figure 9.9).

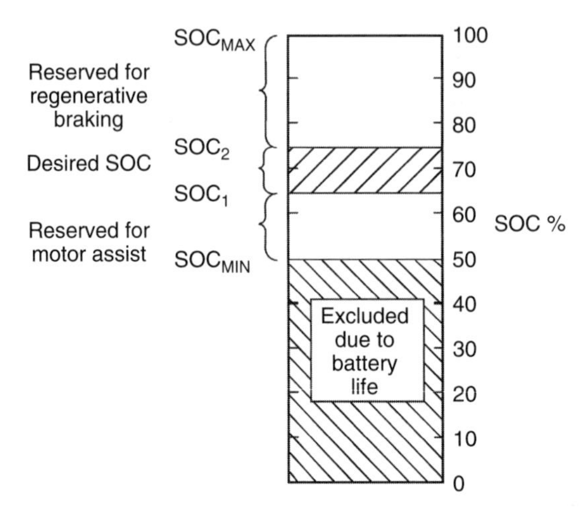

FIGURE 9.9 Bands of SOC for the battery.

Stopping Distance

Two graphs are presented in Figures 9.10 and 9.11. These graphs provide insight into the capability and limitations of regenerative braking. The graphs were calculated using four equations listed below. The equations use several symbols that are now defined.

P is the power of the generator used for regenerative braking, kW
F is the force acting on vehicle, N

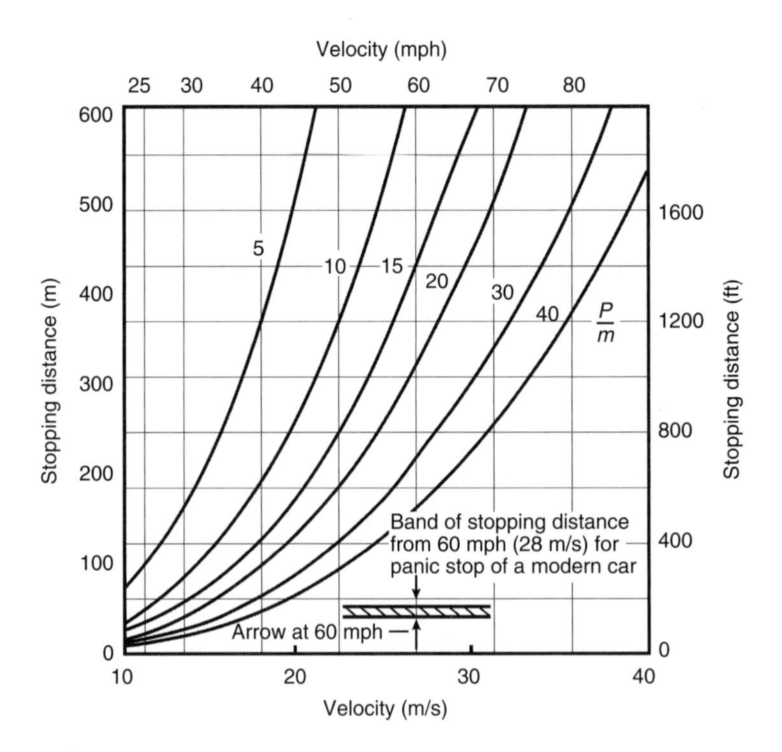

FIGURE 9.10 Stopping distances because of regenerative braking only as a function of hybrid vehicle velocity with power/mass ratio as a parameter. P/m is watt/kilogram.

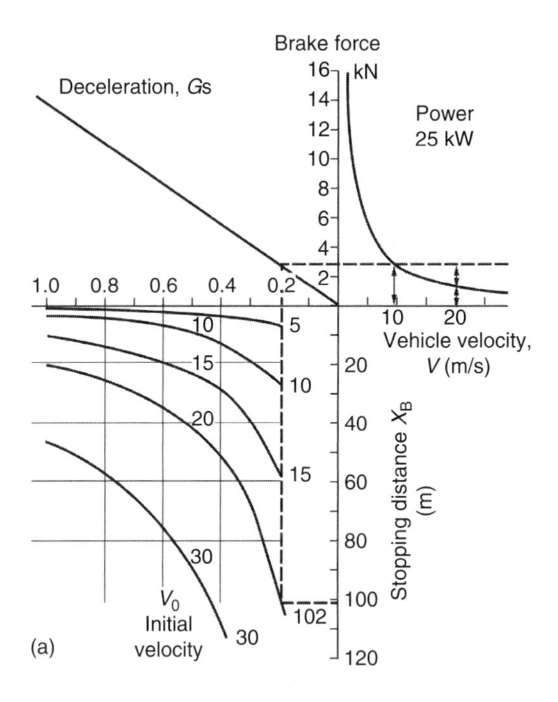

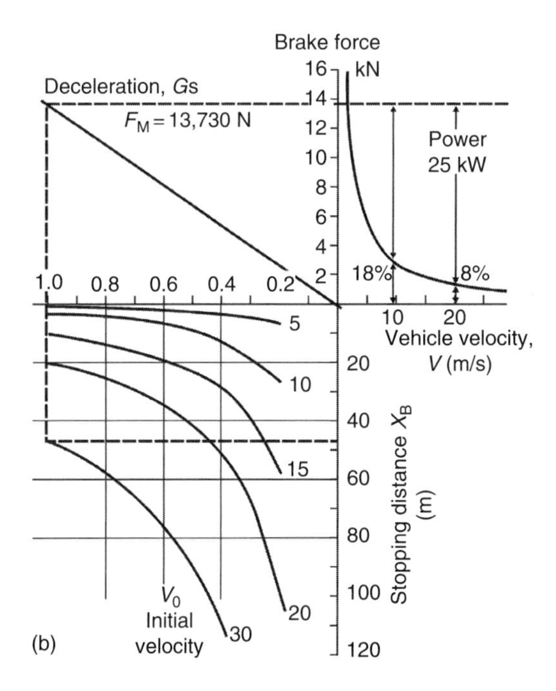

FIGURE 9.11 Three quadrant chart to determine the split between friction and regenerative braking. (a) Braking to achieve 0.2 *G*s deceleration and (b) panic level braking to achieve 1.0 *G* deceleration.

F_F is the braking force due to friction brakes, N
F_R is the braking force due to regenerative brakes, N
F_B is the braking force due to both friction and regenerative brakes, N
F_M is the maximum braking force as limited by tires and road, N
V is the velocity, m/s

V_0 is the initial velocity, m/s
E is the KE, J
E_0 is the initial KE, J
m is the mass of vehicle, kg
a is the acceleration, m/s^2
g is the acceleration of gravity, m/s^2
G is the ratio a/g called Gs, nondimensional
X_F is the stopping distance due to friction brakes, m
X_R is the stopping distance due to regenerative brakes, m
t_F is the time to stop using friction brakes, s
t_R is the time to stop using regenerative brakes, s

Time and Stopping Distance: Friction Brakes

First, the time to stop

$$t_F = \frac{mV_0}{F_F} \tag{9.16}$$

Second, stopping distance

$$X_F = \frac{m}{2F_F} \quad V_0^2 = \frac{E_0}{F_F} \tag{9.17}$$

Time and Stopping Distance: Regenerative Brakes

First, the time to stop

$$t_R = \frac{mV_0^2}{2P} = \frac{E_0}{P} \tag{9.18}$$

Second, stopping distance

$$X_R = \frac{m}{3P}\left(V_0^3\right) = \frac{1}{3}\left(\frac{1}{P/m}\right) V_0^3 \tag{9.19}$$

The stopping distance for friction brakes depends on velocity squared. For regenerative braking, the dependence is on velocity cubed. For a given vehicle velocity, the stopping distance for regenerative braking is greater. An important parameter for hybrid vehicles is the ratio of regenerative braking power, P, to vehicle mass, m. The symbols have been rearranged in Equation 9.19 to explicitly show the ratio P/m. The smaller the P/m ratio, the larger X_R. The regenerative braking power is the same as the power that can be generated by the M/G when operating in generator mode (G-mode).

The stopping distance is plotted as a function of vehicle velocity with the parameter P/m for each curve. Look at the stopping distance for $V_0 = 40$ mph. Along this vertical line pick off values for stopping distance, X_R, for different values of P/m. For $P/m = 5$ kW/1000 kg, $X_R = 1200$ ft. For $P/m =$ 40 kW/1000 kg, $X_R = 160$ ft. For $V_0 = 40$ mph and $P/m = 40$ kW/1000 kg, regenerative braking is adequate and falls within the panic stop band shown. Note the band of stopping distances for panic stop of a modern car at 60 mph. However, for $V_0 = 60$ mph and $P/m = 40$ kW/1000 kg, regenerative

TABLE 9.3
Ratio of Traction Motor Power (kW) to Vehicle Mass (kg)

Vehicle	P/m (kW/1000 kg)	Level of Regenerative Braking
Chevrolet Silverado	2	Low
Toyota Crown	4	Low
Honda Civic	5	Low
Honda Accord	8	Medium
Honda Insight	11	Medium
Nissan Tino	12	Medium
Toyota Camry	21	High
Toyota Prius I	22	High
Ford Escape	30	High
Toyota Prius II	38	High

braking is far from adequate for a panic stop. Since P/m is such an important parameter, values are shown in Table 9.3.

Figure 9.11 is a composite (hybrid) of three curves that appear in three quadrants. The lower left-hand quadrant has a plot of stopping distance as a function of deceleration in Gs. The equation for this quadrant is

$$X_B = \frac{V_0^2}{2gG} \tag{9.20}$$

The upper left-hand quadrant with a straight line for a curve is given by the equation

$$F_B = m\,a = mgG \tag{9.21}$$

Finally, the upper right-hand quadrant is a plot of the equations

$$P = F_R V_0 \tag{9.22}$$

Simple manipulation of Equation 9.22 leads to

$$F_R = \frac{P}{V_0} \tag{9.23}$$

For the curves of Figure 9.11, the values used were $m = 1400$ kg and $P = 25$ kW. This hybrid electric vehicle has $P/m = 18$ kW/1000 kg that can be compared with the discussion of Figure 9.10.

Figure 9.11a has a modest braking at 0.2 Gs. For $V_0 = 20$ m/s, $X_B = 102$ m. Projecting the 0.2 Gs upward, the brake force is

$$F_B = (0.2\ Gs)\ (13{,}734\ \text{N}) = 2747\ \text{N}$$

At $V = V_0 = 20$ m/s, the value of regenerative braking is $F_R = 1250$ N while

$$F_F = (2747 - 1250)\,\text{N} = 1497\ \text{N}$$

Hence at $V = 20$ m/s, 46% of braking is regenerative and 54% is friction. At $V = 9.1$ m/s, 100% is regenerative braking.

Figure 9.11b has a full panic stop at 1.0 *G*s. For $V_0 = 30$ m/s, $X_B = 46$ m. The maximum braking force is $F_M = 13{,}730$ N. As can be seen from the graph in the upper right quadrant, at $V = 20$ m/s, regenerative braking is only 8% of overall braking force. Even at 10 m/s, regenerative braking is only 18% of overall braking force. Friction brakes dominate the panic stop.

Efficiency of Regenerative Braking

As noted above, only a fraction of KE can be returned to the battery. The bigger the power of the M/G in G-mode, the larger the recovery of KE. Figure 9.12 has two braking experiences. One is a panic stop for which about 10% of the energy is returned to battery. The other 90% appears as heat in the friction brakes. The other example is a gentle stop at a traffic light. For this condition about 60% of energy is sent back to the battery. The other 40% heats the brakes. Input values are given in Figure 9.12.

Most of the stops in normal driving are not panic stops. The distribution of stops between panic and mild depends on the driving cycle. Subsequent discussion covers driving cycle and its influence on regenerative braking. For example, Prius recovers 30% of KE during braking ([1], p. 36). Battery losses reduce the energy that can be returned for acceleration. For more details on battery losses, see Chapter 6.

Typical Driving Cycle

The driving cycle shown in Figure 9.13 has vehicle speed as function of time for US06 drive cycle. To place the US06 cycle in context, see Table 12.5, the Environmental Protection Agency (EPA's) Five Test (Driving) Cycles. Looking at the slopes of the curves indicates acceleration and braking. A positive slope of curve gives acceleration of vehicle; an example occurs between 40 and 60 s. A negative slope of curve gives deceleration (braking) of the vehicle; an example occurs between 90 and 120 s. A zero slope, that is, a flat curve, indicates zero acceleration and therefore constant speed. This driving cycle has very little at constant speed. For about a minute between 200 and 280 s, velocity is constant. Toward the end of the cycle from 500 to 560 s, the vehicle is in a traffic jam with alternating slowing to 5 mph and growing to 25 mph. Anyone who drives a freeway at rush hour recognizes this aspect. The cycle has six stops counting the initial stop; however, each stop is less than 10 s. For idle-off of the engine, to improve gas mileage, longer stops are desirable. The driving cycles have a pronounced, not slight, effect on hybrid design.

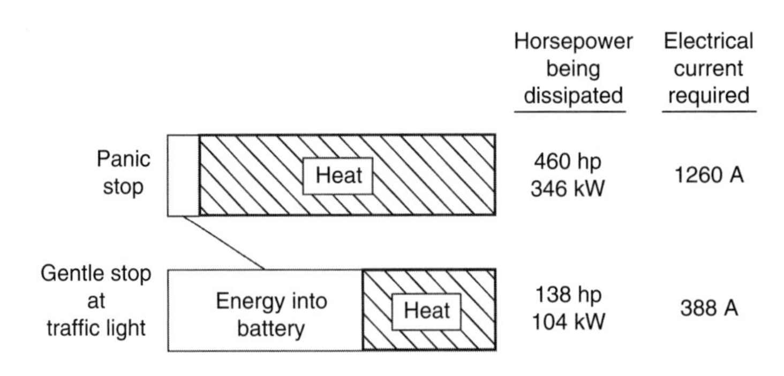

FIGURE 9.12 Efficiency of regenerative braking.

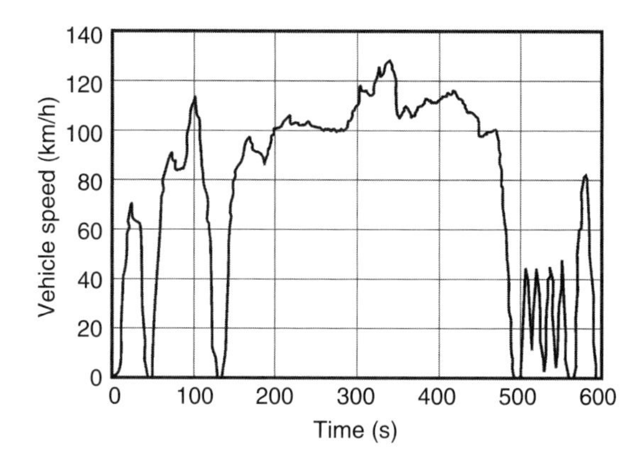

FIGURE 9.13 Vehicle speed variation for the US06 driving cycle. (Reprinted with permission from SAE Paper *Hybrid Electric Vehicles*, SP-1633, 2001 SAE International.)

Braking Force from Electrical Generator

Using the US06 driving cycle, the required brake force was determined for points along the driving cycle [2]. When braking was required for deceleration dictated by the driving cycle, the magnitude of braking force was found. In Figure 9.14, the dark asterisk points are derived from braking during the driving cycle; this shows required brake force for various velocities during cycle. The ratios 2:1 to 6:1 show different generator designs. The ratio 2:1 has constant maximum torque up to 100 km/h; above 100 km/h, the generator has constant maximum power. Constant torque translates into constant brake force. Using the slopes of the velocity curves from Figure 9.13, points in Figure 9.14 can be found.

Some of the dark asterisk points are at zero brake force. These points are for times when the vehicle is not braking. Points with zero braking force correspond to zero slope in the driving cycle curves. For braking at different velocities, the dark asterisk points are distributed throughout the graph. Look at the generator design with 2:1 ratio. Most of the points are below the maximum braking

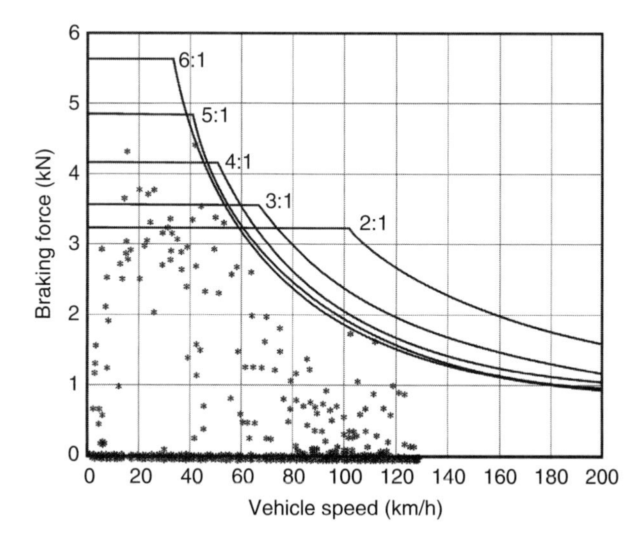

FIGURE 9.14 Distribution of braking forces for the US06 driving cycle. An electrical generator with 5:1 ratio can cover every demand of the driving cycle. (Reprinted with permission from SAE Paper *Hybrid Electric Vehicles*, SP-1633, 2001 SAE International.)

force curve at 3200 kN. This fact means that for this driving cycle, most of the braking can be accomplished by regenerative braking. Friction brakes are needed only when the points are above the maximum braking force curve. Only 12 points exceed the capability of the 2:1 generator design.

Two Design Approaches for Regenerative Braking

The two design approaches for regenerative braking are brake-by-wire shown in Figures 8.3 and 8.4. An alternate approach uses hydraulic pressure from the master cylinder as an input variable for brake control. In this case, regenerative braking is integrated with the conventional hydraulic system. Figure 9.15 shows the latter example.

Regardless of how regenerative braking is implemented, the key problem is the smooth split between friction and regenerative brakes. In today's hybrids, some drivers are able to detect erratic behavior as trade-off is made between friction and regenerative brakes.

Regenerative Braking Integrated with Conventional Hydraulic System

Each component is discussed along with its function (Figure 9.15a and b). To avoid confusion, only the braking for a front wheel is shown. Look carefully at the direction of vehicle motion and the direction of generator torque. Torque opposes the vehicle motion.

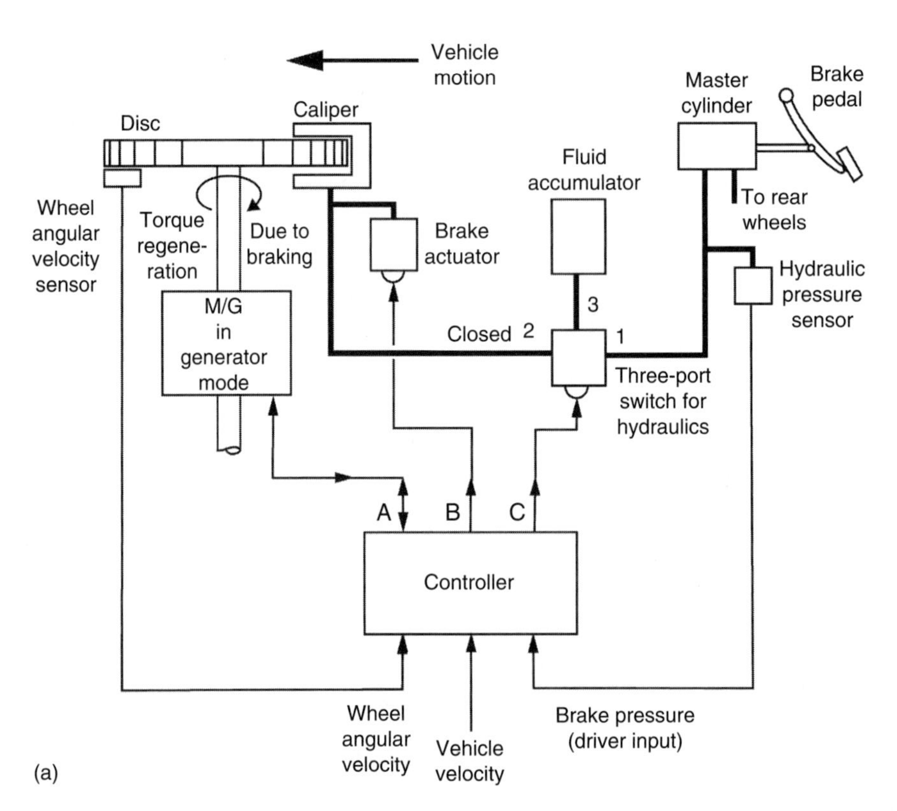

FIGURE 9.15 Regenerative braking integrated with conventional hydraulic braking system. (a) Braking force because of both generator torque and friction by brake pads against the rotor disk and

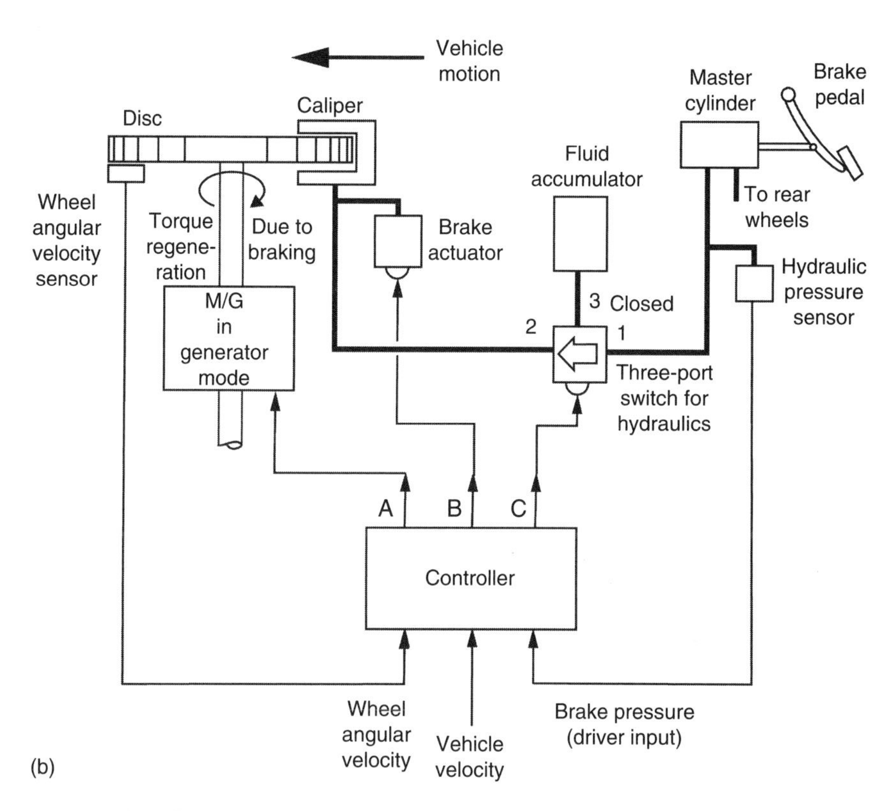

FIGURE 9.15 (continued) (b) fail-safe mode. Master cylinder directly connected hydraulically to caliper. Note the arrow from port 1 to port 2. Both ports are open. Port 3 is closed.

DRIVER INPUT

By stepping on the brake pedal, the driver intends to stop, and the amount of pressure applied on the brake pedal determines the urgency of the situation. The hydraulic pressure sensor communicates braking commands to the controller.

BRAKE ACTUATOR

The brake actuator is an electromechanical device that develops hydraulic pressure. The brake actuator is commanded by the controller. Hydraulic pressure causes the caliper to squeeze the disk brake. Note that two brake tubes supply hydraulic pressure to the caliper. One tube can be traced back to the master cylinder.

THREE-PORT SWITCH

- Three-port switch receives commands from controller.
- In normal operation, ports 1 and 3 are open and port 2 is closed isolating the master cylinder from the brake actuator and caliper (Figure 9.15a).
- In fail-safe mode, port 3 is closed and ports 1 and 2 are open (Figure 9.15b).
- Fail-safe mode. Master cylinder directly connected hydraulically to caliper.

Note the arrow from port 1 to port 2.

CONTROLLER: INPUTS

The controller has several inputs. Four inputs of wheel angular velocity (how fast is wheel spinning?) from each wheel. Vehicle velocity is used along with inputs of wheel spin rate to determine tire slip and/or incipient brake lock. The ABS is activated by incipient brake lock.

Brake hydraulic pressure, which is a driver input, determines the split between the friction and regenerative brakes. For low pressures, braking will likely be regenerative alone. For high pressures, the controller must determine the split between regenerative and friction braking.

CONTROLLER: OUTPUTS

The connection A–A is a data bus; it is not an on–off switch. A–A provides two-way communication between M/G and controller. Also, A–A is used to command the split between regenerative and friction braking. B transmits the commands to the electrically controlled brake actuator that develops hydraulic pressure to activate the caliper. C transmits the commands to open and close ports in the three-port switch for hydraulics. C controls the fail-safe mode.

M/G: GENERATOR MODE

Note the direction of vehicle motion shown at the top of Figure 9.15. The torque due to the generator constitutes regenerative braking. Note the direction of the torque, which opposes vehicle motion; hence, braking action. The battery is not shown in Figure 9.15.

The regenerative braking integrated with the conventional hydraulic system will likely change significantly with the adoption of the new electronic wedge brake (EWB), which is discussed in Chapters 8 and 12. By using EWB, a more satisfying pedal feel may be easier to attain without Herculean control efforts.

SUMMARY

Regenerative braking recovers KE to enhance mpg. Brakes, either friction or regenerative braking, must avoid unsafe conditions such as locked wheels. Fail-safe mode for regenerative braking requires full-size, 100%, friction brakes.

Braking forces are transmitted to the road via the tires. To understand braking, tire behavior must be understood. Regenerative "coasting" from high speed or going downhill recovers a much greater amount of energy compared to that recovered by regenerative braking. Coasting recovers kinetic energy while going downhill recovers potential energy.

Stopping distances for friction and regenerative brakes are presented. Stopping distance for friction brakes varies as V_0^2 whereas for regenerative brakes the variation is as V_0^3. Stopping distances for pure regenerative braking are longer because of the square and cubic dependence. An important parameter for regenerative braking is the generator power to vehicle mass ratio, P/m. Graphs show the influence of P/m on stopping distances.

Driving cycles are important for hybrid design. The distribution of the braking forces for a typical driving cycle is given; for this driving cycle, most of the braking can be accomplished by means of regenerative braking.

Two drawings illustrate regenerative braking integrated into a conventional hydraulic system. The fail-safe mode is shown.

APPENDIX 9.1: EQUATIONS FOR REGENERATIVE AND FRICTION BRAKES

This appendix has the derivation of Equations 9.17 through 9.19. The power being dissipated by the brakes is

$$P = F_B V \tag{A.9.1}$$

The symbols are the same as previously defined. KE is given by

$$E = \frac{1}{2} m V^2 \tag{A.9.2}$$

The derivative of KE assuming velocity as the variable is

$$\mathrm{d}E = mV\,\mathrm{d}V \tag{A.9.3}$$

The force because of deceleration, that is, braking, is

$$F = ma = m\left(\frac{\mathrm{d}V}{\mathrm{d}t}\right) \tag{A.9.4}$$

Equation A.9.1 can be rearranged to give

$$V = \frac{P}{F_B} \tag{A.9.5}$$

The braking equations for regenerative brakes are derived first.

REGENERATIVE BRAKES

Combine Equations A.9.4 and A.9.5 to get

$$\mathrm{d}E = m\left[\left(\frac{P}{m}\right)\left(\frac{\mathrm{d}t}{\mathrm{d}V}\right)\right]\mathrm{d}V = P\,\mathrm{d}t \tag{A.9.6}$$

which also follows from the definitions for energy and power. Equations A.9.1 through A.9.5 were not really needed; however, the results will be used later. Integrate A.9.6 to obtain the time to stop.

$$t_R = \frac{E_0}{P} \tag{A.9.7}$$

Since

$$\mathrm{d}X = V\,\mathrm{d}t \tag{A.9.8}$$

From Equation A.9.6

$$dt = \frac{dE}{P} \tag{A.9.9}$$

Eliminate dt between Equations A.9.8 and A.9.9

$$dX = \frac{V}{P}(dE) \tag{A.9.10}$$

Combine Equations A.9.3 and A.9.10 to eliminate dE

$$dX = \frac{mV^2}{P}(dV) \tag{A.9.11}$$

Integration of Equation A.9.11 yields

$$X_R = \frac{m}{3P}(V_0^3) = \frac{1}{3}\left(\frac{1}{P/m}\right)V_0^3 \tag{A.9.12}$$

As noted earlier, the power to mass ratio, P/m, is important.

FRICTION BRAKES

The braking force developed by friction brakes is F_F and is related to deceleration and vehicle mass by

$$F_F = ma = m\left(\frac{dV}{dt}\right) \tag{A.9.13}$$

This can be rearranged to the form

$$F_F\, dt = m\, dV \tag{A.9.14}$$

Integration yields the time to slow to velocity, V

$$t = \frac{m}{F_F}(V_0 - V) \tag{A.9.15}$$

For the time to stop completely, that is, $V = 0$, the result is simply

$$t_F = \frac{m}{F_F}(V_0) \tag{A.9.16}$$

From Equation A.9.14, the differential equation for stopping distance is

$$t = \frac{m}{F_F}\left(V_0 - \frac{dX}{dt}\right) \tag{A.9.17}$$

Integration of Equation A.9.17 leads to the stopping distance

$$X = V_0\, t - \left(\frac{F_F}{2m}\right) t^2 \tag{A.9.18}$$

Combine Equations A.9.15 and A.9.18 to eliminate time, t, and to obtain distance traveled when the velocity has the value, V

$$X = \frac{m}{2F_F}\left(V_0^2 - V^2\right) \tag{A.9.19}$$

The distance to a complete stop means $V = 0$, and the resulting equation is

$$X_F = \frac{1}{2}\frac{mV_0^2}{F_F} = \frac{E_0}{F_F} \tag{A.9.20}$$

These equations were used to calculate graphs in Figures 9.10 and 9.11.

REFERENCES

1. L. Tchobansky, M. Kozek, G. Schlager, and H. P. Jörgl, A purely mechanical energy storing concept for hybrid vehicles, *Hybrid Electric Vehicle Technology Developments,* SP-1808, Society of Automotive Engineers, Warrendale, PA, 2003.
2. H. Gao, Y. Gao, and M. Ehsani, Design issues of the switched reluctance motor drive for propulsion and regenerative braking in EV and HEV, *Hybrid Electric Vehicles*, SP-1633, Society of Automotive Engineers, Warrendale, PA, 2001, pp. 63–68.
3. H. L. Husted, A comparative study of the production applications of hybrid electric powertrains, *Hybrid Gasoline-Electric Vehicle Development,* PT-117, Society of Automotive Engineers, Warrendale, PA, 2005.
4. S. Birch, Powertrain: Prodrive leads new hybrid project, *Automotive Engineering International*, Society of Automotive Engineers, Warrendale, PA, January 2008, p. 38.

10 Narrow Operating Band for Gasoline Engine

What is surprising, not that the rpm band is so narrow, but that it is so broad.

INTRODUCTION

Two words flow through this chapter: torque and rpm. The driving cycle establishes the required torque and rpm; however, the values are those at the wheels. Of the two variables, rpm is special because of the engine and fuel consumption. For a band of rpm values along a line in the engine map, fuel consumption is a minimum. One frequent goal of hybrid control is to operate on the correct line in the map. The "minimum fuel consumption line," also known as the "basic operating line" is discussed.

Values of the torque and rpm at the wheels are not at the desired location. In order to specify engine operation, values for torque and rpm are needed at the engine. Transfer of torque and rpm from drive wheels to the engine is considered next. Having gained insight into the minimum fuel consumption line, engine operation can be managed so as to yield best mpg and at the same time meet torque and rpm demands at the drive wheels.

DEMANDS OF DRIVING CYCLE

Figure 9.13 shows the vehicle speed variation for the US06 driving cycle. At every operating point, the driving cycle determines the required axle torque and wheel speed. The vehicle, be it hybrid or conventional, must translate the required axle torque and wheel speed into gear ratios and throttle settings. Starting at the tire patch on the road and working upstream to the motor through the powertrain, the engine rpm and torque can be determined. Figure 10.1 focuses on the conventional vehicle (CV).

The situation is much more complicated for a hybrid. The nontrivial decision must be made concerning the split between the motor and engine. Once that decision has been made by the hybrid control, then the issue of gear ratios for the powertrain and throttle settings for the engine can be addressed.

DRIVING CYCLE

Many important factors affect the driving cycle. These factors are important because the mpg attained by a vehicle depends strongly on the particular driving cycle. This fact is obvious from the EPA mileage results which vary widely for the city and highway driving cycles.

ROADS

The three main road types are city streets, country highway, and freeway or interstate. The road types are defined by road width, allowable inclines, frequency of traffic lights or stop signs, speed limits, and similar characteristics.

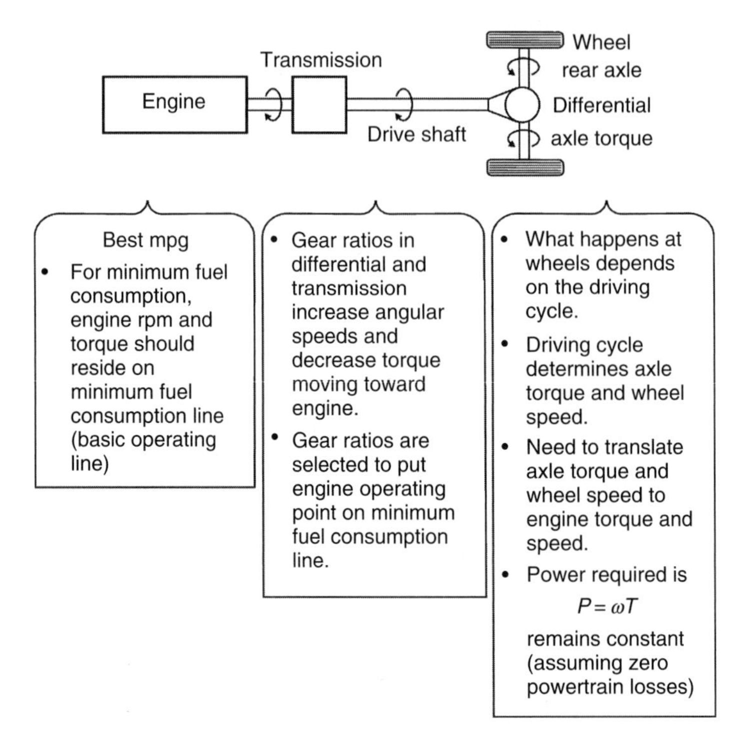

FIGURE 10.1 Demands of driving cycle and placement of engine operating point on the minimum fuel consumption line (basic operating line).

Intensity of Traffic

Just as sailors wish each other fair winds and smooth seas, commuters wish each other light traffic and smooth roads. The traffic can be described as light, moderate, heavy, etc. The traffic can also be described with four letter cuss words. Beyond heavy traffic is the condition of being gridlocked. Obviously, fuel consumption is less for light traffic than for heavy traffic.

Terrain

Terrain is yet another factor affecting performance including mpg. Three different terrain types are flat, mountainous, and hilly. Within each terrain type, other factors come into play. For example, in flat land, how frequent are curves, and what is the average radius of curvature? Tire drag is slightly more going around a curve. The more the drag, the lesser the mpg.

In mountainous terrain, the radius of curvature of the curves dictates the maximum speed both posted and within the laws of physics for your particular car. In hilly country, what are the inclines and length of hills? Very long, steep hills can cause the hybrid battery to go "dead." (See Chapters 4 and 5 for a brief mention of dead battery and Chapter 11 for a more detailed discussion.) Chapter 11 has a section entitled "Residual Performance with a 'Dead' Battery." Chapter 6 has, as you would expect, information about dead batteries.

Road Quality

On a smooth road you sail along with the minimum of motion. The easy and effortless progress down the road is also efficient. Energy is required to create extraneous motion, like jostling up and down over an undulating surface. Energy used is gasoline used.

Rough roads do absorb energy! For more information about rough roads, see Ref. [1]. If energy can be extracted from a bouncing car, an even larger amount has been expended.

Unpaved roads deserve comment. In addition to the uneven surfaces, unpaved roads can become muddy. Muddy roads evidently have more drag on a vehicle. To reiterate, the more the drag, the lesser the mpg.

WEATHER

Blame ice and snow on the weather. Blame muddy roads on the weather. Both by-products of the weather adversely affect mpg. The temperature determines the need and usage of air-conditioning (A/C). As discussed elsewhere in this book, use of A/C is detrimental to attaining high mpg.

Some people win, some people lose. We are talking about head winds and tail winds. A tail wind not only reduces noise inside a vehicle, it also increases mpg by 10%–20%.

One shortcoming of EPA mileage estimates, which is being corrected, is to test vehicles without recognizing the effect of weather on mpg. On hot days, the effect of A/C on mpg is of special importance. A vehicle started from cold-soak, for example overnight on streets in Minnesota in winter, consumes more fuel (see Chapter 12).

MISSION OF VEHICLE: WIDESPREAD USAGE

Driving cycles differ according to the designed task or purpose of the vehicle. Vehicles with widespread usage include commuter cars. Passenger cars used for local shopping constitute another example of a vehicle that is widely used.

MISSION OF VEHICLE: SPECIALTY

Innumerable specialty vehicles exist. Many are produced in small quantities. The vehicle size ranges from tiny to huge. Refuse (trash) trucks have a very distinct driving cycle. About 70,000 trash trucks operate within the United States. Taxicabs also have characteristic driving cycles. New York City auctioned 18 medallions for hybrid taxicabs. Operators of several medallions will use Ford Escape SUV; all are painted yellow.

Chapters 4, 11, 16, and 17 discuss trucks and heavy duty vehicles. Also, Chapter 3 presents information about hybrid truck and bus technology. Buses and trucks have unique driving cycles that determine the economic viability of a hybrid version. Large 18-wheeler trucks may or may not offer an opportunity for a hybrid version. Auxiliary power may be the foot-in-the-door for a hybrid 18-wheeler.

REAL-WORLD DATA

Collecting data for a driving cycle is more than hopping in a car and taking a spin down the road. To collect real-world data, extensive use is made of statistical theory, instrumentation science, satellite technology, geography, and climatology. Acquiring good data for a driving cycle is more difficult than making a political survey.

To acquire real-world data, begin by instrumenting a vehicle with global positioning satellite, recorders, speed sensors, distance sensors, accelerometers, temperature sensors, and wind sensors. Drive actual routes specified by the mission of the vehicle; drive the routes not once but several times. Conversion of the actual test drive numbers to representative numbers is necessary. In brief, conversion of the data is

$$\text{Sets of real data} \rightarrow \text{Typical or representative data}$$

Sampling theory applies as well as other esoteric theories.

Various countries, for example, the United States, Japan, and Europe, have defined typical driving cycles, for example, urban, highway, etc. See Figure 9.13 which shows the US06 driving cycle which combines relatively high speed (up to 120 km/h) with large acceleration and heavy braking. Also, consult Ref. [2].

VARIABLES FOR DRIVING CYCLE

To be of use to the hybrid engineer, the driving cycle needs to provide certain data that are reported using the variables cited below. Most are self-explanatory; however, some comments are made.

- Trip duration (min)
- Trip distance (km)
- Number of stops
- Stop rate (stops/km)
- Idle time (min)
- Idle rate (% of path having an idle event)
- Average speed (km/h)
- Average running speed (km/h)
- Acceleration (m/s^2)
- Deceleration (m/s^2)
- Vehicle load (N)
- Terrain profile (grade % vs. time)

Average speed includes time at stops. Average running speed excludes time at stops and includes only the time that the tires are rolling. For a light delivery truck, or a postal delivery van, the vehicle load varies from point to point in a driving cycle. For buses, as passengers get on and off, the load changes. Changing load changes mpg.

MINIMUM FUEL CONSUMPTION LINE

ENGINE MAP

Many hybrid control strategies drive the engine operation to the minimum fuel consumption line. This is true for most, but not all, control methods. At this juncture, it is appropriate to know what the minimum fuel consumption line may be. One explanation uses the engine map which was also discussed in connection with Figures 8.7 and 8.9. These two figures were generic maps. Figure 10.2 is an engine map with numbers and real data [3]; it is for the second generation Toyota Prius, the hatchback model or Prius II.

The engine map has torque (N·m) plotted as a function of engine speed (rpm) with contours of specific fuel consumption (SFC). Units for SFC are grams per kilowatt-hour (g/kWh). The basic operating line is also the minimum fuel consumption line. The dark line at the top where the SFC lines end is the line of maximum torque for this engine.

SUPERIMPOSE ENGINE POWER CONTOURS

How is the line of minimum fuel consumption determined? The following discussion serves as an explanation. Knowing torque and rpm, power can easily be calculated.

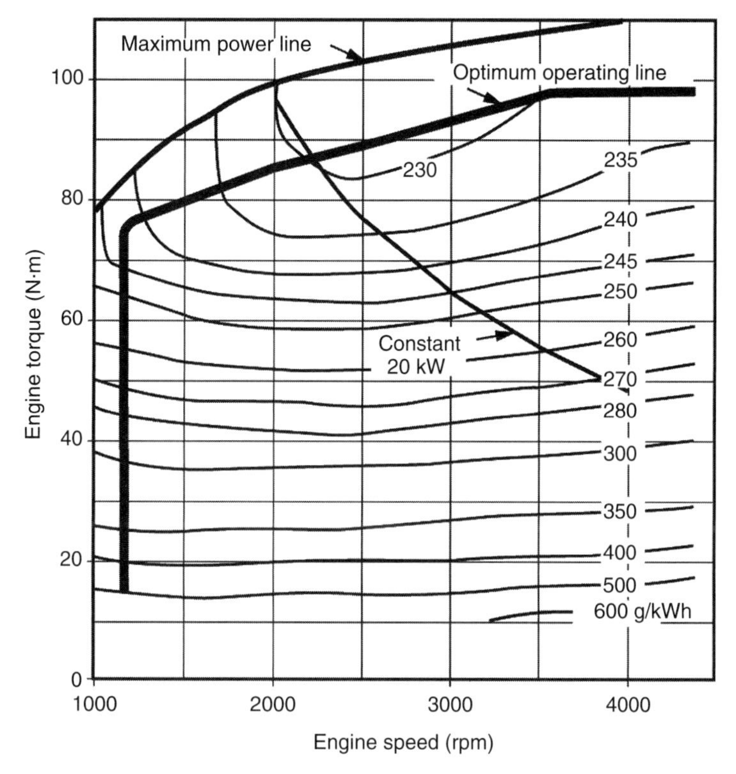

FIGURE 10.2 Engine map for second generation Prius II (hatchback model). (Reprinted with permission from SAE Paper *Hybrid Gasoline–Hybrid Vehicle Development*, PT-117, 2005 SAE International.)

$$P = \omega T = \frac{2\pi NT}{60} \tag{10.1}$$

where

P is the power (W)
ω is the engine angular speed (rad/s)
T is the engine torque (N·m)
N is the engine angular speed (rpm)
The factor 60 is 60 s/min
The factor, 2π, is the number of radians/revolution

A sample calculation illustrates the use of Equation 10.1:

$$P = \frac{2\pi(3000 \text{ rpm})(64 \text{ N·m})}{60} = 20{,}100 \text{ W} = 20.1 \text{ kW} \tag{10.2}$$

Curves of constant power are superimposed on Figure 10.2. These curves were calculated using Equation 10.1 (Figure 10.3).

Basic Operating Line: Minimum Fuel Consumption Line

Any point where a contour of constant engine power is tangent to an SFC contour is a locus on the minimum fuel consumption line; this line is also called the basic operating line. For SFC = 230 g/kWh,

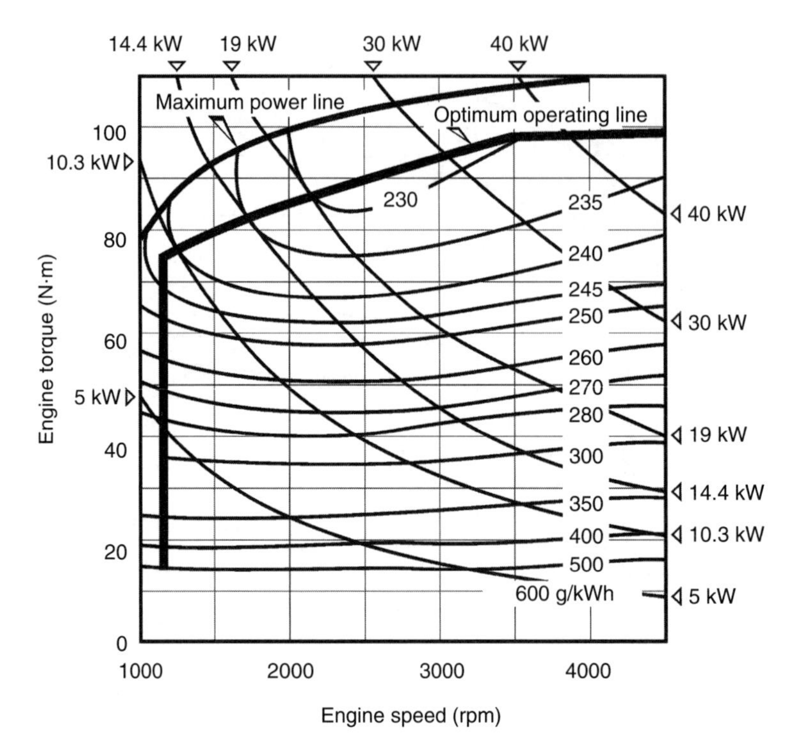

FIGURE 10.3 Engine map for Prius II (hatchback model) with contours of constant power superimposed. (Reprinted with permission from SAE Paper *Hybrid Gasoline–Hybrid Vehicle Development*, PT-117, 2005 SAE International.)

the 19 kW contour is tangent as shown. For SFC = 235 g/kWh, the 14.4 kW contour is tangent as also shown. The tangent point for SFC = 240 g/kWh is 10.3 kW contour. At 1200 rpm, the minimum fuel consumption line (basic operating line) arbitrarily drops in engine torque. The engine does not normally operate on that part of the line except while starting and stopping.

As was done with the discussion on Figure 5.2, the width of the bucket for the lowest fuel consumption needs to be defined. The choice is arbitrary. From Figure 10.3, the following inequalities are found:

$$15 \text{ kW} < P < 40 \text{ kW}$$
$$228 \text{ g/kWh} < \text{SFC} < 235 \text{ g/kWh}$$
$$1800 < \text{rpm} < 4000$$

The band of P was selected to give a band of very small SFC. The corresponding engine speeds, rpm, are included. Table 10.1 gives engine performance for a broader engine speed band along the basic operating line extending from 1200 to 4000 rpm.

What is amazing about the Prius II engine is that from 1500 to 4000 rpm, the deviation from least SFC is less than 2.4%. For the range 1800 to 4000 rpm, the deviation is less than 2%. Variable valve actuation accounts for a portion of this amazing performance.

Narrow Operating Band for Gasoline Engine

Engines are designed specifically for hybrids. The upper bar in Figure 10.4 is for the Prius II engine. Compare with the lower bar which is for a typical, high performance, four-cylinder engine used in a CV. The CV engine operates from idle rpm to the maximum design rpm that yields maximum power.

TABLE 10.1
Engine Performance on the Minimum Fuel Consumption Line (Also Called the Basic Operating Line)

Torque, T (N·m)	Speed, N (rpm)	Power, P (kW)	SFC (g/kWh)	Percent Deviation from Minimum SFC (%)
76	1200	9.5	242	6.1
78	1500	12	237	2.4
85	2000	18	232	1.8
88	2500	23	228	0
92	3000	29	228	0
98	3500	36	230	0.9
98	4000	41	232	1.8

Because of an upper limit of 4000 rpm, the Prius II engine is lightly stressed compared to a high revving (6500 rpm) engine. Later models Prius II have a redline at 4500 rpm. Less stress allows weight reduction. Lower weight is beneficial (see Chapter 7). Even after downsizing, additional weight can be saved due to reduced mechanical stress (Figure 5.2).

Matching the Wheels to the Engine

The angular speed (rpm) and torque are shown in Figure 10.5 for a conventional, rear-wheel drive vehicle. Two gear ratios are illustrated. Low gear is used for vehicle launch, climbing up very steep hills, traversing gooey mud, and descending very steep inclines. High gear is used for freeway cruise. Most freeway cruisers have an overdrive gear not shown here.

The transmission gear ratio is N_T. N_T varies either with a manual or an automatic transmission. The differential gear ratio is N_F and equals 2.92 in Figure 10.5. For low gear, N_T equals 4.28. In low gear, the engine turns (4.28)(2.92) = 12.5 times faster than the rear wheels. In high gear, engine and driveshaft turn at the same speed; the wheels turn 2.92 times slower than engine. Since, in high gear the engine is turning a few 1000 rpm, the driveshaft is also turning a few 1000 rpm. Therefore, the driveshaft must be carefully balanced. Different gear ratios allow matching vehicle and wheel speed with the desired engine speed. Operation near the minimum fuel consumption line is one basis for selection of gear ratios and shift points.

Equation 10.1 is repeated here for convenience.

$$P = \omega T = \frac{2\pi NT}{60} \tag{10.1}$$

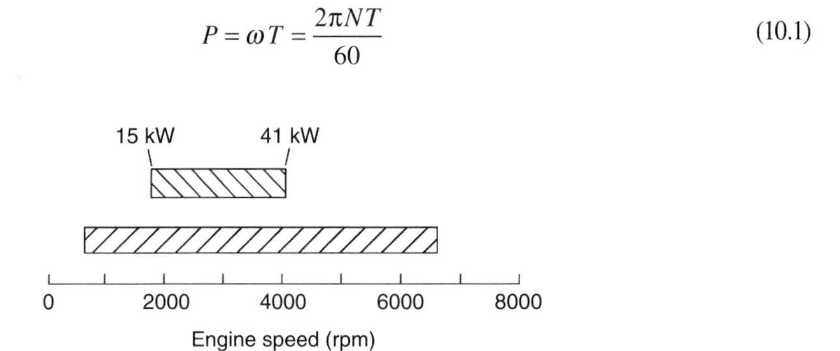

FIGURE 10.4 Span of engine speed illustrating the narrow operating band for an engine designed specifically for hybrid application. Top bar is for Toyota Prius II. Bottom bar is for a typical, conventional, high performance, four-cylinder, gasoline engine.

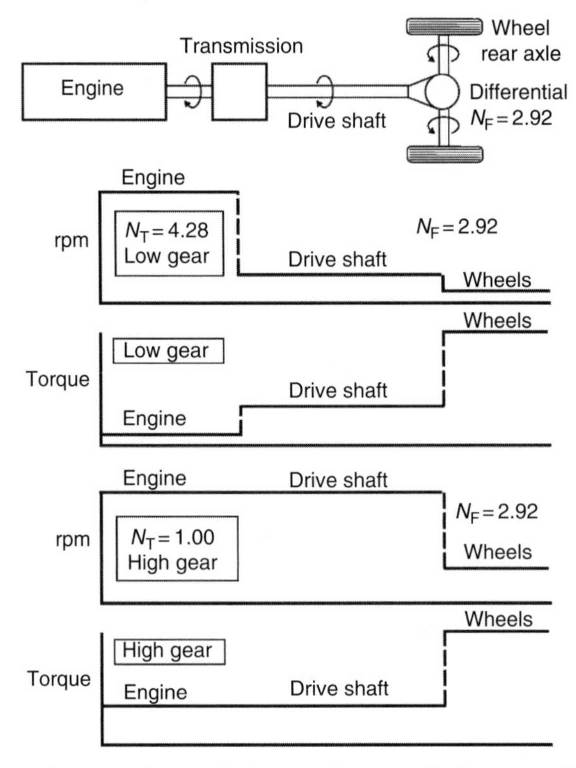

FIGURE 10.5 Engine and powertrain speeds (rpm) and torques for low gear, N_T = 4.28, and high gear, N_T = 1.00, i.e., direct drive through transmission.

For purposes of discussion, powertrain losses are neglected. As a consequence, power is constant at every point along the powertrain. As a result of Equation 10.1, when torque goes up, rpm goes down and vice versa. This inverse relationship is illustrated in Figure 10.5. In low gear, the torque is multiplied (increased) due to the transmission and differential.

How is all of this relevant to an HEV? For a parallel hybrid, the hybrid control commands a split between the torques from the motor and the engine (Figure 8.1). Once the hybrid engine torque and power are specified, the gear ratios in the transmission, the three-way gearbox, and the differential must be selected to put the engine on the minimum fuel consumption line of Figure 10.2.

Effect of Number of Transmission Speeds on Fuel Economy

The number of speeds in an automatic transmission affects fuel economy favorably. Chapter 5 also discusses this factor. Multiple speed automatic transmissions begin to rival the manual shift transmission in efficiency. Automatics with five-speeds, six-speeds, and even eight-speeds are beginning to appear in new cars. The 2007 Ford Expedition SUV changed to a six-speed automatic transmission that increased fuel economy by 1 mpg in spite of the fact that the vehicle weight increased by 250 lb (114 kg) [4].

Effect of Number of Transmission Speeds on Control of Engine Speed

The number of gears in the transmission affects the width of the engine rpm band within which the engine can be operated. See Figure 10.6 that has three graphs of engine rpm as a function of wheel rpm; each graph is for a different number of gears in the transmission. Each graph also has horizontal

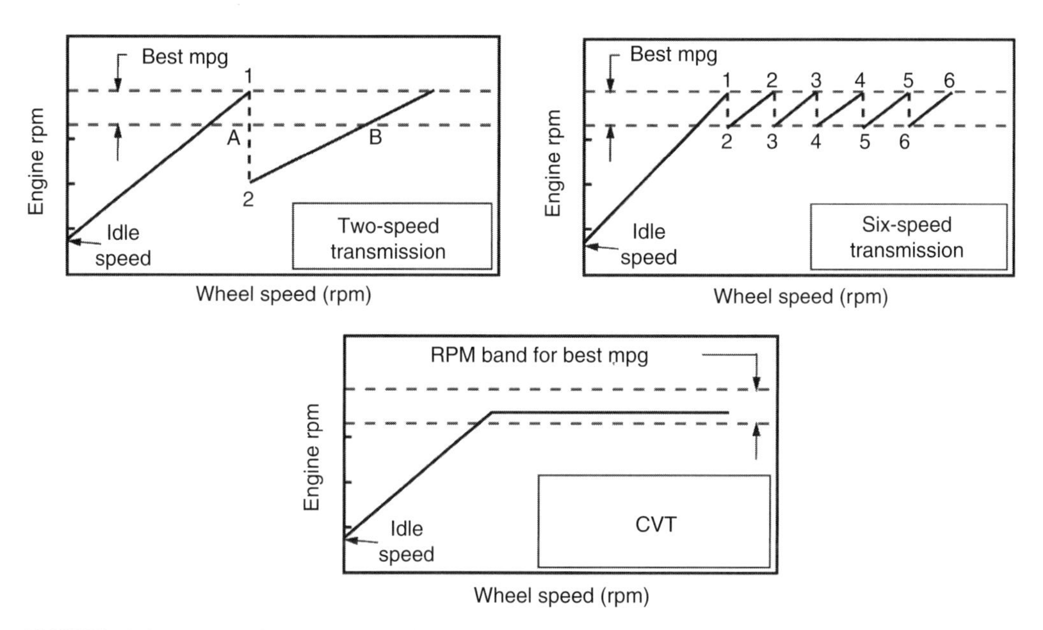

FIGURE 10.6 Range of engine speeds for three different transmissions.

dashed lines defining the rpm band for best mpg. Shift points are shown as vertical dashed lines, for example, 1 $\rightarrow$ 2, for the two-speed transmission at the top, left-hand. Two-speed transmissions have been used. The last two-speed transmissions were the automatics of the 1950s.

The two-speed transmission has wide swings in engine rpm. From point A to point B, engine rpm falls outside the rpm band for best mpg. The six-speed transmission gives a narrow range of engine rpm. The engine can operate within the rpm band for best mpg. The continuously variable transmission (CVT) has an infinite number of gear ratios between two rpm limits. The CVT has an extremely narrow range of engine rpm once it attains a certain speed. CVTs are popular for hybrids because of their ability to match input/output rpm's. The CVT offers seamless acceleration, which some drivers find disconcerting. CVT can improve fuel economy.

Locating engine operation within the rpm band for best mpg is only part of the solution for the hybrid control. As Equation 10.1 shows, the torque and rpm at the wheels determine a power, *P*. For a conventional car (CV), the engine must supply the required power while operating inside the rpm band for best mpg. For a CV, the required torque may not permit operation along the minimum fuel consumption line. However for an HEV, the motor assist almost always enables operation of engine for best mpg.

Role of Electric Motor in a Hybrid

When the power required at the wheels exceeds the engine power, the electric motor can provide motor assist. Motor assist is used so that the engine can operate on the minimum SFC line. Motor assist is explained elsewhere in this book. Periods of motor assist are shown in Figure 4.2. The curve in Figure 5.2 has a much narrower rpm range than the engine of Figure 10.4.

SUMMARY

The driving cycle determines the axle torque and wheel spin rate that, in turn, specifies the power required. The transmission and differential gears adjust wheel torque and speed to that required by the engine to operate on the minimum SFC line. For a hybrid, when the power required exceeds the engine power, motor assist is used to compensate for the deficit of power.

REFERENCES

1. R. B. Goldner, P. Zerigian, and J. R. Hull, A preliminary study of energy recovery in vehicles by using regenerative magnetic shock absorbers, *Hybrid Gasoline-Electric Vehicle Development*, SAE PT-117, Society of Automotive Engineers, J. M. German (Ed.), 2005.
2. C. L. Goodfellow, P. S. Reverault, L. P. Gaedt, D. Kok, T. Hochkirchen, M. Neu, and C. Picod, Simulation based concept analysis for a micro hybrid delivery van, *Advanced Hybrid Vehicle Powertrains 2005*, SAE SP-1973, Society of Automotive Engineers, R. McGee, M. Duoba, and M. A. Theobald (Eds.), 2005.
3. K. Muta, M. Yamazaki, and J. Tokieda, Development of new-generation hybrid system THS II—Drastic improvement of power performance and fuel economy, *Hybrid Gasoline-Electric Vehicle Development*, SAE PT-117, Society of Automotive Engineers, J. M. German (Ed.), 2005.
4. K. Buchholz, New Expedition accents capability, *Automotive Engineering International*, July 2007, p. 10.

11 Hybridness: A Basic Design Decision

Hybridness provides insight into the overall character of the vehicle.

DEFINITION OF HYBRIDNESS

The definition of hybridness, H, is

$$H = \frac{\text{Sum of power of all traction motors}}{\text{Sum of traction motors} + \text{Engine power}} \tag{11.1}$$

Some hybrids have more than one motor/generator (M/G). Hybrids with motor-in-the-wheel and all-wheel-drive (AWD), have more than one motor. The definition uses the sum of all traction motors. The name, hybridization, is occasionally used for H.

As an example of hybridness consider a light delivery van with the propulsion:

Diesel engine: 110 kW at 3000 rpm
Electric motor: 23 kW; maximum torque 243 N·m at 500 rpm

$$H = \frac{23\ \text{kW}}{23 + 110\ \text{kW}} = 0.17 = 17\% \tag{11.2}$$

As will be seen, $H = 17\%$ is a mild hybrid. As a note of caution, the sum of component power 23 + 110 kW = 133 kW is not the maximum hybrid power. The maximum electric motor torque and engine torque occur at different rpm.

What are merits of hybridness, H, as a definition? H defines micro, mild, and full hybrids. The domain of the plug-in hybrid is defined by a range of values of H. Morphing of series hybrids, which is done by varying H, leads to mixed hybrids. H can be an independent variable in an equation for hybrid performance. One example, which is range extension, is discussed below.

STORY

Some auto magazines define a mild hybrid as one that does not have any electric-only propulsion. For example, the 2005 Civic Hybrid does not have any mode which is electric-only propulsion. Hence, by that definition, the 2005 Civic Hybrid was a mild hybrid. The 2006 Civic Hybrid does have an illusive but existent electric-only mode. Hence, the 2006 Civic Hybrid can no longer be classified as a mild hybrid. This chapter unravels the mystery of mild hybrids and provides rational definitions for hybrids.

Based on the definition above, the 1936 Ford with a manual transmission (the only one in 1936) was better than a mild hybrid. In some repair shops of that era, a mechanic might want to move the Ford a few feet. This was done in winter so as not to start the V-8 in a closed space. The mechanic could put the Ford in low gear, leave the clutch out, step on the starter button and move—actually lurch—the desired few feet. This was purely electric-only propulsion. Therefore, according to the above definition for mild hybrid, the 1936 Ford had more hybridization than the 2005 Civic Hybrid! The Ford could be moved 30 or 40 ft before the starter overheated and the battery became weak and disinterested.

HYBRID DESIGN PHILOSOPHY

By considering the major factors for hybrids, an understanding of the various values of H is gained. The basic efficiency of the gasoline engine is low. A typical value is 25%. The efficiency of MGs is higher. Typical values are above 90%. Battery efficiency is moderate; energy is lost putting energy into the battery and again removing energy. Round trip in/out efficiency is typically 70%–80%. Because of the inefficiency, the batteries must be cooled.

Overall hybrid design philosophy has three parts:

- Operate electric motor first (less emissions/less fuel consumed).
- Add gasoline engine only when needed.
- Operate gas engine at the best rpm and throttle setting, that is, operate on minimum fuel consumption line in engine map (see Chapter 10).

HYBRIDNESS: PARALLEL HYBRID

For a parallel hybrid, Figure 11.1 demonstrates the utility of hybridness, H. Some parts are not shown, like the battery. Five different values of H are illustrated. For $H = 0\%$, the vehicle is conventionally powered solely by a gasoline engine. The vehicle has, except for a 12 V system, zero electrical components.

For $H = 25\%$, the hybrid electric vehicle (HEV) has an electrical traction motor with 25 kW and an engine with 75 kW. Both engine and motor shafts are inputs to a three-way transmission. This is

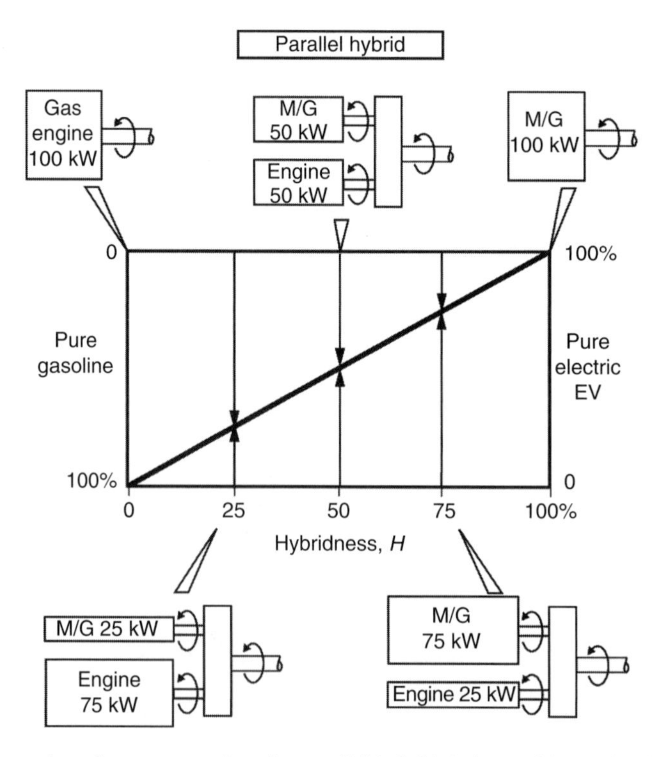

FIGURE 11.1 Illustration of component sizes for parallel hybrid designs with varying hybridness.

the region for a mild hybrid. Mild hybrids are a good solution for certain vehicles. The cost/benefit ratio is highly favorable.

For H = 50%, the HEV has both electrical traction motor and an engine with equal power of 50 kW. As is the case for a parallel hybrid, both engine and motor shafts are inputs to a three-way transmission. This is the region for a full hybrid.

For H = 75%, the HEV has a very large M/G compared to the engine power. To supply the electrical power for the M/G, a large heavy battery is required. This is the region for a plug-in hybrid. Also this is the region for the range extender vehicle. For H = 75%, if the M/G runs for an hour, the energy consumed would be 75 kWh. The engine/generator requires 3 h to recharge the battery.

For H = 100%, the vehicle is a pure electrical vehicle (EV). All electrical power comes from either the battery or regenerative braking. Energy stored in the battery is supplied by charging stations.

Except for H = 0% and H = 100%, each hybrid has the same architecture. The M/G and engine are inputs to a three-way, or three-shaft, transmission.

HYBRIDNESS: SERIES, MIXED, AND RANGE EXTENDER (PLUG-IN) HYBRIDS

Refer to Figure 11.2 for series, mixed, and range extender hybrids.

A series hybrid by its definition has a value of H near 50%. For values of H away from 50%, different classes of hybrids are found (Figure 11.2). The diagram assumes no losses; all components have 100% efficiency. The shafts, which are identified by an ellipse indicating torque, are a mechanical connection between parts.

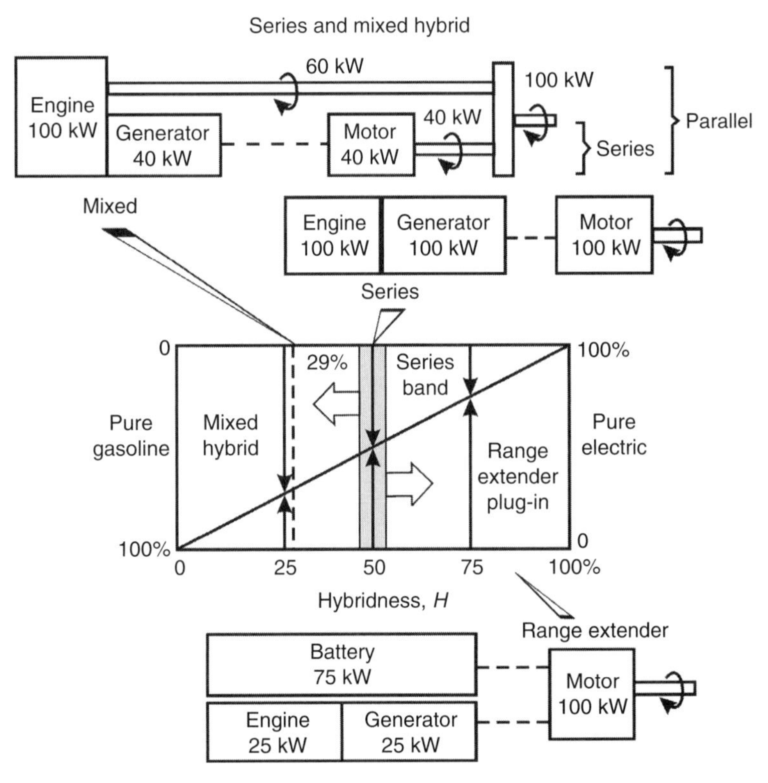

FIGURE 11.2 Series hybrid and its derivatives related to hybridness, H. For H less than 50%, the series hybrid morphs into the mixed hybrid. For H greater than 50%, the series hybrid transforms into the range extender, electric, vehicle. H greater than 50% is also the region for the plug-in hybrid. The dashed line (----) is an electrical connection. For H = 29%, the part attributed to series is shown as well as the part which is parallel. H = 29% is a mixed hybrid.

The series hybrid has motor power approximately equal to engine power; hence, the series hybrid exists in a band near $H = 50\%$. Outside that band, the series hybrid changes into either mixed hybrid or plug-in hybrid. The mixed arrangement shown here is almost the same as the Toyota Prius.

RANGE EXTENDER

The proposed GM Chevrolet Volt (Cadillac Volt) is an example of a range extender hybrid (see Chapter 3). The range extender, which has large value for H, is shown in Figure 11.2.

An almost infinite number of either parallel, mixed, series, or plug-in designs can be made for a hybrid. The equations discussed below apply to one plug-in or range extender hybrid design. The equations have the following assumptions:

1. When it runs, the generator always runs at full power.
2. Power to cruise, which depends on cruise speed, is greater than the generator power.
3. Battery power supplements engine (generator) power.
4. At the end of cruise at maximum range, R, the battery is "dead" and the fuel tank is empty.

Note that when voltage supplied to the motor is equal to the battery open circuit voltage, the battery neither supplies nor absorbs electrical energy (Figure A.6.8). The equations below are derived in Appendix 11.1. The symbols and equations are

$$R = \frac{R_0}{H} \tag{11.3}$$

where

R_0 is the range without engine/generator
R is the range with engine/generator which extends range
H is the hybridness factor

The increment in range, ΔR, is defined by Equation 11.4

$$\Delta R = R - R_0 = R_0 \frac{1-H}{H} \tag{11.4}$$

Input values for power, etc. are given in Appendix 11.1. The numerical results for the example shown in Figure 11.3 are

$R_0 = 240\,\text{km}$
$H = 0.75 = 75\%$
$R = 240\,\text{km}/0.75 = 320\,\text{km}$

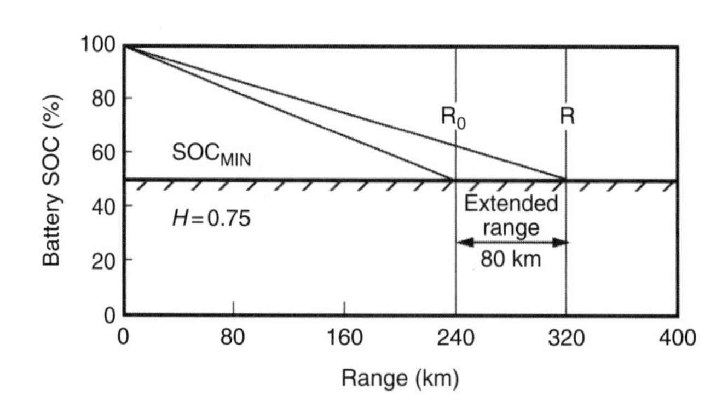

FIGURE 11.3 Extension of range due to a small gasoline-powered generator recharging the battery.

The gain in range is

$$\Delta R = R - R_0 = 240\text{ km }[(1-0.75)/0.75] = 80\text{ km}$$

The required power to move the range extender was $P_R = 20$ kW. Since $H = 75\%$, the gasoline engine is a tiny 5 kW not much larger than a lawn mower engine. According to Equation 11.1, for $H = 1$, $R = R_0$. For $H = 0.5$, $R = 2R_0$.

OPTIMIZATION AND HYBRIDNESS

Chapter 8 discusses optimization of hybrid control. Usually the goal is to obtain best mpg. Since hybrids have two separate propulsion systems, electrical and gasoline, which system is the focus for optimization?

For small hybridness, that is H much less than 50%, the optimum hybrid operates near the best specific fuel consumption line on the engine map. Small H corresponds to mild hybrids. For a band of hybridness near 50%, the efficiencies of both electrical and engine components affect the optimum operating points. Values of H larger than 40% are termed full hybrids. For large values of H, that is, near 100%, hybrid optimization focuses on the M/G, battery, and power electronics.

BATTERY POWER AND ELECTRIC MOTOR POWER

To begin, a few features of batteries are stated. Battery size is determined by

$$\text{Battery energy} = (\text{power of M/G})(\text{run time})$$

This equation assumes the battery power equals the power of the M/G. By a property of the battery known as specific energy (W h/kg), battery energy can be changed to battery mass. Specific energy always has units of (battery energy)/(mass). The run time is the time required for the battery to become dead.

The trends in the variation of battery power and energy are shown in Figure 11.4. The battery power and electric motor power must be matched. For examining the trends, battery and motor powers are assumed equal. Figure 11.4 is divided into two regions with

Mild and full hybrids	$0\% < H < 50\%$
Plug-in hybrids	$50\% < H < 100\%$

The battery variation with increasing H depends on the region.

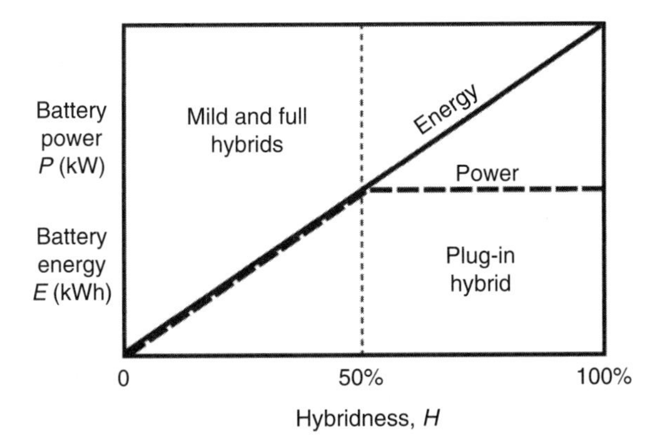

FIGURE 11.4 Variation of battery power and battery energy as a function of hybridness.

TABLE 11.1
Qualitative Aspects of Battery Size

Small	Big
Engine runs too often	Too much weight to haul around
Little loss of mpg due to excess weight	Allows lots of electric-only operation
Insufficient for electric-only operation	Suitable for plug-in operation

For mild and full hybrids, as H increases, the battery power and battery energy increase hand in hand. From the equation Battery energy = (Power of M/G)(Run time), the run time remains fixed in this region. Refer to the definition of H in the opening paragraphs of this chapter. An increase in M/G power increases H. Also a decrease in engine power, that is, downsizing the engine, increases H. Battery energy, E, and power, P, grow to match the growth in M/G. The increase of H in this region is due mainly to growth of M/G power.

For the plug-in hybrid, the M/G power no longer needs to grow. The M/G has sufficient power to move the vehicle. The battery power need not grow; however, battery energy must grow to gain more range. Figure 11.4 shows the parting of the ways of battery E and P in the plug-in region. Battery E continues to increase while battery P remains constant at a value equal to M/G power. In this region, since M/G power is fixed, increases in H are due to shrinking engine power compared to M/G power.

Batteries can be designed so that power remains fixed while energy increases (see Chapter 6).

What is the link that relates the power of the M/G to battery size or weight?

$$\text{Desired run time} = \frac{\text{Energy stored in battery}}{\text{Power of M/G}} \tag{11.5}$$

How long should the hybrid operate before the battery is dead? The answer determines the desired run time. Table 11.1 provides some qualitative comments on battery size.

RESIDUAL PERFORMANCE WITH A "DEAD" BATTERY

Residual performance with a dead battery decreases linearly with increasing H as shown in Figure 11.5.

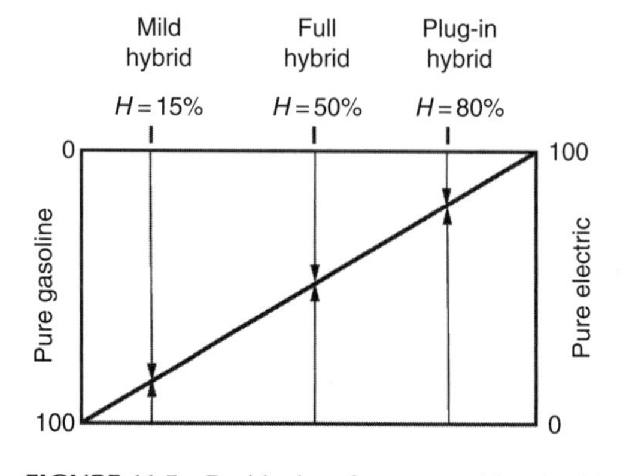

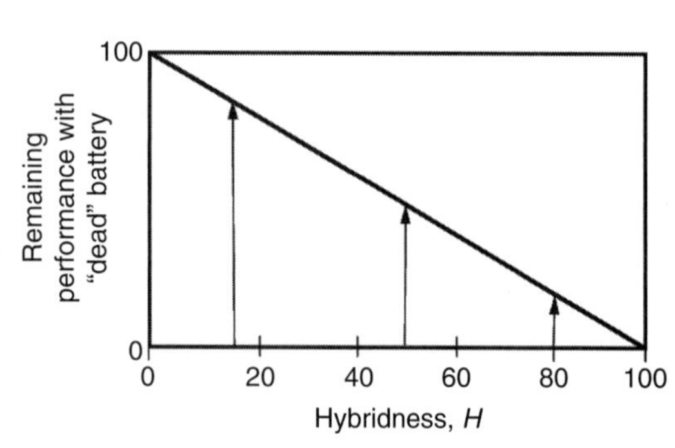

FIGURE 11.5 Residual performance with a dead battery.

Dead does not mean battery state of charge (SOC) is zero; it means SOC = SOC_{MIN} where SOC_{MIN} is the minimum SOC established by designers to protect battery life (Figure 9.9). For a mild hybrid with H = 15%, the 85% remaining power comes from the gasoline engine; robust performance remains. This figure of 85% assumes none of the engine power is diverted to charging the dead battery.

For a full hybrid with H = 50%, the remaining power of the engine is 50% as battery goes dead. A drop of 50% in power is more than a mere inconvenience when climbing a long, steep hill. A full hybrid allows engine downsizing, and a downsized engine cannot provide 100% sparkling performance. For the plug-in hybrid, H = 80%, the engine is very small, and it provides only an excruciatingly slow limp home. For the pure electric car, EV, with H = 100%, when the battery goes dead, the car stops dead also.

EFFECT OF LOW CHARGE

The motor in an HEV contributes to the performance of the vehicle. If the battery cannot contribute as much as when the battery has the design SOC, what is the loss of performance? Table 11.2 provides insight into that question.

POWER-TO-WEIGHT RATIO

Two values for power-to-weight ratio find application for analysis of hybrids. As discussed in connection with Table 9.3, the ratio based only on the power of the M/G is appropriate for regenerative braking. For overall vehicle performance, the "power" in power-to-weight ratio includes both the gasoline engine and the M/G.

Power-to-weight ratio for many hybrids (trucks and pickups excluded) is

$$55 \text{ to } 70 \text{ kW/1000 kg}$$

Power-to-weight ratio affects both stop (regenerative braking) and go (acceleration).

Any performance item that depends on the familiar equation $F = ma$, also depends in a critical manner on power-to-weight ratio.

INTERPRETATION OF RAMPS

Figure 11.7 has a series of ramps below the drawing. The interpretation of the ramps is discussed here.

The example used in Figure 11.6 applies to the capability of hybrids to exploit regenerative braking. For a mild hybrid, H = 15%, regenerative braking is possible but only about 38% of kinetic energy can be recovered. The calculation is 15/40 = 38%, which is the height of the ramp at H = 15%. The limitation is due to the small generator. The ramp ends at H = 40% for which a hybrid has a

TABLE 11.2
Effect of Low Charge on Vehicle Performance

Acceleration	Full Charge	Partial Charge
Time 0–60 mph, s	10.8	12.3
Standing quarter mile (s)	18.3	19.1
Speed at quarter mile (mph)	78	75

Note: Data are for a 2006 Honda Civic Hybrid as compiled from various sources.

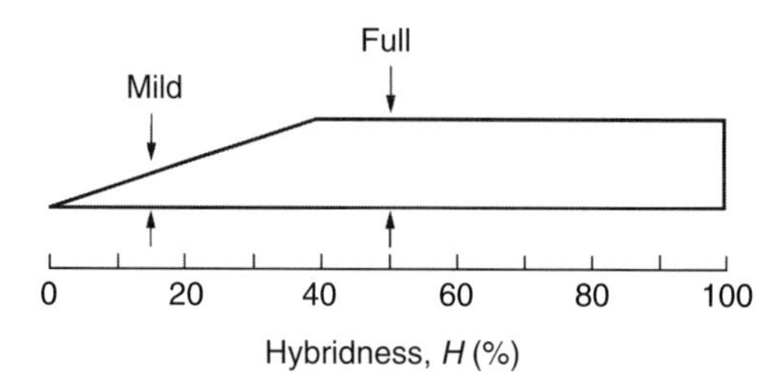

FIGURE 11.6 Interpretation of the ramps used in Figure 11.7.

generator large enough to enable high-efficiency regenerative braking. For a full hybrid, $H = 50\%$, more than enough generating capability exists for regenerative braking. As denoted by the flat bar, the span of H from 40% to 100% allows full regenerative braking. Although numerical values are given for the purpose of illustration, the ramps are qualitative.

VARIOUS TECHNIQUES TO ENHANCE HYBRID PERFORMANCE

As a summary to the hybridness discussion, various techniques to enhance hybrid performance are arrayed with hybridness in Figure 11.7. Each technique is discussed.

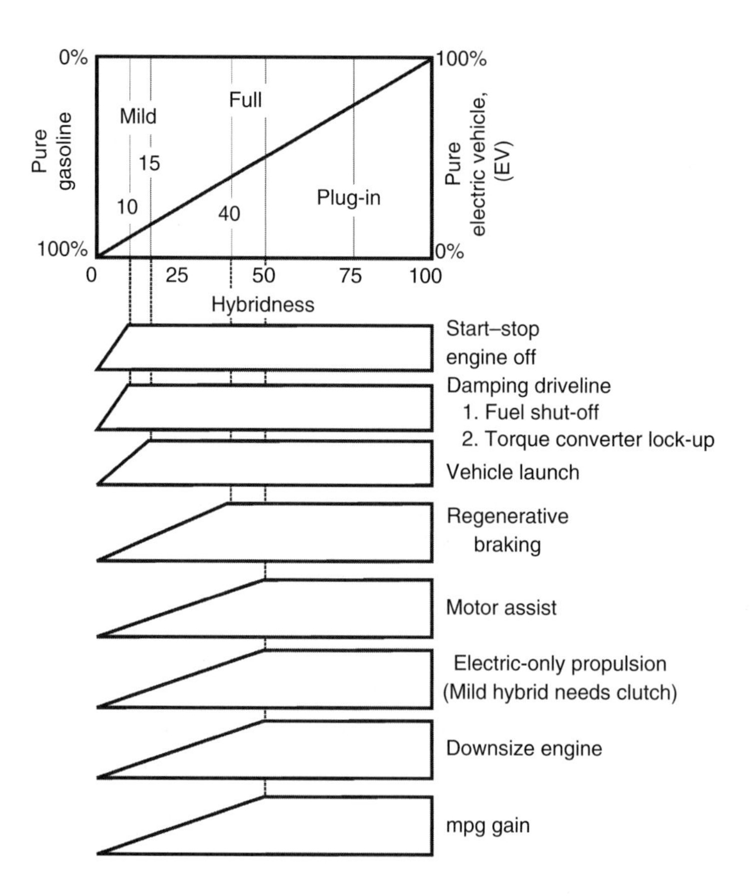

FIGURE 11.7 Availability of various techniques to enhance hybrid performance as a function of hybridness and resulting mpg gain. The bar below the hybridness graph has a ramp which extends from $H = 0\%$ to a value of H for the particular technology. For start–stop, the ramp ends at $H = 10\%$. The flat bar beyond indicates that for all values of $H > 10\%$, that feature is available to the hybrid designer.

Start–Stop

Engine-off during stops in traffic affords a saving in fuel. The usual 12 V starter does not have the power to restart the engine without delay, noise, and vibration. With the more powerful electrical motors, even in mild hybrid, the engine rpm can be quickly increased. Once smoothly and quickly up to starting rpm, the fuel injection can be activated.

Damping Driveline Oscillations

Another way that fuel consumption can be reduced is to shut off fuel flow whenever brakes are applied. Abrupt turn off of fuel can cause shudder and unpleasant oscillations of the engine and of driveline. Damping by the electrical motor can decrease the unpleasantness to an acceptable level.

For a hybrid that uses an automatic transmission, some losses in the torque converter can be reduced by locking the torque converter eliminating slippage. Under some conditions, when the torque converter is locked, driveline oscillations are excited; these oscillations are disagreeable to the customer. Once again, damping by the electrical motor can decrease the unpleasantness to an acceptable level.

Vehicle Launch

An engine at low rpm has little torque. At launch, torque is essential. An electric motor, even a small one, has high torque at low rpm. The motor fills in the torque hole at low rpm. A small motor can contribute significantly to the initial launch.

Regenerative Braking

For small values of H, which implies small generator, the M/G set cannot absorb the kinetic energy of the vehicles forward motion in a rapid stop (see Chapter 9). Although modest regenerative braking is possible and is used at low H, regenerative braking can only be fully exploited when H is about 40%.

Motor Assist

Vehicle launch is part of motor assist, but applies to very low speed. Motor assist covers a broader range of speed and vehicle operations such as hill climbing and driving in snow. More power and a larger electric motor are required. Hybridness, H, of 50% yields enough power from the electrical motor to overcome the power deficiencies of the downsized engine.

Electric-Only Propulsion

Electric-only propulsion means the gasoline engine is shut down and does not consume fuel. Electric-only operation improves mpg. To achieve performance goals, the motor must have adequate power. At $H = 50\%$, the traction motor is as large as the engine. Alone, the traction motor yields the desired performance.

Another reason that electric-only operation is desirable is the fact that emissions are zero or near zero. Stringent emission requirements may be met by electric-only operation. However, cool-down of the catalyst during idle-off is a problem to be solved.

For some mild hybrids, such as Honda Hybrids Accord, Insight, and Civic, the M/G is directly connected to the flywheel. When the motor turns the driveshaft to move the car, the engine is also turned. This robs the battery and motor of considerable power. As a partial solution, the Honda Civic closes all valves which decreases the losses due to the engine.

Downsize the Engine

For $H = 50\%$, the engine and traction motor are of equal power. The engine and motor share equally in propelling the car. The engine can be cut in half so to speak.

Miles per Gallon Gain

As hybridness increases, up to about 50%, mpg also increases. This is a result of a balance between power required and power available. The increase in mpg possible by plug-in is not shown. Plug-in requires energy from charging stations.

MILD OR MICRO HYBRID FEATURES

As a result of being a mild hybrid, certain features follow. The M/G may be belt or chain driven (new Saturn Vue Hybrid). Alternatively, the M/G may be part of the flywheel (Accord, Civic, Insight, Silverado). The M/G serves as the starter/alternator combined.

Mild hybrids have limited regenerative braking. The battery and installed M/G may be large enough to provide low speed motor assist or to provide low speed launch assist. For the rare case of a diesel/hybrid, the M/G in M-mode can provide cold start of the diesel.

For a mild hybrid, other possible design features include fuel cutoff at deceleration, idle shutoff, and torque converter lockup where applicable.

WHY IS *H* ABOUT 50% A GOOD SPOT FOR HIGH MILES PER GALLON?

A driving cycle frequently has required maximum power about twice the average power of the cycle. This suggests that a near equal division between motor and engine is appropriate. The deficiency between required maximum power and average power can be supplied by motor assist. With H near 50%, the engine can be downsized to provide near the average power of that cycle. A value of H near 50% is favorable for electric-only propulsion. There is enough electrical power in the motor and the battery for electric-only operation. Although not too important, excellent (overwhelming) restart capability from engine-off exists.

For $H = 15\%$, fuel savings are 20%–30% while for $H = 50\%$, the fuel savings are 40%–50%. As H increases, the price premium increases. Price premium is the price of a hybrid minus the price of a comparable conventional vehicle. In fact, as H increases, the price premium increases at a fast rate so that recovery through fuel savings is longer for a full hybrid than for a mild hybrid.

PLUG-IN HYBRID

The plug-in hybrid can be viewed as an EV but with a small engine to extend range. Figures 11.2, 11.4, 11.5, and 11.7 show the hybridness domain of the plug-in hybrid. Features of a plug-in hybrid include a large, heavy, expensive battery. The comparison with a full hybrid is a battery of a few 100 lb instead of the typical 100 lb in a full hybrid.

Additional equipment is needed to connect to external "wall plug" electrical source for recharging. Since batteries are high voltage, the voltage of the charging source must be even higher. Inductive rechargers prevent exposure to high voltage. The plug-in will likely have small gasoline engine driven generator for on-board charging; this engine separates the plug-in hybrid from the EV.

For people willing to undertake the recharging chore, the plug-in offers fantastic mpg. To gain the benefits, the range of hybridness for a plug-in is $50\% < H < 100\%$ with H likely to be closer to 100%. The plug-in will limit the filling of the gas tank to each season: spring, summer, fall, winter!

ALL-WHEEL DRIVE HYBRID

For the subsequent discussion, some definitions are necessary:

AWD = All-wheel drive
4WD = Four-wheel drive
2WD = Two-wheel drive
FWD = Front-wheel drive
RWD = Rear-wheel drive

In the discussion to follow, AWD is used for either AWD or 4WD.

The design for an AWD hybrid vehicle depends on whether the starting point is a conversion of an existing AWD vehicle or starting with a clean sheet of paper. With conversion of an existing design, the starting point is called the "legacy design." Many conventional AWD vehicles are sold with the optional choice of either 2WD or AWD. The 2WD is less expensive than the AWD and provides better mpg. The optional 2WD versions may be either FWD or RWD. The 2WD on the left side of Figure 11.8 starts as FWD. The 2WD on the left side of Figure 11.9 starts as RWD.

The legacy design affects the loading for the front and rear tires. With FWD, the front tires have three loads: (1) cornering, (2) braking or traction, and (3) steering. As discussed in Chapter 9, tires have a load limit. Loads are additive. The rear tires carry, at most, two loads: (1) cornering and (2) braking or traction. Too much torque to the front wheels may overload the front tires. An overload adversely affects vehicle handling *in extremis*. To avoid overloading the front tires, a torque split between front/rear is satisfactory with 50/50 or with a bias on the rear wheels of approximately 30/70 F/R. The torque split need not be precisely equal to the numbers 50/50 and 30/70; values near these values are satisfactory.

Table 11.1, which is coordinated with Figures 11.8 and 11.9, shows the front and rear power loading for legacy FWD and legacy RWD. The traction motors are limited in power due to battery

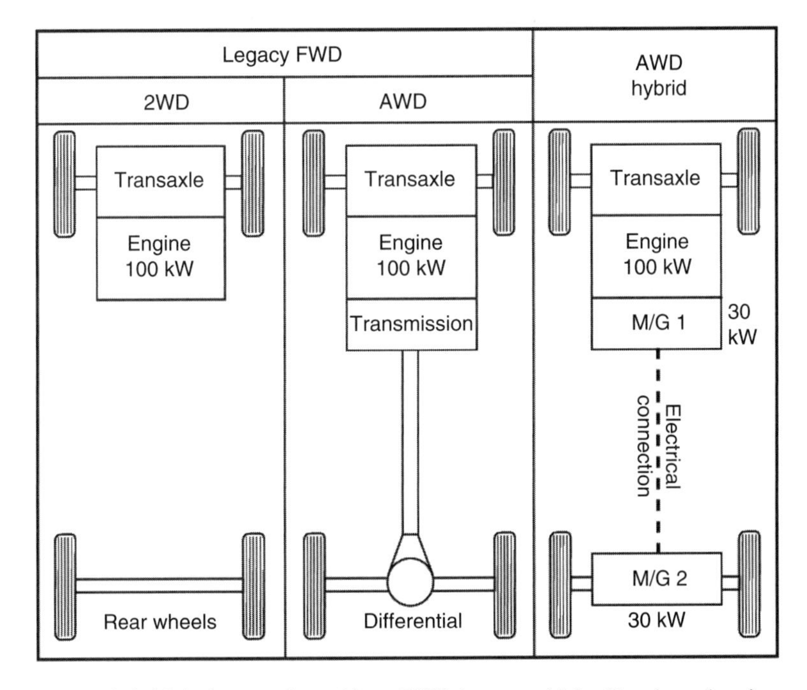

FIGURE 11.8 AWD hybrid design starting with an FWD legacy vehicle. The three drawings are 2WD, the conventional AWD, and hybrid AWD.

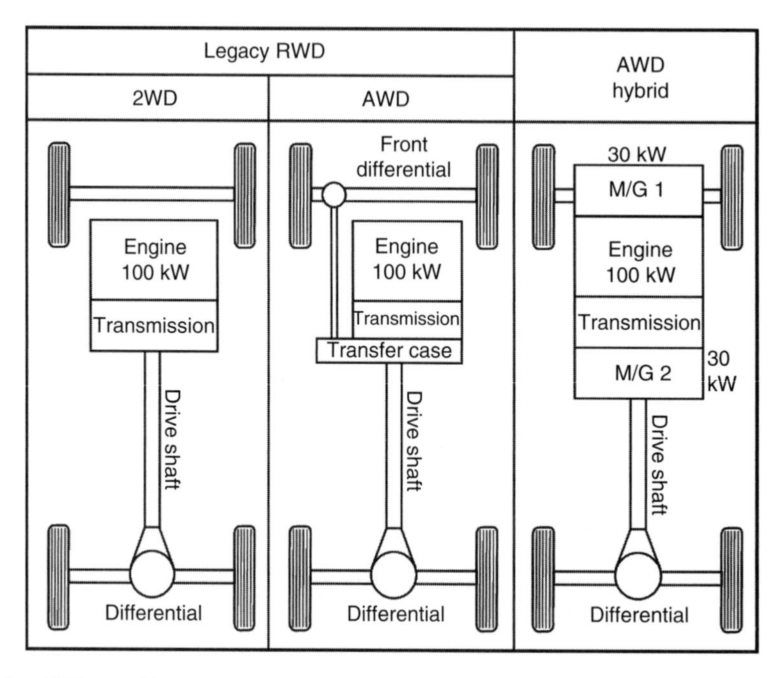

FIGURE 11.9 AWD hybrid design starting with an RWD legacy vehicle. The three drawings are 2WD, the conventional AWD, and hybrid AWD.

limitations. For discussion purposes, each M/G has a realistic 30 kW and the engine is 100 kW. In motor assist, the battery must supply 60 kW, which is 30 kW for each M/G in M-mode (Figure 11.8). With a legacy FWD, the M/G is on the rear axle. This means the traction load on the rear axle is limited to 30 kW (Figure 11.8). With a legacy RWD, the M/G is on the front axle. This means the traction load on the front axle is limited to 30 kW, which is favorable in regard to loading of the tires.

Figure 11.8 shows a legacy design of FWD for the optional 2WD. When the 2WD version of unmodified vehicle is FWD, then a hybrid conversion will undoubtedly have an electric traction motor driving the rear wheels. Front/rear torque bias will likely be reversed 70/30 F/R, which is usually unfavorable. Figure 11.9 shows a legacy design of RWD for the optional 2WD. If the 2WD version of the unmodified vehicle is RWD, then a hybrid conversion will undoubtedly have electric traction motor driving the front wheels. Front/rear torque bias will likely be a favorable 30/70 F/R.

When buying an SUV or any other AWD hybrid, the customer should enquire as to whether the conversion is legacy FWD or legacy RWD. Asking may be a frustrating experience. However, the customer can investigate further on his/her own. Is the optional 2WD, FWD, or is it RWD? A legacy RWD is better from the point of view of tire loading and safety.

For best mpg, many AWD hybrid vehicles operate in the 2WD cruise mode of Table 11.3.

What is the hybridness of the two AWD examples of Figures 11.8 and 11.9? In the motor assist mode, both M/G are in M-mode providing traction. Hence

$$H = 60\ \text{kW}/(100 + 60\ \text{kW}) = 37.5\%$$

TABLE 11.3
Power and Load to Front and Rear Wheels for Two Hybrid AWD Designs Using Either Legacy FWD or Legacy RWD

Operational Mode	M/G 1	M/G 2	Power to Wheels	
			Front	Rear
Legacy Design: FWD				
2WD cruise	Off	Off	100 kW	0 kW
4WD cruise	G-mode, 30 kW	M-mode, 30 kW	70 kW	30 kW
4WD motor assist	M-mode, 30 kW	M-mode, 30 kW	130 kW	30 kW
Legacy Design: RWD				
2WD cruise	Off	Off	0 kW	100 kW
4WD cruise	M-mode, 30 kW	G-mode, 30 kW	30 kW	70 kW
4WD motor assist	M-mode, 30 kW	M-mode, 30 kW	30 kW	130 kW

SUMMARY

Hybridness provides insight into the overall character of the vehicle. Hybridness is applied to parallel, series, and mixed hybrids. Also, the discussion of plug-in hybrids appears throughout the chapter. Typical battery power and battery energy fall into two regions defined by hybridness, H. Mild and full hybrids comprise one region. Plug-in hybrids are in the second region.

Residual performance of the hybrid with a dead battery is a concern and is discussed as a function of hybridness. The ramps of Figure 11.7 summarize the usefulness of H and list various means to improve mpg. Finally, the importance of the legacy design in AWD hybrids is presented. A legacy RWD is better.

APPENDIX 11.1: DERIVATION OF RANGE EXTENSION EQUATIONS

This appendix has the derivation of the Equations 11.1 and 11.2. The symbols used in the derivation are

P_R is the power required to move the vehicle against resistance (kW)
P_B is the power from battery on continuous basis (kW)
P_E is the power of engine (kW)
P_G is the power from the M/G in G-mode (kW)
E_B is the energy from the battery (kWh)
E_G is the energy from M/G in G-mode (kWh)
E is the total energy from battery and generator (kWh)
R is the range with extender (km)
R_0 is the range without extender using battery energy only (km)
t is the battery discharge time with range extender (h)
t_0 is the battery discharge time without range extender (h)
V is the constant velocity (m/s)
H is the hybridness (dimensionless)

Note that $P_G = P_E$. The definition of hybridness for the range extender is

$$H = \frac{P_B}{P_B + P_E} = \frac{P_B}{P_B + P_G} = \frac{P_B}{P_R} \tag{A.11.1}$$

The battery-only range is

$$R_0 = Vt_0 = \frac{VE_B}{P_R} \quad \text{(A.11.2)}$$

With the range extender, the total energy produced is

$$E = E_B + E_G \quad \text{(A.11.3)}$$

The energy produced by the generator is

$$E_G = P_G t \quad \text{(A.11.4)}$$

The range with extender is

$$R = Vt = \frac{VE}{P_D} = \frac{(E_B + E_G)V}{P_D} \quad \text{(A.11.5)}$$

Combine Equations A.11.4 and A.11.5. The result is

$$R = \frac{E_B + (P_G R/V)}{P_D} V \quad \text{(A.11.6)}$$

Rearranging the preceding equation gives the range as

$$R(P_R - P_G) = E_B V \quad \text{(A.11.7)}$$

Total power required is

$$P_R = P_B + P_G \quad \text{(A.11.8)}$$

Combine Equations A.11.7 and A.11.8,

$$RP_B = E_B V \quad \text{(A.11.9)}$$

Using Equation A.11.2

$$R = \frac{E_B V}{P_B} \frac{P_R}{P_R} = R_0 \frac{P_R}{P_B} \quad \text{(A.11.10)}$$

Combine Equations A.11.1 and A.11.10,

$$R = \frac{R_0}{H} \quad \text{(A.11.11)}$$

The increase in range due to range extender follows from

$$\Delta R = R - R_0 \tag{A.11.12}$$

and as a result

$$\Delta R = R_0 \frac{(1-H)}{H} \tag{A.11.13}$$

The value of ΔR is given for two values of H. At $H = 1.0$, $\Delta R = R - R_0 = 0$ since the M/G has vanished. At $\Delta R = 0.5$, $R = 2R_0$. The sample calculation below matches Figure 11.3.

P_R = 20 kW
E_B = 40 kWh
V = 120 km/h
t = 40 kWh/20 kW = 2 h
R_0 = (120 km/h)(2 h) = 240 km
H = 0.75
R = R_0/H = (240 km)/0.75 = 320 km
$\Delta R = R - R_0$ = 320 − 240 km = 80 km

APPENDIX 11.2: RANGE OF A PLUG-IN HYBRID

How far can a plug-in hybrid go in electric-only mode with a large battery? An estimate of the energy required per kilometer of travel can be found from the GM EV-1 which had battery energy of E = 16.2 kWh and a range of 145 km. Hence, the energy/km

$$16{,}200 \text{ kWh}/145 \text{ km} = 112 \text{ W h/km}$$

Assume allowable SOC of 25%. The available energy is (0.25)(16.2 kWh) = 4.05 kWh. Assume the plug-in battery has energy E = 10 kWh. The electric-only range is

$$R = (10 \text{ kWh})(0.25)/(112 \text{ W h/km}) = 22 \text{ km} = 13 \text{ mi}$$

Now assume the allowed SOC = 100%. Then, R = 88 km = 54 mi.

REFERENCE

1. T. Quiroga, What to drive until the perpetual-motion machine arrives, *Car and Driver*, January 2006, p. 65.

12 Mileage Ratings

For good, nay superior, miles per gallon always drive downhill
with a tailwind; never drive uphill with a headwind.

INTRODUCTION

This chapter has three major sections. The first section covers how Environmental Protection Agency (EPA) conducts mileage tests. The earlier (2007 and prior years) shortfall between EPA mpg and real-world driving mpg is addressed. The second section discusses Corporate Average Fuel Economy (CAFE). The third section presents various suggestions and tips for improving fuel economy. Tips to achieve better mpg lurk everywhere; however, some of the information is folklore or old wives-tales. One article indicates a 100 lb load in the trunk of the car will decrease fuel economy by 2%. Another article provides a 4% reduction. Each article makes the statement without any basis for the number quoted. Information is provided here to justify and substantiate the numbers. Although this chapter focuses on mpg, emission measurements are an equally important task of EPA.

EPA mileage ratings and CAFE are federal government's response to various energy crises. The gasoline crisis in 1973 differed from the 2005 jump in gasoline prices. In 1973, gasoline was often not available at any price, and long lines extended down the block from the service station. Congress was stimulated to act now. The new CAFE regulations sailed through Congress. Congress wanted to avoid backlash from not acting. Ever since, CAFE has been a cantankerous issue stirring lots of debate. Since the early 1970s, Congress has wanted to avoid backlash by acting.

Congress is often maligned for what they have or have not done with EPA ratings and CAFE. Congress does respond to the electorate. The electorate (that is you) wanted the choice to buy heavy sport utility vehicles (SUVs). The electorate wanted cheap gasoline, lots more horsepower (hp), and dazzling acceleration. Until recently, Congress' response has been to stiff arm any increases in CAFE. Congress responds to the lobbyists. (Actually, lobbyists are a topic beyond the scope of this book.) Congressmen are told that an increase in CAFE will cause loss of jobs at the automakers. The congressmen are reminded that an increase in CAFE will make cars more expensive and that the voters will be angry.

In defense of Congress, rule making and devising test protocols are not easy. Any regulation needs to be fair and impartial. It must achieve its goals without loopholes or unforeseen consequences. The regulations should encourage the desired behavior. For 90 days, most new or changed rules are exposed to the public, industry, and state governments for comment and review. That is a commendable procedure. For EPA mileage tests, should the testing be done on the highway or inside a laboratory? EPA conducts only a small portion of mileage tests. Automakers conduct the lion's share. If tests are done on the highway, do standard highways exist globally? For outdoor testing, what about wind, temperature (winter and summer), and the condition of road surfaces (wet, dry, snow)? Testing in the laboratory does give reproducible results, which can be duplicated anywhere in the world. EPA has selected laboratory testing.

MILEAGE RATINGS

INTRODUCTION

A single number for fuel economy, mpg, does not exist. The observed mpg depends on many factors including driving techniques. An off-the-shelf Prius II can yield from 35 to 123 mpg. At a mileage competition, one Prius II driver actually obtained 123 mpg! How can such fantastic mpgs be obtained?

TABLE 12.1
Typical Fuel Economy in the United States

	Fuel Economy, mpg	
Vehicle Type	**City**	**Highway**
Midsize sedan	21	27
Large SUV	13	16
Pickup (four cylinder, light)	21	26
Pickup (V-8, full size)	13	15

In spite of the control complexity cited in Chapter 8, some owners of hybrids (who have earned the admiration of this author) gain entry to the control area network (CAN, see Figure 8.5) and modify the allowable boundaries for battery state of charge (SOC). (See also Figure 6.6, which defines the allowable SOC boundaries for a typical hybrid.) The driver modifies the battery ECU, which ordinarily makes 25% SOC available for regenerative or "coasting" braking, and instead, he/she makes 50% available. To utilize these new values of SOC, the driver must maintain real-time control over battery charge. For example, while driving the car, the battery SOC is controlled to be 100% at the bottom of a hill. (Obviously the driver must anticipate the hill to have SOC of 100%.) Ascending the hill, electrical assist is maintained until the SOC falls to 50%. Descending the hill, 50% SOC is available for regenerative "coasting." The amount of electrical operation is doubled with the resultant increased mpg.

Now the question becomes, why don't the hybrid car makers do the same thing? The allowable boundaries on SOC in Figure 6.6 provide safe operation with long battery life for *all driving cycles*. When SOC boundaries are modified, the driver becomes responsible for safe battery operation, and obviously the auto makers do not trust the average driver with this responsibility. Expansion of the SOC boundaries requires detailed information for specific driving situations, e.g., going up and down hills. Of course this information is available from the GPS and may someday be incorporated in future hybrid ECUs. Modification of the hybrid ECU will likely void the warranty.

The United States lags behind the rest of the world in fuel economy. Mileages in other countries are much higher than in the United States (Table 12.1)—Europe, 40 mpg; Japan, 45 mpg; and North America, 20 mpg. For contrast, the 1908 Ford Model T gave 25 mpg. This mileage was attained at low speeds of 35 mph. The old Ford had low compression ratio (low compression hinders good mpg) of 4:1 compared today to 10:1.

Little History

As a response to what was perceived as gouging by a few unscrupulous car dealers, in 1958, the legislation creating the Monroney sticker was passed. Monroney stickers are also known as manufacturers suggested retail price (MSRP) stickers. Congressman A.S. Monroney introduced the bill. In early 1973, the gas crisis caused voter uproar. Shortly thereafter, the legislation creating mileage ratings and CAFE was passed. Gas mileages are determined by the EPA. To extend the usefulness to the car-buying public, the EPA gas mileages were added to the Monroney sticker. The Monroney sticker has been in use for half a century. The EPA mileages have been in use for about 35 years. The two must have served a need since they have been in use for many years. See Figure 12.1 for a sample entry on a Monroney sticker.

By 1984, the EPA mileages had drifted (too optimistic) from the values the typical driver was achieving. Hence, an arbitrary reduction of test values was applied. The reduction was 10% to city

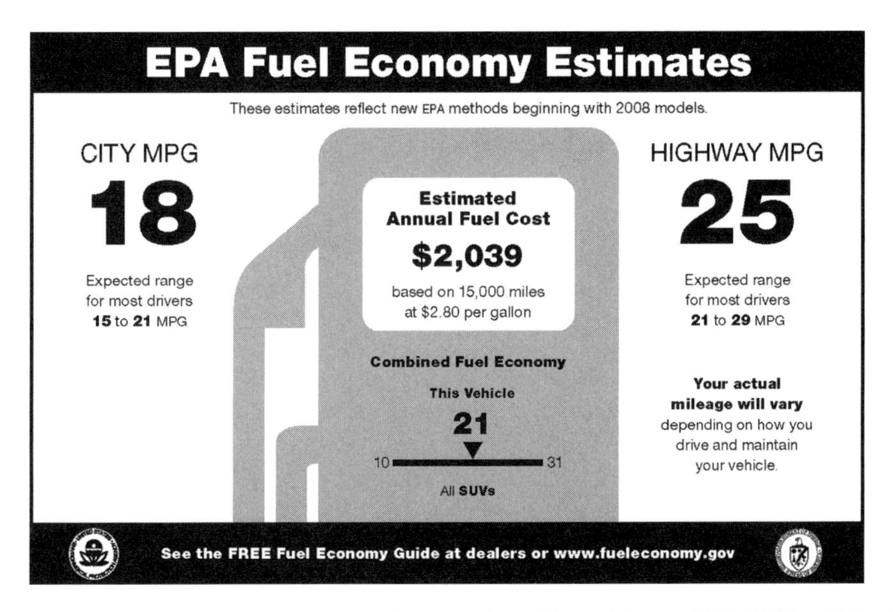

FIGURE 12.1 The new EPA fuel economy ratings starting with model year (MY) 2008 vehicles. This is found on the MSRP, or Monroney, sticker which appears on every new vehicle for sale.

and 22% to highway mpg. The reduced values are known as adjusted values. From 1984, the government has used three different mileages: These are the CAFE required values (to be discussed), the EPA as-tested numbers, and the EPA adjusted values. Table 12.2 shows the history of light-duty Vehicles (LDV) for four MYs spanning more than a quarter century.

Because of technology advances, fuel economy increased in spite of increased power and weight. Table 12.2 shows the dip in weight as well as the big increase in power. The time for 0–60 mph reflects the large increase in acceleration and power. Truck and SUV sales increased to the extent half of the new vehicles sold were large and fuel-thirsty. The number of vehicles equipped with four-wheel drive (4WD) increased markedly. 4WD adds weight and decreases fuel economy. A shift (pun!) from manual to automatic transmissions is also evident. Fortunately, the newer five-, six-, seven-, and even eight-speed automatics yield efficiencies approaching that of a manual transmission. The luxury Lexus LS 460L uses an eight-speed automatic transmission while the equally luxurious Mercedes Benz S550 uses a seven-speed automatic transmission.

TABLE 12.2
Characteristics of LDV for Four MYs

	1975	1987	1997	2006
Adjusted fuel economy	13.1	22.1	20.9	21.0
Weight (lb)	4060	3220	3727	4142
Horsepower	137	118	169	219
0–60 time (s)	14.1	13.1	11.0	9.7
Truck and SUV sales (%)	19	28	42	50
4WD sales (%)	3	10	42	50
Manual transmission (%)	23	29	14	8

PREVIOUS EPA MILEAGE ESTIMATES

WHAT CHANGES AND WHAT DOES NOT CHANGE?

The city and highway tests described below continue intact and unchanged into 2008 and beyond. Hence, a thorough description is warranted. The new EPA procedures add three driving cycles; these are discussed presently.

FACTS CONCERNING MILEAGE RATINGS

What is tested? New passenger cars and light trucks are tested; additional details are presented in subsequent paragraphs. The legal definitions for passenger cars and light trucks are given with the discussion of CAFE.

Who does the testing? The car manufacturer tests at least one sample of all new models and reports the results to EPA. EPA tests 10%–15% of new models to verify the reported results. The test specimen can be a hand-built prototype, pre-production version of a new model. The manufacturer has an opportunity to fine-tune the test specimen. A sample for testing is not selected randomly from the production line. This procedure would delay testing. Instead early testing is used. This accommodates the automaker and allows the automaker to meet the schedules for the introduction of new cars and trucks. Annually, approximately 1250 different vehicle models are tested.

DRIVING CYCLES

Driving cycles are important for achieving high fuel economy (see Chapter 8). Driving cycles are also important for measuring fuel economy. This fact is discussed later in this chapter. Driving cycles form an essential part of hybrid technology and are, therefore, discussed throughout the book (see Chapters 3 through 6, 8, 9, 11, and 16). Two EPA driving cycles continue unchanged from 2007 to 2008 and later years. These unchanged cycles are the city and highway driving cycles shown in Table 12.3.

TABLE 12.3
Details of the EPA City and Highway Driving Cycles Used in 2007 and Prior Years for Estimation of City and Highway Fuel Economy

Test Schedule Characteristics

	Test Schedule	
Driving Schedule Attributes	**City**	**Highway**
Trip type	Low speeds in stop-and-go urban traffic	Free-flow traffic at highway speeds
Simulated distance (mi)	11	10
Time (min)	31	12.5
Average speed (mph)	20	48
Top speed (mph)	56	60
Stops	23	None
Idling time	18% of time	None
Engine temperature at start-up[a]	Cold	Warm
Lab temperature	68°F–86°F	68°F–86°F
Vehicle air-conditioning	Off	Off

[a] Vehicle's engine does not reach maximum fuel efficiency until it is warm.

Standardized Test Procedures

The test procedures are specified in federal law [45]. As stated in the introduction, EPA has selected, indoor laboratory testing. Remember that a few hundred different car factories spread globally must provide reproducible test results. The EPA Web site has useful information [1].

The vehicle is tested using a dynamometer as shown in Figure 12.2. A driver operates the car so as to follow two prescribed driving cycles: one for city and one for highway. Fuel consumption is accurately measured by trapping the exhaust gases. The amount of carbon per gallon of gasoline is known. By measuring the amount of carbon in various carbon-bearing molecules, the gallons of fuel can be determined. Trapping the exhaust gases is essential for emission measurements. Less than 1 gal of fuel is used to complete the city and the highway cycles.

The ducting connecting the exhaust to the test and analysis equipment can be seen in Figure 12.3. The vehicle is inside a closed garage; the closed door can also be seen. Two driving cycles are performed. Tests are at room temperature, 68°F–86°F.

The city driving cycle is intended to duplicate the stop-and-go, creep-and-crawl commuter traffic. The initial part starts with a cold engine. The city cycle is 11 mi. For conventional cars, when the vehicle is stopped, the engine idles. Hybrid electric vehicles (HEVs) may invoke the engine-off mode during stops.

The highway driving cycle represents a mixture of rural country roads with interstate. The engine is at operating temperature. Traffic is not clogged but is flowing freely. The cycle does not have stops; hence, idling is at a minimum. The highway cycle is 10 mi. Maximum velocity is 60 mph. The fraction of overall vehicle operation is 0.55 for city driving, and 0.45 for highway driving. Table 12.3 summarizes the details of the two driving cycles. The highway cycle does not have stops or periods of engine idle. Hybrids are thereby penalized by lack of stops and idle.

Vehicles Exempt from EPA Mileage Sticker

Prior to 2008, vehicles with gross vehicle weight rating (GVWR) over 8500 lb were exempt from EPA mileage ratings and CAFE standards. GVWR includes the empty vehicle plus all fluid (gasoline or diesel fuel) and maximum carrying capacity. Maximum capacity includes

FIGURE 12.2 Vehicle positioned on a dynamometer in an EPA test facility. Tests for fuel economy and emissions are conducted. Note the two rollers for the dynamometer. The two rear drive wheels are on the rollers. (Reproduced from Environmental Protection Agency. Available at www.fueleconomy.gov/feg/fe_test_schedules.shtml.)

FIGURE 12.3 Test vehicle inside an EPA test facility. (Reproduced from Environmental Protection Agency. Available at www.fueleconomy.gov/feg/fe_test_schedules.shtml.)

passengers and cargo. After 2008, heavy vehicles primarily intended to carry passengers must have mileage testing and EPA mileage stickers. A Chevrolet Suburban is mainly a passenger vehicle and must have a sticker. A Ford 350 "dually" pickup is mainly a cargo or towing vehicle and remains exempt.

Mileage Shortfall

In the year 2007 and earlier, the numbers posted on the Monroney sticker did not reflect the mileage attained by the typical driver. Numerous reasons exist for the shortfall. In any event, expectations did not agree with the sticker numbers.

Consequences of Mileage Shortfall

The motivation for the new EPA five-cycle test procedure is to eliminate or reduce the mileage shortfall. A goal of the national energy policy is to decrease dependence on imported oil. Energy is required for the wheels of industry to spin. Energy is necessary for the transportation system to function. Dependence on foreign oil puts the country hostage to the countries that have oil.

Implementation of the national energy policy involves, in part, the CAFE which is a topic to be discussed shortly. CAFE depends on EPA mileages. A discrepancy (overly optimistic) in EPA mileage distorts, weakens, and bypasses the energy policy. From the point of view of the automobile manufacturers, the shortfall enables them to get by with lower fuel economy, F_E. Further, the inflated mileages become a deceptive sales tool. The customer buys expecting to attain the posted mileage. Complaints about the exaggerated mileages can be countered with the statement that the numbers belong to EPA and not the automaker.

From the point of view of the government, the façade can be maintained that high standards have been implemented. The shortfall gives the impression that the government is being tough and is catering to the country's needs. The automobile industry is let off the hook.

From the point of view of the petroleum industry, the mileage shortfall is a case of use more and sell more, but yet be in compliance with national energy policy.

From the point of view of the customer, the shortfall is a case of shortchanging.

EFFECTIVE \$/GAL DUE TO SHORTFALL

A numerical example is used to illustrate the concept of the price the customer pays for gasoline. Assume the EPA mileage is 27 mpg while the observed mileage is 22 mpg. At the observed mpg, the cents/mile is given by

$$\frac{\text{Cents}}{\text{Mile}} = \frac{100(\$/\text{gal})}{\text{miles/gallon}} \tag{12.1}$$

Assume gasoline at \$3/gal. Using the assumed numerical values, the cents/mile is

$$\frac{\text{Cents}}{\text{Mile}} = \frac{100(\$3/\text{gal})}{(22\ \text{mpg})} = 13.6\ ¢/\text{mi}$$

If 27 mpg were actually attained, then what is the effective price for a gallon of gasoline? Manipulate Equation 12.1 to give

$$\frac{\text{Dollars}}{\text{Gallon}} = \frac{(\text{Cents/mile})(\text{miles/gallon})}{(100\,¢/\text{gal})} \tag{12.2}$$

Using the numbers

$$\frac{\text{Dollars}}{\text{Gallon}} = \frac{(13.6\ ¢/\text{mi})\ (27\ \text{mpg})}{(100)} = \$3.67/\text{gal}$$

Due to the shortfall, the effective cost of gasoline is \$3.67/gal.

DEFICIENCIES IN OLD AND NEW EPA METHODS

REVISED TESTS AND STANDARDS

The EPA mileage tests essentially have been unchanged since the 1970s. Prior to 2008, EPA tests failed to take into account the plugged commuter roads and high speed interstate driving. Sections of I-40 in Arizona and New Mexico have posted speed limits of 75 mph. Some drivers automatically add 10% to speed limit and drive at that speed. Since maximum speed is 60 mph for EPA tests, the added fuel to drive 75 mph is not accounted for. See "Optimum Cruise Speed."

The new tests and standards are being introduced. In 2008, the effects of air-conditioning (A/C) are included. Chapter 5 lists many automobile auxiliary and accessory components, which are new or are more widespread since 1973 and which affect fuel economy. Auxiliaries in this category include

- Entertainment systems
- Window defrost
- Electric brakes (coming)
- Electrically heated catalyst
- Electrically heated seats and mirrors
- Power door locks
- Electric valve actuation (it is coming!)
- Electric windows
- Headlights and running lights
- Adaptive cruise control (uses radar)
- Navigation system
- Power sunroof

A typical luxury car has alternator power of 2–4 kW. The new tests will account for the auxiliaries. The mileage ratings are expected to decrease by 10%–15% with the new tests.

Effects of Ambient Temperature

The effect of ambient temperature on fuel economy also is included. See Cold federal test procedure (FTP) in Table 12.5. In 2007, and extending into 2008 and beyond, the FTP tests are conducted when temperature is in the range of 68°F–86°F. The Cold FTP tests are at 20°F. The new tests do introduce new procedures for cold weather.

Testing Four-Wheel Drive

Look at Figure 12.2. Only one set of rollers for the dynamometer is installed. To test a 4WD vehicle, another set of rollers is needed for the front wheels. Since wheelbases vary from vehicle to vehicle, the spacing between roller sets must be variable. EPA does not test 4WD vehicles. Currently, one out of four vehicles on the road is 4WD. The new tests do not introduce any new procedures for 4WD.

Rolling Friction

As discussed in later pages, a source of resistance to motion is tire rolling friction. Chapters 7 and 9 discuss rolling friction and the connection with the tire patch on the road. This resistance originates in the four tire patches on the ground. Figures 12.2 and 12.3 show the front wheels are inactive, and the front wheel contribution is not measured. However, there are four patches. The two rear wheels each have two patches on the rollers, which are partial compensation. Tires have waves, which run in both directions around the circumference of the tire. Abnormal test results may occur if waves from the first roller pile up on the second roller at some critical speed. The new tests do not introduce any new procedures relative to rolling friction.

Aerodynamic Drag

The garage in Figure 12.3 is not a wind tunnel. So, how is the major effect of aerodynamic drag incorporated into the tests? Aerodynamic drag is also discussed in later pages. The effects of aerodynamic drag are calculated and are included by adjusting the torque on the rollers. To calculate the drag, three values for variables are needed. These are ρ, atmospheric density; C_D, aerodynamic drag coefficient; and A, the vehicle reference (frontal) area for drag. Density is easy; values for C_D and A are more difficult. The automaker must supply the numbers. The new tests do not introduce any new procedures relative to aerodynamic drag.

Vehicle Load

What load should the vehicle carry during EPA mileage tests? Should the vehicle be at maximum weight? EPA tests with a load of 300 lb. Consider the effect of load on fuel economy of a minivan, the 2006 Honda Odyssey. The weights of interest are as follows:

Empty	4537 lb	—
EPA test weight	4837 lb	20 mpg EPA city
GVWR	5857 lb	—

When the vehicle is at GVWR, what is the fuel economy? Equation 7.8 provides an answer to the question. The equation is repeated here as Equation 12.3 for convenience. The fractional change of fuel consumption, F_C, and fuel economy, F_E, due to weight change is

$$\frac{\Delta F_C}{F_C} = \frac{\Delta W}{W} = \frac{\Delta m}{m} = -\frac{\Delta F_E}{F_E} \quad (12.3)$$

where
W is vehicle weight, lb
m is vehicle mass, kg

To apply Equation 12.3, the following formula is used:

$$F_E = F_{EO}\left(1 - \frac{\Delta W}{W}\right) \quad (12.4)$$

Numerical values are inserted to find F_E for GVWR; the result is

$$F_E = (20\text{ mpg})\left[\frac{1-(5857-4837)}{(1/2)(4837+5857)}\right] = (20)(1-0.191) = 16.2\text{ mpg}$$

The average weight, (1/2)(4837 + 5857), is used in $\Delta W/W$. When at GVWR, the F_E is reduced from 20 to 16.2 mpg for city driving. The conclusion is that test weight has an important effect on measured mpg. The new tests do not introduce any new procedures relative to load carried by the vehicle.

Aggressive Driving

For purposes of EPA mileage ratings, aggressive driving is defined by the maximum acceleration. In government publications the unit used for acceleration is mph/s. This unit is satisfactory but is somewhat devoid of physical meaning. The acceleration referred to earth's gravity, g, gives the familiar G's. G equals the actual acceleration divided by acceleration of gravity. Required acceleration ranges from 3.3 mph/s (0.15 G's) to 12 mph/s (0.55 G's). Using $g = 32.2\text{ ft/s}^2$ the corresponding G's are as stated in parentheses. Earlier dynamometers could not test high acceleration rates; modern dynamometers can and this fact allows the change in EPA tests. Dynamometers designed specifically for testing hybrids have been developed [53,54].

Driving Cycle

Two aspects of the driving cycle merit attention. Commuting on today's roads frequently involves more stop-and-go with longer idle periods than was the case in 1973. Also, maximum speeds on interstate have increased since 1973. A revised driving cycle can correct these two factors. The existing test, US06, introduces high speed (up to 80 mph) and aggressive driving (up to 0.55 G's). This accounts for the limitations stated above.

A reexamination of urban and interstate driving did not reveal a need to change the 0.55/0.45 split. Currently, and downstream of 2008, the fraction of overall vehicle operation is 0.55 for city driving and 0.45 for highway driving.

A/C and Auxiliaries

The use of A/C and other power-hungry auxiliaries was discussed in connection with revised tests and standards. Operation of A/C and other auxiliaries during tests will reduce fuel economy. The new test, SC03, introduces use of the A/C at 95°F and 45% relative humidity (RH).

Terrain, Curves, and Altitude

In real life, if the road goes uphill, the vehicle is required to go uphill also; obviously. Hilly or mountainous terrain reduces mpg. A slight reduction in fuel economy occurs due to curves in the road. When going along a curve, a drag component is introduced by the tire patch. Any increased resistance (drag) decreases mpg.

Air density varies with altitude and affects aerodynamic drag. Also engine efficiency depends on air density. High altitude tends to increase mpg. The new tests do not introduce any new procedures relative to terrain, curves, and altitude.

Weather

Blame ice and snow on the weather. Blame muddy roads on the weather. Both by-products of the weather adversely affect mpg. Wet or snowy road surfaces decrease mpg.

The temperature determines the need and usage of A/C. As discussed elsewhere in this book, A/C is not a minor problem relative to attainment of high mpg. The hot weather of summer or the desert tends to improve fuel economy while the cold weather of winter tends to decrease mpg.

Some people win, some people lose. We are talking about headwinds and tailwinds. A tailwind not only reduces noise inside a vehicle, it also increases mpg by 10%–20%. Tailwinds help while headwinds decrease fuel economy.

One shortcoming of EPA mileage estimates, which is being corrected, is to test vehicles without recognizing the effect of weather on mpg. On hot days, of special importance is the effect of A/C on mpg. A vehicle started from cold-soak, for example, overnight on streets in Minnesota in winter, consumes more fuel.

NEW EPA TESTS TO ESTIMATE MILEAGE

For 2008 and beyond, EPA estimates are based on five tests or driving cycles. Tables 12.4 and 12.5 summarize the five tests.

Look back at Table 12.3. The city and highway columns are identical to the FTP and highway fuel economy test (HFET), which carry over unchanged from 2007 to 2008. Table 12.4 has five tests that are combined to yield the five-cycle estimate. The method for combining the tests is now discussed.

Five-Cycle Estimate: EPA-Weighted Combination of Tests

A single number appears on the mileage sticker for city mileage and another number for highway mileage (see Figure 12.1). The single number is found by adding together the results from the EPA tests tabulated in Table 12.4. The results are combined by weighted addition. The equations used by EPA, which are lengthy and complex, are much too complicated to discuss here. (It is somewhat analogous to the IRS income tax code!) The equations are given in detail by Ref. [45]. The concepts can be simply illustrated by the following example. The equation is not the same as used by the EPA; however, the easily understood concepts are correct.

$$F_{E1} = W_1 F_1 + W_2 F_2 + W_3 F_3 + W_4 F_4 \quad (12.5)$$

where

F_{E1} is the EPA city mileage
F_1 is the fuel economy from FTP cycle, 24 mpg
F_2 is the fuel economy from US06 cycle, 14 mpg
F_3 is the fuel economy from SC03 cycle, 20 mpg
F_4 is the fuel economy from Cold FTP cycle, 18 mpg

TABLE 12.4

Details of the EPA City and Highway Driving Cycles Used to Estimate Fuel Economy for 2008 and Subsequent Years

Old Tests		New Tests			
City	Highway	High Speed	Air-Conditioning	Cold Temperature	Detailed Comparison
	Test Schedule				
Driving Schedule Attributes	**City**	**Highway**	**High Speed**	**A/C**	**Cold Temperature**
Trip type	Low speeds in stop-and-go urban traffic	Free-flow traffic at highway speeds	Higher speeds; harder acceleration and braking	A/C use under hot ambient conditions	City test w/colder outside temperature
Top speed (mph)	56	60	80	54.8	56
Average speed (mph)	20	48	48	22	20
Maximum acceleration (mph/s)	3.3	3.2	8.46	5.1	3.3
Simulated distance (mi)	11	10	8	3.6	11
Time (min)	31	12.5	10	9.9	31
Stops	23	None	4	5	23
Idling time	18% of time	None	7% of time	19% of time	18% of time
Engine Start-up[a]	Cold	Warm	Warm	Warm	Cold
Lab temperature (°F)		68–86		95	20
Vehicle air-conditioning	Off	Off	Off	On	Off

Notes: The city column is identified as FTP in Table 12.5. Also, the highway column is identified as HFET in Table 12.5.

[a] Vehicle's engine does not reach maximum fuel efficiency until it is warm.

TABLE 12.5

EPA Five Test (Driving) Cycles

Test	Designed to Represent	Average Speed	Maximum Speed, mph	Maximum Acceleration	Ambient Conditions	Primary Use
FTP	Urban stop-and-go, 1970s	21	58	3.3	75°F	Emissions and fuel economy
HFET	Rural driving	48	60	3.3	75°F	Fuel economy
US06	High speeds and aggressive driving	48	80	8.5	75°F	Emissions
SC03	Air-conditioner operation	22	56	5.5	95°F 45% RH	Emissions
Cold FTP	Cold temperature	21	58	3.3	20°F	Emissions

Source: Reproduced from *Federal Register*, Part II, Environmental Protection Agency, Fuel Economy Labeling of Motor Vehicles: Revisions to Improve Calculation of Fuel Economy Estimates; Final Rule, December 27, 2006, p. 77876.

Notes: FTP, federal test procedure identified as city in Table 12.4; HFET, highway fuel economy test identified as highway in Table 12.4; RH, relative humidity.

W_1 is the weighting factor for FTP cycle, 0.62
W_2 is the weighting factor for US06 cycle, 0.06
W_3 is the weighting factor for SC03 cycle, 0.11
W_4 is the weighting factor for Cold FTP cycle, 0.21

Typical numerical values are given for each symbol. The estimated city fuel economy is

$$F_{E1} = (0.62)(24) + (0.06)(14) + (0.11)(20) + (0.21)(18) = 21.7\,\text{mpg}$$

In 2007, the estimated city mileage was simply the fuel economy from FTP cycle equal to 24 mpg. In 2008, using the weighed average, the estimated mileage is 21.7 mpg. The US06, SC03, and Cold FTP tests give lower mileage than the FTP test. The equation includes these lower values leading to a lower fuel economy.

Importance of Weighting Factors

The expensive laboratory tests are designed to yield reproducible results. The estimated mileage depends critically on values assigned to the weighting factors. As an example, change the factors above to an arbitrary new set: $W_1 = 0.50$, $W_2 = 0.40$, $W_3 = 0.06$, and $W_4 = 0.04$. The mpg becomes

$$F_{E1} = (0.50)(24) + (0.40)(14) + (0.06)(20) + (0.04)(18) = 19.5\,\text{mpg}$$

Compare 19.5 with 21.7 mpg; both are based on the same test results but different weighting factors. A strong rationale must exist for determining the values of the weighting factors.

Change in Vocabulary: City and Highway

In the simple days of yore (2007 and before), city mileage was simply based on the second column of Table 12.3 (city). Likewise, highway mileage was simply based on the third column of Table 12.3 (highway). The sticker showing the EPA fuel economy ratings (continuing with MY 2008 vehicles) shows only two numbers—one for city and one for highway. However, these numbers are based on the five-cycle method discussed above. Table 12.6 clarifies the situation.

Relative to EPA mileage ratings, in 2007, city meant something different than city in 2008. In 2007, highway meant something different than highway in 2008.

TABLE 12.6
Tests Used to Determine City and Highway Mileage Estimates

		Tests Included 2007 and Prior Years		Tests Included 2008 and Later Years	
Test	Designed to Represent	City	Highway	City	Highway
FTP	Urban stop-and-go,	Yes	No	Yes	No
HFET	Rural driving	No	Yes	No	Yes
US06	High speeds and aggressive driving	No	No	Yes	Yes
SC03	Air-conditioner	No	No	Yes	Yes
Cold FTP	Cold temperature	No	No	Yes	Yes

Note: FTP, federal test procedure; HFET, highway fuel economy test.

Transition Period: 2008–2010

In the 3-year period, car makers can select one of two ways to estimate mileage for the EPA sticker. One method is based on a scale of adjustments from the five-cycle method. This method is known as mpg-based approach and is not vehicle specific. However, the method does include effects of higher speed, use of A/C, etc. The second choice is the measurement of all five tests (five-cycle method) for a specific vehicle and combining the results in the prescribed weighted manner. After 2010, every new vehicle will be evaluated using the vehicle-specific five-cycle method.

TESTING HYBRIDS

Hybrids and the New EPA Tests to Estimate Fuel Economy

For the FTP cycle, the new EPA procedures to estimate mileage have a different set of weighting factors, equations, and measurements for HEVs. The same five-cycle tests are used. Details are given in Ref. [45].

Hybrid Modes of Operation

Hybrids pose a problem with regard to mileage testing. Each hybrid operates in different modes depending on driver input, battery state of charge (SOC), and the control system commands to maximize fuel economy. For an example of hybrid modes see Chapter 3 and the discussion of the hybrid modes for Saturn Vue Green Line. The modes are accelerating, cruising, decelerating, and full stop. Note that the hybrid has both motor assist and regenerative braking. From the point of view of EPA, the vehicle will be run over the prescribed driving cycle. From the point of view of the automaker, the desire is to achieve the best mpg. The control system of the hybrid will likely be optimized for the EPA test driving cycle.

Electric-only operation yields a significant increase in mpg. This mode needs to be integrated into the test procedure. If not integrated, the HEV is penalized. When stopped, most HEVs turn off the engine. This feature can be readily integrated into the test. Under certain circumstances the engine-off mode is overridden. With cold ambient temperatures and with the heater on, engine-off may not be allowed. Likewise, with hot ambient temperatures and with the A/C on, engine-off is not commanded by the control system. The fuel economy benefits of engine-off are not reaped, and the hybrid is penalized by these conditions. As reported in Ref. [5], FEV Engine Technology, Inc. has added two dedicated test cells for hybrid testing. Dynamometers must be modified to accommodate the high torque and high power of hybrids. In addition to testing of fuel economy, the level of emissions from hybrids can also be measured by FEV.

Driving Technique

Hybrid fuel economy is sensitive to driving technique. Capturing driving technique in a test is very difficult. So, in the real world, some drivers will beat EPA mpg, and others will be deficient.

Regenerative Braking

Regenerative braking affects fuel economy. What is a fair test so that the HEV capability is not downgraded by the test? The answer may be to define a hybrid-specific driving cycle. The HEV operates over that cycle and takes whatever mpg is generated.

Battery State of Charge

The installed battery energy for a typical HEV is 2 kWh. If the test starts and ends with identical SOC, the battery has not contributed to fuel economy. However, if the battery starts the test fully charged and ends the test discharged, the measured fuel economy will be erroneous. How large an effect does battery SOC have on mpg?

The energy of the gasoline consumed during the test is compared to the battery energy. For current EPA test, the city driving distance is 11 mi and highway is 10 mi. Assume city 45 mpg and highway 40 mpg. Gasoline used in city driving is

$$\text{Gallons used} = \frac{11\text{ mi}}{45\text{ mpg}} = 0.24\text{ gal}$$

The gasoline used in highway driving is 0.25 gal. The fraction of the battery energy compared to gasoline energy, Φ, is given by

$$\Phi = \frac{\text{Battery energy}}{(\text{Gallons gasoline})(\text{Gasoline energy density})} \tag{12.6}$$

From Appendix A, the energy density of gasoline is 33.44 kWh/gal. Inserting values into Equation 12.6 gives

$$\Phi_{\text{CITY}} = (2\text{ kWh})/(0.24\text{ gal})(33.44\text{ kWh/gal}) = 0.245$$

For city driving, almost one-fourth of the energy comes from the battery. For highway driving, the result is $\Phi_{\text{HIGHWAY}} = 0.239$. The inflated fuel economy, F_E, is found using

$$F_E = \frac{F_{E0}}{1-\Phi} \tag{12.7}$$

Once again inserting values gives for city driving

$$F_E = \frac{45\text{ mpg}}{1-0.245} = 59.6\text{ mpg}$$

For highway driving, the result is $F_E = 52.6$ mpg. Obviously the potential for major errors exists if the beginning and ending battery SOC is ignored.

Hybrids, Braking, and the Panic Stop

The FTP and HFET cycles have maximum braking of 3.3 mph/s (0.15 G's). In real life, the ordinary automobile is capable of 17 mph/s (0.77 G's). Sports cars are capable of 1.0 G. The aggressive driving cycle, US06, which has decelerations of 0.77 G's, is part of the new EPA fuel economy estimates.

CORPORATE AVERAGE FUEL ECONOMY

Introduction

CAFE is the sales-weighted average fuel economy for a manufacturer's annual production of cars and trucks [38–44]. Fuel economy is defined as miles traveled per gallon of gasoline, mpg. The values for mpg are determined using test protocols and tests stated by the EPA. CAFE is administered by the National Highway and Traffic Safety Agency (NHTSA). Four factors determine the CAFE mileage standards:

1. Technological feasibility
2. Economic practicability
3. Effects of other standards on fuel economy
4. Needs of the nation to conserve energy

Despite its flaws, CAFE did double mpg following the 1970s gas crises. Since then average mpg has fallen due to the sales shift to 50% SUVs and minivans. Looking back, one wishes that CAFE standards had increased at a modest rate of 0.5 mpg per year. Today, we would be driving more passenger cars yielding 40 mpg.

Marked Inconsistency

On one hand, CAFE standards press the car makers to produce more efficient cars and trucks. Yet on the other hand, government policies are oriented toward low gasoline prices. One person stated "In America, people feel it is a God-given right to buy gas at $1.00 per gallon!" Higher gas prices would create a demand for fuel-efficient vehicles. Car makers must sell cars to exist. If the car-buying public demands high mpg, CAFE encouragement is not the only factor improving mpg. The trends of 2006 and 2007 confirm the relationship between high gas prices and sale of fuel-efficient vehicles. High gasoline prices provide an incentive for owners of existing vehicles to drive less. However, the trends of 2006 and 2007 show that high gas prices do not discourage driving.

High gas prices resulting from high demand and low supply are a product of the marketplace. An increase in the federal gas tax is less viable when the market drives prices upward. Even so, arguments can be made in favor of an increased gas tax. Gas taxes are high in Europe and Japan. Why do these countries see merit in high gas tax?

Definitions for Passenger Cars and Trucks

Many people ask "How can an SUV be called a truck?" The CAFE definitions answer that question.

Passenger car is any four-wheel vehicle not designed for off-road use that is manufactured primarily for use in transporting 10 people or less.

Truck is a four-wheel vehicle that is designed for off-road use or has GVWR in excess of 6000 lb, and has physical features consistent with those of a truck. It must have at least four of the physical features:

1. Approach angle of not less than 28°
2. Breakover angle of not less than 14°
3. Departure angle of not less than 20°
4. Running clearance of not less than 20 cm
5. Front and rear axle clearances of not less than 18 cm each

Alternatively, a vehicle is a truck when designed to perform at least one of the following functions:

1. Transport more than 10 people
2. Provide temporary living quarters
3. Transport property in an open bed
4. Permit greater cargo-carrying capacity than passenger-carrying volume
5. Transform to an open bed vehicle by removal of rear seats to form a flat continuous floor with the use of simple hand tools

Usually SUVs qualify as a truck in accordance with item 4 above. Originally light trucks meant pickups and boxy delivery vans. The term now includes SUVs and minivans.

Passenger Car Standards

The current (2007) standard for passenger cars is 27.5 mpg. This value has been in effect since MY 1990. Due to the gas crises of the early 1970s, the goal of CAFE standards was to double the mpg from 14 to 27.5 mpg.

Truck Standards: Unreformed System

Initially, truck standards were different for 2WD and 4WD vehicles. This distinction was eliminated in MY 1992 when 20.2 mpg was set as the standard. New truck standards were issued in 2003 as follows:

MY 2005	21.0 mpg	MY 2008	22.5 mpg
MY 2006	21.6 mpg	MY 2009	23.1 mpg
MY 2007	22.2 mpg	MY 2010	23.5 mpg

These standards are known as the unreformed system.

Truck Standards: Reformed System

For the time period 2008–2010, the light truck manufacturers have the option of meeting the requirements of the reformed or unreformed system. The reformed system is based on footprint, which is the area defined by

$$\text{Footprint} = (\text{track})(\text{wheelbase})$$

Track is the average distance between front and rear tires. Wheelbase is between the centers of the front and rear tires. Each footprint is assigned a target fuel economy as shown in Figure 12.4.

The reformed system for establishing fuel economy for light trucks is based on the curve of Figure 12.4. Instead of weight as the attribute defining target mpg, footprint is used. For a footprint equal to 42 sq ft, the target is 28 mpg. For all footprints greater than 65 sq ft, the target is a constant 22 mpg.

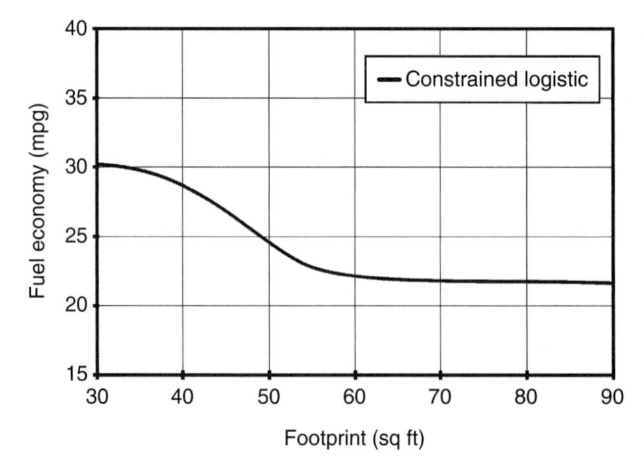

FIGURE 12.4 Target fuel economy, mpg, for light trucks as a function of footprint. (Reproduced from National Highway Traffic Safety Administration, *CAFE Overview*, 2007. Available at www.nhsta.dot.gov/portal/site/nhtsa/template.)

Penalties for Failure to Meet CAFE Standards

The car maker pays $5.50 for each tenth of an mpg the average fails to meet the standard. This is multiplied by the total number of all vehicles produced. Since 1983, car manufacturers have paid more than $500 million in penalties. Some European car makers regularly pay penalties. Since standards have been met, Asian and domestic producers have never paid a penalty.

CAFE Credits

Car makers can earn credits to offset years when they exceed CAFE standards. These credits can be used to offset years when the standards are not met. These are "carry forward" credits. "Carry back" credits can be used. If the standards are exceeded in a year after a year that standards are not met, the credits earned can be applied backwards. Credits can be used for 3 years on either side of current year.

Import Fleets

The rules are different for cars and trucks. Cars follow the two-fleet rule. Manufacturers with both domestic and import production must meet standards for domestic and import. The two fleets cannot be mixed for calculating CAFE. For passenger cars, any vehicle manufactured in United States, Canada, or Mexico is considered to be domestic if 75% of content is domestic. For trucks, the two-fleet rule has been eliminated.

Special Rules for Alternate Fuel Vehicles

To encourage use of alternate fuel vehicles, special mpg standards apply. The fuel economy of an alternate fuel vehicle is determined by dividing actual mpg by 0.15. For example, if the vehicle has 30 mpg, it is calculated as having 30/0.15 = 200 mpg for CAFE. Very generous. The alternate fuel incentive extends to the end of the 2008 MY.

CAFE Exempt Vehicles

In 2007 and earlier, vehicles with GVWR over 8500 lb were exempt from EPA mileage ratings and CAFE standards. GVWR includes the empty vehicle plus all fluids (gasoline or diesel fuel) and maximum carrying capacity. Maximum capacity includes passengers and cargo. After 2008, heavy vehicles primarily intended to carry passengers must have mileage testing and EPA mileage stickers. These passenger-carrying vehicles fall under CAFE. A Chevrolet Suburban is mainly a passenger vehicle and must have a sticker. A Ford 350 "dually" pickup is mainly a cargo or towing vehicle and remains exempt.

National Academies' National Research Council Recommendations

Light-duty trucks (SUVs, minivans, pickups, delivery vans) offer the greatest potential for saving gasoline. New technologies, even though adding cost to a vehicle, can improve mpg. These technologies include variable value timing, five- and six-speed automatic transmissions, dual clutches allowing easy-to-operate manual transmissions, and advanced lightweight materials.

The mpg can be improved by one of two means. First, the engine and power train can be made more efficient. Second, the resistance to vehicle motion can be decreased. Aerodynamic drag can be reduced by sleeker shapes, body underpans, and smaller frontal area. Rolling friction, which is a resistance to motion, can be decreased by lower weight vehicles. Lower weight vehicles introduce a new problem, that is, the safety of the occupants in the vehicle, which may be only partially real. National Research Council (NRC) does feel more analysis is needed with regard to safety and weight.

The system of credits should be expanded to permit tradable credits. The credits could be bankrolled to offset future deficits. The credits could be bought from or sold to other car makers. Tradable credits yield a profit to efficient car manufacturers for each gain in mpg. The profit provides motivation to continue making improvements even if CAFE standards are being met. The price of a credit on the open market would reflect the true cost of improving mpg. Policy decisions can be based on better information.

Currently standards apply to two classes of vehicles: passenger cars and trucks. Standards applied to attribute-based features of a vehicle may achieve desired goals. A weight-based system that encourages car manufacturers to downsize largest vehicles could ultimately reduce the enormous variance between small and large vehicles. Another attribute that is being considered is vehicle footprint. Footprint is obtained by multiplying the wheelbase by track. Wheelbase is the distance between the rear and front wheels.

The two-fleet rule has been rendered obsolete by the global marketplace. The two-fleet rule was passed by Congress to help protect domestic car makers and the number of jobs in the auto industry. No evidence exists indicating either positive or negative impact. Hence, the two-fleet rule should be eliminated.

CAFE standards allow a special credit (see factor of 0.15 discussed above) for dual fuel vehicles. Automakers can use the credits to compensate for less efficient vehicles, and have a negative effect on overall fuel economy. These credits should be eliminated.

Prior cooperative R&D programs between government and industry have been successful. These programs are discussed in Chapter 1. The official name for one of these programs was Partnership for a New Generation of Vehicles. The unofficial name was Supercar. The partnership was between the federal government and domestic car manufacturers consisting of Ford, GM, and Chrysler. Supercar was keenly supported by President Clinton and particularly the "green" Vice President Al Gore. Another ongoing program is Freedom Car which is also discussed in Chapter 1. Further study of hybrid vehicles, fuel cells, advanced engines, and emission control technology is warranted. Some advanced technology, which is common in other parts of the world, is not used in United States because of stringent emission requirements. Diesels and lean-burn gasoline engines are two examples.

CAFE and Hybrid Technology

New CAFE standards for vehicle mileage will certainly increase interest in hybrids [50].

General Motors View on 35 mpg CAFE Standard

The viewpoints of GM on fuel economy, energy conservation, and the 35 mpg CAFE standard are discussed in References 12–56: "Lutz promotes energy conservation, criticizes politicians." In regard to the 35 mpg standard, the huge leap in mileage could cost as much as $6000–$7000 per vehicle. Allegedly, Congress misled the public by giving the impression that the 35 mpg standard is free. There are no free lunches. The CAFE standard will have a major impact on the sales of new cars. Individuals and fleets will keep cars a couple of years longer. Alternate ways exist for achieving the goals. In the short term, going to EV is the best way to lower petroleum consumption.

HOW TO IMPROVE FUEL ECONOMY: TIPS FOR BETTER MILES/GALLON

Debunking Tips for Better mpg: What Is a Myth?

The realm of how to improve fuel economy is a mixture of reality, myths, and folklore [7,8,13, 24–32]. Among the latter category, that is, myths, and folklore, are examples to be discussed in subsequent paragraphs. Here are some examples:

- Negative mpg when idling.
- Buy gas in the cool of the morning.

- Phony test accuracies implied.
- Tire temperature and tire pressure.
- Aggressive driving: rapid acceleration does not require significantly more gas than gentle acceleration!
- Aggressive driving: hard braking has little or no effect on mpg.
- Which uses less gasoline? Windows down? A/C? Opposing views are stated by the media; which is correct?
- Overlooked: even minor speeding causes more loss of mpg than would be expected.

Negative Miles/Gallon

Negative mpg may be below your level of amazement. After discussion of the consequences, you will agree this instance is the poster child of erroneous and misleading statements. The concept, which appears in Ref. [2], is erroneous or misleading. To quote from Ref. [2] "Minimize the time that your car idles—you're getting negative miles per gallon." To have negative mpg, either the numerator (miles) or the denominator (gallons) must be negative. So, what does that mean? Mpg has two numbers as follows:

$$\frac{\text{miles}}{\text{gallon}}$$

miles ← Negative miles must mean you're backing up, yet you are idling

Negative gallons implies car is making gasoline which must be drained from gas tank! → gallon

Idling, of course, implies velocity is zero and the miles are also zero in the numerator of mpg. The negative of fuel consumption is fuel production. Negative mpg is an attempt to be dramatic but is nonsense in the real world of fuel economy. A positive mpg can be had from two negatives; your vehicle is generating gasoline while you back up!

Overview of Analysis

Some of the tips on how to save gasoline are misleading if not erroneous. One method of separating fact from fiction is engineering analysis. Some tips hinge on a driving cycle. A driving cycle can be dissected into distinct operations of the vehicle. Each distinct operation can be viewed as a building block and is then amenable to analysis.

The analysis tools fall into two groups: steady state and transient. A steady-state operation is independent of time. An example is cruise on the freeway. Transient vehicle operations are dependent on time. An example is braking; during braking, vehicle velocity varies from second to second.

One analytical building block may have one, two, or more models. For example, acceleration can be modeled as a constant acceleration event or as a constant power event. A constant power event has (nearly) constant power from the vehicle engine.

As an example of application of analysis, Figure 12.16 has a driving cycle (velocity profile as function of distance) between green and red traffic lights. From the analytical toolbox select acceleration, cruise, and braking. Figure 12.17 has a virtual test track for conducting analytical tests for fuel economy. To analyze the case shown in Figure 12.17, select acceleration, cruise, coasting, cornering, idling, and braking from the appropriate analysis tools.

Fractional Change in Brake-Specific Fuel Consumption, *B*

Later analysis and discussion is abounding with the fractional change in brake-specific fuel consumption (BSFC) which is assigned the symbol B. The fractional change is $\Delta B/B$ and is usually determined by an operating point in an engine map. To locate an operating point, both engine torque and rpm are needed as inputs. Since $\Delta B/B$ is ubiquitous and since reading an engine map is tedious, a comment is inserted here that a value cannot be obtained from a simple equation that is universally (or even narrowly) applicable.

Conventional Vehicle

The suggestions and rules for improving mpg differ from those for a hybrid vehicle. For the conventional vehicle (CV) propulsion is by means of a single engine be it either gasoline of diesel. Every engine has an engine map that incorporates the efficiency of the engine. The map tells how well the engine converts fuel into propulsion.

Hybrid Vehicle

In addition to the engine mentioned above for the CV, the hybrid has an electrical motor. The suggestions and rules for improving mpg must take into account the interplay between the engine and the motor.

Driving Techniques

Driving techniques for improving fuel economy are different for a CV and a hybrid. These differences are discussed in subsequent sections.

PRELIMINARIES

Engine Break-In

The owner of a new car may have expectations for high gas mileage. Note that 3000–5000 mi of driving is necessary before the best mpg is obtained. The engine and power train must be broken-in. Breaking-in means a reduction of friction. For example, the piston rings need to be worn slightly to conform to the grooves in the piston and the circle of the cylinder.

Statistical Variations in Vehicles

Every part in every car has allowable tolerances usually expressed as plus–minus (±) number. Sometimes, the tolerances stack up favorably; at other times they stack up unfavorably. Like a lottery, your car may have all, some, or none favorable (or horrors, all unfavorable). Favorable combinations give better mpg. An example is the combination of transmission seal and shaft. A big shaft with a small seal has more drag and hence fewer mpg. A small shaft with a big seal has less drag and yields more mpg. Another example is the oxygen sensor, which is used to control fuel injection. A low reading for O_2 means less fuel is injected and will likely cause more mpg. An oxygen sensor with a high reading for O_2 means more fuel is injected and will likely cause fewer mpg.

Most Efficient Engine Operating Point

To achieve the most mpg, the engine should operate at, or near, the minimum fuel consumption point. That point is defined by the engine map; see Chapters 5 and 14 for sample maps as well as Figure 12.5. The optimum operating point falls on the optimum operating line shown in the figure. The optimum point (center of the 230 contour) is at 2800 rpm and 92 N·m torque. The corresponding power is 27 kW (36 hp). What engine power is most efficient? At what engine speed? A rule of thumb for optimum (least fuel consumption) power is near 75% of maximum installed hp. The rule of thumb for optimum engine speed, rpm, is near the rpm for maximum engine torque. Engine efficiency and fuel consumption, W_f, are not the same thing. The equation relating W_f and BSFC is

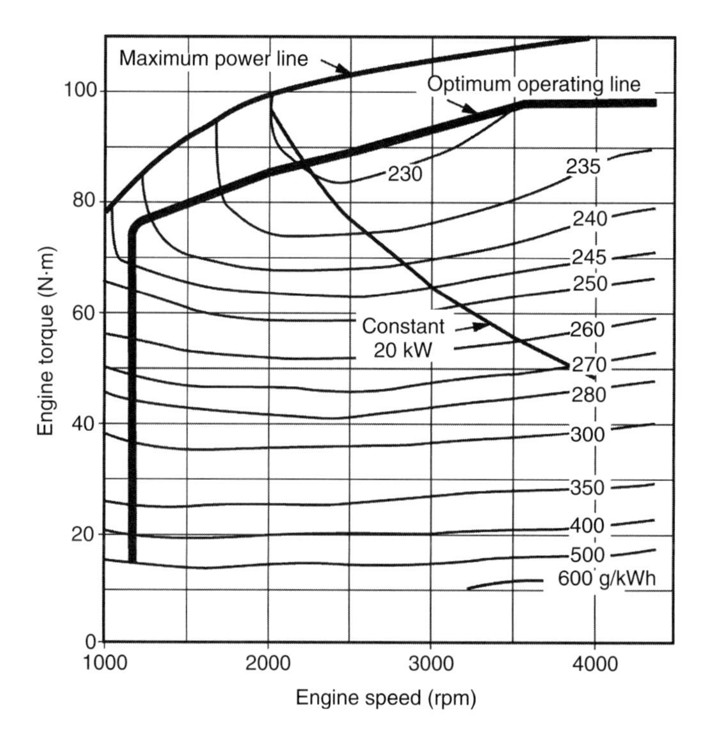

FIGURE 12.5 Sample engine map. Map is for Toyota Prius II. (Reprinted with permission from SAE Paper *Hybrid Gasoline–Hybrid Vehicle Development*, PT-117, 2005 SAE International.)

$$W_f = (\text{BSFC})(\text{power})(\text{time}) = \text{weight of fuel used, lb}$$

where

Power is the engine power
BSFC is brake-specific fuel consumption

The smallest value means the most efficient engine operating point but does not mean the least W_f. The product (power)(time) is determined by the driving cycle and it can be optimized (see Chapter 8).

BUYING GASOLINE

Buy Gas in the Cool of the Morning

Cold gas has a greater density (lb/gal) and hence more energy per gallon (more mpg). A prevailing notion is to buy gas in the morning after the gas at the service station has had a chance to cool down and become more dense. This is a big fallacy. Service station gas tanks are buried a few feet in the ground and, therefore, have nearly constant temperature. The gas will not be denser. Storage tanks above the ground (very rare) may cool during night. A second fallacy is the size of savings.

Let us put some science into this issue. The thermal expansion coefficient for gasoline is 0.0006/°F. Assume a 20°F cooling of gasoline. This gives an increase in density of

$$\text{Increase} = (0.0006/°\text{F})(20°\text{F})(100) = 1.2\%$$

Effectively the price of gas is 1.2% cheaper. You save 60¢ on \$50.00 fill-up. Or you save 3.6¢/gal.

Seventeen oil companies were being sued accusing them of overcharging customers at the pump by failing to compensate for changes in gasoline volumes when the temperature rises. To quote from the article of Ref. [33]

> The consumer fraud suit claims that oil companies fail to take into account the fact that gasoline expands when *the temperature exceeds 60 degrees.*

Italics added for emphasis. Gasoline expands over a very broad range of temperatures from very low minus temperatures on the Fahrenheit scale to much warmer than 60°F. The statement "expands when *the temperature exceeds 60 degrees*" is irrelevant and is phony physics. The defense attorney can show that the change in temperature for gas stored in tanks below ground is less than 2°F giving a savings of 0.36 ¢/gal or 6 ¢ on a $50 fill-up. How will the jury respond to 6 ¢ on a $50 fill-up along with the phony physics at 60°F?

Buy When Tank Is One-Fourth Full

This suggestion makes sense; why? Gasoline has a density of 6.2 lb/gal. The weight of gasoline in the tank can be calculated. Assume 20 gal tank, which gives 20 gal times 6.2 lb/gal yielding 124 lb. At one-fourth full, the weight is 31 lb. Assume a car weight equal to 3600 lb and use Equation 12.3 in the form $\Delta F_E/F_E = -(W_2 - W_1)/W$. Input values are $W_2 = 3600 - (124 - 31) = 3507$; $W_1 = 3600$; $W = 3600$; and $\Delta F_E/F_E = 0.026$. If at 3600 lb with full tank, $F_E = 30$ mpg then at 3507 lb (one-fourth full),

$$F_E = 30(1 + 0.026) = 30.8 \text{ mpg.}$$

Do Not Overfill Tank

Some articles say "If you overfill, gasoline will be spilled from tank." This is no longer true for newer cars; newer cars have sealed gas tanks and a sealed fuel system to avoid pollution of evaporated gasoline. Overfilling may cause gaskets and seals to leak resulting in expensive repairs. To avoid leaks with expensive repairs, do not overfill the tank. If you overfill the tank, you may "bust' or crack the tank.

Do Not Buy Premium Fuel

Motivation to buy premium fuel includes better performance, better speeds, better mileage, and lesser emissions. If the owner's manual says regular 87 octane use 87; do not pay for 91 octane. Higher octane than recommended does not provide any benefit. If owner's manual says premium 91 octane investigate. High octane is to prevent knocking pinging in engine (which can cause damage to engine; usually valves are damaged.) Most cars have a knock sensor. Upon detection of knock, the computer backs off engine timing until the ping goes away; engine power decreases some. Put regular fuel in a premium-fuel car and let the knock sensor do its work. Some loss of that dazzling power will likely occur. Check with your dealer before following this suggestion (let the dealer read this). High octane does not improve mpg.

Different Brands of Gasoline

All gasolines are not the same. Some have a few percent more energy/gallon. It is difficult to verify the gas is really better and find better gas. One source of helpful information is given in Ref. [47], which lists gasoline retailers who provide superior gasoline. The gasoline is better in the opinion of BMW, General Motors, Honda, Toyota, Volkswagen, and Audi. Oxygenated gasoline (for less smog) has 1%–3% loss. Reformulated gasoline (for less smog) has 1%–3% loss. The summer blend is usually 2% better than the winter blend (more volatile). In 2004, California replaced MTBE with ethanol as an additive creating an oxygenated fuel. An oxygenated fuel is gasoline that has been blended with additives that contain oxygen. The blend is 5.7% ethanol and 94.3% gasoline (E5.7). In 2007, California was the largest consumer of ethanol in the United States.

Different Brands of Gasoline: Ethanol E85

From Appendix A, gasoline energy density is 29 MJ/L (megajoule/liter) while E85 is 27.6 MJ/L. (5% less). If mileage on gas is 30 mpg, the mileage using E85 is

$$(30\text{ mpg})(1-0.05) = 28.5\text{ mpg}$$

E85 is 15% gasoline + 85% ethanol.

Credit Card Rebate

Some cards give a generous rebate (5%) on gasoline. Check them out.

How Far to Drive to Save a Penny?

The analysis allows calculation of the breakeven miles.

$$C_M = (G_\$, \text{ \$/gal})/(F_E, \text{ fuel economy}) = \$5\text{/gal/15 mpg} = 33.3\text{ ¢/mi}$$

where
- C_M is the cost per mile to operate car
- $G_\$$ is the \$/gal
- F_E is the fuel economy, mpg

$$D = \frac{(\Delta G_\$, \text{ savings})(G, \text{ gallons to be bought})}{C_M, \text{ cost per mile}}$$

$$D = \frac{(12\text{¢/gal savings})(G = 15\text{ gal to be bought})}{C_M, 33.3\text{ ¢/mi}} = 5.4\text{ mi}$$

where
- D is the maximum distance to drive to save a penny
- $\Delta G_\$$ is the savings in price for gasoline
- G is the gallons to be bought

You can shop for gasoline prices on www.GasBuddy.com or www.GasPriceWatch.com. Gas prices may vary by 15–16 ¢/gal in a local area.

TEST ACCURACY

Some reports and articles give test results with three significant decimal places. A high level of measurement accuracy is required to justify three significant decimal places. A Land Rover was tested in gentle versus aggressive driving with 35.4% better mileage reported for gentle driving. This implies neither 35.3% nor 35.5% are correct. Assume 1 gal of gasoline was used for aggressive driving. Then the gentle case used

$$(1\text{ gal})(1-0.354) = 0.646\text{ gal}$$

It was not 35.5% which would have been (1 gal)(1 − 0.355) = 0.645 gal. Could the experimenters detect accurately 0.646 − 0.645 = 0.001 gal, which is about one thimble full of gas? Did the testers actually have that accuracy to detect a thimble full of gas out of a gallon? If not, what is the credibility of their results?

How Much Gasoline Is Used for a Test?

Accelerate a 4000 lb car to 60 mph at wide-open-throttle (WOT). Acceleration time is 7.5 s. Maximum installed hp may be 300 hp. How much gas is used? The volume of gas used is 0.0217 gal. Assume a 20 gal tank, then

$$\frac{0.0217 \text{ gal}}{20 \text{ gal}} = 0.0011 = 0.1\%$$

An accuracy of 1 part in a 1000 is needed to measure 0.0217 compared to 0.0218.

EPA Tests

In EPA mileage tests, the city, FTP, cycle is 11 mi and for a car yielding 30 mpg, only 0.37 gal are used. The highway, HFET, cycle is 10 mi and at 36 mpg 0.278 gal are used. Total gasoline used for both cycles is 0.65 gal. EPA measures exhaust gases instead of liquid gasoline for greater accuracy.

EPA: Appropriate Implied Accuracy

EPA gives estimates for gains due to the use of ultralow sulfur diesel (ULSD) fuel. To quote:

> "2.6 million tons of smog-causing …"

Comment: To the credit of EPA, the number is not 2,623,851 tons. Another quote is

> "An estimated 360,000 asthma attacks …"

Comment: To the credit of EPA, the number is not 359,403.

PLAN AHEAD

Avoid Unneeded Miles

Drive a Quarter Mile Down the Road

Look at the traffic situation quarter mile in front of you. Look for slow trucks going uphill.

Be alert for red taillights in both lanes; why are the cars stopping or slowing down? Because of the early warning of slowdown, gasoline can be saved and accidents may be avoided.

Just Do Not Drive

The advice "just do not drive" is like the advice "just do not speed" in the book on *How to Avoid a Speeding Ticket*. Some travel is for purpose of meeting someone. Use the telephone. What are alternate means of being there? Telecommute. Use e-mail. Just do not drive; leave your car in the garage.

Alternate Transportation

Car pools may qualify for high occupancy vehicle (HOV) or otherwise known as the car pool lane. Save at least half on commuting costs. Consider using public buses or trains.

Combine Trips

A single trip for each errand uses much more gasoline than multiple errands on one trip; only one warm up of engine and tires is necessary. A single trip for each errand leads to many warm ups of engine, which is bad for engine life. Plan the shortest distance between the stops on a multi-errand trip.

Rush-Hour Traffic

Certain hours (6–10 A.M. and 4–7 P.M.) usually have heavy traffic on commuter roads. Avoid these hours for your commute, if possible. Constipated traffic conditions give lower mpg.

Parking

Do not cruise a parking lot looking for a parking space; take the first open spot, park, then walk some if necessary. Park so that your first move on leaving is to drive forward. Park in the shade for a cooler car upon return; less A/C is needed.

Know Where You Are Going

Avoid searching for a destination. Study a map before leaving home or the office. Driving aimlessly around looking wastes gas and generates frustration. If you have navigation system in your car, use it! (Sometimes the navigation system takes you out of the way with lost gasoline; learn your system.) If a two-car family, use the car giving the greatest mpg.

VEHICLE MAINTENANCE

Keep a Mileage Log

A mileage log will help to detect any changes from the usual mpg. An anomaly provides an alert for corrective action. When your log book reveals poor mpg, have your vehicle checked. Six of the many tests include

- Use (have mechanic do) the onboard diagnostic system check
- Check for dragging brakes
- Check for transmission slippage through all gears
- Check ignition timing
- Check emission control system
- Check vacuum hoses for leaks, proper routing, kinks

The onboard diagnostic system check gives trouble codes for faulty sensors, for example, coolant temperature (radiator boils over); faulty actuators, for example, bad EGR valve; and actual faults, for example, vapor leaks from gas tank (usually for not tightening the gas cap).

Check Engine Light

This light, annoying as it may be, does a similar function as a mileage log. The light alerts you to impending trouble. Pay attention to the “check engine light.”

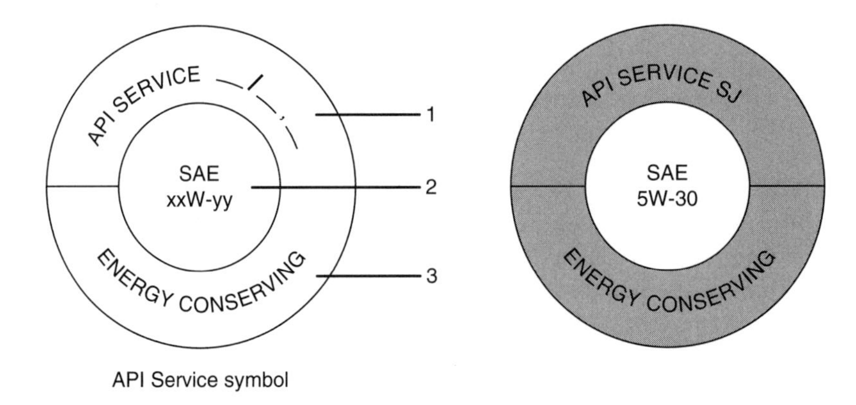

FIGURE 12.6 Marking on oil can showing API marking with identification of "Energy Conserving" when appropriate. (Reproduced with courtesy of the American Petroleum Institute, API.)

Tune-Up

An engine in poor condition uses more gasoline. However, many newer cars can go 100,000 mi between tune-ups. Typically, you gain 4%–5% better mpg after a tune-up. One exception is a faulty O_2 sensor which may cause a large decrease in mpg; up to 40% can be lost. Some tune-up items include spark plugs (100,000 mi), bands for the automatic transmission, low automatic transmission fluid, and low coolant.

Engine Oil Viscosity

The label on the top of an oil can provides three essential facts. Find item 1 in Figure 12.6. Item 1 gives American Petroleum Institute service details, which can be found in an American Institute of Petroleum (API) Web site [55]. For what type of engine (gasoline or diesel) is the oil intended? The API service provides the answer as well as information concerning the severity of service and quality of the oil. Refer to the owner's manual for recommended API service. Item 2 gives the viscosity of the oil. Lower viscosity oils typically have lower friction. Look for "energy conserving" (Item 3 in the figure) on the label of the oil can in API circle; this oil has friction-reducing additives. Some results of mpg loss due to use of the wrong viscosity oil are

Oil Used	Oil Recommended	mpg Loss (%)
10W30	5W30	1–2
5W30	5W20	1–1.5

Plugged Air Filter

A gain of 10% better mpg is typical by replacing (cleaning) the filter. This is true only if it is dirty. A clean filter helps to protect the engine from foreign particles. Follow owner's manual on timing.

To understand why a clean filter affects mpg, look at Figure 17.6, which shows pressure as function of volume within the cylinder as the piston moves from TDC to BDC. A plugged air filter makes area II larger and increases pumping losses. To quote from Chapter 17, since the test equipment that yields the pressure–volume curves is known as an indicator, the energy of area I is termed the "indicated energy." Area II involves moving the gases in and out of the cylinder; this is called pumping. Hence, the term "pumping loss" is applied to the motion in area II. The net output of energy from the engine is

$$\text{Net energy} = \text{indicated energy} - \text{pumping loss}$$

A plugged air filter makes area II larger thereby decreasing net energy. Considerable confusion exists in the popular press about pumping loss.

Wheel Alignment

The drag due to incorrect alignment (an angle, θ) is some fraction of the lateral force. This misalignment force is a drag. The calculation must account for number of tires that are at an angle θ.

$$\text{Alignment drag} = (\text{lateral force/tire})(\text{number of wheels})(\sin\theta)$$

See Figure 9.7 and the adjacent discussion of cornering stiffness. Typical numbers are

$$\text{Alignment drag} = [100\ \text{lb/(tire)(degree)}](1°)(2\ \text{tires})(\sin 1°) = 3.5\ \text{lbt}$$

adds about 3.5% to rolling friction and decreases mpg by the same amount.

For a larger value of θ,

$$\text{Alignment drag} = [100\ \text{lb/(tire)(degree)}](3°)(2\ \text{tires})(\sin 3°) = 31.4\ \text{lb}$$

adds about 30% to rolling friction and decreases mpg by the same amount. An angle of $\theta = 3°$ is very large and would severely decrease mpg. Also with $\theta = 3°$, very severe tire wear occurs. To avoid severe tire wear, wheel alignment requires adjustment to a fraction of a degree accuracy.

Wheel Balance

Unbalance causes wheel vibration or shimmy. Motion is kinetic energy (KE); hence, wheel unbalance soaks up energy from gasoline. The question is how much energy? The energy loss due to wheel unbalance is less than suspension motion. See "Generator in Shock Absorbers" in Chapter 16 for some insight. Rough roads absorb power. See Ref. [1] of Chapter 10, which discusses methods to recover some of that energy. See also "Road Quality" in Chapter 10. As shock absorbers move in and out on rough roads, power can be generated. One implementation would have a magnet on the plunger and a coil on the outside of shock absorber. A few kilowatts could be generated using all four wheels.

Tire Inflation

Have your own accurate tire gage; service station tire pressure gages are notoriously bad for accuracy. To have a meaningful estimate for gain by proper inflation, the following three elements of data are essential:

1. Number of tires under inflated; only one or all four or
2. Pressure of the under-inflated tire(s)
3. Pressure of properly inflated tire(s)

The information above is rarely or never given. Wong's book has data on tire inflation and curve of rolling friction coefficient, f, as a function of tire pressure (Ref. [23], pp. 12–13). A typical value quoted in reports is 3% gain in mpg by proper inflation. One report states 3.3% gain but does not give the pressure of under-inflated tire.

Wong's book (pp. 12–13) has a graph of rolling friction coefficient, f, versus tire pressure. Data from the graph:

Tire Pressure	Rolling Friction Coefficient, *f*
15	0.020
20	0.014
30	0.011
40	0.0091
50	0.0082

From 25 to 50 psi, f varies only by a small amount. Below 25 psi, big changes in f occur.

For 27→34 psi for Wong's radial tire, $\Delta f/f = 0.11 = 11\%$ change. This is not the change in mpg, but can be used to find mpg change. Assume that F_E is proportional to engine power and in turn is proportional to rolling friction coefficient, f. For one under-inflated tire with the vehicle moving at slow speeds, the loss of mpg is 0.11 divided by 4, which is 2.8%. The influence of added tire drag on engine operating point is also ignored. One report for a change 27→34 psi gives 4% better mpg. Another report gives 0.4% gain in mpg for each tire. Proper inflation gives better handling car and hence safer car. Proper inflation gives longer tire life.

Change in Tire Pressure with Temperature

A cold tire has less pressure than a warm tire. Many (incorrect) rules of thumb have been stated for change in pressure with change in temperature. A common rule is for every 10°F change, pressure changes by 1.0 psi. Most are wrong except for a single specific case. The perfect gas law applies:

$$p = \rho RT \tag{12.8}$$

The fractional form is

$$\frac{\Delta p}{p} = \frac{\Delta \rho}{\rho} + \frac{\Delta R}{R} + \frac{\Delta T}{T} \tag{12.9}$$

Since $\Delta\rho/\rho$ and $\Delta R/R = 0$, $\Delta p/p = \Delta T/T$. Note: the temperature in the equation is the absolute temperature °R = °F + 459°. °R denotes degrees Rankine while °F stands for degree Fahrenheit. Assume $T_2 = 90°F$ and $T_1 = 65°F$. The average temperature is 1/2(90 + 65) equals 77.5°F. Absolute temperature is 536.5°R = 77.5°F + 459°. Then

$$\frac{\Delta T}{T} = \frac{90°\text{F} - 65°\text{F}}{536.5°\text{R}} = 0.0652 = \frac{\Delta p}{p}$$

If $p_1 = 28$ psi, then $p_2 = 28$ psi $(1 + 0.0652) = 29.8$ psi. The rule of thumb (for this specific case only) would be

1.8 psi gain for 35°F increase or 0.5 psi for a 10°F gain

Since the formula is so simple, use it instead of a rule of thumb.

Tire Purchase and Construction

When replacing tires, many choices are possible. Rolling friction coefficient, f, affects gas mileage. The rolling friction coefficient, f, depends on tire construction and tire materials.

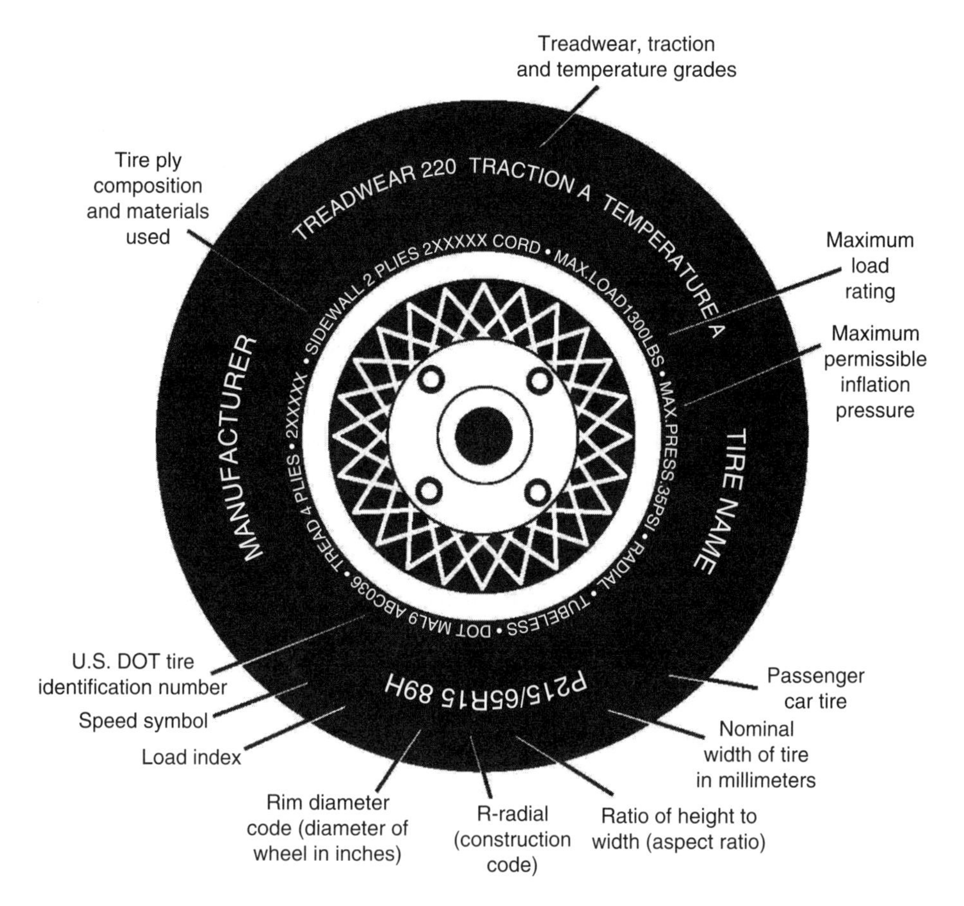

FIGURE 12.7 NHTSA tire sidewall markings.

Unfortunately, the NHTSA does not require data on rolling friction coefficient, f, to be on the sidewall. NHTSA should require the rolling friction coefficient, f, on the tire sidewall. See Figure 12.7 for a tire with sidewall markings. With regard to mpg, rolling friction coefficient, f, is important but not the only factor to consider when buying a tire (see Ref. [14] for other factors to be considered).

Tire Purchase and Construction: Speed Rating

Tires have speed ratings [14]. Standing waves on a tire are discussed in Ref. [23], p. 10. A tire has a critical rotational speed, ω_W, at which standing waves are formed. Due to forward motion of vehicle, the tire has a rotational speed $\omega = V/r$, where V is the vehicle velocity and r is the tire radius. When $\omega = V_C/r = \omega_W$, standing waves appear in the tire as shown by Figure 12.8. V_C is a critical speed for the tire. The speed rating of tire, V_R, must be less than the critical speed, V_C, that is $V_R < V_C$.

Rolling Friction Coefficient: Other Factors: Temperature

Cold rubber has more loss than warm rubber; that is, f is larger. A cold tire may have 20% larger rolling friction coefficient. At least 20–30 mi of driving is necessary to warm up tires (see "Combine Trips" under "Plan Ahead"). Cold tires combined with low inflation causes large loss of mpg.

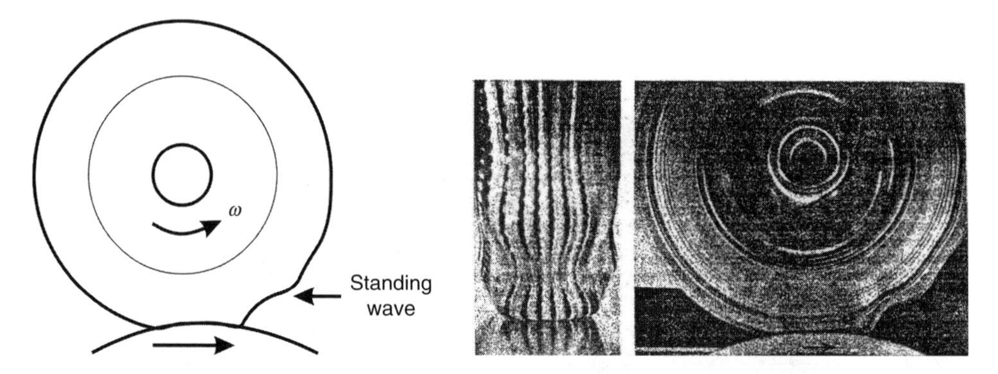

FIGURE 12.8 Formation of standing waves of a tire at high speed. (Reproduced from Pacejka, H.B., Analysis of tire properties, Chapter 9, *Mechanics of Pneumatic Tires*, Clark, S.K. (Ed.), National Highway Safety Administration, Washington, DC, 1982.)

Rolling Friction Coefficient: Other Factors: Aspect Ratio

Aspect ratio (AR) is height of tire section, H, divided by tire width, W.

$$\text{Aspect ratio, AR} = \frac{H}{W}$$

Gillespie's book has a formula for rolling friction coefficient, f, as function of AR [34]. The formula is

$$f \sim \left(\frac{H}{W}\right)^{1/2} \sim (\mathrm{AR})^{1/2} \tag{12.10}$$

To quote from a report

> "But if they (fancy wheels/tires) are wider they'll create more rolling friction and decrease fuel economy."

According to Gillespie's formula the effect of AR is just the opposite from the quote above. Low AR reduces f. Figure 12.9 shows the tire patch for low and high AR tires along with location of CP for the patch. The displacement causes a retarding torque. In Figure 12.9, the wheels are moving up the page. The torque for the high AR tire is

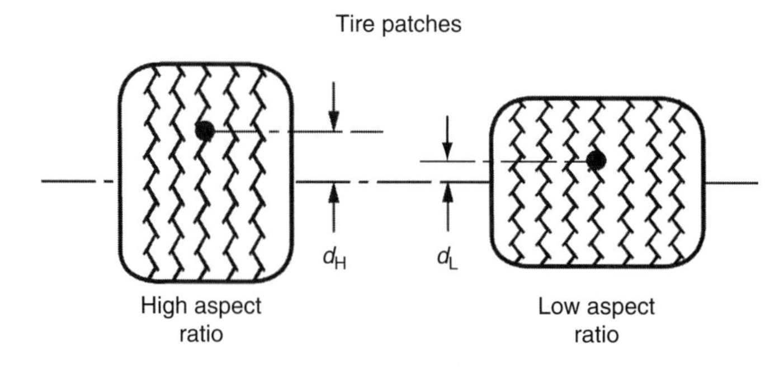

FIGURE 12.9 Tire patch for low and high AR tires along with location of center of pressure (CP) for patch. CP is the black dot.

$$\text{Torque} = (d_H)(\text{weight on tire})$$

The displacements, d_H or d_L, are related to rolling friction coefficient, f. Since $d_H > d_L$, the retarding torque for the high AR tires is greater.

Rolling Friction Coefficient: Other Factors: Vehicle Velocity

Figure 4.31 in Ref. [34] has a graph of rolling friction coefficient, f, as a function of velocity. From 0 to 60 mph, $f \approxeq 0.014$. From 60 to 100 mph, f rapidly climbs to 0.045. Wow! For high speeds, safety demands a tire rated for the high speeds.

Snow Tires

Snow tires typically have larger rolling friction coefficient, f. Take off the snow tires when the first robins of spring declare winter is over.

Aerodynamic Drag: Wheel Wells, Hubcaps, and Wheels

Fancy looking wheels or hubcaps may have more aerodynamic drag. See Figure 1.6, Burt Rutan's Supercar, a hybrid, built under contract to GM. Note the hubcaps for the front wheels. Also note the almost fully enclosed rear wheels. Take a clue from most hybrids; look at the design of hubcaps and wheels on hybrids. These wheels are selected for low drag and good mpg.

Features not evident in Figure 1.6 are the edges of the wheel well. Figure 12.10 illustrates the edges of the wheel well to control the boundary and shear layers. The sharp edge upstream of the tire provides a well-defined detachment point. The rounded edge allows different reattachment points while still defining the location of the shear layer. Control of the shear layer location prevents large disturbances of the outer flow and gives low drag.

GETTING BETTER MPG CRUISING ALONG THE INTERSTATE

Introduction

What is cruising? Cruising involves long distance traveling, which is best done at constant velocity. At constant velocity, the best fuel economy can be attained. Cruising is steady state. Because of the long distances and hours of driving, considerable fuel is used. Consequently, savings here yield big benefits. As stated in the section on CAFE, the two main approaches to better fuel economy are, first, improved engine and power train efficiencies, and, second, reduced resistance to motion.

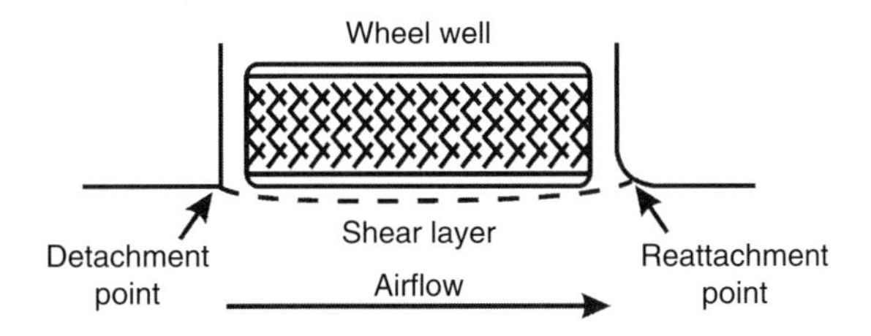

FIGURE 12.10 Wheel well design for minimum aerodynamic drag.

Cruise Control

If you have cruise control, use it. Reference [6] indicates 5%–6% savings on the highway by using cruise control. Reference [12] reports gas savings using cruise control:

- Land Rover at 70 mph saved 14%
- Ford Mustang at 70 mph saved 4.5%

Use of cruise control avoids speed creep and speeding tickets. See Figure 12.11 for an assumed driving cycle without cruise control. For the driving cycle, the average speed is 64.3 mph. Without cruise control, the average fuel economy is 21 mpg. With cruise control set for 60 mph, the fuel economy is 24 mpg, which is a savings of 12.5%. Also note that the average speed of 64.3 mph is greater than the desired speed of 60 mph with cruise control. Speed creep has occurred.

Story: Optimum Speed

What is the optimum speed? Typically, most vehicles have increasing mpg as speed increases. At higher speeds, the mpg begins to drop. The change from better mpg to worse mpg occurs at the peak of the curve. The peak of the curve yields both optimum speed and optimum mpg. Appendix 12.1 provides an extensive discussion of optimum cruise speed. Equations are derived. An example is calculated to illustrate concepts and provide insight into various factors affecting the appropriate driving speed to achieve best mpg.

Optimum Cruise Speed

Long trips on the freeway or interstate involve many hours of driving. Driving at the optimum speed offers big savings in fuel. The fuel economy for a 4,000 lb sedan is shown in Figure 12.12; from the peak of the curve, the optimum cruise speed is 48 mph for this vehicle. The curve is for a fixed transmission gear ratio. From 20 to 48 mph, the mpg is almost constant. At 48 mph, the mpg begins to fall, and, at 60 mph, mpg is 8% less. At maximum speed, 92.5 mph, the fuel economy has been cut to less than half. The facts to remember are that at lower speeds, the mpg varies little with speed. Above the optimum cruise speed of 48 mph, fuel economy falls rapidly.

The effect of transmission gear ratio on the optimum cruise speed is shown in Figure 12.13. The best gear for a particular speed is indicated. As speed increases, the best gear ratio moves upward

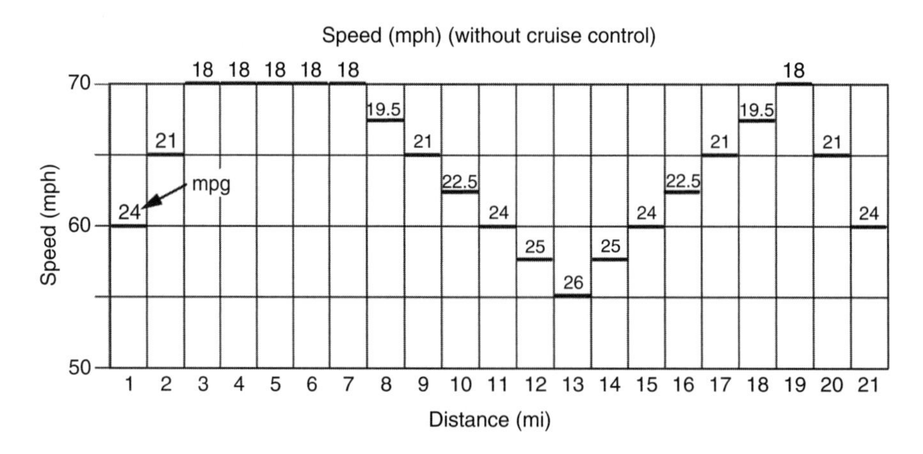

FIGURE 12.11 Driving cycle (velocity profile) for fuel economy without use of cruise control.

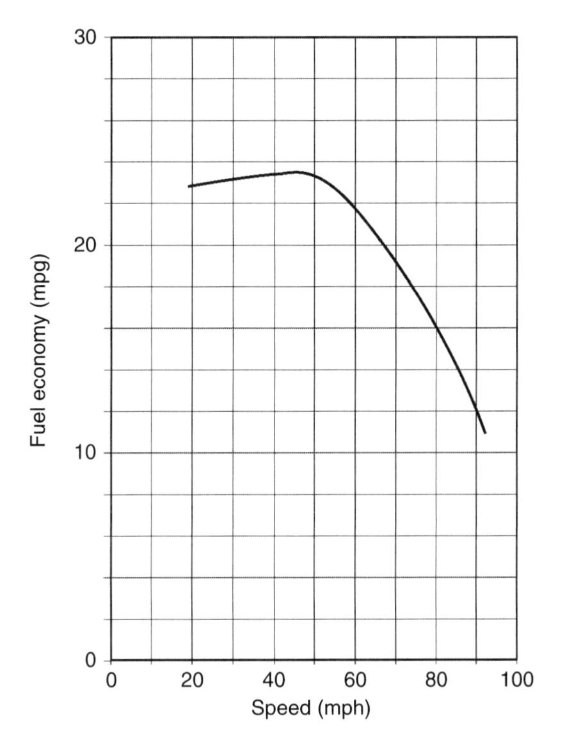

FIGURE 12.12 Graph of fuel economy as a function of vehicle velocity. The optimum speed is 48 mph. Curve is for a large V-8 sedan.

from first to fifth gear. The optimum cruise speed for this vehicle, which is in fifth gear, is lower than the previous vehicle being 40 mph. At 40 mph, the mpg falls from 46 mpg, and, at 60 mph, fuel economy is 34 mpg. This is a 24% drop.

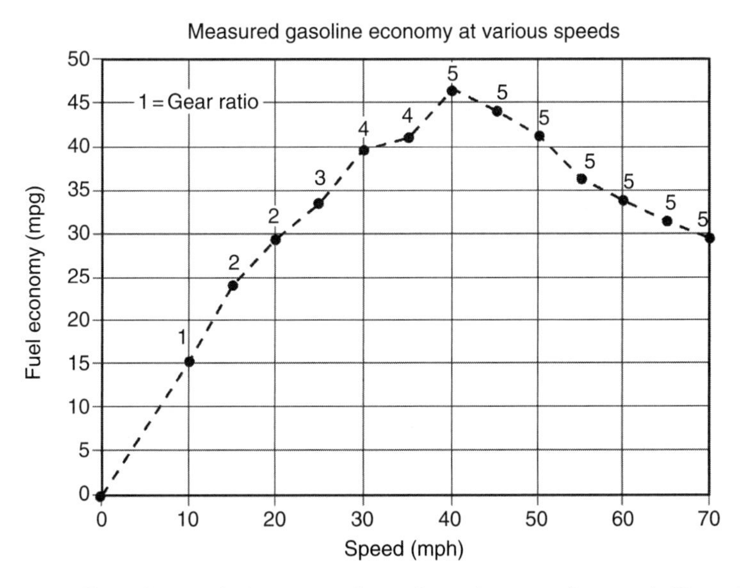

FIGURE 12.13 The effect of transmission gear ratio on the optimum cruise speed. (Reproduced with permission from Parker, D. S., Energy efficient transportation for Florida, Revised 2005. Available at www.fsec.ucf.edu/Pubs/energynotes/en-19.htm.)

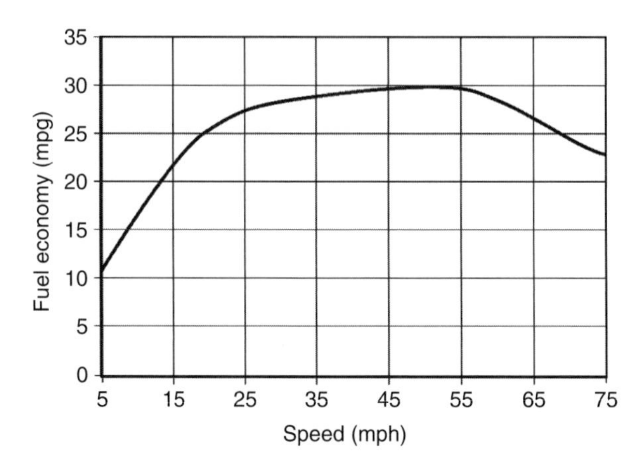

FIGURE 12.14 Graph of fuel economy, mpg, as a function of vehicle velocity, mph. Curve is for V-6 sedan. (Reproduced from Environmental Protection Agency, Driving more efficiently. Available at www.fueleconomy.gov/feg/driveHabits.shtml.)

Figure 12.14 shows another curve of fuel economy as a function of vehicle speed. The three curves in Figures 12.12 through 12.14 are each distinctive. Since the curves are different, the fact is emphasized that each class of car has different fuel economy characteristics. The shape of the curve in Figure 12.12 is similar to that in Figure 12.14.

The curve of fuel economy is found by superimposing a curve of required power on an engine map. The map provides data on fuel consumption. The details for determining the optimum cruise speed are developed in Appendix 12.1. The next topic is resistance to motion.

Resistance to Vehicle Motion: Crossover Velocity

Fuel economy, mpg, depends on the resistance to the forward motion of the vehicle. The two main resistances are rolling friction, F_R, due to the tires; and the aerodynamic drag, F_D. Assume the retarding force due to rolling friction is constant. A brief discussion of aerodynamic drag, F_D, appears in Chapter 5. F_R and F_D are discussed in Appendix 12.1. Total force resisting motion, F, is the sum of F_R and F_D.

Figure 12.15 has a plot of the three forces, F_R, F_D, and F, discussed above. Note that

$$F_R = fW = (0.025)(4000 \text{ lb}) = 100 \text{ lb}$$

In the plot, the aerodynamic drag, F_D, grows from a value of zero at zero velocity. Introduce a reference velocity called the crossover velocity, V_{CO}, with units km/h or mph. The crossover velocity is defined as the point where $F_R = F_D$, that is, the point where the F_D curve crosses over the F_R curve. The equation for crossover velocity, which is derived in Appendix 12.1, is

$$V_{CO} = \left(\frac{2fW}{\rho C_D A} \right)^{1/2} \tag{12.11}$$

A nondimensional velocity is convenient for later equations; it is defined as

$$X = \text{velocity ratio} = \frac{V}{V_{CO}} \tag{12.12}$$

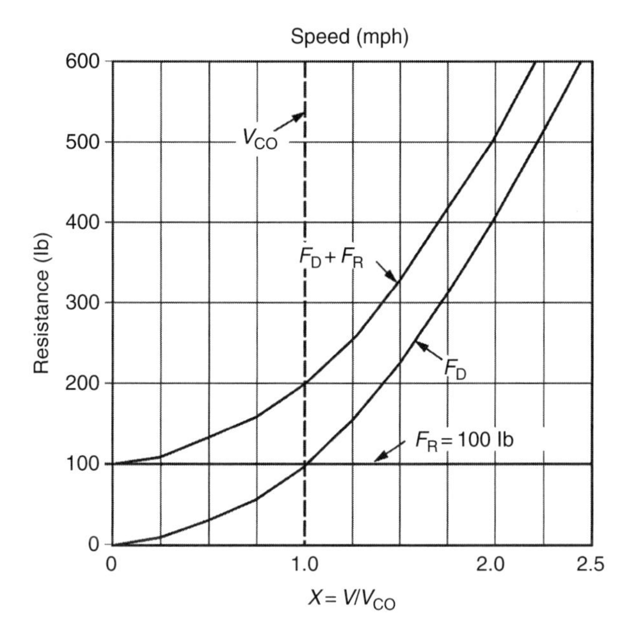

FIGURE 12.15 Curves of rolling friction, F_R, due to the tires and aerodynamic drag, F_D, as a function of nondimensional velocity $X = V/V_{CO}$.

The abscissa of the plot in Figure 12.15 is X.

A sleek sedan is likely to have low aerodynamic drag (meaning of sleek in this context is low aerodynamic drag). Also sleek may imply a certain sportiness with high lateral forces. The tire rolling resistance coefficient, f, is large. Looking at Equation 12.11, as a result V_{CO} is large and aerodynamic drag sets in later.

A blunt, square-box SUV has large aerodynamic drag. Also the knobby tires have a large tire rolling resistance coefficient, f. Consequently, V_{CO} is modest in size. Aerodynamic drag sets in earlier than for the sleek sedan.

Reference [22] gives data for a small European car of a mid-1970s design. These data are $m = 1000\,\text{kg}$, $C_D = 0.6$, $A = 1.65\,\text{m}^2$, ρ = value for standard atmosphere, and $f = 0.01$. Using these values as input to the equation for V_{CO}, the value of V_{CO} is 26 mph.

What is the usefulness of crossover velocity? The use of X simplifies the equation for required power. The size of V compared to V_{CO} indicates which is larger, F_R or F_D. Table 12.7 summarizes the results.

TABLE 12.7
Use of Crossover Velocity to Define Various Regions for the Resisting Forces

$X = V/V_{CO}$	Comparison of Forces, F_R and F_D
$X << 1.0$	F_R dominates F_D. F is nearly constant and equal to F_R
$0 < X < 1.0$	F_R and F_D are nearly equal
$X = 1.0$	$F_R = F_D$ and $F = 2\,F_R = 2\,F_D$
$1.0 < X$	F_D dominates F_R. F varies as V^2 and power as V^3

Speeding: In Cruise Condition

Figures 12.12 through 12.14 show different fuel economy to be expected from three diverse vehicles. At speeds less than optimum, the vehicles of Figures 12.12 and 12.14 are flat with little slope. The vehicle of Figure 12.13 has a sharp increase. For speeds greater than optimum, all vehicles have a sharp fall. Mpg loss for 55 → 70 mph varies in the range 15%–18%. For the three vehicles, the optimum speeds are from 40 to 52 mph.

Tailgating: Behind a Truck on Interstate

This article does not advocate tailgating. Forget safety for a moment; driving behind a truck can appreciably increase mpg. This is called "drafting"; drafting is also used by race car drivers on the race track. The wake of a truck (or any vehicle) is moving at nearly the same speed as the truck. However, the velocity drops. Behind a truck, the drop of wake velocity is as follows:

200 ft	15%
310 ft	13%
400 ft	12%
800 ft	9%
1000 ft	8%

Hence, the V in aerodynamic drag $D = 1/2\ \rho C_D A V^2$ is somewhat reduced. Directly a few feet behind the truck, V is almost reduced to zero.

See Ref. [4] for an example of mpg increase due to tailgating. Driving 3 s (310 ft) behind truck at 70 mph increased mpg during 30 min tests.

- Hyundai Sonata: from 24 to 34 mpg at 70 mph
- Hyundai Sonata: from 22 to 30 mpg at 77 mph

Safety aspects are mixed. The tailgating car is safer; the truck clears the way of obstacles. Crashing into the back of the truck must be considered. The truck cannot stop at zero distance; this fact gives you reaction and stopping time. The driver of the car must be absolutely alert every second. Some new (expensive) cars have adaptive cruise control; radar in the car senses when to apply brakes; good for drafting trucks! Caution: truck tires tend to pick up rocks, which, when released by the tire, can break your windshield.

Tailgating: Analysis

The equation for power to overcome aerodynamic drag is

$$P = \frac{1}{2}\rho\left(V - V_W\right)^2 C_D A V \tag{12.13}$$

The velocity of the car and truck is V. The velocity of the wake of the truck is V_W. Using this equation, an equation for the fractional change in fuel economy, F_E, can be derived. It is

$$\frac{\Delta F_E}{F_E} = -\frac{\Delta B}{B} + \frac{\beta(2-\beta)X^2}{\left(1+X^2\right)} \tag{12.14}$$

where

β is the (wake velocity)/(car velocity), that is, 0.13 for 310 ft behind the truck
B is the BSFC lb fuel/[(hp)(h)]
X is equal to V/V_{CO}, 1.4

For values shown, the fractional change is $\Delta F_E/F_E = +0.16$. This ignores the value of $\Delta B/B$, which was not calculated. For the Sonata discussed above, $\Delta F_E/F_E$ is about 40%.

Tailgating: Good and Bad

Since tailgating has been introduced, another case exists and is now discussed. In stop-and-go traffic, tailgating requires use of brakes more often. Use of brakes is always wasteful, which means mpg suffers. Allowing space in front of your car allows for a more constant speed and less braking, which equals better mpg. Reference [12] points out that trucks in stop-and-go traffic allow extra distance for more constant speed and less braking. In creep-and-crawl traffic, tailgating uses more gasoline, is hard on nerves, and is more dangerous.

AGGRESSIVE DRIVING

Fuel Economy Meter (Instantaneous mpg)

The fuel economy meter helps to improve mpg. If you have one, use it. Many hybrids have a fuel economy meter. Drive with an eye (and foot on gas pedal) on the meter. Drive to get big numbers for mpg.

Aggressive Driving: Definition

As used here aggressive driving is not road rage. It is just people in a hurry to get somewhere, and people in a hurry to empty their gas tanks. The three components usually associated with aggressive driving are

1. Rapid acceleration
2. Speeding
3. Hard, abrupt braking

A surprise is that only one of the three really causes a significant decrease in mpg.

Aggressive Driving versus Gentle Driving

Frequently reported and repeated in the media is the lost 33% mpg in highway due to aggressive driving and 5% in city. Another case reported involved a Land Rover, which gave 35.4% better mileage gentle versus aggressive driving. A Mustang gave 27.1% better mileage. Aggressive driving may not save that much time; a study in Europe showed a saving of 1 min out of a 30 min trip.

Aggressive Driving: Rapid Acceleration

The weight of fuel, W_f, used to accelerate is

$$W_f = (\text{BSFC})(\text{power})(\text{time}) = (\text{B})(\text{power})(\text{time}) = \text{pounds of gasoline}$$

The symbol B is used for BSFC. The power required to accelerate, P, is

$$P = FV = maV \tag{12.15}$$

where
m is the mass of the car, kg
a is acceleration, m/s^2
F is inertial force, N
V is vehicle velocity, m/s

This power, P, ignores that needed to overcome drag and resistance. The increment of fuel used in time $\mathrm{d}t$ is $\mathrm{d}W_\mathrm{f}$ given by

$$\mathrm{d}W_\mathrm{f} = BmaV\mathrm{d}t = BmaV\left(\frac{\mathrm{d}V}{a}\right) = BmV\mathrm{d}V \tag{12.16}$$

At *any time* during the acceleration, the *acceleration cancels*! This means that the fuel used solely to overcome inertia is the same regardless of the value of acceleration, a.

A few comments are appropriate at this point. First, the value of BSFC will depend on the aggressiveness of the acceleration. This variation in BSFC causes minor deviations from the conclusion. Second, the drag caused by rolling friction and by aerodynamics will be a small fraction of the inertial force, $m\,a$. However, drag does alter the conclusion in a minor way as is shown below.

The question of the influence of drag on fuel consumed is now addressed. The equation, W_f = (BSFC)(power)(time) = (B)(power)(time), has three factors. The BSFC is found from an engine map. Power is given by

$$P_\mathrm{Q} = P_\mathrm{RCO}\left[X\left(1+X^2\right)+2X\left(\frac{\mathrm{d}X}{\mathrm{d}t'}\right)\right] \tag{12.17}$$

The term involving time, t', is due to inertia and acceleration. The time is given by the differential equation, which must be integrated to find time

$$t' = \int \frac{2X\,\mathrm{d}X}{\Lambda - X\left(1+X^2\right)} \tag{12.18}$$

where
t' is the nondimensional time
X is the nondimensional velocity, V/V_CO
Λ is the nondimensional power, $P_\mathrm{Q}/P_\mathrm{RCO}$
P_Q is the total power due to resistive forces, kW or hp
P_RCO is the power due to rolling friction coefficient, f, evaluated at V_CO

Some things to notice:

1. The denominator of integrand is a cubic equation in X.
2. Roots are one real and a conjugate complex pair.
3. Looks like numerical integration is easiest.
4. When denominator of integrand is zero, the integral blows up.
5. When denominator of integrand is zero, the vehicle approaches its maximum velocity. In fact

$$\Lambda - X(1 + X^2) = 0 \tag{12.19}$$

can be used to find maximum speed.

Aggressive Driving: Rapid Acceleration: Neglecting Aerodynamic Drag

Since the cubic equation $\Lambda - X(1 + X^2) = 0$ cannot be readily solved or integrated, neglect the aerodynamic drag, which gives rise to the troublesome $(1 + X^2)$ term. The integral for time becomes

$$t' = \int \frac{2X\,dX}{\Lambda - X} \tag{12.20}$$

This is easily integrated. Neglecting aerodynamic drag causes only a few percent error so long as $X \le 1.0$. For $X \ge 1.0$ the error rapidly grows. The equation for time is

$$t' = \Lambda \ln \frac{\Lambda}{\Lambda - X} - X \tag{12.21}$$

Some calculations were made for weight of fuel required and are summarized in Table 12.8. The vehicle is accelerated from 0 to 60 mph. The amount of fuel depends slightly on the magnitude of acceleration. The weight of fuel includes that to accelerate plus the amount to cover the cruise distance. For acceleration time equal to 42.2 s in Table 12.5, the car travels 0.35 mi. For acceleration time equal to 23.7 s, the car travels 0.20 mi. For the 23.7 s case, the car needs to cruise 0.35 − 0.20 = 0.15 mi. The fuel used to go this distance is included in weight of fuel. The reference quantity is the case of infinite acceleration for which 100% is assigned. Note that the difference in fuel consumed between very gentle acceleration (42.2 s) and very hard acceleration (0 s) is only 10%. Conclusion: as power increases, fuel to accelerate actually goes up only slightly from 0.188 to 0.208 lb. The fuel used to overcome inertia is precisely independent of the magnitude of acceleration.

In summary, the reason the fuel consumed is nearly independent of time is that the product of power and time is a constant. If power is very large (hard acceleration), time is very short. If power is very small (very gentle acceleration), time is very long. In either case, multiplication of power and time gives the same numerical value.

TABLE 12.8
Results of Calculations for the Weight of Fuel Required to Accelerate

Nondimensional Power, Λ	Acceleration Time, t, s	Weight of Fuel, W_f, lb	Percent Value, %
2	42.2	0.188	90
3	23.7	0.196	94
5	12.6	0.201	97
8	7.5	0.205	99
∞	0	0.208	100

Another Quote

> Slamming down the gas pedal pushes more fuel into engine while it also keeps the engine running faster.

The implication from the quote is that acceleration has a large effect on mpg. In the equation, W_f = (BSFC)(power)(time), the author is forgetting time.

Reconcile Fuel Economy Meter (Instantaneous mpg) and Previous Conclusions

When the gas pedal is pushed all the way to the floor, the fuel economy meter drops from 25 to 6 mpg. Is not that evidence that sharp acceleration decreases mpg? The meter says so, and the meter is correct. Also the amount of fuel burned is very high. However, the time is short. Once again time (or distance) is being ignored. Turn mpg upside down to gallons/mile. At high power acceleration with meter reading 6 mpg or 0.16 gal/mi; the distance traveled during acceleration is quarter mile. The gas used is

$$(0.16 \text{ gal/mi})(1/4 \text{ mi}) = 0.04 \text{ gal}$$

At low power acceleration with meter reading 20 mpg or 0.05 gal/mi the distance traveled during slow acceleration is 0.8 mi. The gas used is

$$(0.05 \text{ gal/mi})(0.8 \text{ mi}) = 0.04 \text{ gal}$$

which is the same.

Yet Another Quote

> Roaring off to brake at the next light is a disaster for fuel economy.

The report is correct but not for the reason implied. "Roaring off…" is usually associated with speeding, which is the real culprit. See the explanation under "Speeding—In Acceleration."

Aggressive Braking: Braking Hard

When the brake pedal is pushed, the engine control shuts off the fuel. See Figure 12.16, which shows fuel cutoff. The engine becomes an air pump and absorbs power. The engine creates a torque in opposition to vehicle motion. As vehicle speed decreases, engine speed also decreases. When engine rpm approaches idle rpm, fuel flow resumes. Except for the final few moments of stopping, braking consumes zero fuel. With that thought in mind, huge differences between hard and gentle braking does not seem possible. The level of braking has a minor effect on mpg.

One cause for heavy braking to consume more fuel is the fact more distance is traveled in cruise mode before brakes are applied. A numerical example will clarify that concept. Both the gentle and hard braking cars travel the same distance. Assume the following reasonable inputs:

Initial velocity = 20 m/s = 45 mph	BSFC = 0.4 lb/[(hp)(h)]
Gentle deceleration = 1.0 m/s^2	Hard deceleration = 5.0 m/s^2
Cruise power = 10 hp	Time to stop gentle = 20/1.0 = 20 s
Time to stop hard = V/a = 20/5.0 = 4 s	Cruise time = 20 − 4 = 16 s

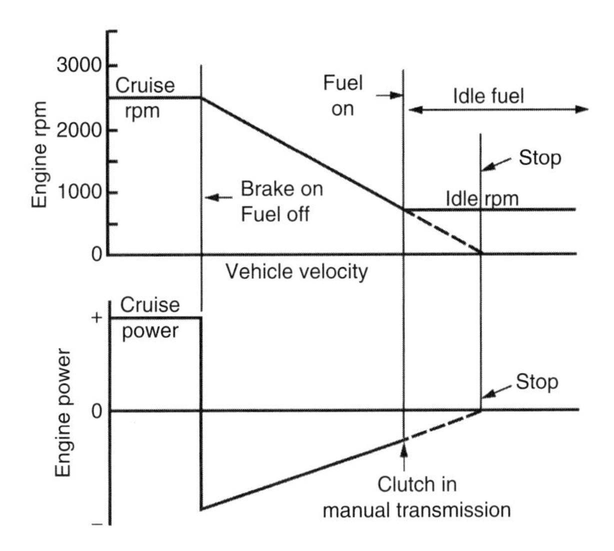

FIGURE 12.16 Fuel consumption during braking. Fuel flow is turned off most of the time during braking.

The fuel consumed by the hard deceleration car is

$$(\text{BSFC})(\text{power})(\text{time}) = (0.4)(10)\left(\frac{16}{3600}\right) = 0.0178 \text{ lb}$$

The 3600 factor converts seconds to hours. This is 0.0178/6.2 = 0.0029 gal of gasoline. The car using gentle brakes can drive 378 ft farther on the same amount of gasoline.

Speeding: In Acceleration

Refer to Figure 12.17, which shows aggressive and gentle driving between traffic lights. Both cars travel the same distance in cruise. The gentle car cruises at 30 mph and the aggressive car at 40 mph. Both cars use the same fuel braking. However, the aggressive car uses more fuel accelerating and cruising. The amount of fuel can be calculated.

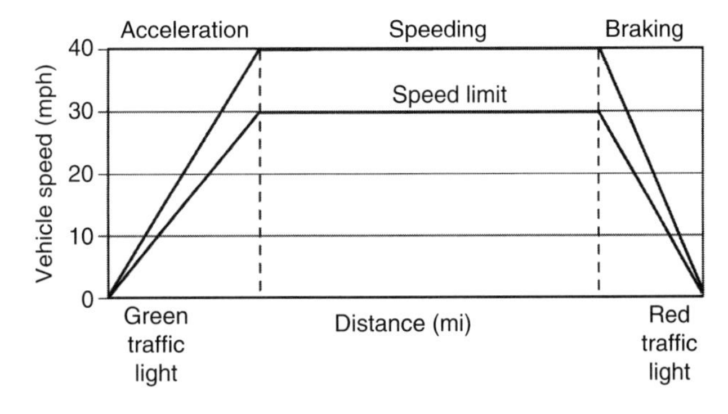

FIGURE 12.17 Figure showing driving cycle (velocity profile as a function of distance) between green and red traffic lights.

Use a KE approach; KE = 1/2 mV^2, kWh or J. Overall vehicle efficiency while accelerating is η. The energy density of gasoline is E_D, kWh/L or kWh/gal, which has a value of 33.44 kWh/gal. Some symbols are

Gallons to accelerate, G_A
Gallons to cruise, G_C
Distance in cruise, D
Fuel economy, F_E, mpg

The equation for the gas to accelerate is

$$G_A = \frac{\text{KE}}{\eta E_D} = \frac{mV^2}{2\eta E_D} \tag{12.22}$$

The gas to cruise is

$$G_C = \frac{D}{F_E} \tag{12.23}$$

The total gas used is

$$G = G_A + G_C = \frac{mV^2}{2\eta E_D} + \frac{D}{F_E} \tag{12.24}$$

Use subscript 1 for the gentle car and 2 for the aggressive car. Added gallons to speed is

$$\Delta G = (G_A + G_C)_2 - (G_A + G_C)_1 = G_{A2} - G_{A1} \tag{12.25}$$

Combining Equations 12.24 and 12.25 gives

$$\Delta G = \frac{m}{2\eta E_D}\left(V_2^2 - V_1^2\right) = \frac{mV_1^2}{2\eta E_D}\left(V_{21}^2 - 1\right) = G_{A1}\left(V_{21}^2 - 1\right) \tag{12.26}$$

Note that V_{21} is shorthand for V_2/V_1. ΔG varies (almost) as the square of V_{21}^2.

$$\frac{\Delta G}{G_{A1}} = \left(V_{21}^2 - 1\right) \tag{12.27}$$

As an example, let $V_1 = 30$ mph and $V_2 = 40$ mph, which are the same values in Figure 12.17. Then

$$\frac{\Delta G}{G_{A1}} = \left[\left(\frac{40}{30}\right)^2 - 1\right] = 0.78 = 78\% \text{ increase in gasoline used}$$

Conclusion: Fuel economy is sensitive to slight speeding; speeding only from 30 to 40 mph gives 78% increase in gas used.

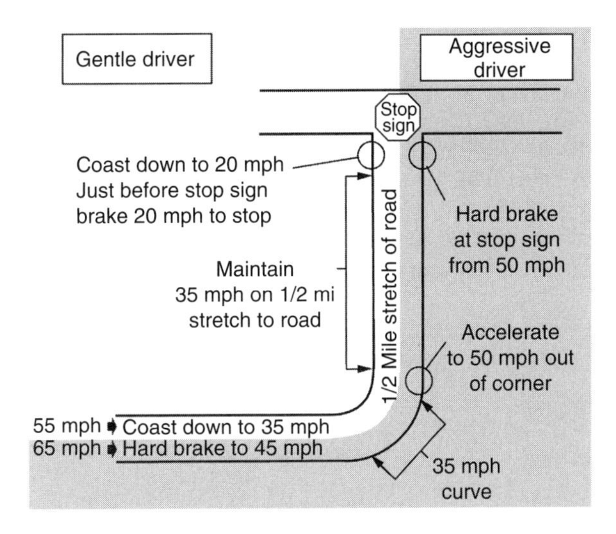

FIGURE 12.18 Virtual test track for conducting analytical tests for fuel economy.

Virtual Test Track

Now we have some fun determining the extra fuel used by an aggressive driver following the virtual test track (Figure 12.18). Many reports lump everything together without definition. A step-by-step method is not used. The virtual test track has gentle driving in various, well-defined steps. The virtual test track has aggressive driving in various contrasting steps. Each step can be calculated to determine fuel consumed. Aggressive driving minus gentle driving gives fuel saved. Each step can be used to find which distinct operation gives the most or least fuel consumed. Each segment of the driving cycle was calculated to yield fuel consumed. The results are shown in Table 12.9.

Using the totals from Table 12.9, the mpg can be calculated. For the gentle driver the values are 6.00 mi/0.255 gal = 23.5 mpg. For the aggressive driver the values are 6.00 mi/0.323 gal = 18.6 mpg, which is 21% fewer mpg than the gentle driver.

TABLE 12.9
Summary of Fuel Consumed by Gentle and Aggressive Drivers

		Gallons Used	
Segment of Virtual Track	**Distance, miles**	**Gentle**	**Aggressive**
Accelerate, 55 mph	0.26	0.023	—
Accelerate, 65 mph	0.19	—	0.032
Straight (gentle)	5.0	0.214	—
Straight (aggressive)	5.07	—	0.251
Coast to 35 mph	—	0.0	—
Brake to 45 mph	—	—	0.0
Corner	0.244	0.010	0.019
Accelerate	0.05	—	0.004
Cruise	0.40	—	0.017
Cruise	0.30	0.013	—
Coast/brake	0.20/0.05	0.0	0.0
Total	6.00	0.255	0.323

DRIVING TIPS

Overdrive and Higher Gears with Manual Transmission

Reference [11] indicates that use of overdrive gears on either manual or automatic transmission improves mpg. Reference [12] suggests shifting the manual transmission up as soon as possible without lugging the engine. Use of overdrive places the engine near its most efficient operating point. The most efficient operating point for an engine is near 75% of maximum installed hp and near maximum engine torque. See "Most Efficient Engine Operating Point." An equation for engine power is

Power = (torque)(rpm)

High	Low	←Overdrive
Low	High	→Low gear

Two combinations of torque and rpm are below the equation; both give the same power. For overdrive, engine rpm is low and torque is high, which places engine near the optimum operating point for least fuel consumption.

In addition to being a sample engine map, Figure 12.5 has a constant power curve plotted to emphasize the preceding paragraph. Power is 20 kW (26.7 hp). Fuel economy, mpg, is

$$F_E = \text{mpg} = \frac{\rho_G V}{(\text{BSFC})(\text{hp})} \tag{12.28}$$

Since ρ_G, V, and P are constant, the ratio of fuel economies at two different rpm's is

$$\frac{F_{E2}}{F_{E1}} = \frac{B_1}{B_2} \tag{12.29}$$

Where the constant power curve crosses a BSFC contour for a fixed rpm, the value of B can be determined. For example, at 4000 rpm, B equals 275 g/kWh. Values of B were obtained at other rpm's. Table 12.7 summarizes the results. The high rpm of 4000 is used as the reference. The fuel economy at 4000 rpm is assumed to be 25 mpg. The value of B at 4000 rpm is assigned subscript 1 so that B_1 equals 275 g/kWh. Based on the values of B_2 the value of the ratio B_2/B_1 was determined as well as F_{E2}/F_{E1}. As the rpm decreases from 4000, the mpg increases. The results shown in Table 12.10 confirm the statements above concerning overdrive.

TABLE 12.10
Change of Fuel Economy with Engine Speed at Constant Engine Power

Engine Speed, rpm	Brake-Specific Fuel Consumption, B, g/kWh	F_{E2}/F_{E1}	Fuel Economy, mpg
2000	231	1.19	29.8
2500	233	1.18	29.5
3000	245	1.12	28.1
3500	260	1.05	26.7
4000	275	1.00	25.0

Aerodynamic Drag: Pickup Truck Tailgate

An item that is discussed from time to time in the magazines is the relative drag with the tailgate up or down. Experiments show that up or down does not really matter.

Aerodynamic Drag: Sunroof

An open sunroof increases aerodynamic drag. The open sunroof increases airflow through the passenger cabin. The more the airflow, the more the ram drag. Ram drag occurs when air is ingested from the incoming airstream. Air going out the sunroof interacts with the external flow with either good or bad results. A tilt-up sunroof also interacts with the external flow to change drag.

Aerodynamic Drag: Deformation of Convertible Tops

If you drive a convertible, this is interesting. Reference [18] discusses the flow and drag changes due to the deformation of the convertible top. The shape of the deformed top depends on pressure—pressure depends on shape of top. This is a coupled problem that is not easily solved. Drag may or may not increase due to deformation.

Aerodynamic Drag: Roof Rack

Two questions are first, is a roof rack necessary and, second, how much penalty is incurred by having a roof rack? With regard to the first question, Ref. [11] discusses car selection

$$\text{Small car} + \text{occasional roof rack use} = \text{needs}$$

$$\text{Large car (no rack)} = \text{needs}$$

A small car has good fuel economy and with occasional drag penalty due to roof rack may be the best selection. A large car has less fuel economy and, with more interior space, eliminates the need for a roof rack.

For improved mpg, remove flags, bike racks, kayaks, cargo boxes, canoes, etc. Reference [19] indicates up to 5% loss of mpg due to roof racks. Analysis provides better insight to the loss of fuel economy, F_E. The equation is

$$\frac{\Delta F_E}{F_E} = -\frac{\Delta B}{B} - \frac{\beta X^2}{1 + X^2} \qquad (12.30)$$

The symbol, β, is ratio of the increment of added drag coefficient, ΔC_D, to the drag coefficient without a roof rack, C_{D0}. That is, $\beta = \Delta C_D/C_{D0}$. Suppose $\beta = 0.25$. Hence, for $X = 0.2$, the term $\beta X^2/(1 + X^2)$ is $(0.25)(0.04)/(1 + 0.04) = 0.0096$. Neglecting the $\Delta B/B$ term, the change in fuel economy would be 0.96%. For $X = 1.0$, the term $\beta X^2/(1 + X^2)$ is $(0.25)(1.00)/(1 + 1) = 0.125$. Neglecting the $\Delta B/B$ term, the change in fuel economy would be 12.5%. As always, perturbations to the reference vehicle are affected by the $\Delta B/B$ term.

Brake Drag

In this day of great interest in fuel economy, it is surprising that the common disk brake design allows the brake pads to rub against the rotor causing a small loss of mpg. Disk brakes allow the brake pad to rub the rotor and cause drag. Also, Refs. [9,10] point out that a foot riding on the brake pedal may waste gas and wear out the brakes. Most concept cars for Supercar, discussed in Chapter 1, had some means of eliminating the drag of disk brakes. Disk brakes do not have any mechanism to retract pad from rotor after braking except the stretched rubber seals; in the future, cars need some means for retraction of brake pads to be added. New electric wedge brakes (EWB) have no drag of the pad on the rotor [16,51]. Reference [52] reports the EWB has been tested on ice yielding a stopping distance of 64.5 m (212 ft) compared to 75 m (246 ft) for a conventional hydraulic brake. Test speed was 80 km/h (50 mph). Expect the EWB to appear on new vehicles starting in 2010.

Reduce Use of Brakes

KE of the car becomes useless heat by the friction brakes (not so with regenerative braking). Braking is equivalent to throwing away useful energy (gasoline). Braking is equivalent to having a hole in your gas tank. Anticipate red traffic lights. Do not tailgate; notice how a car that is tailgating is constantly turning on red stop lights by braking. Pace yourself in traffic to avoid braking for red traffic lights. You must brake for stop signs. Arrive late at the stop sign not as car in a long line but as the first car ready to take your turn. With cars in front of you while you approach a stop sign allow plenty of space to coast to stop and not brake.

Idling

Minimize idling. Avoid idling the car just to run the A/C. At a fast food place, bank, or ATM, park, go in, and avoid the outside service line for cars.

Idling: Engine-Off in Conventional Vehicle

For reasons of safety, do not turn off engine in traffic (hybrids excepted). At road construction when the flagman stops you turn off the engine. At railroad crossings when blocked by a train, once again turn off the engine.

Idling: Engine-Off in Hybrid Vehicle

Hybrids are a different breed. Engine turnoff is done automatically by the hybrid control system. Hybrids have powerful electric motors to use as starters (see Chapter 3).

Power-Hungry Accessories and Auxiliaries

The alternator takes power from the engine and may absorb 2 kW (can be 20% of cruise power if everything is turned on). The alternator provides the electrical energy for many accessories and auxiliaries; see Chapter 5 for a list of accessories and auxiliaries. Reference [9] states that running accessories and auxiliaries reduces mpg. New EPA mileage tests incorporate the effects of the power-hungry accessories and auxiliaries.

Two-Wheel and Four-Wheel Drive

If you have an option, use 2WD instead of 4WD for better gas mileage. 4WD requires more power and, thus, more gasoline. 4WD has additional parts, which in makes it heavy. Added weight causes mpg to decrease. Also, 4WD causes more power train losses and results in lower mpg. The extra weight of 4WD is hauled around wherever you go, there is no vacation from the extra weight and lost mpg. Do you really need the 4WD?

4WD is used for two primary reasons. One reason is to have off-road capability. The other reason is increased handling performance. The comments above apply to both uses.

Extra Weight in Vehicle

This is an important enough subject for an entire chapter (Chapter 7) to be devoted to it. Chapter 7 provides the equation for fractional loss of fuel economy (repeated here for your convenience)

$$\frac{\Delta F_C}{F_C} = +C\left(\frac{\Delta W}{W}\right) = +C\left(\frac{\Delta m}{m}\right) = -C\left(\frac{\Delta F_E}{F_E}\right) \tag{12.31}$$

where

F_C is the fuel consumption (proportional to BSFC)
m is the mass, kg
F_E is the fuel economy, mpg
W is the vehicle weight, lb
C is a constant, that is, 1.0 in some cases of fuel economy and weight correlations

Assume $C = 1.0$ and Equation 12.31 becomes

$$\frac{\Delta F_E}{F_E} = -\frac{\Delta W}{W} \tag{12.32}$$

Assume extra weight, like junk-in-trunk, is $\Delta W = 150$ lb. Assume the vehicle weight is $W = 2500$ lb (case 1) or 4500 lb (case 2).

- Case 1: $\Delta W/W = 150/2500 = 6\%$ loss of F_E = fuel economy, mpg
- Case 2: $\Delta W/W = 150/4500 = 3.3\%$ loss of F_E = fuel economy, mpg

Conclusions: Loss of mpg depends on the vehicle weight. For the same added weight, heavy cars have less loss of fuel economy, F_E, due to the added weight.

An equation that takes into account the influences of rolling friction drag and vehicle velocity can be compared with Equation 12.31.

$$\frac{\Delta F_E}{F_E} = -\frac{\Delta B}{B} - \frac{1}{1+X^2}\left(\frac{\Delta W}{W}\right) \tag{12.33}$$

The two equations are nearly identical if $C = 1/(1 + X^2)$. Driving around town, X is small. On the highway, X may be larger than unity. Large X decreases the effect of added weight.

Anticipate Traffic Conditions

This section reiterates what has been said in the section "Drive a Quarter Mile Down the Road." Pace yourself to avoid using brakes, if possible. Avoid slowing down; if not possible, coast as much as possible. Of course, safety is more important. Do not become involved in a collision just to avoid use of your brakes.

Anticipate Red Lights

With regard to red traffic lights, two cases can be envisioned. These are (1) stop at red light and (2) time yourself to coast through a just-green light. The second case is preferable. Case 1, which is to stop at the red light, forces you to stop. All KE is lost plus the idle penalty.

Case 2, which is to time oneself to coast through a just-green light, retains some of the expensively bought KE and also has no idle penalty.

Towing a Trailer

Towing a trailer is a double whammy on gas mileage. The tires on the trailer add to the rolling friction. The body of the trailer adds aerodynamic drag. One minor point in the analysis is tongue weight, which is a shift in load from the trailer to the tow vehicle (TV). Obviously, trailer towing decreases mpg [21]. Many different kinds of trailers exist: house, boats, etc. These different trailers can be categorized in a meaningful way by the analysis to be discussed.

Aspects of trailers or variables for an analysis include

- TV's rating for towing trailers
- Tongue weight
- Trailer total weight
- Frontal area (drag area) for added aerodynamic drag due to trailer
- Trailer shape as it influences drag coefficient, C_D
- Center of gravity in the trailer
- Dynamics whi le towing; instability

The equation for fuel economy for trailer towing is

$$\frac{\Delta F_E}{F_E} = -\frac{\Delta B}{B} - \frac{\alpha + \beta X^2}{\left(1 + X^2\right)} \tag{12.34}$$

where

α is the ratio of the trailer's rolling friction to that of the TV
β is the ratio of the trailer's drag area $(C_D A)_T$ to that of the TV, that is, $\beta = (C_D A)_T/(C_D A)_{TV}$

Since both α and β may be large and even more than unity, a ratio of the fuel economies should be used instead of the fractional form above. This ratio is based on

$$F_E = \frac{\rho_G V \eta_{PT}}{BP_Q} \tag{12.35}$$

For the ratio, the velocity, V, power train efficiency, η_{PT}, and gasoline density, ρ_G, are assumed constant. With that assumption the ratio becomes

$$\frac{F_{E2}}{F_{E1}} = \frac{B_1}{B_2}\frac{P_{Q1}}{P_{Q2}} \tag{12.36}$$

Subscript 1 is the TV without a trailer and subscript 2 is the TV with a trailer. Both B_1 and B_2 are obtained from the engine map for the TV. The ratio becomes with X_1, α, and β as variables

$$\frac{F_{E2}}{F_{E1}} = \frac{B_1}{B_2} \frac{1 + X_1^2}{(1+\alpha)+(1+\beta)X_1^2} \tag{12.37}$$

As a numerical example use $X_1 = 0.5$, $\alpha = 1.5$, and $\beta = 2.0$. $F_{E2}/F_{E1} = 0.56$. If the TV gives 20 mpg without the trailer, the fuel economy with the trailer is

$$F_{E2} = (20 \text{ mpg})(0.56) = 11.2 \text{ mpg}$$

A very large decrease in mpg results from trailer towing.

The ratio of α/β defines three classes of trailers. These classes are designated I, II, and III. Class I, $\alpha/\beta \ll 1.0$ has features as follows:

- Boxy, large frontal area, and lightweight
- Mpg sensitive to speed
- Example is a house trailer

Class II, $\alpha = \beta$ yields a simple equation for fuel economy ratio, which is

$$\frac{F_{E2}}{F_{E1}} = \frac{B_1}{B_2} \frac{1}{(1+\alpha)} = \frac{B_1}{B_2} \frac{1}{(1+\beta)} \tag{12.38}$$

Features of the trailer are

- Change in fuel economy is independent of speed.
- Getting values for B complicates the analysis.
- Large changes in B are to be expected.
- Example is a horse trailer.

Class III, $\alpha/\beta \gg 1.0$ has features as follows:

- For TV, cargo is limited.
- Streamlined, very heavy trailer.
- Small trailer with small drag area, A_2.
- Examples are a flat bed trailer with load of cement sacks and a transport for heavy, valuable metals.

More details on the analysis and the interpretation of trailer towing are found in Ref. [35].

NEVER "REV-UP" THE ENGINE

According to Ref. [12] "revving up" the engine wastes gas. If your car is stalling (engine dies at idle), you may need a tune-up or need to have the engine control computer checked.

AIR-CONDITIONING AND WINDOWS DOWN

AIR-CONDITIONING: BACKGROUND

In the "advice" literature, many different ideas are given concerning the loss of fuel economy due to A/C. A/C is subject to analysis, which can clear up many differing pieces of advice. Even though the A/C has an on/off switch, the power absorbed by A/C is intermittent. The main power absorber is the A/C compressor, which cycles on/off according to cooling demands. The fraction of on-time affects the power used and mpg. A big fraction of cycling on-time means more loss of mpg. Low fraction of cycling on-time means less loss of mpg. Can the driver control the compressor on-time? Only indirectly; there is no cycle control knob or on/off switch for the compressor. Only through setting of fan speed and temperature control is cycling controlled. What determines the cycle on-time besides outside temperature and humidity? The setting of fan speed and temperature control regulates cycling.

Air-Conditioning: Effects of Humidity

As background for relating humidity to A/C operation some definitions are presented. An endothermic process absorbs heat while an exothermic process releases heat. Humidity relates to the amount of water vapor in the air. The water absorbs or releases heat as follows:

$$\text{Water vapor} \underset{\leftarrow \text{Endothermic}}{\overset{\text{Exothermic} \rightarrow}{\quad}} \text{Water droplets}$$

Water vapor from the outside air condenses on the evaporator coils (inside the cabin). This is an exothermic reaction, which means heat is released. This heat adds to the load due to hot air. The A/C for a car has a drain for the water droplets that are formed on the evaporator coils (inside cabin). A humid day causes a greater loss of fuel economy. Dry, hot air in the desert Southwest of United States has an advantage of low humidity and less load on the A/C.

Air-Conditioning: Control

Control may be manual or thermostat. Also, thermostat control usually has zone temperature control (driver/passenger). The three types of A/C control are

1. Thermostat control (luxury cars)
2. Cold modulate with heat
3. Manual, pure cold only (economy cars)

Air-Conditioning: Zone and Recirculation

Husbands—let your wife set the temperature for her zone anyway she wants; sometimes harmony is more important than better mpg. Almost all cars offer the choice between recirculate the air inside the car (for this setting, some outside air is added) or circulate the outside air. Recirculation of the air inside the car improves mpg.

Air-Conditioning: Thermostat Control

Settings of the A/C can be adjusted for less loss of mpg. Set the thermostat a few degrees less than outside temperature. An example

Outside temperature	94°F
Set thermostat	88°F (seems dumb)

Direct the air vents to your face; the car seems cooler when your face is cool. A/C conditioned air has low humidity and on long trips can dry out eyes, lips, and skin. Wear sunglasses for protection of eyes. These settings, which recycle the compressor less often and use less gasoline, do not cause you to suffer in the heat. The setting beats open windows at an outside temperature of 94°F. If the thermostat is set at 70° instead of 88°F, the compressor will cycle often and decrease mpg.

Air-Conditioning: Cold Modulated with Heat

The heat control knob has a curved red triangle. The A/C knob has a curved blue triangle. When you turn the A/C knob to any position, the A/C cools the air to very cold. Then heat is added to modulate the temperature to just cool (to match A/C knob setting). Wasted power (fewer mpg) is spent to run the compressor. Hence, set knob at full cold, which avoids added heat to waste gas. Put the fan on low setting. If the air is too cold, just let the air glance by your face instead of blowing directly onto your face.

A different, but related, topic is fog on the windshield and other interior windows. Fog gets on the windows of a car whenever the window temperature is less than the dew point of the air inside the car. The windshield defrost system operates with the cold air (removes water vapor) modulated with hot air (heat warms windows).

Air-Conditioning: Pure Cold

This control has no modulation with heat. Position the A/C knob between zero cool and full cold for comfort. A colder setting causes larger loss of mpg. Use as low a setting as is comfortable. Consider the suggestions above.

Air-Conditioning: Reported Losses

Different levels of fuel economy loss due to A/C are reported. Reference [9] reports that on max setting the A/C loss is 5%–25%. Reference [10] suggests using internal air recirculate to avoid dumping cold air overboard. Even in recirculate, some fresh air is added to cabin.

Air-Conditioning: Equation for Loss of Fuel Economy due to A/C

Define symbols:

P_E is the engine power, hp
P_Q is the required power to move the vehicle with A/C on or off, hp
$P_{A/C}$ is the A/C power averaged over the compressor cycle, hp
P_{RCO} is the power required for rolling friction evaluated at the crossover velocity, V_{CO}
F_E is the fuel economy, mpg
F_{E0} is the fuel economy, mpg, with A/C turned off
V is the vehicle velocity, mph
V_{CO} is the crossover velocity, mph or m/s
B is the BSFC, lb/(hp h)
η_{PT} is the power train efficiency
ρ_G is the density of gasoline, 6.2 lb/gal
$\gamma_{A/C}$ is the ratio of power = $P_{A/C}/P_{RCO}$

Power required to move the vehicle, P_Q, and engine power, P_E, are related by the equation

$$\eta P_{\mathrm{E}}' = P_{\mathrm{R}} + P_{\mathrm{D}} + \eta P_{\mathrm{A/C}} = P_{\mathrm{Q}} + \eta P_{\mathrm{A/C}} \tag{12.39}$$

Introducing $\gamma_{\mathrm{A/C}} = P_{\mathrm{A/C}}/P_{\mathrm{RCO}}$ gives

$$\eta P_{\mathrm{E}}' = P_{\mathrm{RCO}}\left[X\left(1 + X^2\right) + \eta\, \gamma_{\mathrm{A/C}}\right] \tag{12.40}$$

The fuel economy, F_{E}, mpg is

$$F_{\mathrm{E}} = \frac{\rho_{\mathrm{G}} V \eta_{\mathrm{PT}}}{B P_{\mathrm{E}}} \tag{12.41}$$

A check of units shows that the equation does, in fact, give mpg. F_{E} = fuel economy, mpg, with A/C turned off or on. In fact, Equation 12.41 applies to any case of steady driving.

$$\eta_{\mathrm{PT}}\left(P_{\mathrm{E}}' - P_{\mathrm{E}}\right) = \eta_{\mathrm{PT}}\, \gamma_{\mathrm{A/C}}\, P_{\mathrm{RCO}} \tag{12.42}$$

Fractional engine power is

$$\frac{\Delta P_{\mathrm{E}}}{P_{\mathrm{E}}} = \frac{\eta_{\mathrm{PT}}\, \gamma_{\mathrm{A/C}}\, P_{\mathrm{RCO}}}{P_{\mathrm{RCO}}\, X\left(1 + X^2\right)} \tag{12.43}$$

The fractional fuel economy is given by

$$\frac{\Delta F_{\mathrm{E}}}{F_{\mathrm{E}}} = -\frac{\Delta B}{B} - \frac{\Delta P_{\mathrm{E}}}{P_{\mathrm{E}}} = -\frac{\Delta B}{B} - \frac{\eta_{\mathrm{PT}}\, \gamma_{\mathrm{A/C}}}{X\left(1 + X^2\right)} \tag{12.44}$$

Air-Conditioning: Sample Calculations for Fuel Economy Loss

Equation 12.44 is used for the calculations.

Typical A/C compressor hp: 6 hp (4.5 kW)
Typical compressor cycle time: 35% of time compressor is compressing

Example 1: Cruise on freeway with small car

P_{RCO} is the rolling friction power with A/C on/off, 8 hp (6 kW)
$P_{\mathrm{A/C}}$ is the A/C power averaged over the compressor cycle, (6)(0.35) = 2.1 hp
$\gamma_{\mathrm{A/C}}$ is the power ratio = 2.1/8 = 0.263
η_{PT} equals 1.0 $\quad \Delta B/B = -0.02$
$X = 1.2$ $\quad \Delta P_{\mathrm{E}}/P_{\mathrm{E}} = -0.090$
Fuel economy loss (due to A/C) = –(–0.02) – 0.090 = 7% loss.

Example 2: Cruise on freeway with large car

P_{RCO} is the rolling friction power with A/C on/off, 14 hp (10.5 kW)
$P_{\mathrm{A/C}}$ is the A/C power averaged over the compressor cycle, 2.1 hp
$\gamma_{\mathrm{A/C}}$ is the power ratio = 2.1/14 = 0.150

TABLE 12.11
Improved Fuel Economy due to Cabin Insulation with Reduced A/C Loading

Insulation Package	A/C	mpg	Loss Due to A/C (%)	Gallons Used	$ A/C Penalty	Cost of Gasoline ($)
No	Off	18.5	0	649	—	1946
No	On	15.4	17	779	391	2337
Yes	On	16.1	13	745	290	2236

η_{PT} equals 1.0 $\Delta B/B = -0.02$
$X = 1.2$ $\Delta P_E/P_E = 0.051$
Fuel economy loss (due to A/C) = −(−0.02) − 0.051 = 3.1% loss.

Conclusions: For a fixed A/C power, $P_{A/C}$, and smaller power to move the vehicle, P_{RCO}, gives greater loss of fuel economy. See the two examples above, which have 7% loss for small $P_{RCO} = 8$ hp and only 3.1% loss for the larger $P_{RCO} = 14$ hp.

Air-Conditioning: 2005 Cadillac STS

Tests were conducted using a Cadillac STS with heat rejection technology and a standard Cadillac STS (see Ref. [17] and Table 12.11). Insulation saves \$391 − \$290 = \$101 annually and decreases emissions.

OPEN WINDOWS

Open Window (Sunroof) Ventilation: Comments

An open window is an alternative to air vents on dash. Air vents usually receive air from the outside grill just in front of the windshield. Windows down as a rule increases aerodynamic drag more than air vents on dash. If an open window is your choice, Reference [12] suggests that wind noise be your guide; the open window with least noise will likely have least drag increase. A pulsating sound inside the car can be created by partially open window; this is called a Helmholtz resonator. An open sunroof also causes an increase in aerodynamic drag.

Open Window Ventilation: Loss of Fuel Economy

From the analysis point of view, windows down, a head/tailwind, and tailgating a truck are all similar. Each involves a change in aerodynamic drag coefficient or relative velocity of the air. Define symbols:

C_{DU} is the vehicle drag coefficient with windows up.
ΔC_D is the increase in aerodynamic drag due to windows down.
β_{CD} is the ratio of increase in drag coefficient, $\Delta C_D/C_{DU}$.

The fractional change in fuel economy is

$$\frac{\Delta F_E}{F_E} = -\frac{\Delta B}{B} - \frac{\beta_{CD} X^2}{1 + X^2} \tag{12.45}$$

This equation can be used to calculate the loss of fuel economy due to the added drag from windows down.

A/C versus Open Windows

The question of whether A/C or windows down causes the greatest loss of fuel economy is a perennial discussion item in magazines, on the Web, and in various reports. Some state that A/C gives less loss than windows down. Others state just the opposite, that is, A/C gives more loss than windows down. Both viewpoints cannot be correct, can they?

A/C versus Open Windows: Sample Calculations

For the loss of fuel economy due to A/C, Equation 12.44 was used. For the loss due to extra aerodynamic drag caused by open windows, Equation 12.45 was used. The input values are

rpm and V values serve as independent variables
$P_{A/C} = 3.8\,\text{hp}$, $\beta_{CD} = 0.20$
$\gamma_{A/C}$ = power ratio = 3.8/4.8 = 0.80 $P_{RCO} = 4.8\,\text{hp}$
$\eta_{PT} = 1.0$ $V_{CO} = 40\,\text{mph}$
B are the values from the engine map (Figure A.12.1)

The results of the sample calculations are graphically summarized in Figure 12.19, which plots percent fuel loss as a function of vehicle velocity.

The two loss curves cross at the velocity V_E. The crossing in Figure 12.19 is at 43 mph. At this point the two losses are equal, which is recognized by the subscript E on V_E. V_E is the dividing value between which loss dominates. Reference [12] reports that at highway speeds the same loss occurs for either A/C or windows down. This is essentially true for speeds near V_E. Reference [15] indicates that at lower speeds, windows down without A/C results in least loss. At higher speeds, the A/C may be more efficient than the wind resistance due to open windows and sunroof. Reference [15] does not give numbers for lower or higher speeds. Reference [2] states that at freeway speeds, windows down versus windows up causes a loss of fuel economy (3%). Reference [2] also reports data at 67 mph for a 1986 VW GTI. These data are summarized in Table 12.12. The conclusions from the data of Table 12.12 are the opposite of Figure 12.19. The two results are inconsistent for $V > V_E$.

Figure 12.19 answers the question posed above, "Both viewpoints cannot be correct, can they?" The answer is yes. For low velocity, A/C loss dominates. At high speed, windows down dominates

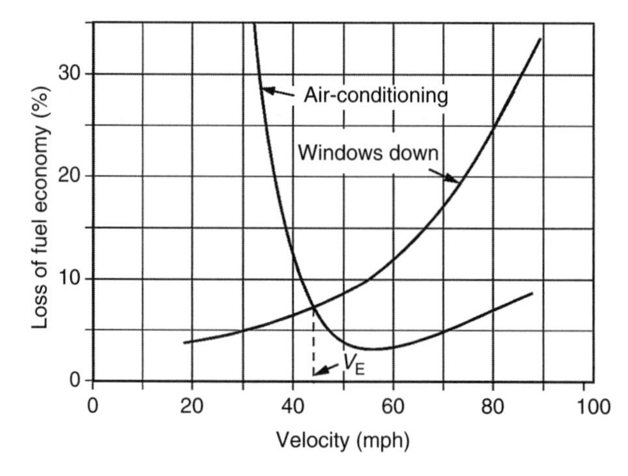

FIGURE 12.19 Percent fuel economy loss as a function of vehicle velocity for A/C and added aerodynamic drag due to windows down for ventilation. Also, illustration of the equal velocity, V_E.

TABLE 12.12
Measured Change of Fuel Economy due to Windows Down or Use of A/C

Velocity (mph)	A/C	Windows	mpg	Loss Percent
67	Off	Up	30.5	0%
67	Off	Down	29.7	2.5%
67	On	Up	26.7	12.5%

Note: Results are for 1986 VW GTI.

the losses. References [10,12] indicate that the value of V_E is about 40 mph. For city driving, use windows down for minimum loss of fuel economy.

Analytical Result for V_E

At V_E, the fractional fuel economies are equal. Using this fact, an equation for V_E is created as shown below.

$$\frac{\Delta F_E}{F_E} = -\left(\frac{\Delta B}{B}\right)_{CD} - \frac{\beta_{CD} X^2}{1 + X^2} = -\left(\frac{\Delta B}{B}\right)_{A/C} - \frac{\eta_{PT}\gamma_{A/C}}{X(1 + X^2)} \quad (12.46)$$

With the two unequal $\Delta B/B$ terms, a simple solution for X = velocity ratio = V/V_{CO} is not possible. For the special case where the two $\Delta B/B$ terms are equal, the equation for V_E is

$$V_E = \left(\frac{2P_{A/C}}{\rho C_D A \beta_{CD}}\right)^{1/3} \quad (12.47)$$

The dependence on the various variables is weak due to the one-third power. Increasing A/C power, $P_{A/C}$, increases V_E. Increasing atmospheric density, ρ, decreases V_E. Likewise increasing drag coefficient, C_D, and frontal area, A, decreases V_E. Since β_{CD} is the ratio of increase in drag coefficient, $\Delta C_D/C_{DU}$ appears in the denominator, an increase of β_{CD} causes a decrease in V_E.

WHERE YOU DRIVE

Headwind and Tailwind

Let V_W be the velocity of wind, and V be the velocity of vehicle. Then for headwind,

$$D = \frac{1}{2}\rho\left(V + V_W\right)^2 C_D A \quad (12.48)$$

Power to overcome the headwind is

$$P = \frac{1}{2}\rho V\left(V + V_W\right)^2 C_D A \quad (12.49)$$

Drag and power for tailwind

$$D = \frac{1}{2}\rho\left(V - V_{\mathrm{W}}\right)^2 C_{\mathrm{D}} A \tag{12.50}$$

With a tailwind, power is

$$P = \frac{1}{2}\rho V\left(V - V_{\mathrm{W}}\right)^2 C_{\mathrm{D}} A \tag{12.51}$$

Drag varies as the square of the relative velocity. Power varies as the cube of the velocity. However, with a head/tailwind, power is the product of vehicle velocity and the square of relative velocity.

The equation for fractional fuel economy for either a tailwind (top algebraic sign) or headwind (bottom algebraic sign) is

$$\frac{\Delta F_{\mathrm{E}}}{F_{\mathrm{E}}} = -\frac{\Delta B}{B} \pm \frac{\beta X^2 (2 \mp \beta)}{1 + X^2} \tag{12.52}$$

Assume V_{W} is the velocity of wind = 10 mph, and V is the velocity of vehicle = 60 mph. For these values, $\beta = 0.167$. Also assume $X = 1.0$. For a headwind, the second term on the left-hand side of the equation has a value of −0.18. If $\Delta B/B = 0$, then mpg is reduced by 18%. For a tailwind, the second term on the left-hand side of the equation has a value of +0.15. Mpg is increased by 15%.

Warming Up the Engine

Many commonly held automotive wisdoms are a holdover from an earlier time when cars were less complex but more demanding. Warming up the engine is in this category. With modern fuel injection, engines provide faultless power. After starting the engine, drive away. The engine will warm up faster while being driven. Almost universal agreement prevails on this advice. Reference [37] states that start driving at moderate speeds. References [10,12] agree; after starting engine wait 30 s or so, and drive away slowly. Reference [2] states "Start up, drive away."

Cold and Hot Weather

Cold weather tends to decrease mpg. Due to low temperature of the air being sucked into the engine, the engine efficiency declines. Hot weather tends to increase mpg. Why?

Curves in Road

Where you drive determines how many curves you will encounter. To go around a corner, the tires must provide a lateral force equal to the centrifugal force. A component of that lateral force is in the drag direction. For this reason, more power is needed to go around a curve, and more fuel is consumed (Figure 12.20). Mpg is decreased. The road has zero bank angle. The symbols are

F_{C} is the centrifugal force or lateral force on the vehicle, N or lb
m is the vehicle mass, kg
V is the vehicle velocity, m/s
R is the radius of curvature of the road in the curve, m

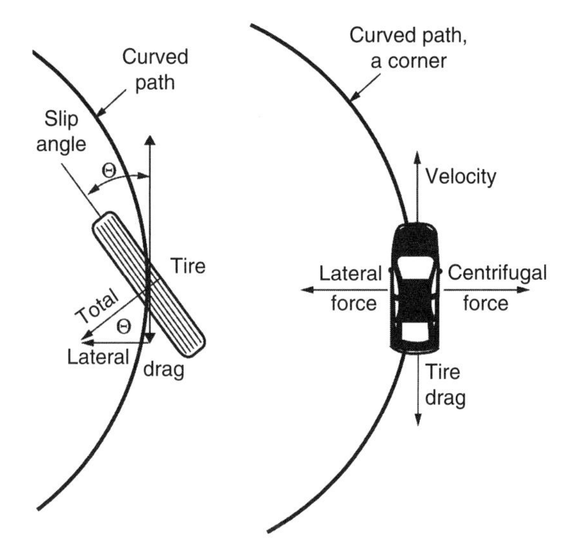

FIGURE 12.20 Tire drag during cornering.

The equation for centrifugal force or lateral force on the vehicle

$$F_C = m\,\frac{V^2}{R} \tag{12.53}$$

Refer to "Brakes and Tires Primer" in Chapter 9. Chapter 9 also discusses lateral tire forces, which are important for both steering and cornering. In Figure 9.7, the angle, θ, is defined by the sketch on the left-hand side. Both locked and rolling tires are shown. The drag component of the lateral force, F_D, is

$$F_D = F_C\ \sin\theta \tag{12.54}$$

The drag component adds to the rolling friction, F_R, which increases power required and, in turn, increases fuel consumption. The added power necessary to go around a curve is

$$P_\theta = V\left(\frac{m\,V^2 \sin\theta}{R}\right) \tag{12.55}$$

Define a ratio, β

$$\beta = \frac{2m\sin\theta}{R\rho C_D A} \tag{12.56}$$

The fractional change in fuel economy, F_E, due to cornering is

$$\frac{\Delta F_E}{F_E} = -\frac{\Delta B}{B} - \frac{\beta X^2}{1+X^2} \tag{12.57}$$

Assume a highway curve that gives 0.1 G's lateral acceleration at 45 mph. The radius of that curve is 411 m. A car is now driven on that curve at the speed of 45 mph. What is the change in fuel economy? Input values are

$m = 1816\,\text{kg}$	$\theta = 1°$	$R = 411\,\text{m}$
$\rho = 1.2\,\text{kg/m}^3$	$C_D = 0.3$	$A = 3\,\text{m}^2$

For these inputs, assuming $X = 1.0$, the second term on right-hand side has a value of 0.051. If $\Delta B/B = 0$, then the fuel economy is decreased by 5% due to the corner.

Ice and Snow

Ice and snow stuck under the car and in the wheel wells is heavy and increases weight, which decreases mpg. The heavy ice and snow becomes a safety hazard. Range of motion of the wheels may be restricted. Humps and bumps of ice and snow on the car body upsets smooth flow over the car and will likely increase aerodynamic drag.

High Altitude

High altitude decreases air density, ρ. Decreased air density has two effects. One effect is decreased aerodynamic drag

$$D = \frac{1}{2}\rho V^2 C_D A \tag{12.58}$$

At high altitude, drag is less. Another effect of altitude is on the engine operation.

Terrain: Climbing Hills

Reference [9] states hills reduce mpg. This statement is obvious. Chapter 7 discusses the added power to climb grades or hills. The equation is

$$P_H = mgV\sin\theta \tag{12.59}$$

where

P_H is the power to climb hill, kW
m is the vehicle mass, kg
g is the acceleration of gravity, 9.81 m/s^2
θ is the angle of the road relative to the horizontal, degrees
P_{Q0} is the power required on level ground, that is, $\theta = 0°$

Performance up a grade is determined by two quantities. One is the velocity, V, going uphill. The other factor, of course, is the steepness of the hill or percent grade. A road that climbs 6 ft while going 100 ft along the road has a grade of 6%. The grade can also be defined as the angle of the road relative to the horizontal, θ. Look at the analysis for A/C; with minor modifications, the same equations apply. Power required to climb a hill becomes

$$P_Q = P_{RCO}\,X\left(X^2 + 1\right) + P_H = P_{Q0} + P_H \tag{12.60}$$

Define a nondimensional ratio, which is hill climbing power compared to rolling friction power

$$\alpha = \frac{P_H}{P_R} = \frac{mgV \sin\theta}{fWV} = \frac{\sin\theta}{f} \tag{12.61}$$

The symbol, f, is rolling friction coefficient. The fractional change in fuel economy is

$$\frac{\Delta F_E}{F_E} = -\frac{\Delta B}{B} - \frac{\alpha}{1 + X^2} \tag{12.62}$$

The analysis is the same as for A/C. For the case of large α, the fractional form may not be accurate. The ratio of fuel economies is appropriate to use for the case of large α.

$$\frac{F_{E2}}{F_{E1}} = \frac{B_1}{B_2} \frac{1 + X^2}{1 + \alpha + X^2} \tag{12.63}$$

Suppose a vehicle is climbing a 6% grade, which is an angle $\theta = 3.43°$. The rolling friction coefficient $f = 0.02$. The alpha ratio, α, is equal to 3. The ratio of fuel economies is the correct equation to use. Assume $B_1 = B_2$. Also, the nondimensional speed, X, is 0.6. The fuel economy ratio has a value $F_{E2}/F_{E1} = 0.405$. If F_{E1} is 25 mpg on flat land, then the value climbing the hill is $F_{E2} = (25\,\text{mpg})(0.405) = 10.1\,\text{mpg}$.

Dirt and Gravel Roads

Besides getting your car dirty, dirt and gravel roads increase rolling friction. Increased rolling friction increases power required and decreases mpg.

Snow and Water Puddles on Road

Besides getting your car dirty, snow and water puddles on the road increase rolling friction. Power is needed to push the water out of the way. Increased rolling friction increases power required and decreases mpg. Remember, when motion is imparted to snow, mud, water, etc. (splash from a puddle) that motion is KE ultimately paid for by gasoline from your tank.

Rough and Undulating Roads

Rough and undulating roads cause the vehicle to move up and down. Remember, when motion is imparted to the vehicle, that motion is KE ultimately paid for by gasoline from your tank. Rough roads absorb power. See Ref. [1] of Chapter 10. Also refer to "Generator in Shock Absorbers" in Chapter 16. A few kilowatts could be generated using all four wheels.

MILEAGE SCAMS

Mileage Scams: Detecting Scams

April showers bring May flowers. High gas prices bring a proliferation of mileage boosting gadgets. These gadgets are always a scam that fails to use the laws of physics to deliver the promise of greater mpg. A gadget that actually boosts mpg is very rare. How do you recognize a scam?

Excessive claims are made for the improvement in mpg; not just 5% improvement but 30%–40% (or even 200%) improvement in mpg. The conspiracy theory is invoked in some form or other. Automakers and oil companies want to squelch the device as it could ruin their business. The government also wants to squelch the device as it could ruin some government regulations, for example, smog. The company making the device will be your knight-in-shining-armor to rescue you from the various conspiracies (see above). The company making the device through great sacrifice on their part makes this device available to you.

The mileage-enhancing device uses some technology previously unknown to mankind (magnetic fields; rarely, if ever, electrostatic fields; magnetic monopoles; exotic beads). Devices to enhance combustion by adding whirl to the incoming air are popular. Other devices claim to exploit miracle magnets, vortex generators, engine ionizers, fuel vapor injection, and water injection to improve mpg. Countless testimonials are used that state "I was skeptical (as you should be); however having tried *Super-X Magic Nuclear Potion*, I'm now convinced it is a miracle." The device does not have EPA test results (if submitted, probably failed tests) to back up claim. Credible engineering tests, which show that the device actually does do what is claimed, are nonexistent. Pressure to buy is used as a technique, that is, 40% off if you buy now; special offer good for the next 30 s. Free installation (only if you buy now). These devices use some irrelevant adjectives like laser, electronic, turbo, nuclear, digital, or nanotech in the name of the gadget.

Mileage Scams: Consequences of Using Scam Device

The promise of greater mileage may not only be dangerous to your pocketbook but may damage your car. Installation of the gadget may void the warranty on your car, may damage the engine, may damage the fuel system, may increase emissions, and may damage the catalytic converter. When *Popular Mechanics* conducted tests on several gadgets, none improved mpg, and some caused a 20% reduction in mpg. One device melted and caused a fire that terminated the test [36].

Some gadgets are flimsy in construction. When placed in the intake manifold, the chances are high that a piece will break off and be sucked into the engine. The valves can be damaged. You may be stranded along the side of the road with a car that will not start and with an expensive valve overhaul to be done.

Mileage Scams: EPA Tests

Your best protection from these scams is to be wary. Apply the criteria above to screen for and to detect a scam. The government provides some assistance with regard to bogus mpg gadgets [20]. EPA does test mileage-enhancing devices. EPA tests of gadgets that are submitted for testing almost universally have negative results. (Is EPA part of the conspiracy? This just shows EPA is part of the conspiracy!) EPA tests of gadgets find in some cases very minor change in mpg. Go to Web site www.epa.gov/otaq/consumer to check for tested products. Examples of devices that have been tested include mixture enhancers, fuel line magnets, and PCV scrubbers.

Mileage Scams: A Perspective

Most of the gadgets to enhance mpg sell for $40–$200. Some are much more expensive at $400–$500 each. The automobile makers face intense competition. Better mpg provides an immense competitive edge. Consequently, the car manufacturers spend a millions on research and development (R&D) to increase mpg. Output from R&D includes such technology as variable valve timing and displacement on demand, DoD. See Figure 17.4 for information on DoD. See also Tables 17.1 and 17.2 for new technologies for reducing fuel consumption applicable to the gasoline engine or diesel engine. Review the tables with the thought in mind of learning how much improvement in mpg is anticipated.

When a technology has been proven, it shows up on the production line. Added cost to the customer for the new technology to yield 15% gain in mpg may be a few hundred dollars. If the car maker could buy one of the magic mpg -enhancers for $40 that gives 30% better mpg, why would they suppress the mpg -enhancer? Of course, the answer is the mpg-enhancer would not be suppressed but would be welcomed. Since a verified mpg-enhancer costing $40 and yielding 30% increase in mpg does not exist, that gadget is not an option.

Mileage Scams: Unsubstantiated Calculations

An example is the plug-in hybrid. See "Electrical-Only—Plug-In Hybrid Fuel Economy" in Chapter 5. The sample calculation in that section gave 2440 mpg for certain conditions.

SUMMARY

Part of our national energy policy involves mileage ratings as measured by EPA and applied by NHTSA through CAFE. Hence, the early (2007 and prior years) mileage shortfalls in EPA were not a mere annoyance but had a serious impact on status of energy usage and availability. Hence, the early EPA tests and procedures are discussed for two reasons. First, the discussion provides the background necessary to understand the current situation. Second, the old city and highway tests carry on unchanged into the new EPA test methods. The new tests solve some of the existing deficiencies but not all. As a result, the discussion of test deficiencies is appropriate. The shortfall in mileage ratings was the motivation for the new EPA test procedures.

Logically, next in the flow of discussion is the description of the new EPA five-cycle test procedure. Two old tests remain intact and three new tests are introduced. The three new tests are high speed, A/C, and cold weather. The results from the five tests are combined by weighted addition yielding one number for city mileage and one number for highway mileage to be placed on the newly designed Monroney sticker for all new car buyers to see. If the new procedures have a weak point, it is the assignment of values to the weighting factors.

The testing of hybrid vehicles is briefly presented. A major source of error is the SOC of the battery at the beginning and at the end of the test. Very significant change in fuel economy occurs if the SOC is not identical at the beginning and at the end of the test.

CAFE provides the teeth for a meaningful energy policy. Despite its flaws, CAFE did double mpg following the 1970s gas crises. Since then average mpg has fallen due to the sales shift to 50% SUVs and minivans. CAFE has its weak points. These are enumerated by a National Academy report, which is discussed. One suggestion incorporated into CAFE is the use of footprint as an attribute to specify fuel economy.

This chapter is long enough to be a minibook. A lengthy portion is devoted to how you can improve your fuel economy, that is, how you can stretch your mpg. Magazines, Web sites, and talk-shows abound with tips for better mpg. Some of these tips fall into the category of folklore and myths quite separated from reality. Analysis of several mileage-increasing tips leads to equations, which are presented along with numerical examples.

Two appendices present an in-depth analysis. Appendix 12.1 discusses optimum cruise velocity and derives equations to calculate the optimum. The original concept of the crossover velocity as a reference velocity is presented. Appendix 12.2 derives the fleet and vehicle averages as used by CAFE.

APPENDIX 12.1: OPTIMUM CRUISE SPEED

This appendix provides an extensive discussion of optimum cruise speed. Equations are derived. An example is calculated to illustrate concepts and provide insight into various factors affecting the appropriate driving speed to achieve best mpg. The necessary symbols for the equations are

P is the power, kW, hp
P_Q is the required power, kW, hp
P_R is the power due to rolling friction, kW, hp
P_{RCO} is the power due to rolling friction at velocity, V_{CO}, kW, hp
P_E is the engine power, kW, hp
P_D is the power due to aerodynamic drag, kW, hp
F_R is the force due to rolling friction, N, lb
F is the force resisting motion of vehicle, N, lb
F_D is the force due to aerodynamic drag, N, lb
W is the vehicle weight, 4,000 lb (17,920 N)
f is the tire rolling resistance coefficient, 0.025
ρ is the atmospheric density, 1.0 kg/m^3
ρ_G is the density of gasoline, 6.2 lb/gal
C_D is the aerodynamic drag coefficient, 0.305
A is the vehicle reference (frontal) area for drag, 4.0 m^2
rpm is the revolutions/minute engine speed
a is the (engine angular speed)/(vehicle speed), that is, rpm/V = 40
V_{CO} is the crossover velocity, km/h, mph
X is the velocity ratio = V/V_{CO}
BSFC is the brake-specific fuel consumption, (lb of fuel)/[(hp)(h)]
mpg is the fuel economy, miles per gallon

The list of symbols also has typical values for the variables to be used in the example and generate Figure A.12.1.

Fuel economy, mpg, depends on the resistance to the forward motion of the vehicle. The two main resistances are rolling friction due to the tires and aerodynamic drag. Assume the retarding

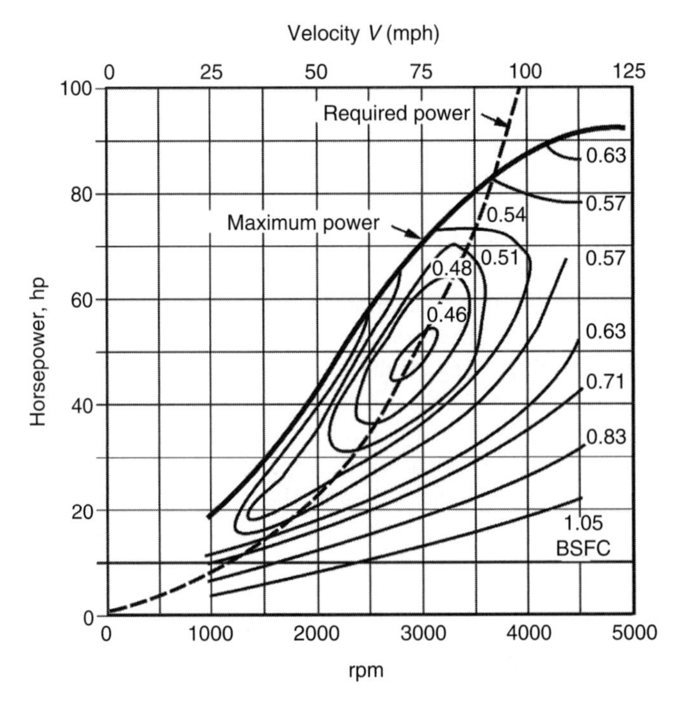

FIGURE A.12.1 Power required overlaid on an engine map using engine power as a function of engine speed, rpm. Engine map has contours of BSFC.

force due to rolling friction is constant. From Chapter 7, the retarding resistance, F_R, due to tire hysteresis is given by the equation

$$F_R = fW \tag{A.12.1}$$

where f is the rolling resistance coefficient with value between 0.01 and 0.04. The value depends on tire construction, tire pressure, road surface condition (sand has larger value than asphalt), and weakly on speed. The resistance increases linearly with weight, W. The power, P_R, required to overcome rolling resistance is

$$P_R = F_R V = fWV \tag{A.12.2}$$

where V is vehicle velocity. Power for V = 60 mph (26.8 m/s), f = 0.01, and W = 3000 lb (m = 1362 kg) is P_R = 3.6 kW (4.8 hp). Note that the power scales directly as vehicle mass (weight).

To determine the fuel economy as a function of vehicle velocity, the power required, P_Q, curve is overlaid on the engine map as shown in Figure A.12.1. The dashed curve is the power required and is calculated using Equation A.12.3.

$$P_Q = fWV_{CO}X\left(1 + X^2\right) = P_{RCO}X\left(1 + X^2\right) \tag{A.12.3}$$

The graph of engine hp as a function of engine angular speed, rpm, with BSFC contours, is known as an engine map. The contours are BSFC, (lb of fuel)/[(hp)(h)]. For an input engine speed (2500 rpm), at the intersection with P_Q curve, the values for hp (34 hp) and BSFC (0.49 lb/hp h) are obtained.

The fuel economy is given by

$$F_E = \text{mpg} = \frac{\rho_G V}{(\text{BSFC})(\text{hp})} = \frac{\rho_G V}{BP} \tag{A.12.4}$$

ρ_G is density of gasoline, 6.2 lb/gal, and rpm and V are connected by

$$a = \frac{\text{Engine angular speed}}{\text{Vehicle speed}} = \frac{\text{rpm}}{V} = 40$$

The symbol a depends on the power train gear ratios.

To understand the shape and trends of the fuel economy versus vehicle velocity curve in Figure 12.1, the fractional form of Equation A.12.4 is needed; it is

$$\frac{\Delta F_E}{F_E} = \frac{\Delta V}{V} - \frac{\Delta B}{B} - \frac{\Delta P}{P} \tag{A.12.5}$$

Equation A.12.5 will now be applied to find mpg for several different velocity regions defined by $X = V/V_{CO}$. See Table A.12.1.

For small X (<0.3, or so), according to Equation A.12. 2, $P_R = F_R V = fWV$ and $\Delta V/V$ is equal to $\Delta P/P$. Consequently, these two terms cancel in Equation A.12.4. Only $\Delta B/B$ affects fuel economy.

TABLE A.12.1
Table of Fractional Values for the Three Terms in Equation A.12.5 for Fuel Economy, mpg

						mpg	
Rpm	Range of X	$\Delta V/V$	$\Delta B/B$	$\Delta P/P$	$\Delta F_E/F_E$	Calculated	Predicted[a]
800–1200	0.33–0.50	0.400	−0.115	0.512	0.003	22.9	—
1200–1600	0.50–0.67	0.286	−0.161	0.425	0.021	23.1	23.0
1600–2000	0.67–0.83	0.222	−0.151	0.379	−0.005	23.5	23.5
2000–2400	0.83–1.00	0.182	−0.087	0.348	−0.079	23.4	23.4
2400–2800	1.00–1.17	0.154	−0.070	0.316	−0.090	21.6	21.6
2800–3200	1.17–1.33	0.133	0.032	0.294	−0.190	19.7	19.7
3200–3600	1.33–1.50	0.118	0.163	0.271	−0.319	16.2	16.0
3600	1.50	—	—	—	—	11.7	10.9

Note: Because of space limitations, the lengthy BSFC has been replaced by B. Also, the lengthy mpg has been replaced by F_E and hp replaced by P.

[a] Predicted using $mpg_{N+1} = mpg_N (1 + \Delta mpg/mpg) = F_E (1 + \Delta F_E/F_E)$.

APPENDIX 12.2: CAFE FLEET AVERAGE FUEL ECONOMY AND VEHICLE AVERAGE

Fleet Average Fuel Economy

Suppose you have a fleet of $N_T = 200$ trucks and $N_C = 800$ cars. The mpg for each are $M_T = 12.5$ for the trucks and $M_C = 14.5$ for the cars. The values for mpg are from Figure 7.2 for 1976. Each vehicle receives 1 gal of gasoline. The distance traveled by the fleet on 200 + 800 gal of gasoline is

$$D_C = (800 \text{ gal})(14.5 \text{ mpg}) = 11{,}600 \text{ mi}$$

$$D_T = (200 \text{ gal})(12.5 \text{ mpg}) = 2{,}500 \text{ mi}$$

Total miles driven are 14,100 on 1000 gal of gasoline. Hence, the fleet average is 14,100 mi/1000 gal = 14.1 mpg

Define a fraction

$$X_T = \frac{N_T}{N_C + N_T} = \frac{200}{800 + 200} = 0.20 \tag{A.12.6}$$

A similar definition applies for X_C and has a value of 0.80 for this example. In the form of a formula, the fleet average is given simply by

$$M_{FLEET} = X_T M_T + X_C M_C \tag{A.12.7}$$

which, for the example, is

$$M_{FLEET} = (0.2)(12.5) + (0.8)(14.5) = 14.1 \text{ mpg}$$

Using the data for 2005 from Figure 7.2, $N_T = 500$ trucks and $N_C = 500$ cars. Consequently, $X_T = X_C = 0.5$. The fleet average is

$$M_{FLEET} = (0.5)(18) + (0.5)(25) = 21.5\text{ mpg}.$$

This value agrees with Figure 7.2.

Vehicle Average

Suppose the driving cycle gives $D_C = 900$ mi in the city and $D_H = 100$ mi on the highway. The fraction of city miles is

$$X_C = \frac{D_C}{D_C + D_H} = \frac{900}{900 + 100} = 0.90 \tag{A.12.8}$$

$X_H = 1-0.90 = 0.10$. The mpg for highway is $M_H = 50$ mpg and for the city $M_C = 10$ mpg. The gallons of fuel used in the city are $G_C = D_C/M_C = 900$ mi/10 mpg = 90 gal. G is the reciprocal of M; $G = 1/M$. The gallons of fuel used on the highway are $G_H = D_H/M_H = 100$ mi/50 mpg = 2 gal. Total fuel consumed is 92 gal while driving 1000 mi. Therefore, the vehicle average fuel economy is 1000 mi/92 gal = 10.87 mpg. An equation for the average mpg is

$$G_{AVG} = \frac{X_C}{M_C} + \frac{X_H}{M_H} \tag{A.12.9}$$

For the example,

$$G_{AVG} = (0.10)/(50\text{ mpg}) + (0.9)/(10\text{ mpg})$$

$$G_{AVG} = 0.002 + 0.090 = 0.092\text{ gal/mi}$$

The average mpg, M_{AVG}, is the reciprocal of G_{AVG}. For this example,

$$M_{AVG} = 1/G_{AVG} = 1/0.092 = 10.87.$$

Compare the calculation procedures for fleet average and vehicle.

APPENDIX 12.3: CHANGE IN FUEL ECONOMY EPA ESTIMATES: OLD VERSUS NEW TEST METHODS

These curves are of intense interest in 2008 while EPA changes tests. After 2008, the data become history with less and less interest. The percentage is calculated by

$$\%\text{ Change} = \frac{\text{mpg}_{NEW} - \text{mpg}_{OLD}}{\text{mpg}_{OLD}} \tag{A.12.10}$$

The first observation is that the changes have considerable scatter for both city and highway mileage. In Figures A.12.2 and A.12.3, the word label is used synonymously with the words MSRP, Monroney sticker. For city driving, Figure A.12.2, none of the data points show a gain for either the CV or the HEV. For CV with low mpg (10–12 mpg), the change is less than for higher mpg vehicles (30–35 mpg). For HEV, significant change occurs.

The changes for highway mpg have even more scatter than for city mpg. As a surprise, some CVs have a gain in mpg! For low mpg vehicles, the change ranges from +5% to −5%. For high mpg vehicles, the change is from 0% to −18%. All HEVs have a change from −9% to −21%.

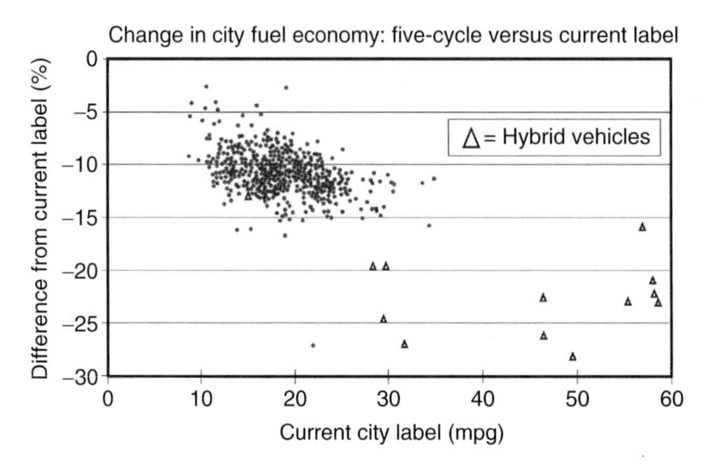

FIGURE A.12.2 Change in city fuel economy, F_E, between current MSRP, Monroney sticker (old EPA) and the five-cycle (new EPA) mileage estimates. (Reproduced from *Federal Register*, Part II, Environmental Protection Agency, Fuel Economy Labeling of Motor Vehicles: Revisions to Improve Calculation of Fuel Economy Estimates; Final Rule, December 27, 2006, p. 77876.)

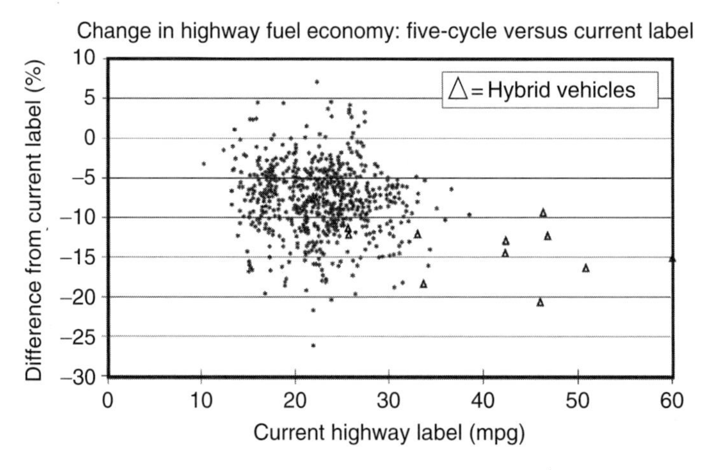

FIGURE A.12.3 Change in highway fuel economy, F_E, between current MSRP, Monroney sticker (old EPA) and the five-cycle (new EPA) mileage estimates. (Reproduced from *Federal Register*, Part II, Environmental Protection Agency, Fuel Economy Labeling of Motor Vehicles: Revisions to Improve Calculation of Fuel Economy Estimates; Final Rule, December 27, 2006, p. 77876.)

REFERENCES

1. Environmental Protection Agency. Available at www.fueleconomy.gov/feg/fe_test_schedules.shtml.
2. D. S. Parker, Energy efficient transportation for Florida, Revised 2005. Available at www.fsec.ucf.edu/Pubs/energynotes/en-19.htm.
3. Environmental Protection Agency, Driving more efficiently. Available at www.fueleconomy.gov/feg/driveHabits.shtml.
4. Fuel economy in automobiles, Wikipedia Encyclopedia, September 2006. Available at http://en.wkipedia.org/wiki/Fuel_economy_in_automobiles.
5. Environmental Protection Agency, Driving more efficiently. Available at www.fueleconomy.gov/feg/maintain.shtml.

6. Uniroyal, Simple tips that could add up to $900 in annual savings on gas. Available at www.uniroyal.com.
7. *Business Week*, Primer on fuel economy. Available at www.businessweek.com/autos/content/jul2006/.
8. Fuel efficient driving, Wikipedia Encyclopedia. Available at http://en.wkipedia.org/wiki/Fuel_efficient_driving.
9. Environmental Protection Agency, Your MPG will vary. Available at www.fueleconomy.gov/feg/why_differ.shtml.
10. Pennsylvania Independent Automobile Dealers Association, Gas saving tips; drive farther for less. Available at http://piada.org/Consumer%20Pages/gas.
11. U.S. DOE Energy Efficiency and Renewable Energy, Driving and car maintenance. Available at http://wwwl.eere.energy.gov/concumer/tips/.
12. P. Vaides-Dapena, Take it slow and save big on gas. Available at http://money.cnn.com/2006/05/01/Autos/.
13. Toyota, Maximum mpgs. Available at http://www.toyota.com/html/.
14. National Highway Traffic Safety Administration, Tire safety; everything rides on it. Available at www.nhsta.dot.gov/cars/rules/TireSafety/.
15. A. Gold, Top 10 fuel saving tips, *About Cars*. Available at http://cars.about.com/.
16. G. Kable, Super binders; electronic wedge brakes will change the way we stop, *AutoWeek*, November 13, 2006, p. 10.
17. P. Weissler, NREL looks at reduced A/C loading, *Automotive Engineering International*, October 2006, p. 22.
18. J. J. Knight, A. D. Lucey, and C. T. Shaw, On the aerodynamic loading and deformation of convertible tops, *Vehicle Aerodynamics*: *Design and Technology*, SP-1600, Society of Automotive Engineers, Warrendale, PA, 2001, p. 325.
19. Environmental Protection Agency, Planning & combining trips, Available at www.fueleconomy.gov/feg/planning.shtml.
20. Federal Trade Commission, FTC consumer alert, good, better, best: How to improve gas mileage, September 2005, 1-877-382-4357.
21. Environmental Protection Agency, Many factors affect MPG, Available at www.fueleconomy.gov/feg/factors.shtml.
22. W. H. Hucho, The aerodynamic drag of cars current understanding, unresolved problems, and future prospects, *Aerodynamic Drag Mechanisms of Bluff Bodies and Road Vehicles*, G. Sovran (Ed.), Plenum Press, New York, 1978, p. 1.
23. J. Y. Wong, *Theory of Ground Vehicles*, Wiley-Interscience Publication, New York, 1978, 330 pp.
24. Federal Trade Commission, FTC facts for consumers, The low-down on high octane gasoline. Available at www.ftc.gov.
25. Environmental Protection Agency, Frequently asked questions, FAQ. Available at www.fueleconomy.gov/feg/info.shtml.
26. Environmental Protection Agency, Heavy-duty highway diesel program; ultra-low sulfur diesel. Available at http://www.epa.gov/cgi-bin/epaprintonly.cgi.
27. H. Pacejka, Analysis of tire properties, Chapter 9, *Mechanics of Pneumatic Tires*, S. K. Clark (Ed.), National Highway Traffic Safety Administration, 1982, 931 pp.
28. CAFE overview—frequently asked questions, National Highway Traffic Safety Administration. Available at http://www.nhtsa.dot.gov/Cars/rules/CAFE/overview.
29. Corporate average fuel economy, Wikipedia Encyclopedia. Available at http://en.wikipedia.org/wiki/Corporate_Average_Fuel_Economy.
30. Staff, Fuel economy: Why you're not getting the MPG you expect, *Consumer Reports*, October 2005, p. 20.
31. United States Environmental Protection Agency, Wikipedia Encyclopedia. Available at http://en.wikipedia.org/wiki/Enovironmental_Protection_Agency.
32. Environmental Protection Agency, EPA Proposes New Test Methods for Fuel Economy Window Stickers, Regulatory Announcement, January 2006, EPA420-F-06–009.
33. New York Times News Service, Oil, gas firms sued, *The Herald*, Monterey, CA, December 16, 2006, p. E1.
34. T. D. Gillespie, *Fundamentals of Vehicle Dynamics*, Society of Automobile Engineers, Warrendale, PA, 1992.
35. A. E. Fuhs, *Improving Fuel Economy; Folklore, Myths, and Reality* (being edited).
36. M. Allen, Looking for a miracle: Bogus MPG gadgets, *Popular Mechanics*, September 2005, pp. 104–108.
37. J. Dunne, Your dad was wrong: 10 auto myths debunked, *Popular Science*, January 2007, pp. 80–81.

38. R. Bamberger, *Automobile and Light Truck Fuel Economy: The CAFE Standards*, Issue Brief for Congress, Library of Congress, March 12, 2003.
39. P. R. Portney, Chair, Federal fuel economy standards program should be retooled, *National Academies*, Washington, DC, Media Release, July 31, 2001.
40. The facts about raising auto fuel efficiency, *National Environmental Trust*, September 2006. Available at www.net.org.
41. S. Barlas, CAFE safety implications debated, *Automotive Engineering International*, Society of Automotive Engineers, Warrendale, PA, June 2006, p. 77.
42. M. Ross and T. Wenzel, Losing weight to save lives: A review of the role of automobile weight and size in traffic fatalities, *American Council for an Energy-Efficient Economy*, July 2001. (Report submitted to National Research Council, March 2001.)
43. Fuel efficiency standards and the laws of physics, *Source Watch*, February 2006. Available at www.sourcewatch.org.
44. Environmental Protection Agency, Light-duty automotive technology and fuel economy trends: 1975–2006, Executive Summary, EPA42-S-06–003, July 2006.
45. *Federal Register*, Part II, Environmental Protection Agency, Fuel economy labeling of motor vehicles: Revisions to improve calculation of fuel economy estimates; final rule, December 27, 2006, p. 77876.
46. National Highway Traffic Safety Administration, *CAFE Overview*, 2007. Available at www.nhsta.dot.gov/portal/site/nhtsa/template.
47. BMW, General Motors, Honda, Toyota, Volkswagen and Audi, Sponsors: www.toptiergas.com.
48. H. B. Pacejka, Analysis of ire properties, Chapter 9, *Mechanics of Pneumatic Tires*, S. K. Clark (Ed.), National Highway Safety Administration, Washington, DC, 1982.
49. H. B. Pacejka, *Tire and Vehicle Dynamics*, Society of Automotive Engineers, Warrendale, PA, 2002, 627 pp.
50. AW Special Report, The state of performance, *AutoWeek*, November 5, 2007, p. 21.
51. J. Christoffel, Continental refines modular hybrid technologies, *Automotive Engineering International*, Society of Automotive Engineers, Warrendale, PA, November 2007, p. 26.
52. M. Newton, Siemen's wedge brake hits the ice, *Automotive Engineering International*, Society of Automotive Engineers, Warrendale, PA, July 2007, pp. 26–30.
53. M. Monaghan, Handling the surge; the pressure facing the automakers and suppliers to meet ever-more-stringent emissions regulations has fueled a boom in business for emissions testing providers, *Automotive Engineering International*, Society of Automotive Engineers, Warrendale, PA, April 2007, pp. 76–80.
54. Staff, Dynamometer systems, *Automotive Engineering International*, Society of Automotive Engineers, Warrendale, PA, January 2008, p. 70.
55. American Petroleum Institute. Available at www.api.org/certifications/engineoil/categroies/upload.
56. T. Costlow, Lutz promotes energy conservation, criticizes politicians, *Automotive Engineering International*, Society of Automotive Engineers, February 2008, pp. 30–31.

13 Enhancing the Sales Brochure

The big print giveth, and the fine print taketh away.

INTRODUCTION

Different levels of information can be provided in the sales brochure for a new vehicle. A superficial level would be limited, almost, to colors and fabrics. The next level is core data helpful to make a purchase decision. To move up one level, additional data are provided and are beyond the nice-to-know category. These additional data, which enable a more intelligent decision, are typical of information found in Consumer Reports. The final level of information, which some persons may label off-the-deep-end, requires some knowledge of engineering. Of course, readers of this book qualify as interpreters of the final level. The sales brochure should provide the information at the final level [1].

Contrast the sales brochure with the owner's manual. The sales brochure is printed on glossy paper in multiple colors showing exotic backgrounds. The owner's manual, which you receive after purchase of the car, is typically black and white on standard paper. Some owner's manuals have two colors if you count black as a color.

Later in this chapter, recommendations appear. Some of the tests associated with the recommendations require careful definitions. Standards define the test protocols and permit comparisons between vehicles. An example is SAE Standard J1349, which specifies procedures for determining horsepower and torque of gasoline engines (SAE is the Society of Automotive Engineers). Who should define the standards? SAE? Or some agency of the federal government?

Some sales brochures give data in the form of percentages. To be truly meaningful and informative, both the numerator (the number on top) and denominator (the number on the bottom) must be completely defined.

FUNCTION OF SALES BROCHURE: CONSUMER VIEWPOINT

As stated above, the primary function of the sales brochure is to provide data. The data help in making an intelligent decision concerning the purchase of a vehicle. A secondary function is analogous to the menu in a restaurant. Various selections can be made for a good meal or for an appropriate car that meets the customer's needs.

FUNCTION OF SALES BROCHURE: CAR MANUFACTURER VIEWPOINT

The sales brochure highlights the competitive features of the vehicle. Exclusive features are given prominence. Deficiencies are ignored. Remember, it is, after all, a sales brochure.

HYBRID COMPONENTS

What information should be provided in the sales brochure? Data about the components identified in Figure 4.1 are needed.

Battery

A key component is the battery because of its effect on performance, customer satisfaction, purchase costs, and operating costs. Chapter 6 discusses the numerous quantities that characterize a battery. The battery warranty is vital to the customer. Also the battery weight affects payload. Battery size affects passenger space and cargo volume. The electrochemistry of the battery, that is, battery type, determines cost and electrical performance.

Electrically, the battery needs to be specified. The capacity of the battery in ampere-hours and the stored energy in kilowatt-hours determine the electric-only range. As discussed later, the battery energy is important for regenerative braking. Battery efficiency determines the cooling load for the battery. Battery voltage is significant relative to safety. Battery voltage affects the size and weight of electrical motors. Battery maximum power influences the performance of the hybrid. In summary, the specification sheet should give facts about

Battery warranty
Size (dimensions)
Capacity, Ah
Efficiency
Maximum power, kW
Weight
Battery type, e.g., NiMH
Stored energy (kWh)
Volts

The sales brochure for the 2005 Toyota Prius states the battery is a sealed NiMH with voltage of 201.6V and a power of 28 hp (21 kW). The sales brochure for the 2006 Honda Civic Hybrid gives the NiMH battery capacity as 5.5 Ah. Information for the NiMH battery for the 2006 Honda Accord hybrid was capacity of 6.0 Ah with power output of 13.8 kW.

Electrical Motor/Generator

Different types of motor/generators (M/Gs) are discussed in Chapter 14. The sales brochure should state the number and types of M/Gs used in the vehicle. The maximum power that the M/G can provide in motor-mode and the maximum power that the M/G can absorb in generator-mode affects hybrid performance. The battery, of course, must be capable of providing the maximum power for the motor and absorbing maximum power from the generator.

Motor torque and power are a function of motor speed (rpm). Curves of motor torque and power should be provided in the sales brochure.

Gasoline (Diesel) Engine

In addition to the usual and expected displacement, number of cylinders, compression ratio, and variable valve timing, novel engine cycles should be identified. For example, the Atkinson cycle, which is discussed in Chapter 17, is popular for HEV. Engine torque and power are a function of motor speed (rpm). Curves of engine torque and power should be provided in the sales brochure.

Combined Engine and Motor Torque and Power

With regard to torque, the engine and motor complement each other as discussed in Chapter 14. Combined engine and motor torque and power curves are useful in estimating performance. An example of the combined curves appears in Figure 14.11, which compares 2006 and 2005 generation powertrain models for Honda Civic Hybrid.

Transmissions

Transmissions have an effect on powertrain efficiency, and hence on mpg. In recent years, a variety of transmissions have been updated or newly introduced. The newer transmissions include the continuously variable transmission (CVT) based on the cone and vee-belt design. The CVT appears in the Honda Civic Hybrid. The planetary gear set, which was used in the Ford Model T, has been updated for use in the Toyota Prius. For the Prius application, the transmission is identified as the electronic continuously variable transmission (ECVT). The ECVT provides torque split between the motor and engine.

The automatic transmission using the torque converter is still widely used. The sales brochure should identify the type of transmission being used.

Accessories and Auxiliaries

With the advent of the HEV, more and more of the accessories or auxiliaries are electrically powered. Some items, which were formerly belt driven off the engine, are now powered by an electric motor. Examples are power steering and the compressor for air-conditioning (A/C). Electrically driven water pumps and electrically actuated friction brakes are also in the pipeline.

For a hybrid, a problem with A/C occurs when the engine is turned off at stop. Having an electrically driven A/C compressor helps solve that problem. Items that have traditionally been engine driven but that are now motor driven should be identified in the sales brochure.

HYBRID SYSTEM

Conversion or Clean Sheet Design

Only three production hybrids start with a clean sheet of paper design. These are the Honda Insight and Toyota Prius I and II. All other production hybrids are conversions (modifications) of an existing product line. An example is the Honda Civic and the Honda Civic Hybrid. When hybrid production starts with an existing car design, added weight of hybrid components may increase braking distance, decrease storage space, increase weight, and affect handling characteristics, when compared to nonhybrid cars. Added weight likely subtracts from the allowed payload. The potential customer should verify the design changes that were made during conversion to a hybrid to offset the effects of added weight.

Conversion or Clean Sheet Design: Impact on AWD and 4WD

The impact on AWD or 4WD vehicles is the torque split between front and rear wheels. The torque split front/rear is usually best with 50/50 ranging to a bias on rear wheels 30/70 F/R. Too much torque to the front wheels may overload front tires, which affects handling capability. This information is discussed in Chapter 3.

Regenerative Braking

The sales brochure should not be limited to saying "Yes, we have regenerative braking." Some regenerative braking systems have limited ability to recover the kinetic energy (KE) of motion or the potential energy (PE) of elevation. The maximum power and energy that can be absorbed are useful data. Another insight would be the fraction of KE that can be recovered at some specified speed. The recovery ratio, R, is

$$R = \frac{\text{Energy into battery}}{\text{KE at 35 mph}} \tag{13.1}$$

A set of standards are needed for recovery ratio, *R*. Another criterion for the capability of the regenerative braking system is the percent of battery state of charge (SOC) nominally dedicated to the regenerative braking system.

During coasting, and especially downhill, the battery is recharged. In the case of going downhill, the battery may be limited on the number of feet elevation change that can be absorbed. A statement in the sales brochure would provide useful information for the consumer. Once again a standard needs to be defined so that the number given has meaning.

Hybrid System Layout

Chapter 4 discusses the layout (arrangement of components) for both the series and parallel hybrid designs. Also the various operating modes are discussed. These hybrid operating modes include

- Acceleration, climbing hill, deep snow; motor assist
- Normal cruise
- Battery charging
- Regenerative braking
- Electric-only propulsion

Usually, the sales brochures provide details about the various operating modes. However, a layout of the hybrid system is typically missing.

Some hybrids have the motor/generator bolted to the engine where the motor rotor serves as the flywheel for the engine. Hence, the engine and motor must turn together at the same rpm. During regenerative braking, the braking torque is divided between the M/G and the engine. Whatever torque absorbed by the engine is not available to charge the battery. Further, for this case in electric-only operation, the motor must supply power not only to drive the wheels but also to turn the engine. The power to turn the engine is wasted. A layout diagram shows this situation. A diagram should show the location of clutches.

Hybridness or Hybridization

In Chapter 11 the value of hybridness, *H*, is defined and discussed. *H* helps define the characteristics of an HEV. Although the consumer can calculate *H* for himself/herself using data in the sales brochure, why not have the car manufacturer do the calculation and present the results in the brochure?

Plug-In Hybrid

For the plug-in hybrid, details on charging are necessary information. The charging time at the required voltage and current is the essential information. Some plug-in vehicles offer two options for charging: for 110V AC and for 220V AC.

Navigation systems are becoming popular. For the plug-in, locations of charging stations should be part of the navigation software. This fact should be stated in the sales brochure.

Starting the Engine

HEVs typically have two batteries. One is the 12V battery for the radio, lights, power windows, etc. This battery can be called the hotel battery. The other battery is the propulsion battery. Some hybrids can be started only with the propulsion battery. Other hybrids can be started with either the propulsion (primary) or the hotel battery (secondary). The sales brochure should state the starting capabilities of the hybrid.

HYBRID PERFORMANCE

ELECTRIC-ONLY

Of interest in regard to fuel economy is the electric-only range for the HEV. The electric-only range depends on two factors: battery SOC and vehicle speed. Figure 13.1 shows a map of distance as a function of speed with SOC as a parameter. For the HEV of the figure, at slow speeds and with 100% SOC, the distance in electric-only mode is about 1.4 mi. At 20 mph, again with 100% SOC, the distance is 0.8 mi.

The sales brochure should provide a map showing the electric-only range. Standards need to be developed so that the map is comparable between car makers.

FUEL ECONOMY DURING CRUISE

Chapter 12 has extensive discussion of fuel economy during cruise. Three different graphs of mpg versus speed were presented for three different vehicle types in Figures 12.10 through 12.12. Figure 12.12 is reproduced here for your convenience. A curve similar to Figure 12.12 would greatly enhance the sales brochure.

FUEL ECONOMY

The sales brochure should show the EPA mileages for city, highway, and combined. Most sales brochures do show these numbers since fuel economy is a major selling point for hybrids. However, the brochures do not show a curve of fuel economy as a function of vehicle speed. From the curve, the optimum cruise velocity can be determined. It would be still better if the manufacturer simply stated the optimum cruise speed as 50 mph.

Three quantities affect the curve in Figure 12.12. First is frontal area, A, as defined for aerodynamic drag. Drag coefficient, C_D, is the second quantity. The third is rolling friction coefficient, f, defined in Chapter 12. Some sales brochures report the drag coefficient, C_D. All sales brochures should report A, C_D, and f. Both A and C_D are reported to EPA for use in fuel economy testing; the same values are appropriate for the sales brochure. The rolling friction coefficient, f, depends on

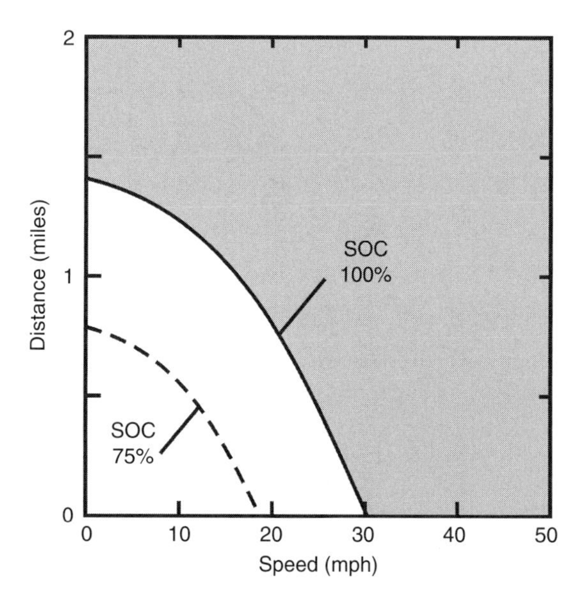

FIGURE 13.1 Map of electric-only range as a function of speed with SOC as a parameter.

TABLE 13.1
Acceleration Times with and without a Fully Charged Battery

Time to Change Speed (s)	Full Charge	Partial Charge
0–30 mph	4.0	4.4
0–60 mph	10.8	12.3
0–100 mph	32.6	41.4

many factors such as type of road surface, tire pressure, and tire temperature. A standard is necessary for reporting f in the sales brochure.

Performance with a "Dead" Battery

"Dead" is in quotes since the definition varies. Dead does not necessarily mean SOC equal to zero. Dead does mean partially discharged (see Chapter 6).

Two tests may be appropriate for revealing performance with a dead battery. The first test is a hill climb. Maximum hill climb speed on a 6% grade with motor assist might be 65 mph. Without motor assist (dead battery), the maximum speed up the same hill may fall to 46 mph. The two hill-climbing speeds tell the customer the level of performance to be expected.

The second test is acceleration with a charged and discharged battery. Table 13.1 shows some results. *Car and Driver* is one of the few magazines to report the effect of a dead battery on acceleration times. Without motor assist (due to dead battery) the times are stretched from 10% for 0–30 mph to 27% for 0–100 mph.

Reporting hill climb and acceleration in the sales brochure is a good idea to keep the customer fully informed.

MISCELLANEOUS

Crossover Velocity

A factor that involves all vehicle quantities for vehicle weight and both rolling friction and aerodynamic drag is the crossover velocity defined in Chapter 12. The formula for crossover velocity is repeated here for your convenience.

$$V_{\mathrm{CO}} = \left(\frac{2fW}{\rho C_{\mathrm{D}} A} \right)^{1/2}$$

Reporting crossover velocity in the sales brochure adds to the depth of knowledge that the consumer attains. The crossover velocity, which is discussed in Chapter 12, adds to the customer's knowledge and is appropriate for the sales brochure.

Optimum Cruise Velocity

The optimum cruise velocity for the vehicle shown in Figure 12.12 is 48 mph, which yields 30 mpg. The customer can determine the optimum cruise velocity by reading the graph; however, showing the optimum cruise velocity explicitly has merit.

Instrumentation and Displays

A gage showing instantaneous fuel economy (mpg) is useful and should be included with the list of options in the sales brochure. An explanation of the diagrams showing flow (if installed on vehicle) of energy during different operating modes is proper for the sales brochure. Most sales brochures highlight the flow diagrams. An old-fashioned ammeter for the propulsion battery helps the driver. If ammeter is installed, tell the customer in the sales brochure.

SUMMARY

In the introduction, the sales brochure is described from both the customer and car makers' viewpoints. The contrast between the glossy sales brochure and the owner's manual is noted.

Desirable data for each of the important hybrid components are identified. This includes the battery, electrical motor, and engine. Next, the components assembled to create the HEV are discussed from the point of view of necessary information. Reporting in a sales brochure the performance with a dead battery is discussed. Electric-only performance is another topic presented.

REFERENCE

1. T. Quiroga, 2006. What to drive until the perpetual-motion machine arrives? *Car and Driver*, January, p. 68.

14 Torque Curves: A Match Made in Heaven

At vehicle launch, gobs of torque are needed. Sadly, the gasoline engine has a torque "hole" at launch. Enter the electric motor. A motor with ¼ the power has as much torque as the engine at low rpm. Motors and engines—truly a match made in heaven!

HISTORY OF ELECTRICAL ALTERNATORS, MOTORS, AND GENERATORS IN AUTOMOBILES

The history of automobile electrical systems starting a century ago divides into low power, a few kilowatts, and high power, tens of kilowatts (Figure 14.1). The years for the introduction of two different transistors are shown on the dateline. The historical landmarks for conventional, low-power systems are divided into starter and generator. Solid-state physics and the diodes made possible the alternator in the early 1960s. With the aid of solid-state physics, an irritating, costly, maintenance problem was overcome when the old generator was replaced by the alternator. Commutators of DC generators failed often. Hybrids today would not be possible without the advances in solid-state physics.

For the low-power portion of electric vehicle (EV) and hybrid electric vehicle (HEV), a 12V bus fed from the hybrid higher voltage system is likely. The 42V system shown in the 1990s is not mainstream; most activity is limited to R&D. The 42V is needed for CV due to increased electrical loads. To avoid shock the upper limit for voltage is about 42V; use of this voltage is a safety item. The higher voltage cuts current (amperes) by one-third compared to the 12V system.

Some mild hybrids use 42V due to costs and availability of data from R&D. The starter and alternator of CV are usually replaced by the M/G in the HEV.

The high power region, which appears at the bottom of Figure 14.1, has the old-style equipment and the modern M/G. A gap in time exists between the two regions; for EV and HEV, note the gap from 1924 to the 1970s. Years ago, the tools to tailor the M/G torque and rpm were limited. The bag of design tricks consisted of commutation, relays, and polyphase windings. To be compatible with the battery, all M/G were DC. In modern times, tools to tailor the M/G torque have increased in number. Sample tools are listed.

FUNCTIONS OF THE MOTOR/GENERATOR

The battery, fuel cell, or engine-driven generator provide the electrical power. The power electronics massage the power to provide correct voltage and current. To drive the wheels, torque is needed on a shaft. The motor converts electrical power to shaft torque. In reverse, the generator converts torque to electrical power.

ORIGIN OF FORCES AND TORQUES

A force creates a torque when acting on a pivoted lever arm. An example is the common door. Pulling on the door handle with a force causes a torque about the door hinges. Torque is a twist about an axis. Opening a screw-top jar requires a torque to be applied to the lid. Torque at the axle of a car causes the motion. Torques are frequently associated with spinning motions.

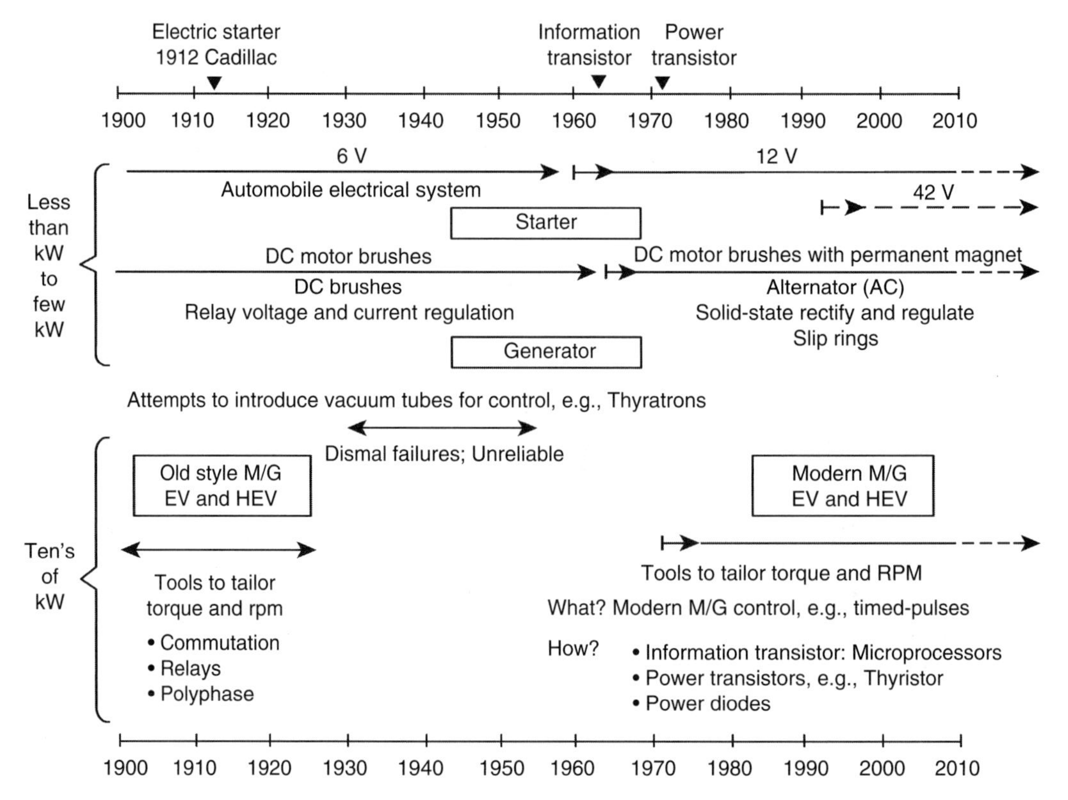

FIGURE 14.1 Electrical motors and generators and control technologies for CV and EV, and HEV. Note the thyratron (vacuum tube) and thyristor (transistor); both have "thyr…."

How are forces and torques created in an electrical motor? The forces are due to either magnetic forces or electrostatic forces. Although Benjamin Franklin invented an electrostatic motor, this class of motors remains a curiosity. All M/Gs depend on electromagnetic forces to develop torque.

Forces are developed between magnets, ferromagnetic (iron) materials, and electrical currents. As shown by magnet–magnet interaction in Figure 14.2a, unlike magnetic poles attract and like poles repel. The magnet–magnet interaction is the basis for the DC brushless motor. As shown in Figure 14.2b, ferromagnetic material is attracted by a magnet. This magnet–ferromagnetic interaction is called reluctance and forms the foundation for reluctance motors.

Figure 14.2c shows the magnet-electrical current interaction, which is called *inductance*, and forms the starting point for inductance motors. For the eddy currents to develop, the conductor must

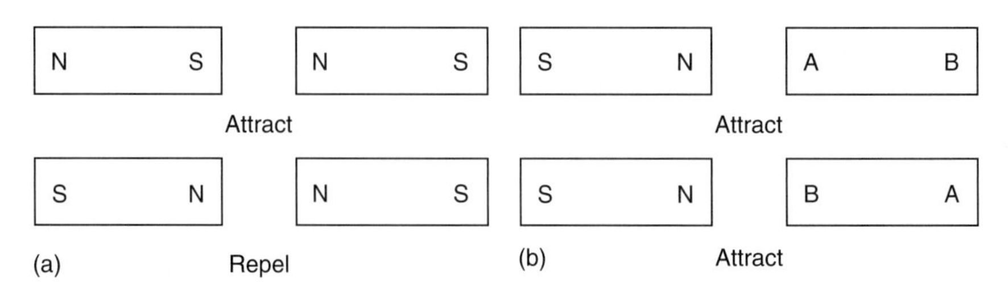

FIGURE 14.2 Origin of forces and torques in motors and generators. (a) Unlike magnetic poles attract each other; like magnetic poles N–N and S–S repel each other. An example using forces of attraction between N–S and S–N poles is a DC brushless motor. (b) Object A–B is ferromagnetic (iron) and unmagnitized. When A is adjacent to N, a force of attraction occurs. Reverse A and B, and the force remains attraction. An example is the reluctance motor.

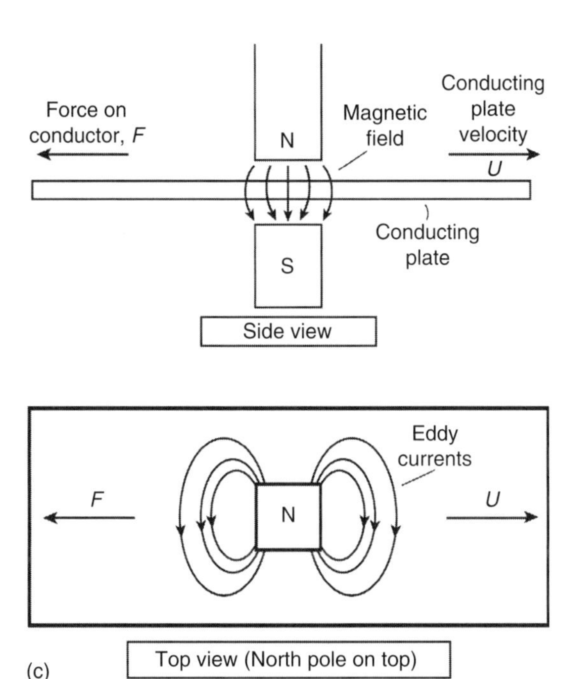

FIGURE 14.2 (continued) (c) Eddy currents are created (induced) by motion of the conducting plate through the magnetic field. The eddy currents interact with the magnetic field and cause a force in a direction opposing conductor motion. An example using this force is an inductance motor.

be moving relative to the magnetic field. In motors or generators, this relative motion is called "slip". No slip, no inductive force.

Forces are created in the M/G by one of the three mechanisms discussed above: magnet–magnet, reluctance, or inductance. Torque is due to a force (N) acting on a lever arm (m) giving units Newton meter (N·m). Torque describes the action on a rotor to make it spin. All electrical M/Gs use one or more of the three forces.

TYPES OF MOTORS AND/OR GENERATORS

The earliest and simplest M/Gs were the DC commutating machines. Induction electrical machines have a rotating magnetic field and require "slip" to operate. Both synchronous reluctance and DC magnet machines have a rotating magnetic field; however, neither needs slip to operate. An alternator is an AC generator that usually has a DC output via a rectifier. Stepping motors include motors that can be commanded to move the rotor to a specific angle and stop. Stepping motors may have rotors shaped like gears. The teeth of the gears serve as *salient* poles. This stepping is not suitable for an M/G used in an HEV; however, continuous angular motion is possible making stepping motors a candidate for M/G. Stepping motors are also known as switched reluctance motors (SRM). The advances in the control of motors and generators make M/Gs based on complex designs, such as stepping motors, possible for use in HEVs.

PERMANENT MAGNETS

Permanent magnets have contributed to the high efficiency of motors and generators. Some motors have magnets placed in position by hand (very expensive method) or robots. Reference [13] of Chapter 3 states General Motors has developed a permanent magnet that can be cast into the motor and then magnetized.

DEFINITIONS FOR MOTORS AND GENERATORS

Five ubiquitous words are used to describe motors and generators. The word field is used as an adjective in terms such as field windings or field poles. The field windings produce the magnetic field, which interacts with the magnetic field of the armature to create the forces and torques described earlier. Permanent magnets can be used and replace separate field windings. Field windings typically have much less electrical current (A). The armature handles the heavy power. The armature windings have the heavy current.

The rotor spins and may incorporate either the field or armature. The stator is stationary and may encompass either the field or armature. Different M/G designs use different combinations of the rotor as an armature, etc. The word salient is used as an adjective to describe the geometry of the poles on either the rotor or armature. Salient poles stick out like gear teeth.

SIMPLE DC MOTOR/GENERATOR

The simplest of all motors is the DC motor; simplicity may be the reason for its early invention. The DC M/G was used in the golden age of EVs. The DC motor is used even today as the starter on almost every conventional vehicle (CV). Because of cost and simplicity, DC motors are the selection of choice for niche EV such as golf carts.

Field coils are stationary and provide the DC magnetic field. Permanent magnets may replace the field coils as is the case for modern starters. The armature coils rotate within the magnetic field created by the field coils. Because of commutation, AC in the armature coils becomes DC at the output terminals.

In an application for an HEV, three aspects of the M/G must be changed: torque, rpm, and status as M or G. Torque from a DC motor is varied by the current in the armature. Recall that

$$(\text{Power}) = (\text{Torque})(\text{rpm})$$

Motor rpm becomes faster if motor power is greater than load power. If motor power is less than load power, the motor slows. For HEV, the M/G must switch from M to G and vice versa. How does the DC M/G switch? One method is to reverse the direction of rotation, which is an action completely incompatible with regenerative braking. Alternatively, the switch can be done by reversing the direction of current in the field coils. This is feasible for regenerative braking.

ROTATING MAGNETIC FIELD

Many advanced M/G designs use a rotating magnetic field (Figure 14.3). Each pole has a coil wound around the pole. Current in the pole windings creates a magnetic field. Poles directly across from each other, at 180°, have field coils that create N–S pairs. In sketch 1, the top pole "a" has current with direction to make a north pole, N. Both "b" and "d" have zero current and, hence, zero magnetic field. Bottom pole "c" has current with direction to make a south pole, S. The currents fed to each pole change and thereby change the N–S poles as shown. The effect is a rotating magnetic field. The rotating magnetic field is important in many M/G designs. To progress beyond the simple DC motor, rotating magnetic fields are essential.

With a rotating magnetic field, two things are spinning in a motor. These are the rotor and the field. The two spin rates must be coordinated, which is a process called synchronization. Synchronous motors come in two flavors relative to synchronization: slip and no-slip. Induction motors, as discussed in the pages to follow, require slip, which is a slight difference in rotational speeds:

$$\text{Slip} = (\text{Rotating magnetic field speed} - \text{Rotor speed})$$

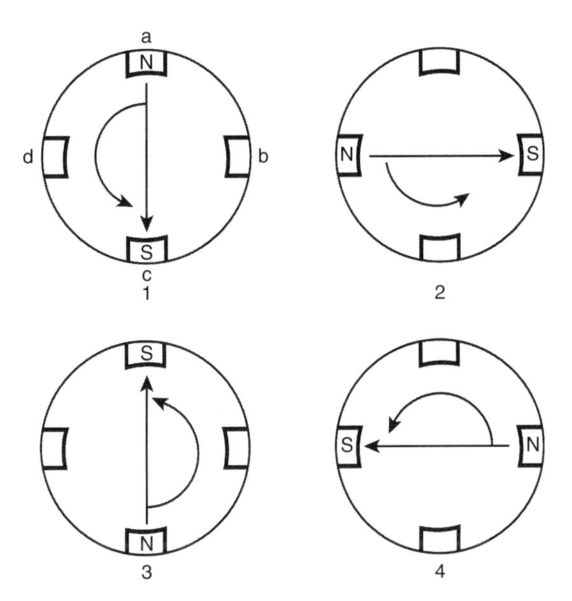

FIGURE 14.3 Rotating magnetic field created by properly phased currents at poles. Each pole has a separate field coil.

Positive slip gives motor action. Negative slip gives generator action. Slip appears in Figure 14.2c; the velocity, U, causes the slip.

The other flavor of synchronous motors appears in DC brushless motors and reluctance motors. The fundamental origin of force and torque for a DC brushless motor is magnet–magnet interaction (Figure 14.2a). The fundamental origin of force and torque for a reluctance motor is reluctance illustrated in Figure 14.2b. Both DC brushless and reluctance motors can operate only with zero slip. Slip causes both types of motors to cease to run; the motors "drop-out" and stall.

DC BRUSHLESS MOTOR

The DC brushless motor is brushless but is not DC. The DC brushless motor is a synchronous machine. The angular position of the rotor must be synchronized with the angular position of the rotating magnetic field (Figure 14.4). The N-poles of the rotating field of the armature must match with the S-poles of the rotor. Permanent magnets are frequently used for the rotor. The stator is the armature. AC is needed to create the rotating magnetic field. The DC brushless M/G is used in Honda Insight, Accord, and Civic hybrids.

As stated earlier, three aspects of the M/G must be changed to be suitable for an application in an HEV. DC brushless motor varies the torque by the current in the armature, which creates the

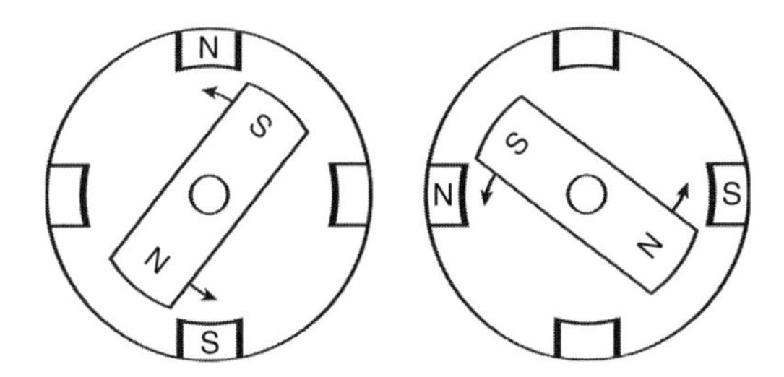

FIGURE 14.4 This DC brushless motor uses a permanent magnet as the rotor. Unlike poles, N–S and at 180° S–N, have force of attraction and chase each other as magnetic field rotates. The action is analogous to dog at racetrack chasing the rabbit.

rotating magnetic field. The rpm is changed by the angular rate of the rotating magnetic field, which in turn is varied by the frequency of the AC driving the armature. Switching from M to G and vice versa is done by changing the angular position of the rotor compared to the rotating magnetic field. In Figure 14.4, suppose the phase angle between the rotor and rotating field were shifted by 180°. Keep the direction of rotation fixed. The armature poles are switched from N to S. The rotor is repelled by the armature poles. The torque is reversed. The M/G has been switched from the M-mode to the G-mode.

ALTERNATOR

An alternator has a rotating DC field. The DC in the rotor allows use of slip rings instead of commutation rings. Current in the rotor is relatively low compared to the armature. The armature is stationary and generates AC. The output of an alternator is AC. The frequency varies with the alternator rpm.

Some mild hybrids use an alternator with higher power and voltage than a CV. The alternator also serves as a motor to start the engine.

INDUCTION MOTOR AND ASYNCHRONOUS GENERATOR

The induction motor is analogous to a transformer. Energy is transferred from the primary (input) windings to the secondary (output) windings of the transformer by means of electromagnetic induction. Direct electrical connections between primary and secondary do not exist. The stator of an induction motor is analogous to the primary. The rotor of an induction motor is analogous to the secondary. In a sense, an induction motor can be viewed as a transformer in which the primary creates a rotating magnetic field and the secondary is a rotor.

Because the rotor and armature are coupled only by induction, commutators, slip rings, and brushes are not needed. Absence of commutators or brushes is a good feature for low maintenance. Except for the shaft, the induction M/G can be sealed, which helps to keep out dirt and muck.

Some features of induction M/G include being rugged combined with simplicity. The rotor can have windings in the shape of a squirrel cage. An induction M/G requires AC. The operating characteristics of an induction M/G are favorable for constant speed operation. For application in HEV, the M/G needs modification for variable speed operation. The invention and development of the high-power, variable frequency inverter makes the use of an induction M/G possible in an HEV.

The word asynchronous means, of course, not synchronous. What is not synchronous? An asynchronous generator is actually an induction motor caused to run at rpm above its synchronous speed. Recall the definition of slip, which is

$$\text{Slip} = (\text{Rotating magnetic field speed} - \text{Rotor speed})$$

Positive slip gives motor action and is called synchronous. Negative slip gives generator action and is called asynchronous (for more discussion see Figure 14.10).

SWITCHED RELUCTANCE MOTOR: STEPPING MOTOR

The second hand on digital clock jumps each second. This is an example of a stepping motor. Some features of stepping motors, also called SRM, include no magnet in the rotor and simple construction. The rotor does not have windings, magnets, or a squirrel cage. Rugged and economical design is possible. Other good aspects include high power density (kW/volume), good efficiency, high-speed capability, and fault tolerant operation.

Problem areas involve high frequency torque ripple with the associated noise and vibration. Customers loathe both noise and vibration. Noise suppression techniques can be used [4,5].

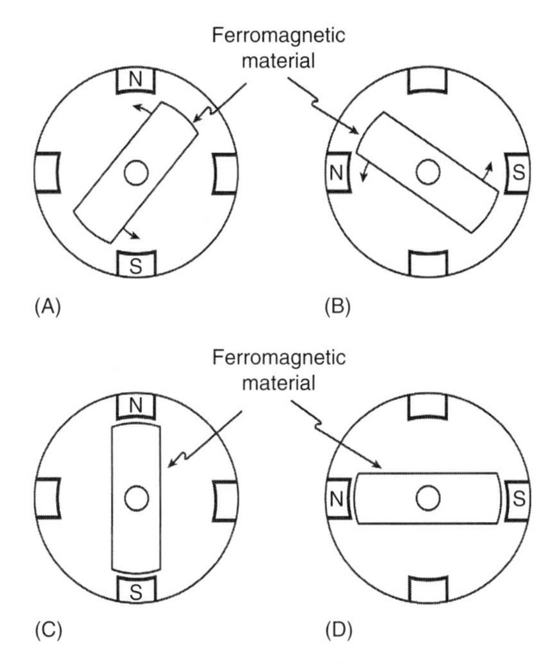

FIGURE 14.5 Reluctance motor with ferromagnetic material as the rotor.

The rotor of Figure 14.5 is the same as Figure 14.4 except the permanent magnets have been replaced by ferromagnetic material (iron). This change of rotor changes the DC brushless motor into a reluctance motor. Compare Figure 14.3 with Figure 14.2a, which illustrates forces between magnetic poles. Also, compare Figure 14.5 with Figure 14.2b, which illustrates forces due to magnetic reluctance.

In Figure 14.5, sketches A and B show the rotation of the iron rotor with a slight lag behind the rotating magnetic field. Sketches C and D show the iron aligned with the N–S poles. In either position C or D, the torque on the iron is zero due to symmetry. Torque is required to displace the iron rotor from the aligned position.

The rotor of Figure 14.6 is made of ferromagnetic material. The rotor has six salient poles. The poles on rotor are identified as A, B, and C. Pairs A–A, B–B, and C–C are symmetric relative to the

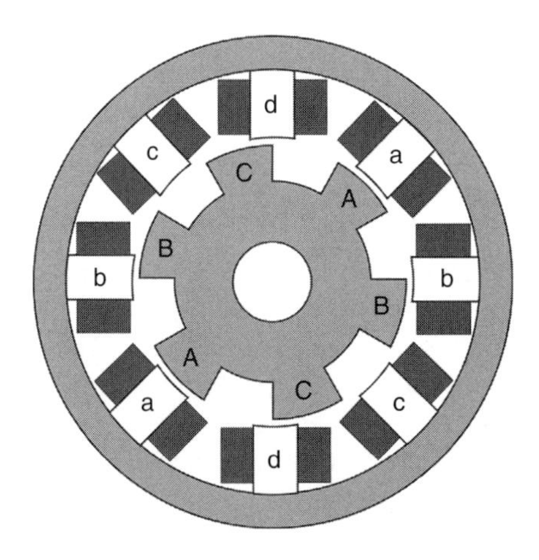

FIGURE 14.6 Switched reluctance motor with eight poles for stator and six poles for rotor.

rotor axis. The stator has eight poles identified as a, b, c, and d. Pairs a–a, b–b, c–c, and d–d are symmetric relative to the rotor axis. Stator poles directly across from each other have windings in series. The current to the windings is not applied sinusoidally but as pulses.

LOSSES AND MOTOR EFFICIENCY

All motors have at least some windings and input/output circuits. The wires give rise to $I^2 R$ losses; energy is converted to heat. The symbols are I equals current (A), and R equals resistance (Ω). Eddy currents in laminations cause a loss. Hysteresis in magnetic materials is another loss. Windage, which is aerodynamic drag on rotor as it spins, absorbs energy. Smooth rotors help reduce windage losses.

The losses are a cause of motor inefficiency. Further, losses appear as heat, which requires cooling of the M/G. The logic is

High efficiency → Low losses → Lower cooling requirements → Better mpg

Even seemingly minor design details can have an important influence on losses.

COOLING

Cooling is a consequence of losses [6]. Cooling is necessary to protect insulation strength especially on rotors. Cooling helps maintain higher electrical conductivity of the wires. The electrical resistance of wires increases as temperature increases. The power rating of an electrical M/G is limited by the allowable temperature rise.

Provision for cooling is an added cost item. Hybrids already have a price premium.

SELECTION OF M/G

Many factors affect the selection of the M/G. Dominant factors are cost, size, and mass. Other factors to be considered are efficiency both at key points and the overall efficiency map. Electrical specifications are involved such as the power capability both steady and surge. Voltage and electrical current must match the power electronics. Almost all applications for HEVs use a variable frequency for motor drive; fixed frequency is not suitable. For design of the power electronics, knowledge of motor resistance (Ω), motor inductance (μH), and electrical lead/lag properties is essential.

Mechanical specifications are involved such as the maximum speed, rpm, and the map of torque and rpm curves. M/G mass or weight expressed in kilograms or pounds, and the size with dimensions are two additional factors. For thermal management, the temperature limits must be known. Reliability expressed as mean time to failure affects warranty costs and customer satisfaction. In this regard, corrosion resistance for the wheel-in-rotor M/G affects reliability. As has been pointed out before in connection with the switched reluctance motor, vibration production and noise generation by the motor are important. This comment applies to all M/Gs.

A major advancement made since the golden age of EV is the variable frequency inverter. Three features of the inverter are important. First, of course, is the ability to change DC to AC. Second, is the variable frequency, which is essential for regenerative braking. Third, is the bidirectional aspect of power flow in an HEV. During motor assist the power flow is DC → AC. During regenerative braking, the power flow is reversed, AC → DC. As illustrated in Figure 14.7b, the variable, bidirectional inverter opens up use of AC motors in HEV.

Six of the selection factors discussed so far are summarized in a selection chart in Table 14.1. The relative merits of each choice are shown. For example, the switched reluctance M/G earns the high scores shown in Table 14.1. The cost for the switched reluctance M/G is favorable because of construction and design features discussed above [9].

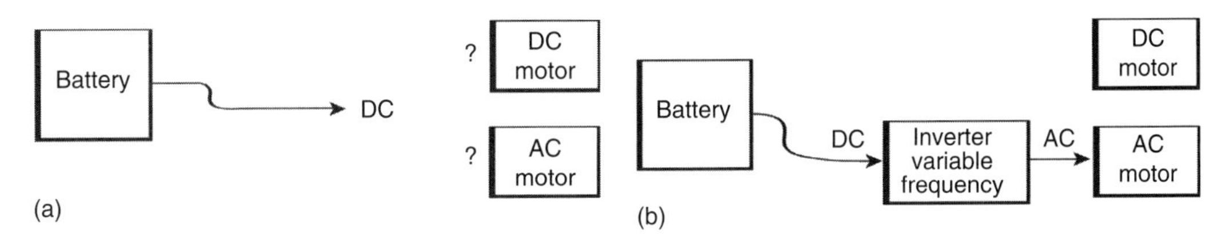

FIGURE 14.7 Modern power electronics opens usage of AC motors, which expands design options. (a) Which motor to select? DC–DC used in past and will be in future. DC–AC not possible without electrical matching of some kind. (b) Advent of variable frequency inverter makes AC motors a viable, and likely attractive option.

DESIGN AND CONTROL ASPECTS

Both motor speed and torque need control and regulation. Control is used in the same sense as in Chapter 8. Regulation indicates a situation where the command for speed or torque is constant for a relatively long time, for example, cruising the freeway at constant speed.

A vehicle must have the ability to back up. Usually, the motor is not reversed to back up. In the rare case where motor reversal occurs, control of the motor spin direction is necessary. Transient conditions require control of M/G acceleration and deceleration. Starting requirements for the motor differ according to motor design. Synchronous motors must increase in rpm to match synch speed. In some designs, braking of the motor is desirable. Sensors and methods for overload protection assure long life. Some motor designs have a propensity to runaway, for example, increase speeds to a dangerous level. The runaway condition, if any, must be avoided.

DESIRED AND ATTAINABLE TORQUE CURVE: MATCH MADE IN HEAVEN

Gasoline engines have a torque hole at low rpm. Motors provide high torque at low rpm. See Figure 14.8a for an example of high torque at low motor rpm. The curves are for a DC shunt motor. Figure 14.12 has data for the engine and motor torques of the 2006 Honda Civic Hybrid. The speed range of the curves extends from 1000 to 6250 rpm.

TABLE 14.1
Relative Merits of Various Motor Designs

	Efficiency		Size		Cooling	
	↓	Maximum rpm	↓	Weight	↓	Cost
DC: shunt, series, compound	2	2	2	2	2	4
Synchronous: permanent magnet	8	6	8	8	8	4
Synchronous: B field by current	6	6	6	6	6	4
Asynchronous	6	8	6	6	6	8
Brushless DC	6	4	6	6	6	6
Transverse flux	8	2	4	6	6	2
Switched reluctance M/G	6	8	6	6	8	8

Notes: The transverse flux motor is used for motor-in-wheel designs. Although numbers are assigned, this is a semi-quantitative evaluation of motors suitable for HEV. The higher the number, the better the choice.

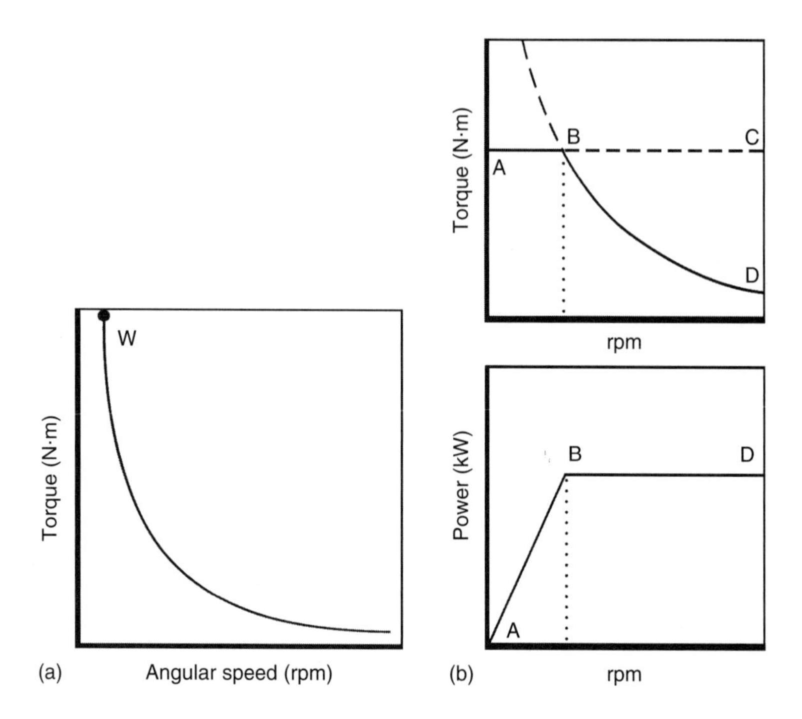

FIGURE 14.8 Two examples of torque–speed characteristics. (a) Torque–speed characteristics of a shunt DC motor. (b) Torque–speed characteristics to best match battery.

If high torque is desirable, why not point W in Figure 14.8a? Torque is proportional to armature current. Point W has very high armature current that exceeds current limits of battery. Wanted or not, the battery limits are imposed. Look now at Figure 14.8b. Line A–B is a line of constant torque and constant current equal to battery maximum current limit. If A–B is good, why not use B–C? As well as having an upper limit on current, a battery has an upper limit on power. The line B–C exceeds battery maximum power, P_{MAX}. At point B, $P_{MAX} = P_{MOTOR}$. To move toward point C, $P_{MAX} > P_{MOTOR}$, which cannot happen.

Motor power is given by

$$P_{MOTOR} = T\omega$$

where

T is the torque (N·m)

ω is the motor angular speed (rad/s)

Line B–D is the curve where T is proportional to $1/\omega$ and, hence, is a line of constant motor power. It is a line for $P_{MAX} = P_{MOTOR}$. In summary, line A–B is a line of constant torque dictated by maximum battery current while line B–D is a line of constant power dictated by maximum battery power. In terms of power, line A–B is a line of linearly increasing power. Also, in terms of power, line B–D is a line of constant power. The torque–speed characteristics of the motor match the maximum current and maximum voltage characteristics of the battery.

A motor without an electronic and power controller has torque versus speed curves, which are designated as raw, natural behavior in Figure 14.9a. With modern-day electronic and power controllers almost any torque–rpm performance can be achieved as shown in Figure 14.9b.

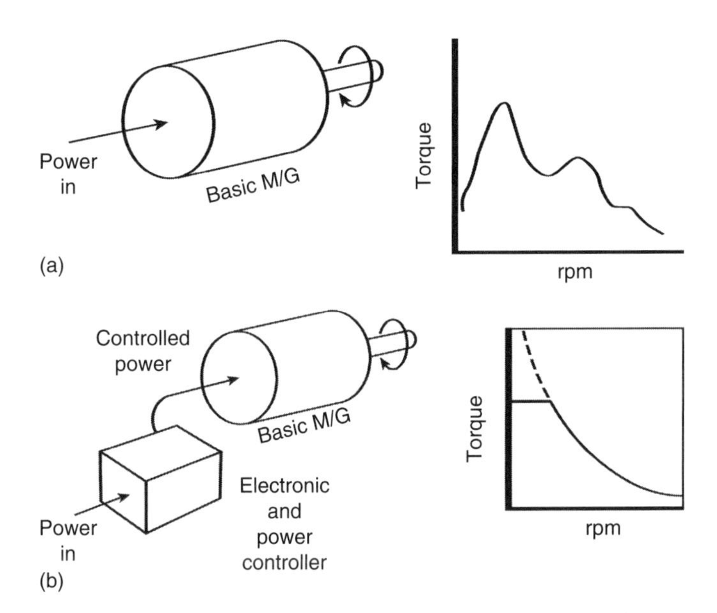

FIGURE 14.9 Modern control techniques allow tailored performance with almost any basic M/G. (a) Raw, natural, pre-modern torque–speed characteristics of the motor. (b) Tailored performance due to electronic and power controller yielding desired torque–speed characteristics of the motor.

SWITCHING FROM MOTOR TO GENERATOR AND VICE VERSA

Improved mpg for a hybrid is critically dependent on regenerative braking. In turn, regenerative braking is critically dependent on ability to switch the M/G from M-mode to G-mode without technical hassles. The switch from M-mode to G-mode has been discussed for some M/G designs in the preceding paragraphs. The method for switching an induction motor is presented in Figure 14.10.

Figure 14.10 has four sketches A–D. Sketches B–D are a plot of the torque from the generator as a function of speed (rpm). Sync speed, which is the rpm of the rotating magnetic field, is identified with a triangle. Sync speed is that of the rotating magnetic field similar to Figure 14.3. The rotor speed, which ranges from 0 to 8000 rpm, is also marked. Recall once again the definition of slip, which is

$$\text{Slip} = (\text{Rotating magnetic field speed} - \text{Rotor speed})$$

As shown in Figure 14.2c, slip is necessary for torque. The line at 45° going to the left upward from the sync speed is a curve of motor torque depending on slip. The more the slip, the more the torque. Sketch A is motor mode.

The line at 45° going to the right upward from the sync speed is a curve of generator torque depending on slip. The more the slip, the more the torque. The 45° curve to the right has negative slip and represents a generator. The switch from M-mode to G-mode has been made between sketches A and B. Sketches B–D are all generator mode but at decreasing rotor rpm.

With the rotor at 6000 rpm and sync speed 6300 rpm, the M/G operates as motor (Sketch A). The driver of the hybrid decides to stop or coast creating a need to switch from M-mode to G-mode. The frequency variable inverter (Figure 14.7) is commanded to decrease sync speed from 6300 to 5700 rpm thereby creating negative slip. As the vehicle speed decreases, the rotor speed also decreases going from sketch B to C to D from 6000 to 4000 to 2000 rpm. The sync speed also decreases to maintain constant negative slip and constant generator torque. Note that the sync speed could vary to modulate slip and torque as rotor speed decreases.

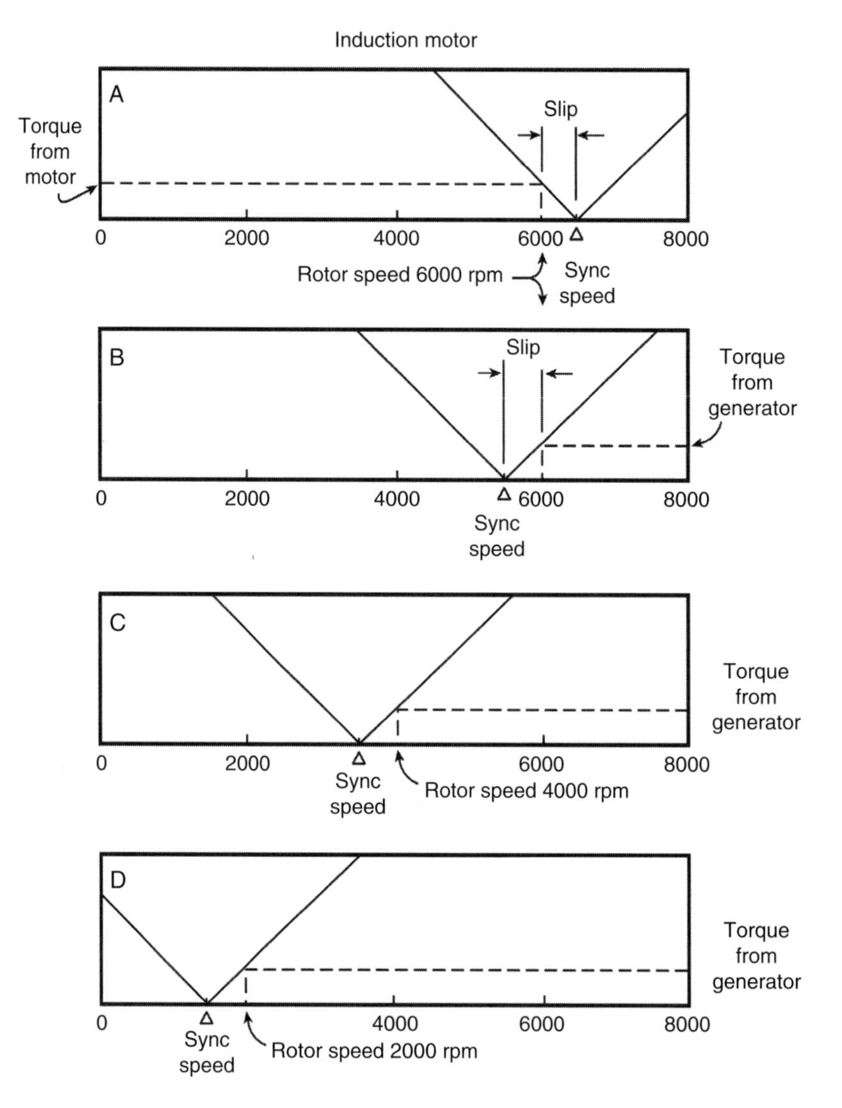

FIGURE 14.10 Switching from motor to generator operation for an induction motor.

2006 Honda Civic Torque Curves

The curves of Figure 14.11 allow calculation of the individual motor and engine torque and power curves. Figure 14.11 shows combined motor plus engine torque and power curves. The motor or engine power is given by

$$P = T\omega$$

where

T is the torque (N·m)

ω is the motor or engine angular speed (rad/s).

The results of reading the curves of Figure 14.11 and making the appropriate calculations are given in Table 14.2. The peak motor power is only 15 kW compared to the peak engine power of 69 kW.

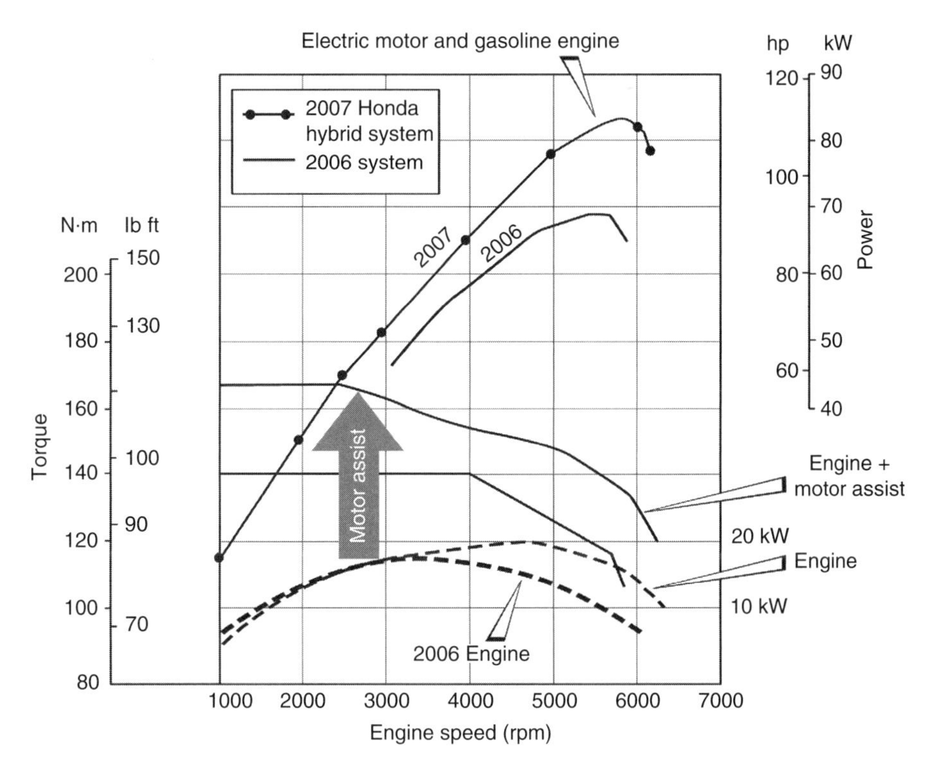

FIGURE 14.11 Comparison of 2006 and previous generation powertrains for Honda Civic Hybrid. (Reproduced with permission of Honda Motor Company.)

Motor power is only 22% of engine power. The peakmotor torque is 76 N·m at 1000 rpm compared to engine torque of 92 N·m at the same 1000 rpm. Motor torque is 83% of engine torque at 1000 rpm.

In summary of Figure 14.11, a small 15 kW motor only provides 22% of power but provides 83% of torque at a low speed of 1000 rpm. The combined engine plus motor torque is flat from 1000 to 2500 rpm. At low rpm, motors provide gobs of torque for very little power, which translates into low energy drain from battery. Truly a match made in heaven!

The data of Table 14.2 have been plotted and appear in Figure 14.12. The motor torque decays with increasing rpm, which is consistent with Figure 14.8. Also, from 2000 to 6000 rpm, the motor torque follows a 14 kW power contour. Once again this is consistent with Figure 14.8.

2006 HONDA CIVIC HYBRID ELECTRIC MOTOR

The motor for the 2006 Honda HEV is a DC brushless motor with the permanent magnets on the rotor. The rotating magnetic field is due to the armature in the stator. The motor is attached directly to the flywheel; hence, motor rpm equals engine rpm. The torque is due to a mixture of forces from reluctance and forces from magnet poles. The rotor has permanent magnets, which are the origin of N–S and S–N pole forces. The rotor has ferromagnetic material, which is the origin of forces due to magnetic reluctance. The motor operates at 159 V. Rectangular wire fills the slots in stator and decreases the ohmic resistance. Square wires eliminate the empty space found between round wires thereby decreasing the electrical resistance. Toyota has also adopted wire with rectangular cross section (see Ref. [13] of Chapter 3). The motor is rated at 15 kW. Motor ratings correspond to point B in Figure 14.8b. The dashed curve in Figure 14.12 shows the Honda Civic Hybrid motor follows a constant power profile from 2000 to 6250 rpm.

TABLE 14.2
Motor and Engine Torque and Power Data for 2006 Honda Civic Hybrid

Motor (rpm)	Engine Torque (N·m)	Motor + Engine Torque (N·m)	Motor Torque (N·m)	Engine Power (kW)	Motor Power (kW)	Combined Power (kW)
1000	92	168	76	9.6	8.0	17.6
2000	110	168	58	23.0	12.1	35.1
2500	118	166	48	30.9	12.6	43.5
3000	116	163	47	36.0	14.8	50.8
3900	119	155	36	48.6	14.7	63.3
4000	121	154	33	50.7	13.8	64.5
5000	121	148	27	63.4	14.1	77.5
5900	111	135	24	68.6	14.8	83.4
6000	108	129	21	67.9	13.2	81.1
6250	104	121	17	68.1	11.1	79.2

ALTERNATE VIEW OF ELECTRIC MOTORS: TORQUE OR SPEED

This chapter has touted the electric motor for its virtue as a torque machine. Yet Getrag–Bosch, as reported in Ref. [8], sees merit in the motor as a "speed" machine rather than a "torque" component. The application would be in a parallel hybrid with the electrical side (motor) independent of the engine. The motor should not be between the crankshaft and the transmission, which does allow

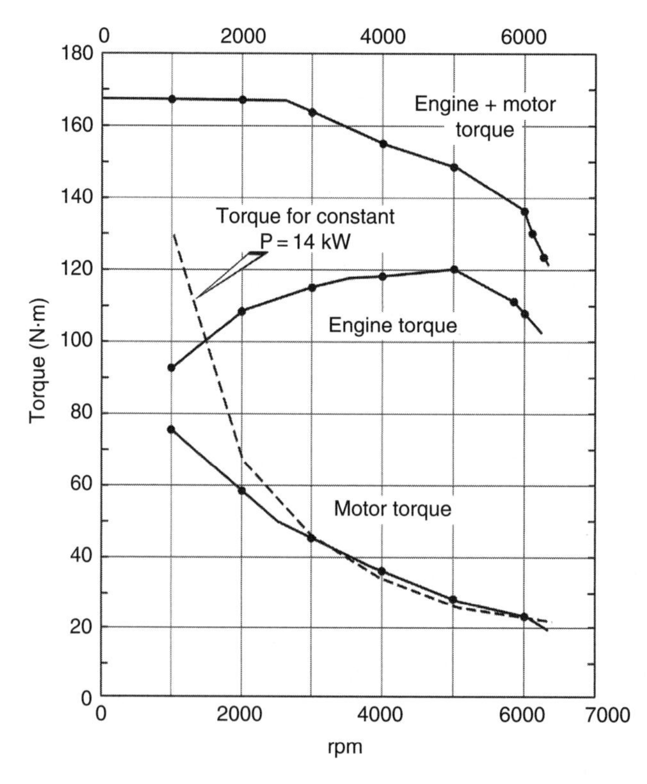

FIGURE 14.12 Engine, motor, and combined engine + motor torque as a function of rpm. Dashed line (---) is torque for a constant motor power of 14 kW.

high torque at low rpm. However, due to packaging and weight, a high-speed motor may be superior. The high-speed motor offers more flexibility in design.

SUMMARY

After a brief historical introduction, the attention shifts to the origin of the forces in an M/G for a hybrid application.

For magnet–magnet interaction, the basics are shown in Figure 14.2a with Figure 14.4 representing a simple motor. For magnetic reluctance, the basics are shown in Figure 14.2b with Figure 14.5 representing a simple motor. Figure 14.6 shows the design of the rotor of a reluctance motor. For electromagnetic induction, the basics are shown in Figure 14.2c with Figure 14.10 illustrating how the switch is made from motor to generator for an induction M/G.

Matching the battery and the M/G requires consideration of maximum current and maximum voltage. DC motors are discussed. To move into AC M/G, the concept of a rotating magnetic field is introduced. This leads to idea of synchronization between the rotor and rotating field. Synchronous motors come in two flavors relative to synchronization: slip and no-slip. The importance of the variable frequency inverter is discussed. Modern electronic and power controllers for M/G open many new design choices for the hybrid engineer.

M/G technology such as losses, cooling, efficiency, size, maximum rpm, weight, and cost are presented. Factors affecting the selection of an M/G are discussed for each type of M/G and are summarized in Table 14.1.

Gasoline engines have a torque "hole" at low rpm. Motors provide high torque at low rpm. Figure 14.12 has data for the engine and motor torques of the 2006 Honda Civic Hybrid. The speed range of the curves extends from 1000 to 6250 rpm.

APPENDIX 14.1: MATCHING POLES FOR A SWITCHED RELUCTANCE MOTOR

The following discussion is based on Figure 14.6, which illustrates a switched reluctance motor with eight poles for the stator and six poles for the rotor. To visualize the alignment of poles, the rotor in Figure 14.6 could be cut out and turned through various angles. However, an equally serviceable method is to lay out a linear scale with 0° to 360° for both the rotor and stator. Put arrows at 360°/8 = 45° for the stator. On another sheet of paper, put arrows at 360°/6 = 60° for the stator. Slide the rotor along the stator and note when rotor and stator poles match. Using this tool and procedure, the following pole matching was found.

Rotor Angle	Poles	Rotor Angle	Poles
0°	A–d	105°	C–a
15°	C–c	120°	B–d
30°	B–b	135°	A–c
45°	A–a	150°	C–b
60°	C–d	165°	B–a
75°	B–c	180°	A–d
90°	A–b	195°	C–c

The rotor angle increases clockwise; hence, the rotor is turning clockwise, CW. The stator field is moving d → c → b → a and appears to be rotating CCW. While the rotor moves 120°, the stator field moves 360°.

REFERENCES

1. I. M. Gottlieb, *Electric Motors and Control Techniques*, 2nd ed, TAB Books, McGraw-Hill, New York, 1994.
2. J. Yamaguchi, Honda unveils intelligent powertrains, *Automotive Engineering*, Tech Briefs, October 2005, pp. 20–22.
3. M. H. Westbrook, *The Electric and Hybrid Electric Car*, Society of Automotive Engineers, Warrendale, PA, 2001.
4. A. Emadi, Low-voltage switched reluctance machine based traction systems for lightly hybridized vehicles, *Hybrid Electric Vehicles*, SP-1633, Society of Automotive Engineers, 2001, p. 41.
5. H. Gao, Y. Gao, and M. Ehsani, Design issues of the switched reluctance motor drive for propulsion and regenerative braking in EV and HEV, *Hybrid Electric Vehicles* SP-1633, Society of Automotive Engineers, 2001, p. 63.
6. K. Muta, M. Yamazaki, and J. Tokieda, Development of new-generation hybrid system THS II—drastic improvement of power performance and fuel economy, *Hybrid Gasoline-Electric Vehicle Development*, J. M. German (Ed.), SAE PT-117, 2005, p. 47.
7. J. Yamaguchi, Hondas unveils intelligent powertrains, *Automotive Engineering*, Tech Briefs, October 2005, pp. 19, 20.
8. N. Palmem, Ecology needs dual power, *Automotive Industries*, Quarter 3, 2007, pp. 24–25.
9. J. Abthoff, P. Antony, M. Krämer, and J. Seiler, The Mercedes_Benz C-Class series hybrid, *Electric and Hybrid-Electric Vehicles*, PT-85, Society of Automotive Engineers, 2002.
10. Marketing staff, press release, October 2005, Honda Corporation America.

15 Economics of Hybrid Ownership

Driving a car—any car—is expensive. Hybrids are even more expensive.

INTRODUCTION

A cold, hard look at the expenses of owning and operating a car reveals just how expensive it is. As discussed in subsequent paragraphs, many of the costs are stealthy. The owner's attention is not focused on the costs all at one time. Many features of a hybrid add to the costs and complexity.

This is a chapter on economics. So, there are many tables and graphs with numbers preceded by "$" signs.

COSTS TO PRODUCE A HYBRID

One factor contributing to hybrid costs is the price premium, which is the added cost relative to a comparable conventional vehicle (CV). The manufacturer's viewpoint on costs is briefly presented next. The cost of added electric drive contributes to the price premium. Having two propulsion systems adds cost and weight. These facts are discussed throughout this book (see Chapters 3, 4, and 7). The manufacturer can start with the modification of an existing vehicle. The Honda Civic Hybrid is an example. Alternatively, the manufacturer can start with clean sheet of paper design. The Honda Insight and Toyota Prius are examples. A clean sheet design requires a larger investment. Determination of the price premium is difficult [3].

Initially, the sales for a newly introduced hybrid will be low. A unit cost penalty for initial low-volume production jacks up the sticker price on the showroom floor. Even to convert an existing CV to a hybrid involves significant program costs. First items are engineering and testing costs. Next, the hybrid has added pieces or components with attendant piece costs. Added components, which largely account for the infamous price premium, include electrical motors and generators. Production investments include integration of the hybrid onto the production line to permit simultaneous production of the CV and the hybrid.

In October 2006, Toyota announced the goals for reducing both price premium and hybrid component weight. Within 3 years, Toyota will reduce the price premium by 50% to $1900 for the next generation Prius (Prius III). The size and weight of the hybrid components (not the complete car) will be reduced to half compared to Prius II. In July 2006, Honda announced that they expect to cut the price premium of its integrated motor assist (IMA), hybrid system to approximately $1700 by 2009.

TABLE FROM THE INTERNET

Table 15.1, which is from various sources on the Internet, introduces several concepts that will be useful in later discussion. Table 15.1, which is dated January 2008, can be updated using the Internet. Update by logging on to J. D. Power, Kelly Blue Book, and Edmunds Auto Internet sites. Data available on the Web include the *Days to Sell*, which allows the potential hybrid customer to judge how willing the dealer may be to offer a discount or sales incentive. Other valuable facts include annual sales, which affect tax credits (see the discussion in later sections). The price premium is of central

TABLE 15.1
Relevant Data for Assessing Purchase Price and Cost of Hybrid Ownership

Vehicle	Base Price ($)	Price Premium ($)	Hybrid Mileage	Conventional Mileage	Observed Mileage
Prius II	21,610	4,535	48/45	26/35	40/38
Camry	25,860	2,200	33/34	21/31	—
Malibu	22,790	2,795	—	—	24/32
Civic	23,235	2,690	40/45	26/34	36/45
Lexus RX400h 2WD	42,045	3,880	27/24	18/23	—
Lexus RX400h AWD	43,445	3,880	26/24	17/22	—
Ford Escape	28,745	1,865	29/27	17/22	—
Saturn Vue FWD	24,795	3,290	—	19/26	—
Chev Tahoe AWD	53,295	13,905	—	14/19	—

importance in determining overall hybrid vehicle expense of ownership. Finally, the fuel economy, mpg, as stated by Environmental Protection Agency (EPA) and as observed in real-world driving affects the number of years to recoup the added expenses of a hybrid.

Economics of Bus and Truck

Whereas hybrid cars may sell at a premium without hope of recovery of added cost from savings on fuel, in the commercial world this is not the case. Total cost of ownership dictates purchase or none; the only issue is the effect on the bottom line. Car owners will buy for reasons other than saving money.

Base Model versus Top-of-the-Line

The manufacturer makes a decision as to what range of trim lines the hybrid will be offered. The decision affects sales. Some hybrids are offered only in top-of-the-line trim and with almost all options. The hybrids presented only in high trim level include Toyota Highlander sport utility vehicle (SUV), Lexus RX 400 h SUV, Lexus 450 h hybrid sedan, and Lexus 600 h hybrid sedan. The hybrids with choice of trim levels from base to fully loaded include Toyota Prius and Saturn Vue Green Line.

BENEFITS OF HYBRIDS

Summary of Benefits

Chapter 3 discusses in more detail the benefits of hybrids to the owner and to society. The opening sentence for Chapter 3 is "The future of hybrid vehicles rests on the twin pillars of enhanced fuel economy and lower emissions." In addition, the hybrid provides special concessions such as solo driver use of car pool lane. In the eyes of the hybrid owner, the benefits balance, or more than balance, the added costs.

PAY AT THE PUMP

Perspective on Cost of Fuel: Pay at the Pump

Each time the gas tank is filled, the customer is reminded of the cost of gasoline. In contrast, the cost of depreciation is stealthy. Even insurance is paid annually and not at the corner gas station. Insurance cost is soon forgotten. Maintenance is sporadic. If the customer were to pay for all car expenses each time at refill, a new perspective on cost of gas would emerge. The feeling about high cost of gas might be revised.

Consider the expenses for a new car during the first year of ownership. Assume an entry luxury car which was purchased for $38,000. First year expenses are

Depreciation	$10,000
Insurance for year	$1,800
Maintenance	$1,200
Capital cost at 5%	$1,900
Miles driven in year	12,000
mpg combined	20
Miles driven between fill-ups	300

Now some calculations:

Number of fill-ups per year = 12,000/300 = 40
Gallons of gas used = 12,000/20 = 600 gal
Gallons to fill-up = 600/40 = 15 gal
Cost of gas at fill-up = 15 × $5.00/gal = $75

The cost of each item at fill-up

Gas	$75.00
Depreciation	$250.00
Insurance	$45.00
Maintenance	$30.00
Capital	$47.50
Total cost at fill-up	$447.50

If all costs were paid at the pump, the service station attendant would say "That will be only $447.50; cash or charge or our easy payment plan?" Compared with a staggering amount of $447.50, $75 for gasoline does not seem so big. And yet, that is what the owner is actually paying to drive a $38,000 entry luxury car for 300 mi! Is that amount really true or is it faulty estimated numbers? Look at Table 15.2 which provides data for price, depreciation, cost per mile, and true cost at each fill-up. The cost over 5 year ownership would be a lesser amount but not less dramatic.

FUEL COSTS

ANNUAL COST OF GASOLINE (DIESEL) FUEL

Define some symbols for the cost equation.

D = miles driven per year (mi/year)
$G_{\$}$ = cost of gasoline ($/gal)
F_E = fuel economy (mpg)
C = annual cost for fuel ($/year)

From Chapter 7 the fuel economy has units of mi/gal (mpg), or km/L, and the symbol for fuel economy is F_E. Note that fuel economy and fuel consumption are reciprocals: $F_E = 1/F_C$. The cost equation is

TABLE 15.2
Data for Price, Depreciation, Cost per Mile, and True Cost at Each Fill-Up

		Depreciation		$/Mile		Cost at
Car	Price ($)	First Year ($)	Five ($)	First ($)	Five ($)	Fill-Up ($)
2006 Eclipse GT	28,000	8,700	17,000	1.14	0.71	342
2006 Ford Explorer	38,700	10,500	21,500	1.25	0.73	375
2006 Chevy Corvette	51,400	10,600	26,400	1.42	0.88	426
2006 Mercedes Benz E350	54,200	7,700	28,300	1.25	0.95	375
2006 Cadillac STS-V	77,100	18,800	44,000	2.43	1.36	729
2006 VW Jetta GLI	26,500	6,400	14,700	0.89	0.60	267
2005 Ford GT	167,000	88,500	135,900	8.11	2.92	2,433
2005 Toyota Tacoma	24,100	5,570	13,000	0.80	0.53	240
2006 Ford Escape Hybrid	30,700	6,900	17,700	0.89	0.56	263

Notes: The depreciation and dollars per mile are for the first year and the sum for all 5 years of ownership. The cost at fill-up assumes 300 mi between fill-up. Data are calculated and/or extracted from various Web sites on Internet.

$$C = \frac{(D,\ \text{mi/year})(G_{\$},\ \$/\text{gal})}{F_{\text{E}},\ \text{mi/gal}} = DG_{\$}F_{\text{C}} \quad (15.1)$$

Note that some units cancel in Equation 15.1. The miles in the denominator and numerator cancel. The 1/gal in the denominator and numerator cancel. Remaining is $/year, which, of course, is the annual cost for fuel. Table 15.3, which was calculated using Equation 15.1, provides annual fuel costs, C, for a variety of vehicles.

TABLE 15.3
Annual Fuel Costs for a Variety of Vehicles

	Annual Mileage		
Vehicle	12,000	16,000	20,000
Honda Insight Hybrid (51)	$1,180	$1,575	$1,970
Toyota Prius II (41)	$1,470	$1,950	$2,430
VW Golf TDI Diesel (41)	$1,580	$1,300	$2,535
Toyota Yaris (38)	$1,580	$2,120	$2,630
Toyota Corolla auto (29)	$2,070	$2,750	$3,450
Ford Focus (24)	$2,500	$3,330	$4,170
Jeep Grand Cherokee V-8 (15)	$4,000	$5,330	$6,670
Ford Excursion SUV V-10 (10)	$6,000	$8,000	$10,000

Notes: The number in parenthesis is the combined mpg. Gasoline is $5.00/gal and diesel $5.20/gal.

FUEL SAVINGS

ANNUAL FUEL SAVINGS EQUATION

In addition to the symbols defined for Equation 15.1, add the following symbols:

S = annual savings from cost of gasoline (\$/year)
C_H = cost for fuel for hybrid (\$/year)
C_C = cost for fuel for CV (\$/year)
F_{EH} = fuel economy of Hybrid (mpg)
F_{EC} = fuel economy of CV (mpg)

Both F_{EH} and F_{EC} are based on combined city/highway fuel economy. The fuel savings for an HEV compared to a CV is given by

$$S = C_C - C_H \tag{15.2}$$

Combining Equations 15.1 and 15.2 yields

$$S = DG_\$\left(\frac{1}{F_{EC}} - \frac{1}{F_{EH}}\right) \tag{15.3}$$

As a sample calculation, consider a Honda Civic Hybrid with $F_{EH} = 47.5$ mi/gal and $F_{EC} = 34$ mi/gal. Also $G_\$ = \$5.00$/gal and $D = 15{,}000$ mi/year. Inserting these values into Equation 15.3 gives

$$S = (15{,}000 \text{ mi/year})(\$5.00/\text{gallon})\left(\frac{1}{34} - \frac{1}{47.5}\right) = \$630/\text{year}$$

Equation 15.3 is used to calculate portions of the data for Table 15.4. Tax credits are discussed later. For Table 15.4, a tax credit of \$2000 is assumed.

TABLE 15.4
Fuel Savings, Price Premium, and Years to Recoup Price Premium with and without \$2000 Tax Credit

Vehicle	Mileage CV	Mileage HEV	Fuel Savings, S (\$)	Price Premium, P (\$)	Years to Recoup, Y	Years with Tax Credit, Y_T
Honda Civic	34	47.5	630	2390	3.8	0.6
Honda Accord	24.3	32.1	750	3630	4.8	2.2
Toyota Prius II[a]	29	55.6	1,240	4600	3.7	2.1
Toyota Prius II[b]	29	40.5	730	4600	6.3	3.6
Ford Escape	22	33.6	1,180	3300	2.8	1.1
Ford Escape[b]	22	24.7	370	3300	8.9	3.5
Saturn Vue	21.9	29	830	2000	2.4	0
VW Beetle[c]	26.7	41.2	870	1240	1.4	—

Note: Gasoline in \$5.00/gal.
[a] Compared to Toyota Corolla.
[b] With observed fuel mileage in real-world driving.
[c] Gasoline versus TDI diesel.

Combined EPA Mileage Ratings

Subsequent discussion uses EPA combined mileages. See Chapter 12 on how to combine city and highway mpg. The equation for combined fuel economy, F_E, is

$$F_E = \frac{1}{(0.55/\text{city}) + (0.45/\text{highway})} \tag{15.4}$$

As an example, for city = 29 mpg and highway = 37 mpg, the combined is 32 mpg.

Real-World Fuel Economy and Market Realities

Real-world fuel economy, which usually differs (usually less) from EPA estimates, depends on driving habits and driving style. See Chapter 12 for information on how EPA ratings are determined. Also Chapter 12 has suggestions for driving habits to improve mpg. In 2008, a big improvement was made in the quality of EPA estimates for mpg.

Obviously, real gasoline savings are based on real mpg not fictional mpg. The years to recoup the price premium depend on use of valid mpg data.

Marginal Savings: One More Mile per Gallon

Suppose that a hybrid is rated at F_{E1} = 35 mpg. What is the added annual gasoline savings, S, if F_{E2} were increased to 36 mpg? The increase in savings is known as the marginal savings. Using Equation 15.3, a formula for marginal savings is derived. The two fractions within the parenthesis of Equation 15.3 are combined to a single fraction; thus

$$\frac{\Delta S}{\Delta F_E} = DG_\$ \frac{F_{E2} - F_{E1}}{F_{E2} F_{E1}} \approx DG_\$ \frac{\Delta F_E}{F_E^2} \tag{15.5}$$

Marginal savings is the ratio $\Delta S/\Delta F_E$, where ΔS is the change in savings and ΔF_E is the change in fuel economy. The units are \$/(year)(mpg).

Using F_{E2} = 36 mi/gal, F_{E1} = 35 mi/gal, $G_\$$ = \$5.00/gal, and D = 15,000 mi/year, the marginal savings are \$59.52/(year)(mpg). Table 15.5 shows the decrease in marginal savings as mpg increases by 1 mpg.

TABLE 15.5
Decrease of the Marginal Savings and CO_2 Emissions as mpg Increases by 1 mpg

Mileage (mpg)	Marginal Savings, $\Delta S/\Delta F_E$ \$/(year) (mpg)	Percentage Reduction Carbon Dioxide, CO_2 (%)
10	750.00	−10.0
20	187.50	−5.0
30	83.33	−3.3
40	46.90	−2.5
50	30.00	−2.0
60	20.83	−1.7

Note: Calculated for gasoline at \$5.00/gal.

Very little is saved by better mpg as mpg increases. When $F_E = 10$ mpg, the marginal savings are \$750.00. For $F_E = 60$ mpg, the marginal savings is reduced to a mere \$20.83. Obtaining cost benefits by increased mpg becomes more difficult as mpg increases. The ability to lower mpg becomes harder and harder with increasing mpg.

YEARS TO RECOUP

Years to Recoup Price Premium

The potential hybrid buyer wants an answer to the question "How many years of ownership are required to recoup the price premium through savings from gasoline?" To answer the question, an equation is presented next. Define five more symbols:

P = added cost to the customer for hybrid above a CV; the price premium.
S = annual savings due to cost of gasoline due to better mpg (see Equation 15.3)
Y = years to recoup price premium based on gasoline savings
T = tax credit
Y_T = years to recoup with tax credit

The equations for Y and Y_T are

$$Y = \frac{P}{S} \tag{15.6}$$

and

$$Y_T = \frac{(P-T)}{S} \tag{15.7}$$

values for Y and Y_T are given in Table 15.4. For a graphical interpretation of the years to recover the price premium, see later discussion of Figures 15.4 through 15.8.

TAX SAVINGS

Tax Credit and Tax Deduction

To encourage production and sales of hybrid vehicles, and in the process decrease emissions and dependence on oil imports, governments are offering various tax incentives. The federal government offers tax credits and tax deductions. A tax credit is more valuable than a tax deduction. A tax credit subtracts directly from the income tax due. A tax deduction is less valuable as can be seen from the formula:

$$\$ \text{ value to tax payer} = (\text{Tax bracket\%})(\$ \text{ tax deduction})$$

The tax incentives are in a state of flux year to year. A tax credit ranging from \$650 to \$3150 is available for certain hybrids in the time period 2006–2010. Each hybrid receives more or less depending on annual fuel savings. However, the number of credits is limited to 60,000 cars sold annually by any one manufacturer. The interested reader can look on the Internet for the latest information. Search for tax credits for hybrids.

State Tax Credit and Tax Deduction

The interested reader can look on the Internet for the latest information. Search for (insert name of your state) tax credits for hybrids. For example in Nevada, search Nevada tax credits for hybrids.

Gas Guzzler Tax

The buyers of certain vehicles pay a gas guzzler tax. Decreasing fuel economy, mpg, increases the tax.

21.5–22.5 mpg	$1000 tax
<12.5 mpg	$7700 tax

The gas guzzler tax does not apply to SUVs, pickups, and minivans. Because of the narrow and arbitrary application of the gas guzzler tax, some citizens view this as being grossly unfair and unwise. The erratic application of the tax does not help with the goals of reducing oil imports and reducing emissions.

Car Pool Lane Permits: Two-Tier Value

In some states, counties, and cities, car pool lane permits are limited in number. Hence, a two-tier system is created—those who have and those who do not have permits. When the hybrid is sold, does the permit go with the car? If so, some people are rewarded and others are penalized.

COSTS

Costs: Added Costs as a Result of Price Premium

The customer for a hybrid must pay sales tax on the price premium. Assume 8% sales tax on a price premium of $3000. The added cost is (8%)($3000) = $240. Compare with the annual fuel savings shown in Table 15.4, which range from $223 to $742.

Either additional finance expenses (for the easy payment customer) or added capital costs (for the cash buyer) increase the cost of ownership. The amount will be about the same as for added sales tax. Combining sales tax with added finance charges wipes out from 1 to 3 years of annual fuel savings.

Costs (Savings): Maintenance and Repair

Repairs are either due to a car-part failure or collision damage. For repairs of collision damage, see the discussion of insurance in subsequent paragraphs. Routine maintenance can be divided (split) between independent shops and dealers. In either case, check on mechanic specialized training for hybrids. Specialized training implies a larger hourly rate that will be charged. Independent shops usually have lower rates. The Auto Club may help with a list of AAA certified garages.

See Chapters 20 and 21 for additional information on mechanics and repairs. One special aspect of hybrids concerns brakes. Regenerative braking extends the life of mechanical brakes (see Chapter 9).

Costs: Automobile Insurance

Your automobile insurance policy lists charges for bodily injury, medical payments, uninsured motorist, property damage, comprehensive coverage, and collision. Except for uninsured motorist, each item is affected by being a hybrid. Other chapters have information relevant to the cost of insurance.

Chapter 3 discusses conversion or clean sheet design. Only three production hybrids start with a clean sheet of paper design. All other production hybrids are conversions (modifications) of an existing product line. An example is the Honda Civic and the Honda Civic Hybrid. When hybrid production starts with an existing car design, added weight of hybrid components may increase braking distance, decrease storage space, increase weight, and adversely affect handling characteristics, when compared to nonhybrid cars.

Chapter 20 discusses performance and handling penalties due to added weight. Added weight has an effect on several of the vehicle handling and performance characteristics including acceleration, braking, and steering. Depending on the location of the added weight, the location of center of gravity (CG), is affected. This influences susceptibility to roll over, the weight distribution front-to-rear, and directional stability. Further, added weight increases tire rolling friction; the more the weight, the more the friction.

Consider insurance for bodily injury and medical payments. The statements summarized above suggest that a hybrid may be less safe than a CV. Experience in the coming years will prove or disprove that assertion. In the meanwhile, insurance companies may increase rates.

Consider insurance for property damage, comprehensive coverage, and collision. Increased weight and cost of collision repairs once again may increase insurance premiums. The availability of car parts, as discussed in Chapter 20, affects repair costs. Car parts stores do not stock repair parts for hybrids. The logistics for hybrid parts has not been established. Collision damage is in a separate category of repair. Eventually, hybrids will be involved in collisions that may damage both body and drive train components. Currently, drive train parts are only available from the appropriate new car dealer. The collision repair can perhaps be done only in a dealer's body shop. Competition from independent shops is reduced or nonexistent; hence, prices are higher. Auto insurance companies will likely detect the price discrepancy between HEV and CV. Insurance rates may increase for hybrids.

The comments above focus on the narrow viewpoint connecting insurance premiums with the costs of repair and damage. A much broader viewpoint may motivate insurance companies to offer insurance discounts for hybrid vehicles. In 2006, the Traveler's Insurance Company offered owners of hybrids in California a 10% discount on automobile insurance. Traveler's offered discounts in 41 other states. (Traveler's provides both hurricane and automobile insurance.) The logic is something like this. The number and severity of hurricanes is increasing due to global warming. Consequently, the insurance company is at greater risk due to hurricanes. Global warming is thought to be due to increased levels of CO_2 in the atmosphere. Hybrids spew out less CO_2. Encouraging sales of hybrids will in the long run reduce the very major insurance risks associated with hurricanes.

Costs: Depreciation Takes Effect

Depreciation is a stealthy cost of ownership. It takes effect when the owner sells or trades-in the hybrid. Depreciation is important at lease-end for a hybrid. Depreciation determines cost of ownership. Small depreciation yields smaller costs. After the theft of a hybrid, depreciation influences the amount that the insurance company will pay. A vehicle may be totaled due to collision or flood damage. Again, depreciation influences the amount that the insurance company will pay.

Costs: The Showroom Syndrome

The moment the buyer signs the purchase contract at the new car dealer, the vehicle is transformed from a new car to a used car. The transformation generates a $3000–$5000 instant depreciation. The drop in value is known as the "showroom syndrome." The proud new owner starts his/her new vehicle; the engine is cold. Even before he/she has driven far enough to warm up the engine, the $3000–$5000 depreciation has taken place! Showroom syndrome is not hybrid unique but applies to every new vehicle. The extra large first-year depreciation is due to the showroom syndrome.

What Is a Major Improvement?

This section is a prologue for the discussion on depreciation and the pace of technology. What is an improvement? Any advances that lead to increased mpg are an improvement. Likewise, decreased

TABLE 15.6
Examples of Technological Improvements

Component	Improvement	Benefits
Brakes	Two-wheel → four-wheel	Shorter stopping distance; safety
Brakes	Mechanical → hydraulic	Shorter stopping distance; safety
Brakes	Drum → disc	Shorter stopping distance; safety
Ignition	Breaker points → microprocessor	Greater reliability
Cooling	Fan → thermostatically controlled	Better mpg
Tires	Bias belted → radial	Better mpg; longer tread life
Fuel metering	Carburetor → fuel injection	Better mpg; lower emissions
Fuel injection	Manifold → direct injection	Better mpg; lower emissions
Battery	Vented → sealed	Longer life; eliminate adding water
Lighting	Separate lens → sealed beams	Avoid dirty lens and reflectors
Electrical	Six-volt → 12-volt	Lower current; less weight
Electrical	DC generator → AC alternator	Greater reliability; longer life
Antifreeze	Annually → 5 years	Less expense; fewer trips to garage

emissions and greater reliability are desired quantities. Cost always hovers over the hybrid production scene. If a new gadget or new production technique can simultaneously improve performance and reduce costs, that new technology is especially welcome.

Improvements fall into the following five categories: performance, safety, comfort and convenience, reliability, and styling. Performance hinges on many mechanical features of a vehicle including brakes, engine, and steering. Mechanical features affect fuel economy, which is a dominant factor with regard to hybrids. Safety has many facets. Crash protection (seat belts, crumple zones) is but one aspect. Agility on the road with superior ride and handling may prevent an accident. A short stopping distance enabled by brakes affects safety. Headlamps enable the driver to see better at night with obvious impact on safety. Headlamps are also a dramatic styling feature with jewel-like lenses. The category of comfort and convenience includes fold-down seats in sedans, SUVs, and minivans. Backup TV is a convenience and safety item. Navigation systems help to avoid distractions and driver confusion in unfamiliar locations. Safety is enhanced. Reliability is important. When the car stops in the middle of a freeway, the car owners cringe and vow all kinds of retribution on the car company. Both mechanical and computer components determine reliability. Styling affects a buyer's decision to buy, and, therefore, cannot be ignored. As discussed in Chapter 7 styling and safety interact. Figure 7.9 shows a styling comparison between an original vehicle and a vehicle having added crumple zone.

The modern automobile is a marvel of technology. Some 25–30 years ago, the owner was plagued by frequent (annually or every 10,000 mi or so) maintenance and repairs. This included tire replacement at 10,000–20,000 mi, breaker point settings every 10,000 mi, carburetor float setting every 10,000 mi, and battery replacement after 1 or 2 years of service. Current vehicles run 100,000 mi with minimal service.

These comments and Table 15.6 are an introduction to the pace of technology and the effect on the magnitude of depreciation.

Costs: Depreciation and Pace of Technology

The pace of technology does affect depreciation. Two examples are given to initiate the discussion. An example of the pace of technology appeared earlier in this chapter; call this Example A. Both

Toyota and Honda announced the goals for reducing the price premium and hybrid component weight. The timescale was 3 years. The advances are significant enough so as to qualify as basis for a new model. Most important is that the prior models will be made obsolete. Another example, which is called Example B, appears in Figure 7.3, which gives vehicle weight and performance for the years 1975–2005. From 1976 to 1981, major reductions were made in weight with a slight decrease in acceleration. From Figure 7.2, major improvements were also made in fuel economy. A 1981 car was superior to a 1976 car.

Two factors are relevant. First, the years of ownership denoted by the symbol, Y_O. Second, in the span of Y_P years, a new hybrid has had such major improvements due to advanced technology that earlier hybrids are made obsolete. The subscript P represents the pace of technology. Y_O is well defined. Y_P is not well defined; in fact, Y_P is rather nebulous at this point. Y_P does exist and is not constant. Definition of Y_P is the crux of this discussion. Many influences affect the value for Y_P including the buying public's whims, the strength of the economy, the price of gasoline, and vehicle styling. Our goal is to exclude the outside influences and focus on the pace of technology. Styling is particularly dominant in determining depreciation.

For a more refined definition of Y_P, think of extremes in hybrid know-how. At the far most revolutionary extreme is a perpetual motion machine, which would make all hybrids obsolete. At the barely-no-advance extreme would be a change in paint color. But that is styling, which is excluded. From Table 15.6 select an advance such as the change in batteries from vented to sealed, or the change in antifreeze from annual replacement to a 5 year schedule. Both of these, while important in aggregate, are at the barely-no-advance extreme. To have an impact on Y_P the sum of the contribution from new designs, new components, etc. must be between the extremes at the high end.

The definition for Y_P should incorporate the difference in value of the old hybrid compared to a new hybrid fresh off the showroom. An assumption made by the used-vehicle buyer is that the vehicle mileage today is an indicator of remaining relatively trouble free miles. Another facet is the fact that after several years, the car's mileage and age are less important. An older car having had tender, loving care is more valuable than a neglected car (never changed oil, never washed, never garaged).

The following two sentences comprise a circular argument. "A large depreciation in a few short years, Y_O, implies a small value for Y_P." To complete the circle, the second sentence is, "If Y_P is less than Y_O, then large depreciation results." The more direct definition for Y_P is as follows. A hybrid is purchased in year zero. After Y_P years, the new hybrids offered in the market are in the "must-have" category. The owner will seriously consider trading in his/her old hybrid for a new technologically advanced hybrid. Major improvements in the five vehicle categories—performance, safety, comfort and convenience, reliability, and styling—place a new model in the must-have category.

Using Y_O and Y_P, a qualitative graph is made as shown in Figure 15.1. The two symbols can be related by

$$Y_O = n\,Y_P \tag{15.8}$$

where n is the number of new models. One new model, that is, $n = 1$, is sufficient for obsolescence. $Y_O = Y_P$, which is a 45° line in Figure 15.1a, and is also the line for $n = 1$. The use of equations and numbers on a graph may suggest a more quantitative representation than is justified. However, Examples A and B suggest that Y_P is approximately 5 years for each example. Example A states a goal of 3 years; still, announcements are not made without assurances the goal can be reached. Add 2 years for the prior R&D to the 3 years giving $Y_P = 5$ years.

Figure 15.1a is a plot of Y_P as a function of Y_O with n as a parameter yielding the straight slanted lines. A dotted horizontal line is drawn for $Y_P = 5$ years. After 2.5 years of ownership, point A is located at $n = 0.5$. What is a vehicle with only half the benefits of advanced technology? Because $n = 0.5$, the new hybrids have newly advanced parts and pieces but the overall package does not create a compelling reason to buy. After 5 years of ownership, point B is located on the $n = 1$ line.

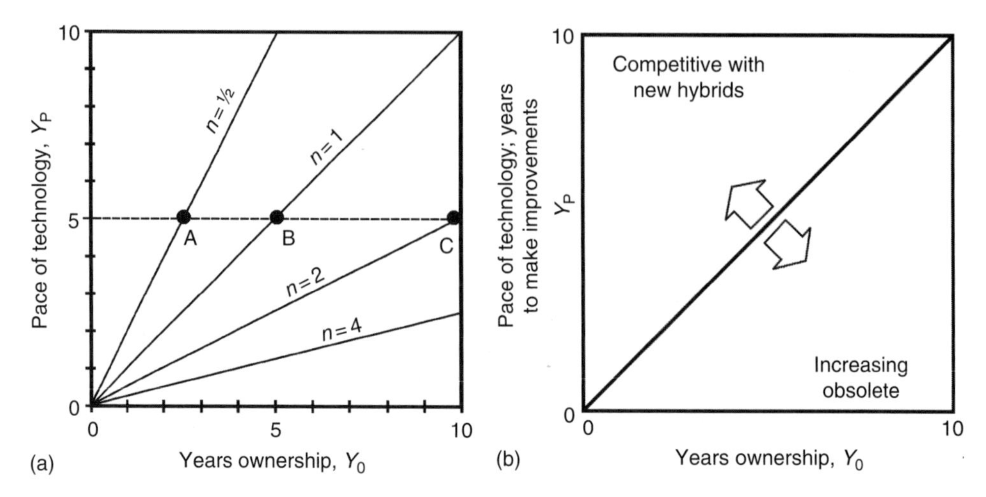

FIGURE 15.1 Effect of technological improvements, a plot of years for technological pace versus years of ownership. (a) Curves of constant n relating obsolescence to years of ownership. (b) Qualitative chart for depreciation.

This means the old hybrid has become appreciably obsolete. At $Y_O = 10$ years, point C is located on $n = 2$ line. The old hybrid is badly out of date with large depreciation.

In Figure 15.1b the 45° line divides the plot into two regions. Above the line, upwards and to the left, the farther the point is from the line the more competitive the used hybrid is with the new models on sale. Below the 45° line, downwards and towards the right, the farther the point is from the line the more hopelessly obsolete the used hybrid is compared with the new models on sale.

Costs: Amount of Depreciation for Various Vehicles

Data on depreciation for various vehicles appear in Table 15.2. Also, information concerning vehicle price, cost per mile, and true cost at each fill-up is presented in Table 15.2.

Two isolated data points on depreciation were found. One is from a classified advertisement. In October 2006 a 2001 Prius I was offered for sale. The Prius I had 63,000 mi. The offering price was \$14,000. Base price was \$19,995. Depreciation was \$19,995–\$14,000 = \$5,995 = \$6,000 which is \$1,200/year. Very low. Another example was found in *AutoWeek* 5 December 2005 for a 2004 Prius II which retained 89% of its value after 1 year. Mileage was 16,000 in that year. Sticker price was \$22,949. Trade-in value was \$20,400. So, depreciation was \$2,549 = 11% which is very low (also see Appendix 15.1).

Threat of Battery Replacement

As the years pass and the miles on the odometer grow, the value of the hybrid depreciates. The cost of a replacement battery does not decrease. Refer to Figure 15.2, which illustrates the fraction of the Prius value as depreciation takes its toll. When new, the battery represented 16% of vehicle value. After 3 years, the battery is 28% of Prius value. With depreciation, the fraction of vehicle value due to the battery grows and becomes a more significant factor. The size of the rectangles varies with the value. The area within the rectangle for the battery stays the same but becomes relatively larger.

A telephone call to the local Toyota dealer parts room gave an estimate of \$3,000 for a Prius new battery with approximately \$600 for removal and disposal of the old battery and installation of the

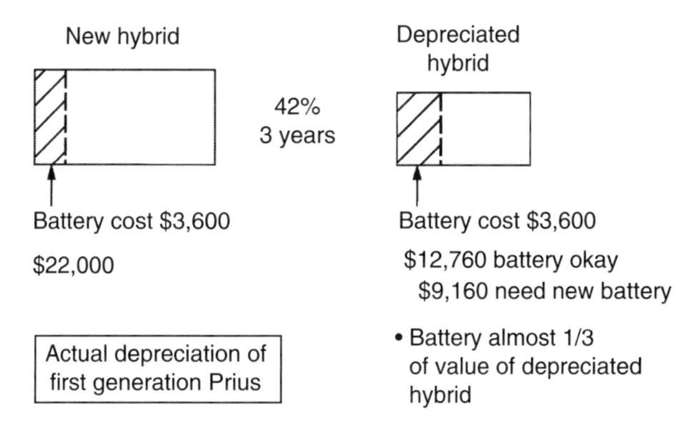

FIGURE 15.2 Depreciation of Prius I after 3 years with the impact of battery replacement on depreciation.

new battery. At 3 years, if the original battery continues to function satisfactorily, the Prius is worth \$12,760 yielding 42% depreciation. However, if the battery is weak and needs to be replaced, then the Prius is worth only \$9,160 which is a 58% depreciation.

Several measures of battery life are used. These are years of service; vehicle miles, which the customer understands; and the number of charge/discharge cycles, which the electrochemist uses (see Chapter 6). Consumer Reports indicates that automakers expect a battery life 150,000–180,000 mi (see Ref. [2]). The sales brochure for the 2005 Prius II states a battery warranty of 80,000 mi. Battery life in terms of cycles is reported in Table 3.6. For a light duty vehicle (LDV) the number of charge/discharge cycles is a few 10,000 over a distance of 124,000 mi (200,000 km). The expected life of a hybrid is 124,000 mi. The outlook looks positive with regard to battery life.

INTERPLAY HYBRID TECHNOLOGY AND PRICE PREMIUM

Graphical Representation of Profit/Loss and Reward/Penalty

A graphical format is developed in this section, which shows where the customer is rewarded or penalized and where the manufacturer profits or loses. The graphical format also solves for values of the recoup time Y_O. Figure 15.3a defines three parts of any graph. The ordinate is the vertical axis while the abscissa is the horizontal axis. The origin is the point where the coordinates of a point are (0,0). Symbols are defined for the discussion; these are

P is added cost to the customer for hybrid above a CV; the price **P**remium.
M is the added cost to the **M**anufacturer for hybrid above a CV.
S is the annual **S**avings due to cost of gasoline due to better mpg (see Equation 15.3).
F is the pro**F**it/loss due to hybrid conversion.
R_N is the financial **R**eward to customer for hybrid compared to CV over N years.
N is the **N**umber of ownership years; integer values.
Y is the **Y**ears to recoup price premium based on gasoline savings.

The letter with bold font suggests what each symbol means.

The net financial reward to the customer is for N years

$$R_N = NS - P \tag{15.9}$$

Negative R_N means a penalty. If R_N is negative, at some value of N, which may not be an integer, the value for R_N switches from negative to positive. At that point R_N is zero and becomes R_Y.

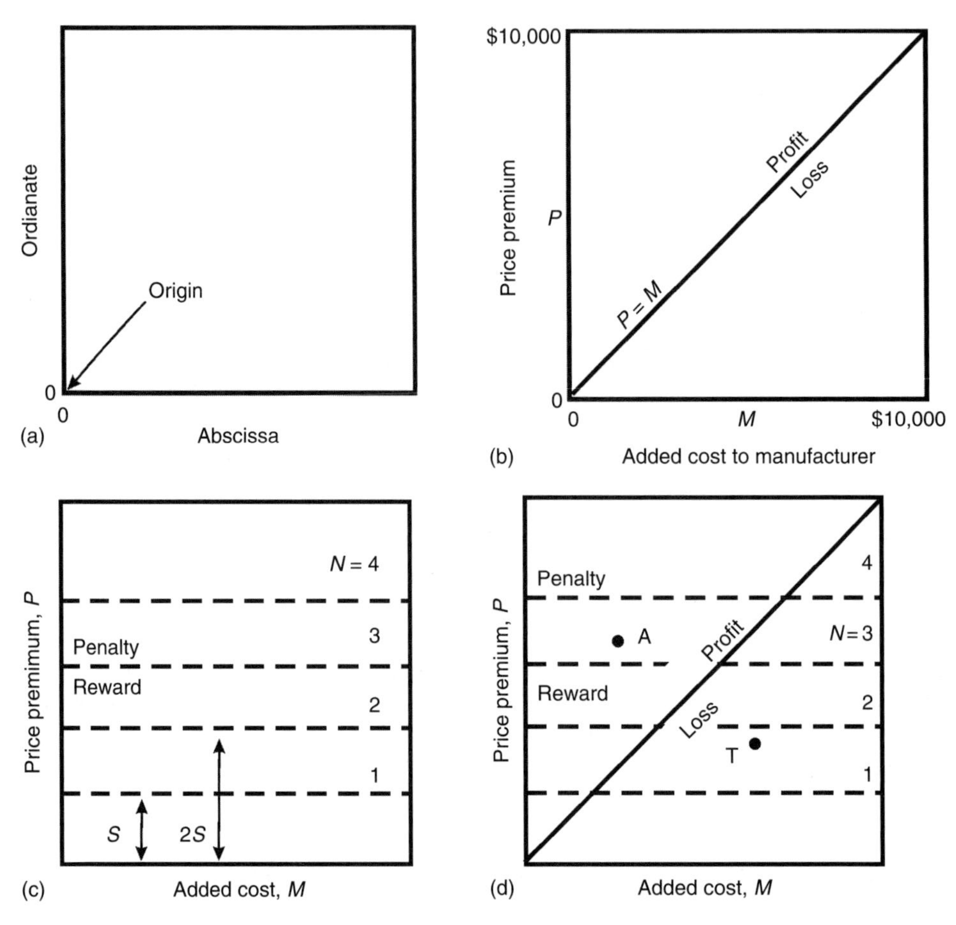

FIGURE 15.3 Graph incorporating profit/loss for the manufacturer and reward/penalty for the customer. (a) Nomenclature for graphs (all graphs throughout the book). (b) Profit/loss line for manufacturer. (c) Reward/penalty line for customer. (d) Points for current hybrids and an introductory hybrid.

$$R_Y = YS - P = 0 \tag{15.10}$$

Table 15.4 has data on the years, Y, to recoup price premium with and without a $2000 tax credit. Equations 15.6 and 15.7 provide values for Y and Y_T.

In Figure 15.3b the ordinate is price premium, P, and the abscissa is the added manufacturing cost, M. Numerical values of $10,000 have been arbitrarily assigned to graph. The zero profit line occurs where $P = M$ and is a line at 45° in the graph. Above the zero profit line is a region of profit. Below the zero profit line is a region of loss.

In Figure 15.3c the attention shifts to the customer. The horizontal dashed lines are at the value of annual gasoline savings, S. Table 15.3 gives the annual fuel costs, C, for a variety of vehicles. Table 15.4 has the annual fuel savings, S. The horizontal dashed line at $2S$ represents the fuel savings for $N = 2$ years, and $3S$ represents the savings for $N = 3$ years, etc. For a given year of ownership, that is, value for N, the region above the dashed line represents the financial penalty to hybrid owner. Financial rewards to the hybrid owner fall in the region below the dashed line.

In Figure 15.3d two points appear. Point A is typical of hybrids for sale in 2007. The point is located in the customer penalty region above the line for $N = 3$. From the graph, the value for Y is 3.4 years. Point T is typical of the loss a manufacturer may be willing to absorb to launch a new hybrid or to gain market share. An example is the Japan-only-market Prius, which reportedly cost $41,000 to produce and sold for $20,000. Net loss was $21,000 (see Ref. [1] of Chapter 1). As a commentary on the loss of $21,000, suppose the manufacturer had a budget of $10,000,000 with a choice of advertising or a sweetheart deal on each car. The manufacturer could sell approximately 5,000 vehicles using the $10 million. The amount of buzz generated, the solid engineering data obtained, and the impact on future sales was considered to be greater than the benefits of advertising. Thus the first Prius hybrids were sold at a loss.

In Chapter 11, H is used to define micro, mild, and full hybrids. The domain of the plug-in hybrid is defined by a range of values of H. Morphing of series hybrids, which is done by varying H, leads to mixed hybrids. Figure 15.4 shows the effect of mild and full hybrid designs on profit/loss for the manufacturer and reward/penalty for the customer. At the origin, $H = 0$, and the vehicle is a CV. Compare Figure 15.4a and b. The annual savings in fuel costs have grown to $750/year due to benefits of a full hybrid. For the full hybrid, R_1 through R_5 are all negative. Finally at $N = 6$, R_6 becomes positive, and the customer is now rewarded. From Figure 15.4a, $Y = 5.3$ years to recoup the price premium.

Shift attention to Figure 15.4b, which is for a mild hybrid. Figure 15.4b has a smaller price premium, P, than for a full hybrid. Of course, the mild hybrid yields smaller annual fuel savings, S. For both the mild and full hybrids, the price premium exceeds fuel savings, and the reward, R_1, is negative. R_1 means the first year with $N = 1$. In fact R_2, R_3, and R_4 are all negative, which means the customer is paying a penalty. Finally at $N = 5$, R_5 becomes positive, and the customer is now rewarded. From Figure15.4b, $Y = 4.4$ years to recoup the price premium based on gasoline savings.

One might have expected Y for a full hybrid to be fewer years than Y for a mild hybrid. Even though the annual fuel savings is greater for the full hybrid compared to the mild hybrid ($750 > $450), the annual fuel savings did not grow as fast as the price premium. This fact stretched out the years, Y. In an earlier section, "Marginal Savings: One More Mile per Gallon," the point was made that the cost to gain one more mpg increases as vehicle mpg grows. The mild and full hybrid examples show this effect.

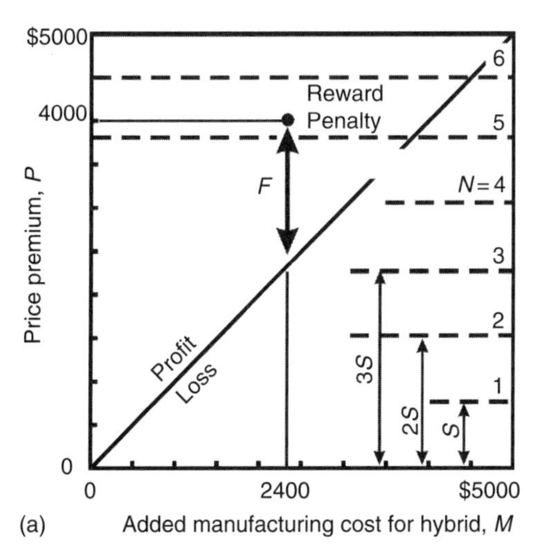

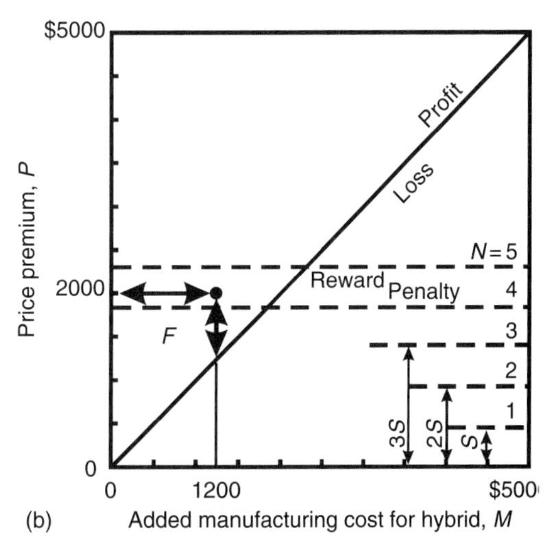

FIGURE 15.4 Effect of mild and full hybrid designs on profit/loss for the manufacturer and reward/penalty for the customer. (a) Point for a full hybrid in the P, M coordinates. Numerical values for the graph are S = $750/year, P = $4000, and M = 0.6, P = $2400. Profit, F, is $1600. (b) Point for a mild hybrid in the P, M coordinates. Numerical values for the graph are S = $450/year, P = $2000, and M = 0.6, P = $1200. Profit, F, is $800.

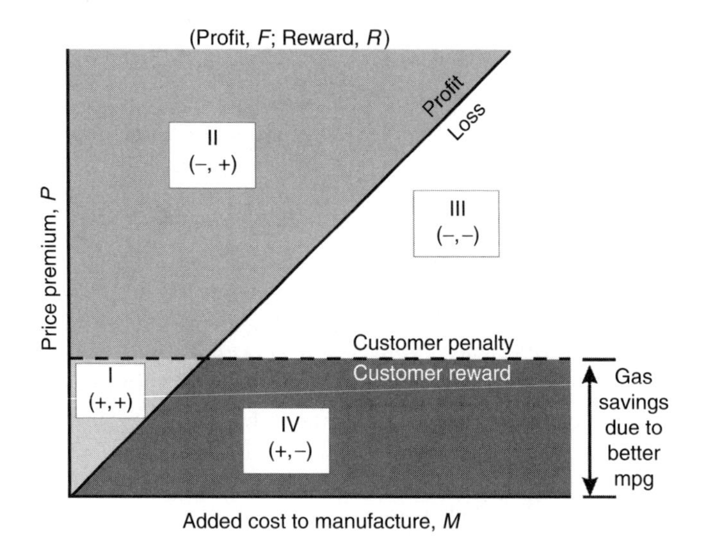

FIGURE 15.5 Four regions showing the influence of price premium and added manufacturing costs on the profit/loss for the manufacturer and reward/penalty for the customer.

As shown in Figure 15.5, the zero profit line and the customer reward/penalty line divide the graph of P versus M into four regions. Recall that the symbol F means profit, and symbol R means reward. In region I where both F and R are positive, denoted in a shorthand as $(F, R) = (+, +)$, everyone wins. The manufacturer is making a profit and the customer is saving money on his/her purchase of a new hybrid. In region II, which is $(F, R) = (-, +)$, the manufacturer's hybrid is profitable; however, the customer is not quickly recovering the price premium. The hybrid is selling on features other than savings due to better mpg. In the long term, this is an untenable position if the desire is to move the hybrid from a niche market to the mainstream (see Chapter 18).

In region III, $(F, R) = (-, -)$, everyone loses. What more can be said? In region IV, $(F, R) = (+, -)$, the customer gains but the automaker is losing. Once again, region IV is unstable and changes must be made.

As shown in Figure 15.6, decreasing the price premium moves point A to A′ and point B to B′. Moving from point A to A′ puts the pricing in favorable region I. However, the profit is very meager. In spite of moving from point B to B′, the price reduction does not move B′ into favorable region I. Reducing price fails to solve the problem. Some aspects of the hybrid must be redesigned.

Profit for hybrid A extends from a → A before the reduction in price premium. After reduction, profit has shrunk to a → A′. Both before and after profits are positive. Before reduction of price premium, the profit for hybrid B was positive, b → B. After reduction of price premium, the profit becomes negative, b → B′.

At point B in Figure 15.7, the customer is paying a price premium while the manufacturer is earning profit. By decreasing content in the hybrid by an amount ΔM, point B is shifted to point B′. The hybridness, H, has been decreased. As a result of the lowered content and smaller H, the annual gasoline savings line falls by an amount ΔS thereby shrinking region I. Reduction of the price premium by an amount ΔP, lowers point B′ from region II to point B″ in region I. The profit has not changed significantly by the reduction of content. However, the amount the customer pays for the price premium is less than the savings, S.

Compare the relative size of ΔM and ΔS. A large reduction in manufacturing cost has been made with minimal loss of fuel savings. It is apparent that the strategy of reducing content is viable only if ΔM can be much larger than lost fuel savings, ΔS.

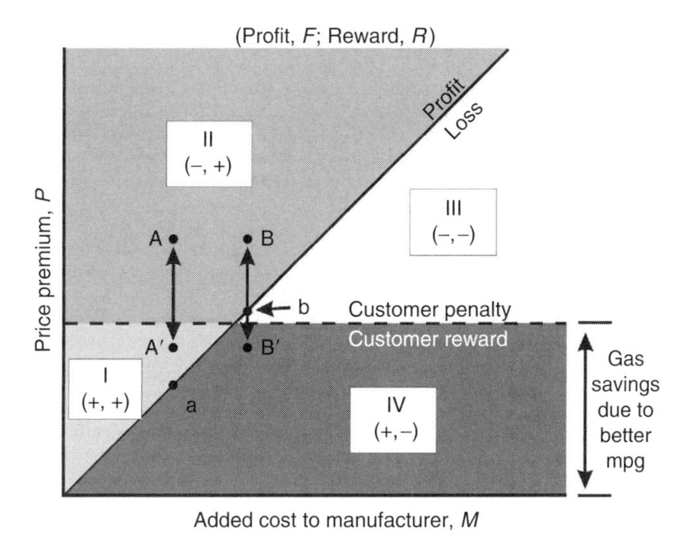

FIGURE 15.6 Effect of decreasing price premium for hybrids in region II with goal of a more marketable vehicle. As indicated above the figure, in parenthesis are Profit, F and Reward, R.

Point B in Figure 15.8 is at the same location as point B in Figure 15.7. Increasing content by an amount ΔM moves point B from region II to point B′ in region III. At point B′, the profit is negative. An increase in price premium, ΔP, restores profit. Because of increased ΔM and increased H, the fuel economy is increased. The increase in the annual savings for gasoline is ΔS as shown in the Figure 15.8. Because of the increased annual fuel savings, the ceiling for region I is moved upward. Note that ΔS is much larger than added manufacturing cost, ΔM.

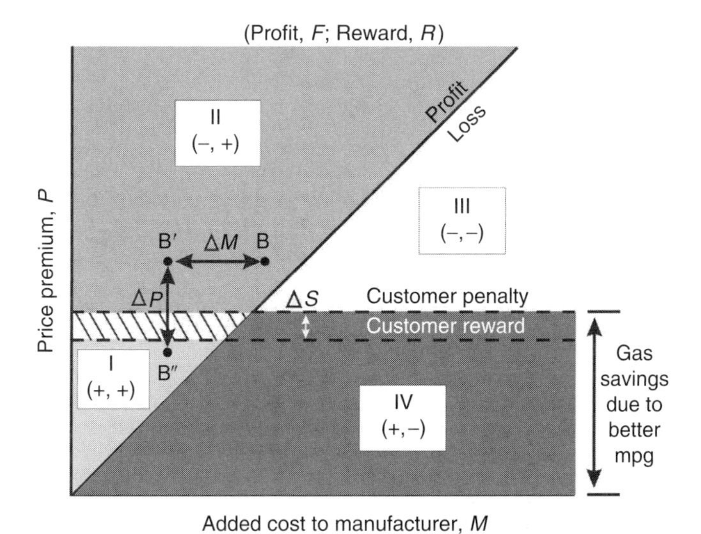

FIGURE 15.7 *Reducing* content, including a reduction of hybridness, H, to obtain a more marketable vehicle. As indicated above the figure, in parenthesis are Profit, F and Reward, R.

SUMMARY

Price premium is introduced early because of its major impact on the economics of hybrid ownership. The reasons for the price premium, P, are briefly explained. Table 15.1 gives the price premium for several hybrids; values for P range from $1,150 to $11,110.

Pay at the pump is intended to emphasize the expense of operating a car and comparing that expense with "high cost of gasoline." The cost to drive a car is about $1.00/mi. If driven for 300 mi, the actual total operating cost is $300. If the $300 were paid at each fill-up, a new perspective on gasoline cost would prevail. The gasoline cost for the 300 mi is $40–$50. The percentage for gasoline is only 13%–17% of overall costs.

Equations are derived for both the annual fuel costs, C, and annual fuel savings, S. For calculation of savings, S, the hybrid must be compared with some other vehicle. Usually, the comparison is with a CV of the same class. Units for C and S are $/year.

Marginal savings, $\Delta S/\Delta F_E$, decrease significantly as F_E increases. An equation for marginal savings is derived. This equation shows that at 10 mpg, the marginal savings is $750/(year)(mpg). At 60 mpg, the amount has fallen to a mere $20.83/(year)(mpg). Marginal savings are available to be applied to hardware costs to further increase mpg. A lot of hardware can be bought with $750/(year)(mpg); $20.83/(year)(mpg) does not buy much new hardware.

Tax credits and deductions for hybrid cars, both federal and state, are intended to encourage production and sales of hybrid vehicles, and in the process decrease emissions and dependence on oil imports. Tax credits are more valuable than tax deductions.

The years to recoup the price premium are determined by simply dividing the premium, P, by the annual fuel savings, S. This was done analytically (Equations 15.3 through 15.7) and later in the chapter, graphically. The graphical technique appears in Figures 15.3 through 15.8. The graphs relate manufacturer profit/loss with customer reward/penalty. Figure 15.4 studies degree of hybridness, H. Figure 15.6 examines a simple reduction of price premium. Figure 15.7 compares a "reduction" of content with an "addition" of content in Figure 15.8. The recoup time, Y, varies from 4.7 to 8 years. With a tax credit, the years to recoup, Y_T, are considerably less.

In addition to the price premium, numerous other costs accrue due solely to the hybrid. These include added sales tax, finance charges, and cost of capital. Mechanics that service a hybrid need more training, which increases hourly rate. Maintenance is higher. Collision repairs are higher.

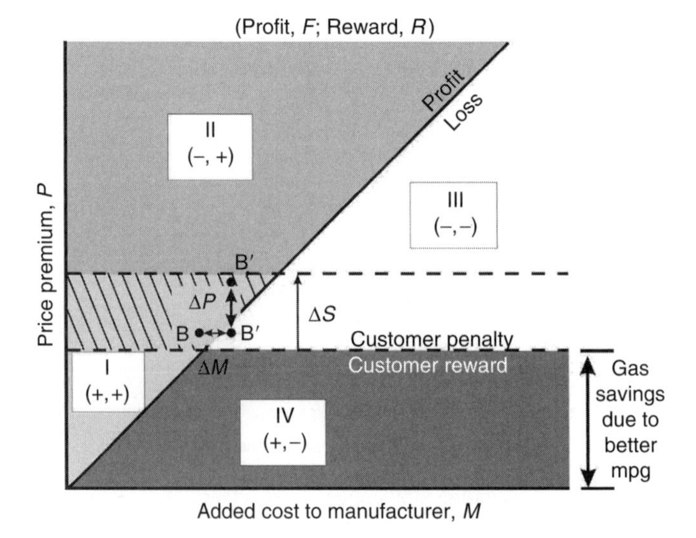

FIGURE 15.8 *Adding* content, including increased hybridness, H, to obtain more a marketable vehicle.

A focus on the narrow viewpoint connecting insurance premiums with the costs of repair and damage yields higher insurance premiums. However, a much broader viewpoint may motivate insurance companies to offer insurance discounts for hybrid vehicles. This motivation may be stimulated by global warming and hurricanes.

Depreciation depends on the pace of technology. This fact can be understood in terms of the years of ownership, Y_O, and the pace expressed in terms of the years, Y_P, between hybrid models. The suggested value for Y_P is 5 years. If the manufacturers are motivated (as is the case for hybrids), Y_P is near 5 years. If motivation is low due to slight market pressure, Y_P exceeds 5 years.

After 3 years, the battery for a Prius II is almost one-third the depreciated value. With an okay battery, depreciation is 42%. If a new battery is needed, depreciation balloons to a huge 58%. Hence, battery replacement looms large as a potential financial pitfall, but fortunately has not occurred in practice. The battery on the Tesla electric (not hybrid) sports car is Li-ion and costs $20,000.

APPENDIX 15.1: CLASSIFIED ADVERTISEMENT FOR TOYOTA PRIUS II

Prior discussion of the Prius II depreciation appears with Figure 15.2. The classified advertisement provides one data point for a used Prius II.

The data from the section "Costs: Amount of Depreciation for Various Vehicles," along with the information from the advertisement of Figure A.15.1, are used to create Table A.15.1. Using these data and Equation 19.1 yields the estimated value for the characteristic depreciation time, τ. The typical CV has a value for τ between 3 and 5 years. The larger the value for τ, the lesser the depreciation. The values of τ for the three examples are 6.6, 8.5, and a spectacular 14 years. The wide variation was affected by such factors as resale value, which was the asking price in two cases. Even with the wide variation, the depreciation of hybrids looks favorable as of the year 2008.

TOYOTA '01 Prius
exclt. cond. green,
new tires/battery
90 Kmi Maintenance
records $11,250
obo call
831-624-WXYZ

FIGURE A.15.1 Classified advertisement for Prius II (October 2007).

TABLE A.15.1
Estimation of the Characteristic Depreciation Time, τ, for Three Different Prius Vehicles

Car	Sticker Price V_0 ($)	Resale Value V ($)	Depreciation Period, t, Years	Characteristic Depreciation, τ	Conventional Vehicle Resale ($)
Prius I	19,995	14,000	5	14.0	7,400
Prius II	22,949	20,400	1	8.5	18,800
Prius II	28,000[a]	11,250	6	6.6	8,400

[a] The value for τ was 5 years.

The far right-hand column gives the resale value of a CV assuming the same depreciation period and sticker price. The value for τ was 5 years. Compared to a hybrid, the CV returns significantly less at trade-in time.

REFERENCES

1. D. Welch, What makes a hybrid hot, *Business Week*, November 14, 2005, p. 41.
2. Editorial Staff, The dollars and sense of hybrids, *Consumer Reports*, April 2006, p. 21.
3. K. Reynolds, The future is now, *Motor Trend*, November 2007, pp. 86–95.

16 New Technologies: Hybrids

Although new technology is expensive, the lack of advanced technology changes a front-runner to a has-been.

OVERVIEW

Suppose you were a high-level government official with a few billion dollars to solve the nation's energy problems. Congress, at least for now, likes your program. So, you can focus on solving energy problems. The first question is who uses a major portion of the petroleum? Transportation—airplanes, trains, buses, trucks, and cars (called light duty vehicles (LDV) in Washington D.C.)—uses a major fraction; so this is a place to concentrate some resources. Ultimately, you filter down the roster of users until you come to trucks and cars.

Trucks, thinking of 18-wheelers as well as smaller delivery trucks, are almost exclusively powered by diesel engines. (Fuel economy of 18-wheelers: many of trucks on road today, 4.5 to 5.0 mpg; new trucks available for purchase today, 5.5 to 6.5 mpg; and future trucks, 9.0 to 10.0 mpg.) So, what can be done to improve efficiency of diesel engines? This is a topic in Chapter 17. Looking at how the giant 18-wheelers operate when the driver is sleeping in his apartment suite at the back of the cab, the engine may be idled all night long. This is a big waste of diesel fuel. Also, at truck stops the driver may relax, have a steak, and shower while idling the diesel for 2 h or more. A fuel cell auxiliary power supply could save barrels of diesel fuel. Can hybrid technology be applied to heavy trucks? Alternatively, truck stops could offer plug-in electrical power. According to *Fortune* [1], a nationwide system is being built to offer truckers plug-in power and other amenities while at a truck stop.

For specialty trucks, such as door-to-door delivery or trash trucks, diesel electric hybrids offer major fuel savings. With technology gains across the spectrum, fuel consumption for fleets of specialty trucks may be cut in half. What components at the 270 hp (200 kW) level offer the best gain in miles per gallon (mpg)?

Continuing onward, the personal transportation vehicle, or LDV, comes under scrutiny. In your role as a high-level government official, you inherited the Partnership for a New Generation of Vehicles, also known as Supercar (see Chapter 1). Considerable technology already exists from Supercar days. Hybrid technology was emphasized for Supercar. What new thrust should be defined to use Supercar data and to advance capability for new hybrids?

The goals are to decrease cost and to increase mpg. The government can cooperate with automobile manufacturers to develop and refine hybrid technology. The government cannot put the technology into production cars on the assembly line. Only the automobile manufacturers can do that. Everyone shares the goals of less cost and more mpg, but government role tapers off at some point.

Increasing mpg cannot be done in isolation looking at only hybrid propulsion technology. One needs the whole vehicle approach. Chapters 16 and 17 focus on hybrid propulsion technology and take an even narrower view of components. The whole vehicle approach must include direct and indirect input technologies. Technologies directly affecting mpg include

- Aerodynamics
- Tires
- Weight
- Engine efficiency

- Powertrain efficiency
- Integration and control

Technologies indirectly affecting mpg include, for example, advanced materials. A better alloy can reduce weight of an engine or M/G. A new semiconductor material may reduce cost, improve efficiency, and decrease weight of a variable frequency inverter. A new semiconductor material may eliminate the need for liquid cooling of the power electronics. A better synthetic rubber with less hysteresis can reduce the rolling friction of tires.

FACTORS AFFECTING NEW TECHNOLOGY

Factors Affecting New Technology: Overall Items

What can be done with existing, known technology? Before investing in R&D, existing technology needs to be identified and screened. Finding a nugget among the vast store of knowledge saves time and money. If a nugget cannot be found, then the deficiencies of existing technology can, in the process, be identified. These deficiencies can form the basis for a well oriented R&D program. Innovation plays a role. Combining existing individual technologies in novel ways may solve a problem.

Many factors influence new technology and the move from the laboratory to the production line. The market and commercialization affect R&D programs and transition out of the laboratory. Government programs and leadership have a pronounced influence. For example, emission requirements are established by governments and have a large impact. Different segments of petroleum users, for example, trucks and cars, have a different bearing on new technology. Very large trucks likely will remain diesel powered for a decade or two whereas passenger cars likely will have large shifts in technology.

Any program to create new technology needs to recognize the effect of possible R&D dead ends. An example is the fuel cell. Perhaps fuel cells may have a serious flaw that is not realized at the moment. Then what? Alternate or replacement technologies would be needed. The effect of possible R&D breakthroughs creates the enthusiasm for new programs.

Elusive Fuel Cell

In the 1950s, the H-bomb was exploded a few short years after the A-bombs were developed. (H is hydrogen, and A is atomic.) The very high pressure and temperature created by fission ignited the fusion reactions. This demonstrated that man could duplicate on earth the nuclear fusion found deep in the core of the sun. These were heady times. However, the release of energy is uncontrolled in the H-bomb. Could controlled fusion be developed? Based on the rapid success with the H-bomb, controlled fusion seemed just around the corner. Some 60 years and several billion dollars later, controlled fusion remains an elusive goal.

By analogy, is the commercially viable fuel cell for hybrids equally elusive? Certainly fuel cells have been used very successfully on Apollo to the moon and the space shuttle. For these applications of intense, cost-is-no-object, national interest, a $1 million fuel cell can assuredly fly (pun intended). However, for a hybrid, cost is the whole crux of the matter, although some may argue reliability is the bottom line. Department of Energy (DOE) states on their Web site [10] that the fuel cell stack cost to be competitive with conventional technology for automotive applications must be $30/kW or less.

Impact of Emission Requirements: ZEV Program

Government requirements have a pronounced influence. The zero emissions vehicle (ZEV), which was not successful in achieving stated goals, is an example. The ZEV emphasis was on the pure electric vehicle (EV). At the time the public was not ready to accept the limitations of the EV. The focus is now on the partial zero emissions vehicle (PZEV). The hybrid electric vehicle (HEV) qualifies as a PZEV. Cleaner gasoline and diesel engines may be acceptable in regard to emissions. In the future, the fuel cell may provide the technology for ZEV.

Infrastructure: Charging EV and Plug-In HEV

For an expanded fleet of EV and plug-in HEV (PHEV), charging the battery becomes a major magnitude problem. If the fuel cell programs were to hit a dead end, however remote that may be, EV and PHEV become very important. An infrastructure to charge batteries is much different than an infrastructure to provide hydrogen.

Battery charging must accommodate a variety of needs. A few different types of batteries may be in service. Each requires different voltages and charge rates. The problem of really knowing the state of charge (SOC) of the battery must be addressed.

Depending on the type of battery, some fast charge, 1 h, capability is needed. Some trickle charge, 8 h, capability is also needed. Home charging in the garage will be popular. Hovering over all the other factors is the question of safety is of paramount importance.

For HEV, battery charging is accomplished while driving. For EV and PHEV, the charging time and charger power are key factors. The energy used per kilometer (or mile) for the GM EV-1 was 160 W h/km (262 W h/mi). For a 400 km range, the battery energy required is (400 km)(160 W h/km) = 64 kWh. The battery charging time depends on the amount of energy. Due to the battery efficiency (putting energy into battery and taking it out) of 80%, 64 kWh becomes 64 kWh/0.80 = 80 kWh for charging energy. Data for Table 16.1 was calculated using

$$\text{(Charging time)} = \text{(Battery energy)/(charger power)}$$

The current from the charger was calculated using

$$\text{(Current)} = \text{(Charging power)/(voltage)}$$

which can be verified using sample numbers from Table 16.1.

Some novel ideas have been stated for solving the problem of a "dead" battery. One suggested approach is to remove the dead battery and install a fully charged battery. This is done with flashlights. Trade batteries, jack up the EV, and install a charged battery. Some questions are: Is this a rental battery? Who owns the battery? Some problems are the heavy battery and the complicated exchange. Due to high voltage, safety is an issue. Finally, obtaining reliable connections every time must be demonstrated.

Another novel suggestion is to have a small trailer to be towed behind the vehicle. The trailer has a battery and may have a small engine to help charge the battery. For a trip that is longer than the normal range of the vehicle, the trailer would be towed. This suggestion is applicable to EV, PHEV, and HEV.

TABLE 16.1
Charging Current at 220 V for a Battery Similar to That in the GM EV-1

Charging Time (h)	Current (A)	Charging Power (kW)
1	364	80
2	182	40
4	91	20
8	45	10

Infrastructure: Hydrogen

The question of the hydrogen infrastructure is discussed in more detail in Chapter 22. An important report is that from the National Academy of Engineering [2].

According to DOE, the transition to hydrogen will take 40–50 years to unfold. Both hydrogen-fueled internal combustion engine (ICE) and fuel cells are affected by the speed of development and breadth of the infrastructure.

One step toward a hydrogen infrastructure is fleet sale or lease of fuel cell vehicle (FCV) and hydrogen ICE. A large fleet operating in a small area can afford to build a hydrogen plant and service station for the fleet. This is in contrast for hydrogen for the public; hydrogen service stations will need to be dispersed and conveniently located.

Impact of Emissions Requirements: Internal Combustion Engine H_2 Powered

Two reasons exist to consider an ICE powered by H_2. First, zero CO_2 is a major advantage relative to global warming. Second, the ICE H_2 is a step along the way to have the necessary H_2 for fuel cells. Motivation to develop the hydrogen infrastructure is increased since a realistic near-term application awaits. Pure H_2 may not be absolutely essential for the fuel cell. Means to reformulate hydrocarbons may find use in fuel cells (see Chapter 6).

Another method for creating demand for hydrogen is to design hybrids that use hydrogen. The FCV comes immediately to mind but the idea here is anything but a fuel cell. An example is reported in Ref. [14]. A Prius was changed from a gasoline ICE/electric hybrid to a hydrogen ICE/electric hybrid. Hydrogen replaced gasoline as fuel. Emissions immediately dropped dramatically. A 60 L storage tank holds 3 kg of H_2 giving a range of 140–150 mi (224–240 km).

Relative Importance of ICE and Fuel Cell: Creation of Hydrogen Infrastructure

Hydrogen can be used by both ICEs and fuel cells. Further, the ICE-powered cars can be dual fuel and use either gasoline or hydrogen. The second column in Table 16.2 gives the percentage of ICE vehicles compared to FCVs. In the year 2010, 95% of the hydrogen users will be ICE with only 5% FCV. By the year 2040, FCV will be more common at 70% with 30% ICE. After gasoline becomes prohibitively expensive, sometime near 2040, the ICE will likely continue as an alternate to FCV.

The third column in Table 16.2 applies only to ICE dual-fuel vehicles. Being dual fuel, the owners have the option of filling up with gasoline or hydrogen. Of course, if hydrogen is not available, then the fill-up will be with gasoline. In 2010, hydrogen will not be universally available as the infrastructure expands. For this reason, in 2010, the dual-fuel ICE vehicle will use 95% gasoline and 5% hydrogen. Later, with more plentiful sources for hydrogen, the percentages shift so that in 2040 gasoline has vanished as a fuel.

TABLE 16.2
Relative Importance of ICE and Fuel Cell during Creation of the Hydrogen Infrastructure

Year	Internal Combustion (%)/ Fuel Cell (%)	Dual-Fuel ICE Gasoline (%)/ Hydrogen (%)
2010	95/5	95/5
2020	75/25	65/35
2030	50/50	40/60
2040	30/70	0/100

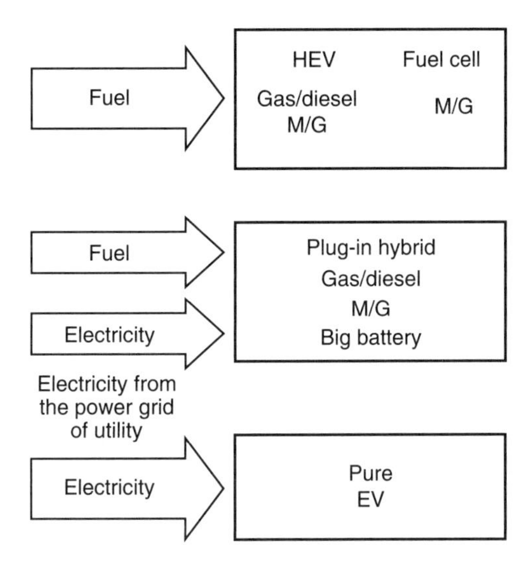

FIGURE 16.1 To be capable of moving down the road, the HEV and EV must have energy supplied from outside.

Three General Vehicles and Associated Infrastructure

Figure 16.1 shows three different vehicles: two HEV and one EV. The box representing a vehicle is blank except for the name; this is to focus attention on supply of energy. How energy is delivered to the vehicle is different in each case. Each vehicle has different infrastructure requirements. The HEV already has the gasoline refineries, pipelines, and service stations in place.

Some designs for the fuel cell need hydrogen production plants, storage, distribution, etc. In other words, these designs require a whole new infrastructure. The plug-in hybrid requires both a battery charging infrastructure and a gasoline infrastructure. The EV needs only charging infrastructure, which is partially in place with home charging in the garage. Each infrastructure (gasoline, hydrogen for fuel cell, and power grid charging) offers a different level of global emission. Analysis of the options should include the effect on global emissions. The power grid for EV and PHEV allows use of a broader range of energy sources such as nuclear, solar, wind, waves, and similar energy. The EV moves generation of electricity out of the vehicle and upstream to utilities. This statement is partially true for PHEV.

Generalized Hybrid

The generalized hybrid has features common to all hybrids (Figure 16.2). Fuel and energy storage are on the left-hand side of the diagram. Fuel is converted to useable energy, which feeds into the powertrain to supply torque and power at the wheels. Table 16.3 summarizes the conversion possibilities.

Energy storage implies a flow of energy into and out of storage. Different storage devices are discussed below.

Shared Components: Fuel Cell and HEV

Fortunately, many technologies for HEV are also relevant to the FCV. Looking at the O–X or X–O combinations in Table 16.4, certain technologies are unique to one or the other vehicle. Combustion is unique to HEV while electrochemistry of fuel cells is unique to, obviously, FCV. HEV are unlikely to use hydrogen as fuel.

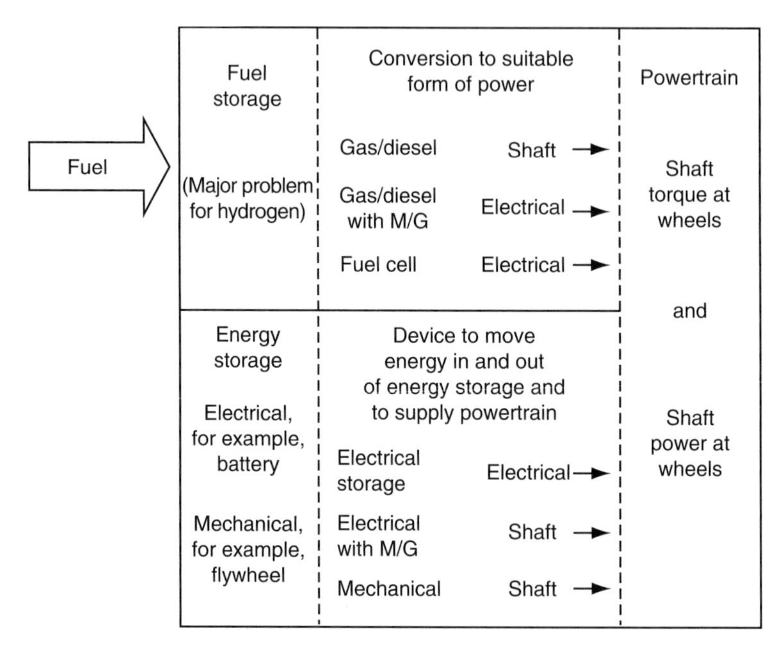

FIGURE 16.2 A generalized look inside a hybrid vehicle. This look provides an indication as to promising technology to pursue. Also, commonality between HEV and fuel cell is evident.

TABLE 16.3
Conversion of Fuel for the Generalized Hybrid

Fuel	Conversion	Output
Gasoline/diesel	ICE	Shaft power
Alternate fuel	ICE	Shaft power
Hydrogen	ICE	Shaft power
Hydrogen	Fuel cell	Electrical power

TABLE 16.4
Commonality between HEV and Fuel Cell Technology

Technology	Hybrid	Fuel Cell
Fuel storage (H_2)	O	X
Electric M/G	X	X
Improved combustion	X	O
Improved electrochemistry	O	X
Battery	X	X
Ultracapacitor	X	X
Lightweight materials	X	X
Power electronics	X	X
Integration and control	X	X

BIGGEST GAINS

What aspects of hybrids will yield the greatest gains? Some insight can be gained from Figure 5.5, which shows the various efficiencies for a hybrid. From Figure 5.5,

$$\eta = \eta_E \, \eta_G \, \eta_B \, \eta_M \, \eta_P$$

Use logarithmic differentiation, which is a mathematical technique for finding fractional quantities. Subsequent discussion will illustrate fractional quantities. The result of logarithmic differentiation is

$$\frac{\Delta \text{mpg}}{\text{mpg}} = \frac{\Delta \eta_E}{\eta_E} + \frac{\Delta \eta_G}{\eta_G} + \frac{\Delta \eta_B}{\eta_B} + \frac{\Delta \eta_M}{\eta_M} + \frac{\Delta \eta_P}{\eta_P} \tag{16.1}$$

From Figure 5.5, values for the efficiencies are engine efficiency, $\eta_E = 0.24$; generator efficiency, $\eta_G = 0.94$; battery efficiency, $\eta_B = 0.60$; motor efficiency, $\eta_M = 0.95$; and powertrain efficiency, $\eta_P = 0.92$. Assume a 1% change in engine efficiency, η_E; what is the fractional change in mpg? Inserting values in the equation above,

$$\frac{\Delta \text{mpg}}{\text{mpg}} = \frac{0.01}{0.24} = 4.2\%$$

Once again, assume a 1% change in generator efficiency, η_G; what is the fractional change in mpg? Inserting values in the equation above,

$$\frac{\Delta \text{mpg}}{\text{mpg}} = \frac{0.01}{0.94} = 1.1\%$$

The conclusion is that if a component has small efficiency, for example, $\eta_E = 0.24$, a small improvement yields a large gain in mpg. Starting with a larger efficiency, the improvement in mpg is considerably less.

The same analysis applies to the effects of poor hybrid control. Good control has the component operating near its peak efficiency (see Chapter 8). If the engine is operating 1% off peak efficiency, mpg suffers by 4.2%. Look for components with low efficiency; these yield the greatest improvement in hybrid performance.

BIGGEST GAINS: SAVING GASOLINE

What class of vehicles is the best place to look for saving gasoline; is it the gas-hogs or the fuel-sippers? The answer is gas-hogs. The annual saving of gasoline in gallons, S, is given by

$$S = \frac{M}{G_C} \left(\frac{f}{1+f} \right) \tag{16.2}$$

This equation is derived in Appendix 16.1. The symbols used in Equations 16.2 and 16.3 are M is the miles per year driven; f is the fractional gain in mpg due to hybrid; G_C is the mpg of the conventional vehicle; and G_H is the mpg of the hybrid. For a sample calculation, the same value for $M = 12{,}000$ mi/year is used. Values for the sport utility vehicle (SUV) and pickup class (the very definition of gas-hog) are $G_C = 18$ mpg and $f = 20\%$. For the gas-hogs, $G_H = 21.6$ mpg. The gasoline saved annually is 111 gal. Values for the small fuel-sipper vehicle are $G_C = 36$ mpg and the same $f = 20\%$. For a small fuel-sipper, $G_H = 43.2$ mpg. The gasoline saved annually is 55.5 gal, which is one-half the saving by the gas-hogs. From Equation 16.2 assuming f is constant

$$\frac{S_2}{S_1} = \frac{G_{C1}}{G_{C2}} \tag{16.3}$$

Use $G_{C1} = 36$ mpg and $G_{C2} = 18$ mpg. Then the gas saved by a gas-hog equals twice the gas saved by the fuel-sipper.

MINING THE OBVIOUS: ENERGY STORAGE

BATTERIES

Chapter 6 discussed options and selection of batteries. Chapter 14 focused on the M/G–battery match from the point of view of the M/G. Here some comments are made from the point of view of the battery.

A battery is composed of individual cells. Each cell has a voltage depending on the electrochemical potential of the chemicals. NiMH has cell potential of 1.2 V. Lead acid has a cell potential of 2 V. To increase voltage, cells are placed in series end-to-end. A 12 V lead acid battery has six cells in series.

Batteries have three attributes of interest to the hybrid designer: maximum current, maximum power, and stored energy. A paragraph or two provides a mental picture concerning battery design and maximum current, maximum power, and stored energy. With regard to battery energy, consider the chemical equation

$$AB + C \xleftrightarrow[\text{Discharge}]{\text{Charge}} A + BC$$

Charging the battery creates large amounts of the chemical BC. Discharge generates the chemical AB. Battery energy is stored in BC. A large mass of BC, when SOC is 100%, means large stored energy. Want more energy? Increase the mass of chemical BC.

With regard to maximum current, the chemical reaction occurs on an electrode of the battery. The larger the electrode area, the greater the amount of current produced by the battery. Only the electrode area at the same voltage is included in the count. Want more current? Increase the electrode area of the battery.

Power of the battery is given by the equation

$$(\text{Battery power}) = (\text{Battery current})(\text{battery voltage})$$

For this equation, the units are

$$(\text{Watts}) = (\text{Amperes})(\text{volts})$$

The battery designer has two choices to increase power. First, the electrode area can be increased to provide more current. Second, the voltage can be increased by putting more cells in series.

What is the next battery likely to be used in HEV? The lithium-ion (Li-ion) battery is on the horizon. The lithium battery comes in two flavors. The Li-ion battery uses a liquid or gel electrolyte. The lithium polymer battery uses a plastic electrolyte. See Chapter 6 for more details.

DOE development goals for batteries suitable for HEV and EV are as follows: a battery with power of 25 kW, which cost $3000 in 1998 will be available with the same power in 2010 for $500.

Ultracapacitor or Supercapacitor

The ultracapacitor and supercapacitor are discussed in Chapter 6. The ultracapacitor and supercapacitor are energy storage devices. Capacitors can be used to supplement or replace batteries.

The ability to absorb high power from regenerative braking makes capacitors an attractive choice. Further, the capacitors do not have chemical reactions on charge and discharge as batteries do. As a result, the capacitors have a much longer life compared to batteries. The capacitors are expensive and have poor energy density, which means considerable volume is needed. However, the capacitors can be stuffed in unused space, for example, in a wheel well. For trucks and buses, the cost is less of a problem because of the desire to have long, trouble-free life.

Flywheel

A flywheel can store an amazing amount of energy. To store the energy, high rotational speeds are necessary. Some designs have considered 100,000 rpm. Tests have been conducted at 42,000 rpm. The rotational energy stored is

$$E = \frac{1}{2} I\omega^2 \tag{16.4}$$

where

I is the moment of inertia of the flywheel, kg m^2

ω is the rotational speed, rad/s

The tests used a titanium flywheel designed to withstand the large centrifugal stresses. The flywheel was magnetically levitated with magnetic bearings. An M/G was located on the same shaft. The flywheel turned in vacuum to eliminate windage, which is aerodynamic drag. The flywheel could be charged at a power up to 90 kW. Energy stored was 2 kWh, which is comparable to a chemical battery. To help isolate the flywheel assembly from vehicle motion, gimbal mounting was used. Due to the high spin rate, gyroscopic forces are significant. In an application, counter-rotating flywheels may be used.

Compared to a chemical battery, flywheels have higher efficiencies. As the tests indicated above, the power absorption can be high, which is good for regenerative braking. Flywheels can perform all the duties for regenerative braking, that is, recuperate, store, and discharge the kinetic energy of vehicle motion. Also, flywheels do not have the safety problems of high voltages. However, if the flywheel disintegrates, shrapnel may pose a threat. Structure to enclose the shrapnel is possible but adds weight and cost. Such containment structures are common for aircraft gas turbines.

Reference [12] describes the *flybrid* which uses a flywheel to store energy. The unit weighs 12 lb (5.4 kg), operates in a vacuum, and has maximum angular speed of 64,500 rpm. The flywheel shaft is connected to a continuous variable transmission (CVT), and angular speeds are reduced to 10,000–12,500 rpm. The *flybrid* may be used in Formula One race cars.

Hydraulic/Pneumatic

The hydraulic/pneumatic energy storage uses pumps to compress gas. High pressures up to 72 MPa (10,500 psi) may be used. Just as for a battery, the words SoC are used to describe the fraction of energy stored. This storage method can be thought of as a mechanical battery. In a 95 kg system, power can be absorbed at 75 kW.

In the same 95 kg system, the energy stored is 0.2 kWh. A 95 kg chemical battery can store 5–6 times more energy. However, when life restrictions are imposed on the chemical battery, the numbers reduce to 1.2–1.5 times more energy.

Some features are better than for a chemical battery. Hydraulic/pneumatic energy storage has higher efficiency of energy storage going in or coming out. The hybrid can use full range $0\% < SOC < 100\%$ from hydraulic/pneumatic storage. Chemical batteries are restricted to useable SOC of about 25%.

As discussed in Ref. [11], two delivery trucks have been developed in a joint project involving EPA, UPS, and Eaton. A pneumatic system replaces the battery for energy storage. The claimed improvement in fuel economy is 70%. Sufficient energy is stored to allow the truck to move several hundred feet with the engine off. This is adequate for some UPS deliveries.

Compressed Air Vehicle

Deletion of the engine from an HEV yields a pure EV. Likewise, deletion of the engine from a hydraulic/pneumatic hybrid creates the compressed air vehicle (CAV). The range of an EV is determined by the energy stored in the battery. The range of a CAV depends on the energy stored in the high pressure air tank. Equation 22.5 can be used to calculate the stored energy. Ranges up to 200 km (120 mi) have been claimed for CAV. Hybrid compressed air/electrical vehicles are possible.

Spiral Springs

Spiral springs, much like those used in the old windup alarm clocks, have been investigated for energy storage. With a mass of 59 kg, the energy stored was small being 30 kJ = 0.0083 kWh. With regard to amount of energy stored, the spiral springs are not competitive to chemical batteries. Maximum torque for the test equipment was 676 N·m. A spiral spring does not spin; the maximum revolutions of the spring shaft was 6.2. Energy density was 0.14 W h/kg. Clever, but complicated, gearing is required to convert 6.2 revolutions into useful shaft speeds and to reverse spring rotation for energy in and out.

Rubber Torsion Springs

Many youth have made rubber-powered model airplanes. Modern flying of model aircraft involves electrical motors using either NiMH or Li-ion batteries. Proposed use of rubber torsion to store energy causes one to think of shades of rubber band-powered model airplanes. The idea of rubber torsion springs for hybrids has been proposed but not developed.

MINING THE OBVIOUS: OTHER HYBRID TECHNOLOGIES

Integration and Control

Overall control of the hybrid systems is as important as, or more so than, any of the other technologies. See Chapter 8 for more details. Both control hardware, for example, computers and microprocessors, and control theory, for example, fuzzy logic control, are becoming mature technologies. However, merging and adapting the hardware and theories to meet the needs of hybrids is in its infancy. Considerable progress is anticipated. Some of the tools for advancing integration and control include simulations, modeling, and hardware in the loop.

Power Electronics

Major breakthroughs are unlikely. Progress will likely be by incremental advances. One path is integration of the electronic control with the overall hybrid propulsion control and the power electronics. The focus is on decreasing costs and improving reliability.

DOE goal for the combined M/G and power electronics is decreased cost. The power electronics and 25 kW M/G that cost $3000 in 1998 will be available for $700 in 2010 for the same power electronics and 25 kW M/G. This assumes success in R&D. Typical performance values for an advanced DC–DC converter are

- Efficiency 96%
- Power density 6 kW/L (8 hp/L)
- Specific power 4 kW/kg (2.4 hp/lb)

Wheel Motors and Electrical M/G

Although not common now, in the near future wheel motors will be appearing in HEVs or EVs. Wheel motors are useful for all-wheel drive and provide savings of space in the vehicle. However, wheel motors typically have large unsprung mass, which affects vehicle dynamics. Space saving benefits accrue due to no transmission, no drive shafts, and no M/G within the body of the vehicle. Various wheel motor designs have been considered including DC brushless, induction, axial magnetic flux geometry, and radial magnetic flux geometry.

Expectations for M/G performance gains consider breakthroughs to be unlikely. Only incremental gains will be in lower costs, less weight, and smaller, that is, less volume. M/G efficiency is already in the mid to high 90%. Big improvements for peak efficiency are not expected. However, in efficiency maps, the region for high M/G efficiency can be greatly expanded and shifted to match required power demands [8].

Energy Conversion

As shown in Figure 16.2, conversion of the fuel energy to either shaft or electrical power is an important aspect of hybrid technology. For more details on the fuel cell, see Chapter 22. For more details and discussion on the future of the ICE, see Chapter 17.

Powertrain

For front-wheel drive (FWD), powertrain components must fit between front wheels. For hybrids, the trend is for integrated design for FWD. Integrated design involves the electric M/G, gear sets, inverter, and controller all in one package. Having one package eliminates high voltage cables and signal (control) wiring exterior to package. This in turn yields high reliability and ease of assembly during manufacture. References [3–7,9] offer more information. Planetary gear sets are popular.

Reference [4] describes a new, improved transmission; see Table 16.5 for the performance gains. The new transmission eliminates internal electrical power circulation that may occur in the Prius

TABLE 16.5
Improvement in mpg with New Transmission Design Compared to Toyota Prius

	mpg		
Driving Cycle	New Transmission	Prius Transmission	Improvement mpg (%)
Urban	51.3	47.0	8.3
Highway	61.7	58.4	5.3
ECE[a]	44.8	43.1	3.7

[a] European driving cycle.

CVT. Lower power demands are made on the M/G yielding higher transmission efficiencies. By means of simulation, the improvements in mpg were determined as reported in Table 16.5. For the urban driving cycle, a significant improvement of 8.3% was achieved.

Savings from New Transmission

Cost plays a dominant role in the introduction of new technology such as the new transmission design. Benefits must be delivered to the customer to justify added cost. The savings in fuel can be readily determined. The annual fuel cost is given by the equation:

$$C_{\mathrm{MPG}} = \frac{(\text{Miles/year})(\$/\text{gallon})}{\text{Miles/gallon}} \tag{16.5}$$

Using the mpg data from Table 16.5 for the urban driving cycle

$$C_{47} = \frac{(12{,}000 \text{ mi/year})(\$3.00/\text{gal})}{47 \text{ mi/gal}} = \$765.96 \tag{16.6}$$

With similar calculation, C_{51} = $701.75. The saving is $64.21 annually. Is that saving enough to pay the incremental costs for a more complex transmission?

Automated Manual Transmission

One newly developed automated manual transmission (AMT) is called Zeroshift as reported in Ref. [11]. The big problem with AMT has been roughness during shift. For the more common automatic transmission, the torque converter absorbs much of the roughness. Zeroshift uses control rings to smooth the operation. AMT, without the usual planetary gear set, with a torque converter offers potential benefits. These benefits include a 2% fuel savings compared to manual transmission, and a 7% fuel savings compared to automatic transmission. Further, AMT offers substantially lower cost and is smaller and lighter than the planetary gear automatic transmission. AMT offers the fuel economy of a manual transmission with the convenience of an automatic transmission.

AMT typically uses double-clutch systems to control roughness of shifting. Both wet and dry clutches are used. Small efficiency gains are possible using dry rather than wet clutches.

Starting the Engine

Starting the engine in a quiet, unobtrusive, way is important for operation at engine-off-at-idle, relative to improved mpg the energy consumed by the starter is very small. Starting energy is typically 1000–2000 J for a four-cylinder engine; this is 0.0003–0.0006 kWh. The energy saved by improved starting is not meaningful. The main drivers for R&D on the use of M/G as starters are lower cost and customer satisfaction.

Honda is developing a start method that does not rely on energy input to an electrical starting motor. Upon stopping, the alternator brakes (stops) the engine so that a selected cylinder stops at mid-stroke. Fuel is injected and ignited thereby causing the engine to run in reverse for a fraction of a revolution. This strange behavior causes compression in other cylinders. Fuel injection and ignition then cause rotation in the correct direction [13].

Solar Cells

Solar cells are frequently suggested; why not use solar cells for hybrid vehicles? The solar irradiance outside the earth's atmosphere is 1 kW/m^2. Putting in the various losses such as atmospheric attenuation, off-normal angle due to latitude, and photovoltaic conversion efficiency, the required area of a solar cell panel is about the same as a double garage door to get 1 kW.

It must be a very clear day at high noon to get 1 kW. Further the solar panel must be clean; dirt on the surface further reduces power. Is your car always clean? With a reasonable area, the power from solar cells is about a factor of 8–10, too small for meaningful hybrid propulsion. However, solar cells have a place. See the solar-powered cabin cooling fan used in the Ford Reflex concept car in Chapter 3.

Assume your hybrid is not at home during the day. Instead of having solar cells installed on the car, put the cells on the roof of the garage or house. During the day, electrical power is generated. What can be done with the daytime-only power? Two choices exit. First, charge batteries at home; these batteries are a second set not installed in the hybrid or EV. Then charge the hybrid batteries at night from the second set of batteries. Not a great idea. Second, sell the daytime power from the solar cells to local utility. Then at night buy the power to recharge the hybrid batteries.

To size the solar cell panel, assume E is the battery energy, 10 kWh. Assume charge time t equals 8 h. (These 8 h must be in the middle of the day.) Power needed from the solar cell panel is 1.25 kW. Assume power out from solar panel is 0.1 kW/m^2. Then area, A, of the panel is

$$A = \frac{1.25\ \text{kW}}{0.1\ \text{kW/m}^2} = 12.5\ \text{m}^2$$

For a square panel, the length of one side is 3.54 m = 11.6 ft. A = 134 sq ft. A double garage door is about 84 sq ft. The panel is 134/84 = 1.6 or roughly two double garage doors.

INNOVATION: SEEKING THE NOVEL

Modular Hybrid Units

Instead of a unique design for each new hybrid vehicle, modular hybrid units are developed. Packages of the hybrid components are assembled for use in a wide range of vehicles. The packages can be adjusted to provide different power levels. From, say, four different transmissions, six different electrical motors, and five different power electronic units, a combination is selected and assembled to meet the hybrid vehicle design needs. As hybrid technology matures, the modular hybrid unit approach becomes feasible.

Generator in Shock Absorbers

Rough roads absorb power. See Ref. [1] in Chapter 10, which discusses how to recover some of that energy. See also "Road Quality" in Chapter 10. As shock absorbers move in and out on rough roads, power can be generated. Onc implementation would have a magnet on the plunger and a coil on the outside of shock absorber. A few kilowatts could be generated using all four wheels.

Generator Driven by Exhaust Turbine

Turbocharged engines use an exhaust gas turbine to drive an air compressor. The new idea is to place an additional item on the same shaft. An exhaust-driven turbine, an air compressor (to compress intake air), and an electrical generator are all on a single shaft. When normally the waste gate would be open, the gate stays closed and the extra torque drives a generator. Up to 10kW can be obtained to recharge the battery. The system is altitude sensitive as you might expect.

Fuel-Cell/Gas-Turbine Hybrid

For stationary electrical power plants, which are used for peaking power (large power, a megawatt or so), the combined cycle fuel cell/gas turbine may yield 60%–70% efficiency. The fuel cell itself offers an efficiency of 40%–50%. The gain in efficiency is due to the added gas turbine. These plants tend to use high temperature fuel cells (not good for hybrid vehicles) so that the gas turbine can exploit the waste heat from the fuel cell. Some work is being done by General Electric and the DOE using the relatively cool proton exchange membrane fuel cells.

Can the combined cycle fuel cell/gas turbine be developed for hybrid vehicles? Certainly the very high efficiency of 60%–70% is a strong inducement to try.

Free Piston Engines

Free piston engines are a form of ICE and appropriately are discussed in Chapter 17. Free piston engines are included here because the engine and motor are a single unit.

Free piston engines do not have a crankshaft. The pistons have linear, back-and-forth motion. Variable compression ratio is easy to obtain. Camless valve trains are used. Lower costs are expected. Free piston engines have higher efficiency than a crankshaft gas engine because of lower friction losses and better combustion control.

At least two different designs have been considered. The first design has direct electricity generation by magnets on the free piston that moves to and fro. The magnets move inside coils and form an induction generator. This design is applicable to HEV. The second design has direct pumping of a hydraulic/pneumatic system. Each end of the free piston geometry has two pistons. One is driven by the combustion of fuel. Another smaller piston compresses fluid. See "Hydraulic/Pneumatic" under "Energy Storage" above.

Transmotor: Rotating Rotor and Rotating Stator

"Trans" comes from transmission and "motor" obviously comes from motor. Hence, the name transmotor. Both the rotor and the stator are supported by the common shaft, and both can turn. This arrangement effectively provides three ports or three input/output ports. The three ports equal two shafts + one electrical.

Some advantages of the transmotor include operation as a motor or a generator. The gasoline engine connected to the transmotor can run at constant rpm while the drive shaft for the wheels has variable speed. The transmotor can replace the traditional mechanical transmission (see Ref. [8] for more details).

Digital versus Analog Control

The computer side of hybrid control is digital. However, control of the M/G is analog. Most M/Gs are not digital. However, motors that are inherently digital in operation do exist. Stepping motors, which are discussed in Chapter 14, are digital in nature. Stepping motors are also known as switched reluctance motors.

Think of the second hand on a quartz clock. The second hand moves by 1 s in discrete steps; it is a stepping motor. A stepping motor may have a rotor that looks like a gear. The gear tooth is a magnetic pole. Motion can be controlled in steps; each step is one tooth on gear. Although motion can be jerky like the second hand on a clock, smooth, continuous motion is also possible like a regular motor.

The attractiveness of a digital motor is that each step can have its own tailored current and magnetic field. Each step can make its own contribution to the average torque of motor. New levels of performance are possible.

Photosynthesis and Other Bioinspired Chemistry

Solar cells, which are 10%–15% efficient, are discussed above. Because of poor efficiency, solar panels are not a viable option for propulsion of an HEV. Photosynthesis is nearly 100% efficient converting sunlight to chemical energy. In addition, photosynthesis has as intermediate step involving chemical dissociation to form hydrogen. Photosynthesis offers two possibilities. First is converting sunlight to chemical energy. The second is production of hydrogen. Perhaps photosynthesis can help produce hydrogen for the FCV [15,16]. Reference [16] has other suggestions for applying biology and chemistry to solving energy problems.

Cold Fusion

If cold fusion were possible, the concerns over energy would vanish or at least be greatly diminished. The author feels cold fusion is in the same category as UFOs!

MOTIVATION FOR NEW TECHNOLOGY

Uncertain Future

As discussed throughout this book, petroleum will run out someday. A complete change in personal and commercial transportation will be necessary. Hybrids will play an important role as the future evolves.

In a shorter time frame, the manufacturers want to sell more cars. Hybrids enhance the manufacturer's image. Technology that provides lower costs, better mpg, and better acceleration helps to sell cars. The government and the public want to be greener with lower levels of emissions. The threat of global warming drives interest in hydrogen ICE and fuel cells.

SUMMARY

A hypothetical government official contemplates which R&D programs deserve sponsorship. Multiple factors affect the choices of R&D programs. These factors include government funding and leadership, emissions regulations, the marketplace, commercialization possibilities, and the impact of the cost of a barrel of oil. Two different infrastructures are discussed. One is for charging of EV and PHEV assuming these vehicles become a significant fraction of the automobile population. Another infrastructure is for hydrogen to support hydrogen as a fuel for ICE and for fuel cells. Either or both infrastructures may evolve.

The focus on a generalized hybrid leads into energy conversion within the HEV, and technology shared by HEV, EV, and FCV. The mathematical tool of fractional quantities provides insight into which HEV components offer the biggest gains.

Mining the obvious launches the discussion of storage devices, which range from the common to the exotic, technically speaking. Continuing with mining the obvious, the various technologies essential for the HEV are assessed one by one.

Innovation, seeking the novel, roams through several new ideas that may offer improved performance. The chapter starts with the observation that although new technology is expensive, it is not optional. The chapter ends with the observation that petroleum will run out someday forcing a completely new mixture of vehicles. R&D on the new vehicles is indispensable.

APPENDIX 16.1: SAVING GASOLINE

The gasoline used by a conventional vehicle is M/G_C. The gasoline used by a hybrid is M/G_H. The savings, S, is the difference

$$S = \frac{M}{G_C} - \frac{M}{G_H} \tag{A.16.1}$$

Define the fractional gain in mpg due to hybrid as

$$f = \frac{G_H - G_C}{G_C} \tag{A.16.2}$$

Combining Equations A.16.1 and A.16.2 leads to Equation 16.2.

REFERENCES

1. Staff, *Fortune*, May 29, 2006, p. 24.
2. National Academy of Engineering, *The Hydrogen Economy: Opportunities, Costs, Barriers, and R&D Needs*, 2004.
3. H. L. Benford and M. B. Leising, *The Lever Analogy: A New Tool in Transmission Analysis*, Society of Automotive Engineers Technical Paper Series, Number 810102, Warrendale, PA, February 1981.
4. R. Nesbitt, C. Muehlfeld, and S. Pandit, Powersplit hybrid electric vehicle control with electronic throttle control (ETC), Society of Automotive Engineers, Warrendale, PA, *Hybrid Electric Vehicle Technology Developments*, SP-1808, 2003.
5. H. Hisada, T. Taniguchi, K. Tsukamoto, K. Yamaguchi, M. Iizuka, M. Mochizuki, and Y. Hirano, AISIN AW new full hybrid transmission for FWD vehicles, *Advanced Hybrid Vehicle Powertrains 2005*, Society of Automotive Engineers, Warrendale, PA, R. McGee, M. Duoba, and M. A. Theobald (Eds.), SP-1973, 2005, p. 55.
6. X. Ai and S. Anderson, An electro-mechanical infinitely variable transmission for hybrid electric vehicles, *Advanced Hybrid Vehicle Powertrains 2005*, Society of Automotive Engineers, Warrendale, PA, R. McGee, M. Duoba, and M. A. Theobald (Eds.), SP-1973, 2005, p. 81.
7. B. Conlon, Comparative analysis of single and combined hybrid electrically variable transmission operating modes, *Advanced Hybrid Vehicle Powertrains 2005*, Society of Automotive Engineers, Warrendale, PA, R. McGee, M. Duoba, and M. A. Theobald (Eds.), SP-1973, 2005, p. 149.
8. S. W. Moore and M. Ehsani, A charge sustaining parallel HEV application of the transmotor, *Electric and Hybrid-Electric Vehicles*, Society of Automotive Engineers, Warrendale, PA, R. K. Jurgen (Ed.), PT-85, 2002.
9. T. Hanyu, H. Iwano, H. Ooba, S. Kamada, Y. Kosaka, S. Komiyama, and K. Takeda, A study of the power transfer systems for HEVs, *Advanced Hybrid Vehicle Powertrains 2006*, SP-2008, Society of Automotive Engineers, Warrendale, PA, Organized by M. A. Theobald, J. Moore, M. E. Fleming, and M. Duoaba, 2006, p. 137.
10. U.S. DOE Fuel Cell Technology Challenges. Available at http://www.eere.energy.gov/hydrogenandfuelcells/fuelcells/.
11 Eaton uses hydraulics for hybrid, *Automotive Engineering International*, Society of Automotive Engineers, Warrendale, PA, December 2006, p. 26.
12. N. Kurczewski, Taking a flyer; hybrid depends on high-tech flywheel, *AutoWeek*, December 3, 2007, p. 9.
13. B. Gritzinger, Smarter stop/start?, *AutoWeek*, December 3, 2007, p. 9.
14. K. Bucholz, ECD fuels a hybrid with hydrogen, *Automotive Engineering International*, Society of Automotive Engineers, Warrendale, PA, December 2003, pp. 40–41.
15. F.M. Harold, *The Vital Force: A Study of Bioenergetics*, W.H. Freeman and Company, New York, 1986.
16. S. Schwartz, T. Masciangioli, and B. Boonycratanakornit, (Eds.), *Bioinspired Chemistry for Energy*, The National Academies Press, Washington DC, 2008.

17 New Technologies: Internal Combustion Engine

The competitive race between the CV and HEV becomes more interesting as new technologies are developed for the ICE.

INTRODUCTION

The internal combustion engine (ICE) is examined from two viewpoints. The first viewpoint considers the ICE to be a component of a hybrid vehicle. The second viewpoint considers the ICE to be the sole source of torque and power for a conventional vehicle (CV).

As new technologies are introduced into the ICE, the competition between the CV and the HEV becomes more interesting. As new technologies are introduced into an ICE, and with the advanced ICE being used in a hybrid, that HEV becomes more fuel efficient. In the competition between CV and HEV, the improved HEV becomes a moving target for the improved CV. Improvements in an ICE enhance both CV and HEV.

Alternatives to the gasoline-powered CV include clean diesels, HEV gasoline/electric, HEV diesel/electric, the hydrogen ICE, and the hydrogen fuel cell. In the United States, diesel engines have not penetrated the market to the extent as in Europe. The BMW H-7 dual-fuel car with an ICE is in limited production with delivery of 100 H-7 in 2007. The two fuels are gasoline and hydrogen. Ford is producing dedicated hydrogen-fueled ICE engine, which is a supercharged, 6.8 L V-10. The engines are for Ford E-450 shuttle bus [20]. Hydrogen-fueled ICEs are a long-term project and, being long term, have elements of uncertainty.

WHAT AND WHERE ARE THE LOSSES?

What aspects of hybrids will yield the greatest gains? The same approach as in Chapter 5 applies. Some insight can be gained from Figure 5.5, which shows the various efficiencies for a hybrid. From Figure 5.5

$$\eta = \eta_E \, \eta_G \, \eta_B \, \eta_M \, \eta_P$$

Use the same equation as was done before in Chapter 16

$$\frac{\Delta \text{mpg}}{\text{mpg}} = \frac{\Delta\eta_E}{\eta_E} + \frac{\Delta\eta_G}{\eta_G} + \frac{\Delta\eta_B}{\eta_B} + \frac{\Delta\eta_M}{\eta_M} + \frac{\Delta\eta_P}{\eta_P} \tag{17.1}$$

The conclusion remains that if a component has small efficiency, a small improvement yields a large gain in mpg. The big gains due to low engine efficiency ($\eta_E = 0.24$) confirm the need for this chapter.

As shown by Figure 17.1 the losses for a gasoline engine depend on rpm and load.

At ¼ load, the two heat losses, engine cooling and exhaust waste heat, add to 63%. Engine output is only 19%. Recall from Chapter 5, a lazy, loafing engine has poor fuel economy. At full

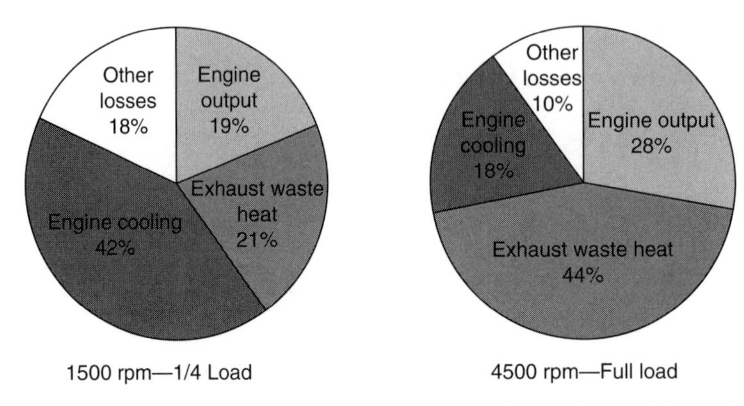

FIGURE 17.1 Losses in a 1.4L SI engine at ¼ load and full load. (Reprinted with permission from SAE Paper *Advanced Vehicle Powertrains 2005*, SP-1973, © 2005 SAE International.)

load, the heat losses remain high at 62%. Engine output increases to 28%. Other losses are 10%. Examine Figure 17.2 to determine the origin of other losses. One loss lumped into other losses is the power to drive the lubricating oil pump.

Although Figures 17.1 and 17.2 are from different sources, the numbers agree. From the diagram in Figure 17.2, auxiliaries account for two-thirds of the losses. Figure 17.2 corresponds to full load engine operation. Chapter 5 considered the gains to be made from powering automobile auxiliary components by electrical motors. Included in the "charge cycle and auxiliaries" is the pumping loss. At full load, pumping losses are small. At partial load, charging cycle losses are dominated by pumping losses. Some of the losses identified within "other losses" can be reduced slightly by better lubricants.

After looking at Table 17.1, the conclusion is that there are lots of ways to improve efficiency and reduce emissions. Due to the numerous avenues to improved efficiency, one can predict with

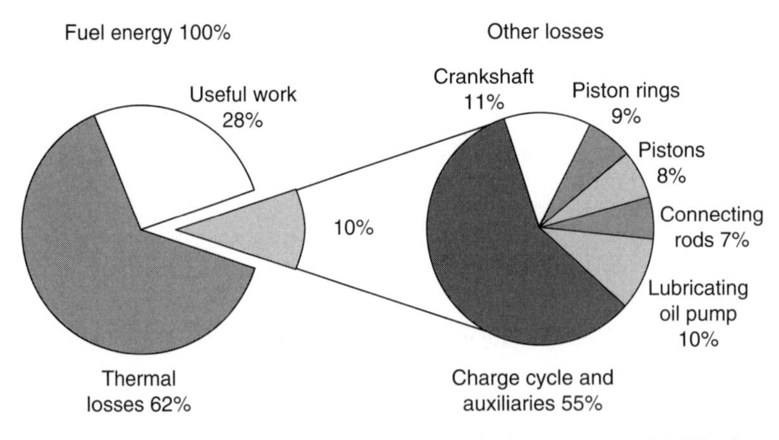

FIGURE 17.2 Expanded information on friction losses denoted by "Other losses 10%" in the previous figure. Although from different sources notice the agreement in values for losses. Charging cycle losses are dominated by pumping losses.

TABLE 17.1
New Technologies for Reducing Fuel Consumption Applicable to the Spark Ignition (SI) or Gasoline Engine

Technology	Efficiency Gain (%)	Remarks
High-precision fuel injection	Few	Allows lean burn
Fully variable valves	10%	Options for different valve train designs Mechanical; costs go up 10% Electrohydraulic Electromechanical
Pulse turbocharging using an air pulse valve	20–30	Increased charge (mass of air and fuel) over entire rpm range
Cylinder cut-off; variable stroke volume	8–20	Useful in partial load range
Dual spark plugs	5	Sequential dual ignition
Crankshaft starter/generator	15–22	Available with 14/42 V electrical system Mild hybrid; permits downsizing engine and engine off at idle
Variable CR	30	Requires supercharging or turbocharging
Reduce friction losses	10	In past 10–15 years friction losses have been reduced by remarkable 20%. An engine with zero friction loss would yield another 20% gain; assume can get ½ or 10%
Thermal management	5–10	Hot cooling under partial load
Fuel and oil optimization	5–10	
Reduced warm up time	5–10	Cold engines burn more fuel
Improved knowledge of sprays and combustion	5	Controlling emissions is a trade-off with efficiency. Better and detailed knowledge allows reduced emissions without sacrifice of efficiency
Ceramic engine	10–15	Out-of-style now. Liquid cooled engines operate at 200°F (98°C). Operate hotter increases Carnot efficiency
Recover waste heat; radiator	1–5	Low grade (low temperature) difficult to convert heat to mechanical or electrical energy
Recover waste heat; exhaust	1–16	Higher grade heat 900°C (1650°F) makes Rankine cycle feasible. Also thermoelectrics

some confidence a fuel reduction of 50% by 2015. Obviously, the efficiencies in the column cannot be added to yield an efficiency in excess of 100%. Some techniques exclude use of other techniques. Table 17.1 focuses on fuel economy; emissions are equally important. Within the automobile industry, finding a refined cost-effective, market-acceptable (market-enhancing) solution to the combined emissions and fuel economy scenario is expensive, time-consuming, and wrought with tough decisions. Complications include the necessity to integrate in unknown proportions alternate fuels, hybrids, and fuel cells [23].

TECHNOLOGY TO INCREASE EFFICIENCY

Advanced Combustion: Gas and Diesels

Within DOE is a program called Energy Efficiency and Renewable Energy (EERE). This program sponsors work in alternate fuels and advanced combustion for engines including emissions controls R&D. The projects are needed to support the development of advanced HEVs.

For diesels, other EERE projects ask what happens in the cylinder of the diesel? The aim is to reduce emissions, improve the effectiveness of exhaust aftertreatment technologies, and to optimize fuel formulation. As discussed later, the two big issues with regard to diesel emissions are particulate matter (PM) and NO_X emissions. One approach in a French laboratory to understanding combustion in diesels is described in Ref. [24]. Using a transparent cylinder head made of quartz and sapphire, the events are visible. Laser probing and spectroscopy are tools that can be applied. The strategy for obtaining low soot and NO_X emissions is low temperature combustion.

Stop and Go: Wear and Tear

Integrated starter/generator systems (ISG) automatically turn off the engine when the car is stopped. Note that ISG is synonymous with M/G. Idle-off improves mpg for either CV or HEV, but do engines suffer extra wear due to the frequent start–stops? Engine idle-off results in changes in engine lubrication compared to continuous running; these changes may cause added engine wear. Stop–starts cause a fluctuating temperature within the engine. Engine idle-off may also cause deposits on the valves. Catalytic converters may be affected by engine on–off. The words "may also cause" appear in the preceding sentences, which suggests R&D on the effects of idle-off is appropriate. Engine-condition monitoring programs can shed light on the question of added wear and tear.

Recover Thermal Energy Dumped by Exhaust and Radiator

For the 1.4 L engine (78 kW) with losses shown in Figure 17.1, the power recovery from the cooling by the radiator is limited by maximum temperature. From cooling, the power being dumped by the radiator is 9–48 kW. Assume recovery of 10%, which is 0.9–4.8 kW.

The exhaust is dumping from 4.6 to 120 kW. The temperature of the exhaust gas can be 900°C. For ideal Carnot efficiency, the recovery is 1.7–45 kW. Assume that 30% of the ideal Carnot efficiency can be recovered. The power recovered is 0.5–13.5 kW. By recovery of the 13.5 kW of the waste heat, the engine efficiency is improved from 28% at full load to 33%.

Both *Popular Science* [14] and *Motor Trend* [15] describe BMW's turbosteamer, which uses the Rankine cycle to recover exhaust heat (Figure 17.3).

Recover Thermal Energy Dumped by Power Electronics

Energy is wasted in the form of heat from the power electronics. The "cold" plate for power electronics is at a relatively low temperature as dictated by the maximum allowable temperature for the electronics. The cold plate is a low grade heat source, which means the temperature is low. This fact limits the amount of energy recovery which is possible.

Variable Valve Timing and Variable Valve Lift

Why is variable valve timing, which is common today, important for improving fuel economy? The flow of fuel and air into and the exhaust gases out of the cylinders can be optimized. Engines operated from several 100 rpm to a few 1000 rpm. Variable valve timing accommodates this broad range of engine speeds. High torque can be obtained over a broad speed range. Today, almost everything about engine control is electronic. One exception is variable valve timing, which is accomplished

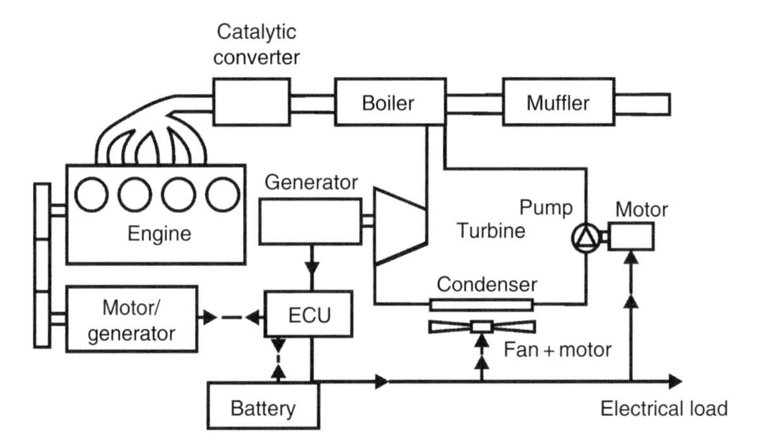

FIGURE 17.3 HEV with Rankine cycle to capture some of the thermal energy flowing out the exhaust with the hot gases. Boiler is located in exhaust pipe. (Reprinted with permission from SAE Paper *Advanced Vehicle Powertrains 2005*, SP-1973, © 2005 SAE International.)

mechanically by old-fashioned cam shafts. The ability to control timing and valve lift by mechanical means is impressive. Dual cams help. The day is coming when each valve will be lifted individually by electromechanical actuators.

Direct Injection Stratified Charge Engines

With direct injection, the fuel is injected directly into the cylinder. An overall lean mixture is possible. Locally, a relatively rich mixture occurs near the spark plug and allows reliable ignition. Direct injection may be combined with turbocharging or superchargers.

Dual Injection

Dual injection is discussed in Chapter 5. A clever combination to improve power, reduce emissions, and reduce fuel consumption is dual injection used by the Lexus V-6 in both CV and HEV applications such as hybrid Lexus GS450 h. Dual injection has injection both into the manifold and directly into the cylinder. Manifold injection is called port injection. Dual injection allows higher compression ratios (CRs) without knocking. A gain of 7% in peak engine power results from dual injection. Engine torque and fuel economy are improved. Also emissions are reduced.

Comments on Diesels

Table 17.2 summarizes various technologies to improve diesel efficiency. The same comment made about Table 17.1 applies here also. The efficiencies in the column cannot be added to get more than 100% efficiency. Diesel engines are already efficient; for this reason, it is harder to improve efficiency.

The efficiency of gasoline engines will grow faster than diesel engines; consequently, the efficiencies will converge with the diesel engine advantage reducing to about 10% (see Table 17.4 and Figure 17.9). With the 10% advantage, the fuel savings may not be enough to pay extra costs associated with a diesel engine. This convergence of gasoline engine and diesel engine efficiencies puts the diesel/electric hybrid for passenger cars in doubt. For buses and trucks, the diesel/electric hybrid is the clear choice. For example, New York City is adding 500 diesel/electric hybrid buses according to *Fleet Owner* [6] (also see Chapter 3).

TABLE 17.2
New Technologies for Reducing Fuel Consumption Applicable to the Diesel Also Known as Compression Ignition Engine

Technology	Efficiency Gain (%)	Remarks
Piezoelectric fuel injectors	Few	Fuel injection profile becomes more precise
Improved knowledge of sprays and combustion	Few	See remarks same item for gas engines, also
Homogeneous combustion, HCCI	Up to 15%	Large reduction in pollution Avoids local temperature spikes Avoid soot formation
Fully variable valves	5–15	Diesels already operate unthrottled (compare remarks in Table 17.1)
Pulse turbocharging	20	Substantial savings of fuel eliminate glow plug or reduce CR
Cylinder cutoff	Almost none	Much less improvement in efficiency than gasoline engine (compare Table 17.1)
Crankshaft starter generator	15–20	Similar to SI engine; mild hybrid (compare Table 17.1)
Variable CR	—	Not recommended due to high CR of diesel engines which have very high loads on pistons, rods, etc.
Reducing friction loses	13	Diesel engines with zero friction losses would gain 26% improvement; assume that ½ can be attained or 13%
Downsizing made possible by higher operating pressures	20–25	Increase peak pressures to 250–300 bar (3700–4500 psi) • High pressure means smaller engine • Very hard to start with high pressure, use electrically supported turbocharger
Recover waste heat	5–10	See remarks Table 17.1
Optimize thermodynamic cycle	5	Wrestle with Carnot efficiency
Turbocompound supercharging	5	Exploit exhaust energy better than simple exhaust gas turbocharger

New Technology: Diesels

Yamaguchi [7] discuss the new diesel technology being incorporated in commercial vehicles in Japan. Intercooled turbocharging is being used. Turbocharging heats the intake air making it less dense. Cooling increases the density and the mass of air going into the engine via the intake. Common-rail fuel injection is at high pressure 180 MPa (26,100 psi). Multiple fuel injections (up to four) replace the simple one-squirt. Electronically controlled, water-cooled precision control EGR assists engine performance.

The two big issues with regard to diesel emissions are PM and NO_X emissions. The problem of near-term NO_X is solved with a catalytic converter. PM (black smoke and soot) is reduced by using a diesel particulate filter. Another method is urea injection. Regulations are getting tougher worldwide for allowable PM and NO_X.

Partnership for a New Generation of Vehicles: Supercar Diesel Engines

Partnership for a New Generation of Vehicles (PNGV), which is discussed in Chapter 1, had projects oriented toward diesels. Direct injection was emphasized. The General Motors concept hybrid for PNGV used a 1.3 L, three-cylinder, dual overhead camshaft CIDI engine. CIDI is compression ignition (diesel) with direct injection. The diesel had an intercooler using water to air. The General Motors concept hybrid for PNGV was known as Precept.

Clean Diesel for SUV: DOE Program

DOE has programs focused on better mpg for sport utility vehicles, SUVs. SUVs are popular in the United States. One goal is to reduce fuel consumption by 50% compared to previous diesels. The power size, 150–188 kW (200–250 hp), has been selected to be suitable for SUV towing, hauling, and all-wheel drive.

Most SUVs use gasoline engines, which provide a certain mpg. Another goal is to provide a diesel that gives 50% better mpg. For example, if the current gasoline SUV gives 20 mpg, the SUV with a new diesel would provide 30 mpg. At the same time, the new diesel meets the future emissions requirements. The new EPA standards for emissions are 90% less NO_X emission and 90% less PM. The diesel must be able to operate 120,000 mi while meeting the new EPA emission standards. This requires the new low-sulfur content diesel fuel [19].

Heavy Truck Diesels

For the foreseeable future, heavy duty trucks, 18-wheelers, will continue to use diesel. These trucks will burn diesel fuel from petroleum. Work on cost-effective, advanced commercial hybrids for heavy trucks continues.

EXAMPLES OF NEW DESIGNS

Closing All Valves

One engine mode of the 2006 Honda Civic Hybrid is to close all valves in the engine during regenerative braking. In this hybrid, the motor is mounted and connected to the engine. The braking torque is split:

$$\text{(Braking torque)} = \text{(Engine torque)} + \text{(Generator torque)}$$

The generator torque is responsible for returning energy to the battery. If engine torque is large, then generator torque must be small. Hence, the goal is to make engine torque as small as possible. How is engine torque, which causes wasted energy, reduced?

All valves are closed trapping gas in the cylinders. The result is that the gas behaves as a spring and mass as shown in Figure 17.4. The heavy black arrow is the direction of motion of the piston or mass, m. The thin black arrow is the force acting on the piston (mass). When the arrows are in opposite directions, as in A, energy is being removed from the engine. When the arrows are in the same direction, as in B and C, energy is being returned to the engine. With the valves closed, no pumping loss occurs. Other dissipation mechanisms, such as turbulence and thermal conduction, cause losses.

Sketch A has the piston moving upward to compress the gas trapped in the cylinder. The spring is being compressed in A. In B, the piston is moving downward in the direction of the force. The force is large due to high pressure of the gas, or, in the case of the spring, the spring is tightly compressed. In C, the piston is still moving downward but is near the end of the stroke. The force is small. After C, the compression stroke A starts again. In one revolution, very little energy is lost, which means the average engine torque is small; therefore, the generator torque is bigger.

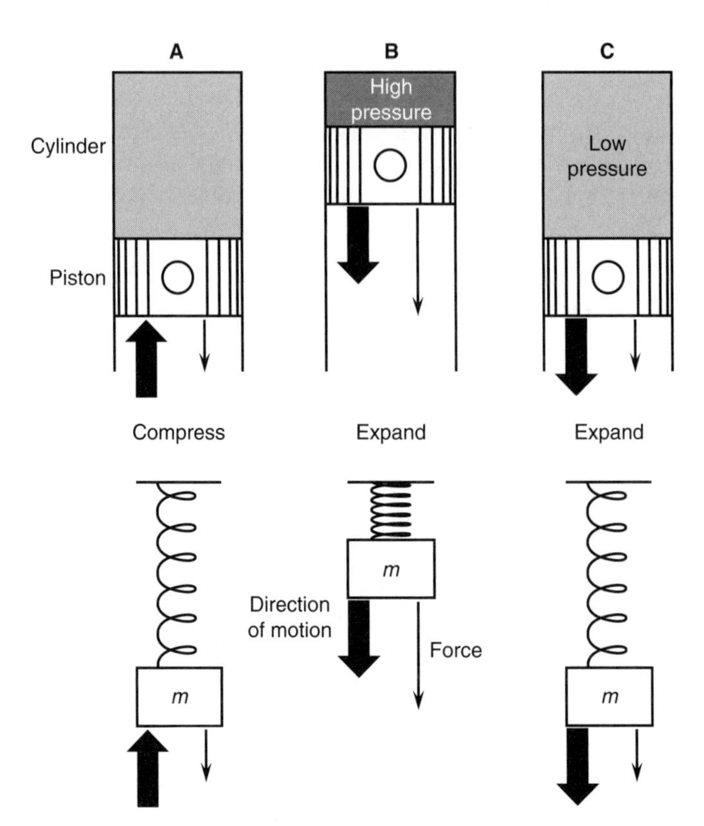

FIGURE 17.4 Closing all valves for improved regenerative braking. Engine operation is analogous to low-loss motion of a spring with a mass, *m*. The minimum loss for an engine is attained with all valves closed.

Cylinder Deactivation

The hybrid application is not the only case where minimum engine torque is desired. The displacement on demand (cylinder deactivation), where three cylinders in a V-6 engine are shut down, also needs less energy loss due to nonworking cylinders. Usually, for displacement on demand, the valves are closed for the inactivated cylinders. Fuel economy increases of 5%–10% are possible.

Atkinson Cycle Engine

The Atkinson cycle engine is definitely not a new design having been invented in 1882. Because the cycle has advantages when applied to hybrids, the 1882 invention is enjoying a period of popularity. See Figure 17.5, which shows the various strokes of a four-stroke engine with intermediate positions as well.

Key features of Atkinson cycle are a long expansion stroke which allows extraction of more energy. The short compression stroke reduces pumping losses. The design allows retaining and designing any compression ratio desired. The results are improved engine efficiency which is provided at the expense of power.

That is the good news; now for the bad news. Due to the reduced charge, discussed presently, the power is reduced compared to the same engine of equal displacement. "Charge" is the maximum mass of fuel plus air in the cylinder; usually this mass occurs when the piston is at TDC and all valves are (nearly) closed ready for expansion stroke.

Some relevant definitions: as crankshaft angle passes through 0°, the piston pauses and stops; hence the word "dead" to describe top dead center (TDC). See sketch E in Figure 17.5. Likewise as

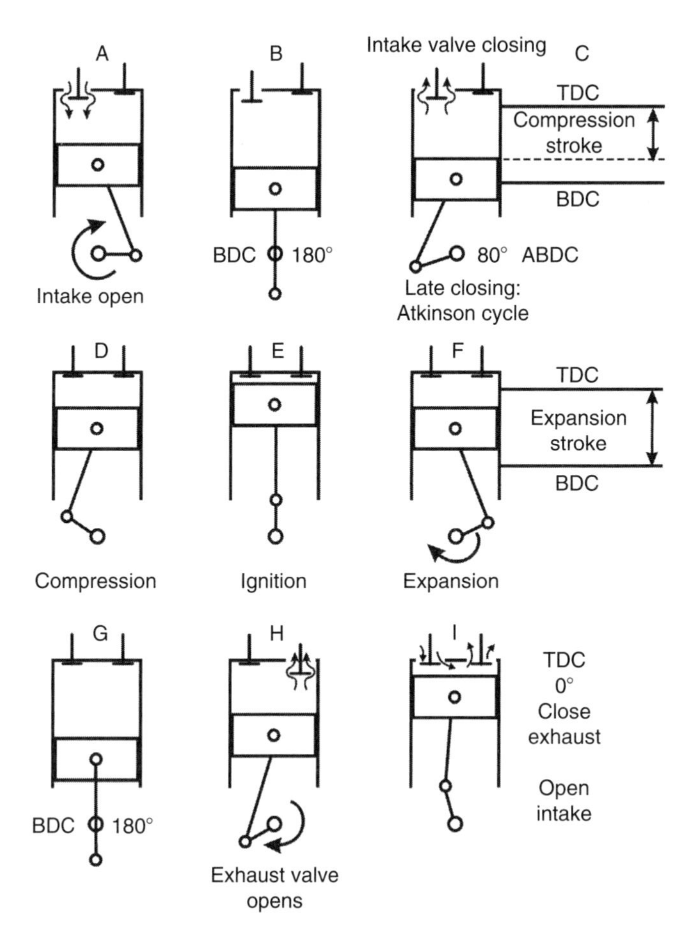

FIGURE 17.5 Illustration of all of the four strokes of a four-stroke engine with some intermediary piston positions to show features of Atkinson cycle.

crankshaft passes through 180°, the piston pauses and stops giving bottom dead center (BDC); this is sketch G in Figure 17.5. In Figure 17.5 an important attribute of Atkinson engine is illustrated. Starting with sketch A, the piston is moving toward BDC. Sketch B is at BDC and normally the compression stroke commences. Sketch C shows an open intake valve; the charge cannot be compressed. An open intake valve is like leaving the door open; the charge leaks out. Compression is delayed creating a shortened compression stroke, which is one feature of the Atkinson cycle (see sketch C). Sketch F shows full expansion stroke, which is another feature of the Atkinson cycle.

Figure 17.6 provides an excellent way to understand reduced pumping losses from the Atkinson cycle as applied to a four-stroke engine. The pressure traces enclose two areas, I and II. Analysis shows that the area enclosed in I is proportional to the energy produced by the engine. Since the test equipment that yields the pressure–volume curves is known as an indicator, the energy of I is termed indicated energy. Area II involves moving the gases in and out of the cylinder; this is called pumping. Hence, the term pumping loss is applied to II. The net output of energy from the engine is

$$\text{Net energy} = \text{Indicated energy} - \text{Pumping loss}$$

Considerable confusion exists in the popular press about pumping loss.

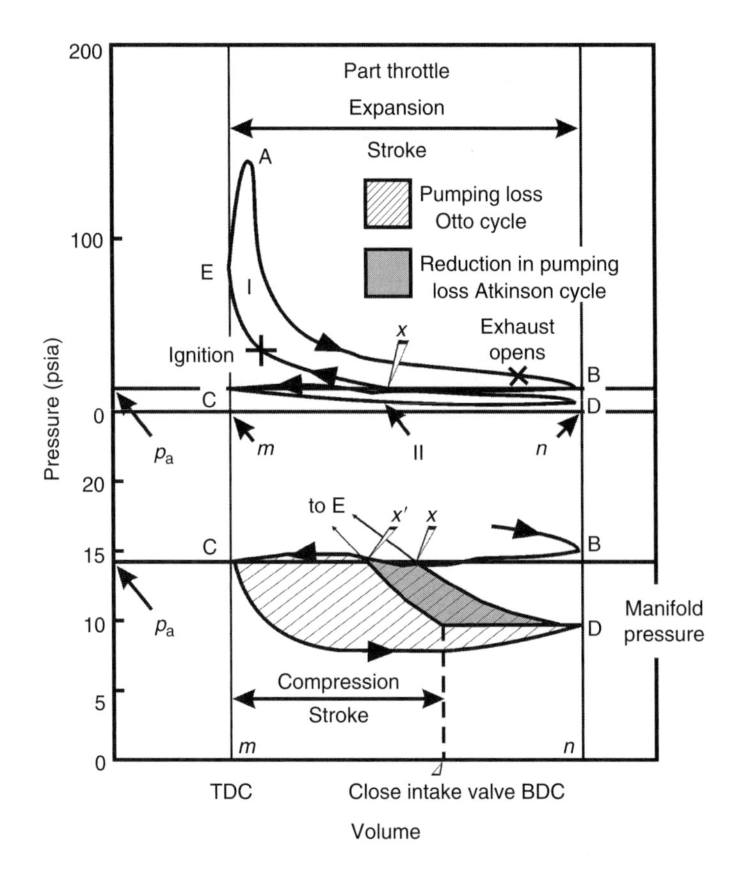

FIGURE 17.6 Pressure shown as function of volume within cylinder as piston moves from TDC to BDC. The four strokes are D–E, compression; E–A–B, expansion; B–C, exhaust; and C–D, intake. Pressures within a cylinder are measured as engine is operating at part throttle.

The magnitude of pumping loss depends on the throttle setting. Consider the ratio of pumping loss divided by net work done by the engine. At full throttle, the pumping loss ratio is 1%–3%. At partial throttle, the pumping loss ratio is much larger being 30%–40%.

In Figure 17.6, p_a is the ambient pressure. Point D, which is below ambient pressure, is a partial vacuum. Also point D is equal to manifold pressure. In the Otto cycle, the intake valve closes at D, and the charge is being compressed. Notice the pressure curve going upward toward point E. However, with the Atkinson cycle the intake valve remains open. As the piston moves toward TDC from BDC, the pressure remains equal to that at point D. When the intake valve closes, the pressure increases and the curve heads off toward E. Area II is reduced in size by the slice, which is gray shaded. Pumping losses are less.

Figure 17.6 also shows the shortened compression stroke and the comparatively long expansion stroke. The long expansion stroke yields a greater extraction of energy from the fuel.

Does Atkinson cycle allow control of CR? The answer is yes as depicted in Figure 17.7. Displacement, D, clearance volume, c, are used to define compression and expansion ratios. CR, r, is defined by Equation 17.2

$$r = \frac{c + D}{c} \tag{17.2}$$

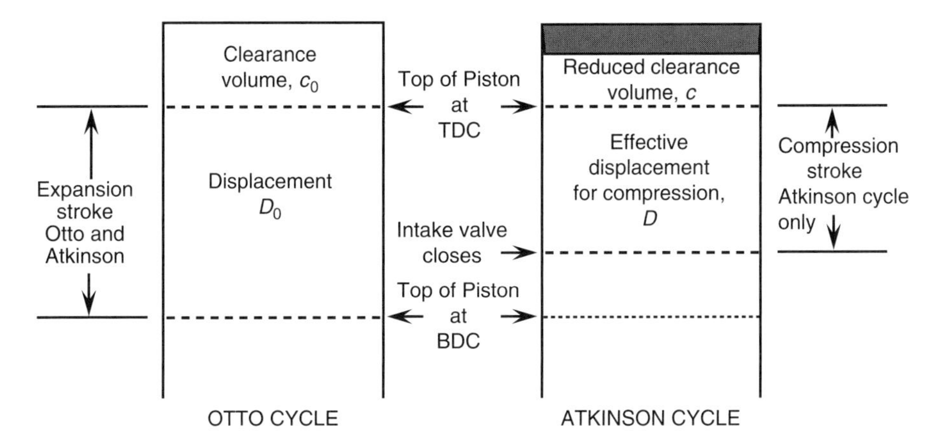

FIGURE 17.7 Displacement and clearance volumes for Otto and Atkinson cycles. Both have the same CR = 4. If displacement D_0 is combined with reduced clearance volume, c, the Otto cycle CR is 5.5. This higher value is occasionally used in sales brochures.

Further define

f = fraction of D_0 swept before intake valve closes; $f = 0$ for Otto cycle
D_0 = displacement when intake valve closes at BDC; value when $f = 0$
c_0 = clearance volume when $f = 0$
r_0 = CR when $f = 0$

To have the same CR at both $f \neq 0$ and $f = 0$, what is the new value for the reduced clearance, c? It is simply

$$c = c_0(1 - f) \tag{17.3}$$

Define a CR, r_{OTTO}, based on reduced clearance volume, c, and on the original displacement, D_0. This is appropriate value for the Otto cycle.

$$r_{\text{OTTO}} = \frac{c + D_0}{c} = \frac{(1-f)+(r_0-1)}{1-f} \tag{17.4}$$

Figure 17.7 is drawn to a scale such that $c_0 = 2$ and $D_0 = 6$. Also $f = 1/3$. With these input values, $r_0 = 4$, $D = 4$, $c = 4/3$, and $r_{\text{OTTO}} = 5.5$. The Atkinson CR is

$$r = (4/3 + 4)/(4/3) = 4$$

which is the same as the original value with $f = 0$.

A hybrid engine with Atkinson cycle can usually be identified by the CR. For CR < 11, the engine is not Atkinson cycle. For CR > 11, the engine likely uses the Atkinson cycle.

The first generation Prius, Prius I, has continuously variable valve timing as shown in Table 17.3 [8]. The delay in closing of the intake valve is from 120° ABDC/80° ABDC. Compare with Figure 17.5, which illustrates all of the four strokes of a four-stroke engine with some intermediary piston positions to show features of the Atkinson cycle.

TABLE 17.3
Continuously Variable Valve Timing for Atkinson Cycle for the Toyota Prius I

Intake valve opens	30° ATDC/10° BTDC
Closes	120° ABDC/80° ABDC
Exhaust valve opens	32° BBDC
Closes	2° ATDC

Note: The phase angle for the intake valves is 40° cam. Only the intake valves are variable; the exhaust valves have a fixed cam angle.

Miller Cycle

The Miller cycle engine is closely akin to the Atkinson cycle engine. Both adjust the valve timing to achieve an enhanced expansion stroke and a reduced compression stroke. In contrast to the Atkinson cycle, the Miller cycle requires a supercharger to be practical. Lower fuel consumption of 10%–15% is expected compared to the Otto cycle. Mazda has a V-6 Miller cycle engine in production. Since Mazda is producing both Miller and Wankel engines, this suggests Mazda is an innovative company. Try Mazda.com for more data on the Miller cycle engine.

Stirling Cycle

The Stirling cycle is not an ICE. It is a heat engine. Heat can be supplied from external combustion, solar power, nuclear power, or any other heat source. For an ICE, the working fluid (air + fuel) flows through the engine. For a Stirling cycle, the same gas remains in the engine. In one form, consider two pistons and cylinders connected at head end by a pipe. When one cylinder is expanding the other is compressing. The gas sloshes back and forth between cylinders. One cylinder is maintained at high temperature and the other at a cooler temperature. As each piston moves through both compression (work absorbed) and expansion (work generated) strokes, overall work occurs. The work for the hot piston exceeds that of the cold, and net work can be removed at the crankshaft. For more information on Stirling cycle, see Ref. [13]. Surprisingly, Refs. [11,12] have little or no information on Miller and Stirling cycle engines.

Alvar Engine

The Alvar engine may or may not have any impact on future hybrids. However, in the spirit of something new and novel, the engine fits. A sketch of an Alvar engine appears in Figure 17.8. The motion of the small piston can be controlled independently of the main piston. Further, during one revolution, the rotation of the small crankshaft can vary; that is, the angular speed can vary. Why bother with the added complexity?

The effect of crankshaft speed, nonstandard cylinder pressure traces, and variable CR on emissions has largely been ignored [9]. The potential exists for decreased emissions and improved engine efficiency. Only by research can the potential be assessed. Figure 17.6 is an example of a standard pressure trace. The Alvar engine can create and test the effects of nonstandard cylinder pressure traces.

Offset Crankshaft

Figure 17.5 illustrates the normal case where the axis of the cylinder intersects the axis of the crankshaft. The crankshaft axis can be displaced laterally so that the two axes no longer intersect. This changes time spent near TDC for longer combustion time. Also, the offset crankshaft affects engine friction.

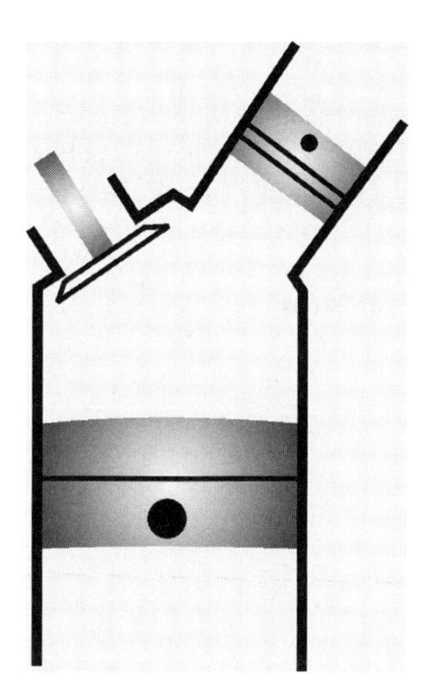

FIGURE 17.8 Combustion chamber of an Alvar engine. The small piston has a crankshaft separate from the crankshaft of main piston. An electrical motor drives the crankshaft for the small piston. (Reprinted with permission from SAE Paper *Advanced Vehicle Powertrains 2005*, SP-1973, © 2005 SAE International.)

Free Piston Engines

Free piston engines are another example of a departure from the ordinary crankshaft incarnation of the Otto cycle engine. These were discussed in Chapter 16. The ICE and electrical devices such as M/G are married as separate components in an HEV. A free piston engine can have the same marriage of functions in one component. For a CV, this marriage may not be feasible.

CONVERGENCE OF GAS AND DIESEL

Advances in mpg in Recent Years

In the years 1985–2010, major gains have been, or will be, made in engine efficiency. Table 17.4 lists the trend in mpg and the technology making the gains possible. Note the convergence of mpg from the gasoline engine and diesel engine.

TABLE 17.4
Comparison of Gas Mileage for Diesels and Gasoline Showing the Effect of New Technology Introduced in the Last Two Decades

Technology	Gasoline Engine (mpg)	Diesel Engine (mpg)
Without direct injection, DI	25	29–30
DI diesel only	25	34
DI both gas and diesel	27–29	34
Gas engine high pressure injection; charge stratification	31	34

Hybrid Engines

Diesel engines offer high torque at low rpm and low specific fuel consumption but have problems with emissions and particulates (soot or black smoke). Gasoline engines have high power per kilogram (or pound) and run smoothly and quietly. Can the best features of each be combined? This is the goal of the hybrid engine. "Hybrid" is used in a different sense when applied to engines instead of vehicles. A hybrid engine combines facets of two cycles. The two cycles are usually Otto and diesel [2,11]. Features of a hybrid engine can be selected from several options.

• Compression:	fuel/air, *f/a*, mixture or pure air
• Mixture condition:	homogeneous or non-homogeneous
• Ignition:	external or auto-ignition
• Control:	qualitative or quantitative
• CR:	low or high
• Fuel:	gasoline or diesel

One demonstration of a hybrid engine is the Mercedes Benz DiesOtto engine also known as the homogeneous charge compression ignition (HCCI) engine. Another is the GM Saturn Aura with HCCI engine.

HCCI Engine

The HCCI engine is the ultimate convergence of the diesel and Otto cycles. The HCCI combines the fuel efficiency of diesel with the power and low cost of a spark ignition (SI) engine. HCCI uses a homogeneous fuel/air mixture which is auto-ignited by the temperature increase accompanying compression. Ignition in a gasoline engine is initiated by the firing of the spark plug. In a diesel engine, the ignition sequence starts at the moment the fuel is injected into the already hot, compressed air. For HCCI, the ignition depends on the chemical nature of the fuel. Iso-octane auto-ignites at a lower temperature than methane. Detailed knowledge of chemical kinetics coupled with the fluid flow within the engine is necessary to design an HCCI engine [16]. Engine speed and power are controlled by the chemical nature of the fuel, the fuel/air ratio, and the temperature and pressure of the mixture.

In a conventional gasoline engine, auto-ignition is bad since it causes knocking. In an HCCI, auto-ignition is good. Auto-ignition is the necessary trigger mechanism for combustion.

What are the advantages of HCCI engine? An especially strong point is that HCCI lowers emissions and particulates. The tough emissions for a diesel are NO_X and soot. HCCI replaces the tough emissions with unburned HC and CO, which are easily controlled by existing catalytic converters. The engine offers 15% better fuel economy. The HCCI offers the potential of 40% peak efficiency compared to 30% for gasoline engine. The technology is scalable from lawn mower engines to mammoth engines for ship propulsion. Lean fuel/air mixtures can be burned. Any fuel that can be injected can be used. However, it must be quickly stated that the fuel and engine design must be carefully matched. The critical triggering of auto-ignition depends on chemical kinetics. The HCCI can use high CR and still avoid pinging. Power output can be attained that is competitive with existing engines.

Table 17.5 compares the combustion processes for different ICE. Looking at the engine attributes, the HCCI has several features in common with the diesel engine. Compare the column under each heading; except for "Ignition trigger" the two are identical. For a discussion of surface and volume combustion, see Appendix 17.1.

What are the problems of HCCI engine? What is the best way to create the homogeneous mixture? The ability to operate over wide range of speeds and loads remains a problem. HCCI has limited operating range with regard to rpm. Control of the start of combustion as well as the rate of combustion is a focus of R&D activities. Another problem is the control of the optimal equivalence

TABLE 17.5
Comparison of Combustion Processes for Different ICE

		Throttle Body Injection		
Process	**Diesel**	**Direct Injection Gasoline**	**Carburetor Gasoline**	**HCCI**
Intake pure air	X	X	—	X
Intake *f/a* mixture	—	—	X	—
Compression	X	X	X	X
Fuel injection	X	X	—	X
Ignition trigger	Injection	Spark	Spark	Auto-ignition
Combustion	Volume	Volume	Surface	Volume

ratio for combustion as a function of engine speed and load. Another minor problem is the minimization of HC and CO emissions, which increase due to low temperatures.

Why does HCCI provide 15% better fuel economy? One reason is downsizing. Here the gains of downsizing differ from "Engine Downsizing" in Chapter 5. (In Chapter 5, downsizing allowed the engine to function at an optimum operating point for minimum fuel consumption.) Of course, downsizing means a smaller engine. The friction drag internal to the engine decreases as size decreases. Also the number of cylinders may be decreased. Going from a V-6 engine to an I-4 engine lowers friction drag. Other features decrease fuel consumption whether or not used in HCCI. These include the benefits of direct fuel injection, higher CRs, and the absence of pumping losses due to throttling.

The Mercedes Benz HCCI DiesOtto engine is a turbocharged four-cylinder with 1.8 L displacement yielding 238 hp (179 kW) and 295 ft·lb$_f$ (400 N·m) torque. The engine uses gasoline and incorporates a homogeneous charge, which provides the necessary compression ignition. Other technologies enhance fuel economy including direct fuel injection, turbocharging, and variable compression. Up to 2000 rpm, the engine uses the diesel-like mode. Above 2000 rpm, it switches to the SI mode. Being the engine component of an HEV, the engine provides 39 mpg. The engine has sufficient power to be used in an S-Class vehicle, which is a large sedan. The test vehicle is the Mercedes Benz F700 Hybrid (F = sedan for test of technology) S-Class sedan. The motor portion of the hybrid is a 20 hp (15 kW) electric motor. Acceleration time from 0 to 60 mph is 7.5 s, which is quite acceptable in a luxury sedan.

The GM Saturn Aura mule uses a 2.2 L Ecotec four-cylinder engine. Computer control plays an important role operating the engine. Sustaining HCCI operation at low and also high loads is a challenge. The engine operates in HCCI mode for engines speeds between 1000 and 3000 rpm [21].

Turbocharging and/or Supercharging

Both turbocharging and supercharging increase the mass of fuel and air inside the cylinder thereby increasing power. The engine can be made smaller, which has two effects. First, the smaller engine means the total vehicle weight is smaller. Second, as noted in the discussion of HCCI, the friction drag internal to the engine decreases as engine size decreases.

Six-Stroke Engine

Although the four-stroke ICE dominates, both two- and six-stroke engines are possible. The two-stroke engine has a power (expansion) stroke every other stroke; one-half of the strokes contribute to power output. The six-stroke engine has two power strokes; one-third of the strokes contribute to power output. The four-stroke has one power stroke; one-fourth of the strokes contribute to power output. The temptation exists to say that a four-cylinder, six-stroke engine will have the same power as a V-8, four-stroke engine. This will only be true, of course, if the second expansion stroke produces as much power as the first. This equality of power is unlikely.

For the same displacement (cubic inches or liters) the engine with the greater fraction of power strokes will have the greater power output. However, the engine with the greater fraction of power strokes will have the greater stresses (primarily thermal stress). Greater stresses may mean shorter life. The six-stroke engine has not been the focus of development until recently. Why are three-stroke, five-stroke, etc. engines impossible?

Several six-stroke engine designs have been invented. The goal is to capture the waste heat. Figure 17.1 shows the fraction of energy input that appears as waste heat. At quarter load, the waste heat loss is 21% while at full load it is 44%. The first expansion (work producing) stroke uses the hot gases from combustion to push the piston. In some engines, the hot gases to push the piston for the second expansion stroke have been heated by the waste heat. For the design of Ref. [22], fuel is added for the second expansion. Ignition occurs due to oxygen-free radicals. In other designs, water is injected, and the waste heat creates steam. The steam pushes the piston during the second expansion stroke. Exhaust valve timing is very important for the second expansion stroke.

Advantages attributed to the six-stroke design are a cooler running engine and increased efficiency. Less engine cooling is needed since some heat is converted to engine power. Efficiency is improved since added power is extracted by the second expansion stroke. Also, emissions are less. Claims are made that the six-stroke engine yields 40% better fuel economy than a four-stroke engine. The anticipated gain in efficiency is, of course, one compelling motivation for research on six-stroke [11,12,22].

Trends in MPG for Gasoline and Diesel Propulsion and HEV

In Europe, diesels are extremely popular. Two factors contribute to this result. First, emission standards have been lower than those in the United States. Second, gasoline (and diesel) are taxed so that the selling price is $3–$7 per gallon. The high fuel price stimulates high fuel economy. In France, 65% of light duty vehicles (LDV) are diesels. One-half of all LDV in Europe are diesels. The diesel engine was invented as an alternate to Otto cycle. In the United States, the diesel earned a reputation for being unreliable in the 1970s and 1980s. Can diesels be sold in the United States? The high sulfur fuel prior to October 2006 could foul catalysts. The low sulfur fuel currently on sale permits clean diesels. Advanced diesels may use sequential turbochargers, that is, turbo for low rpm and another turbo for high diesel rpm [10].

A Lexus GS 450 h sedan, which is an HEV, is compared with the Mercedes Benz E320 BlueTec diesel. The MB BlueTec offers first-class motoring on an economy class fuel budget (29 mpg combined). The Lexus GS 450 h has been designed for performance instead of fuel economy and delivers 23 mpg combined (for more information, see Ref. [18]).

Assume a 1400 kg (3000 lb) vehicle. The Freedom Car Program of DOE has as a goal of improved efficiency for a gasoline ICE from $\eta_E = 30\%$ to $\eta_E = 50\%$. Since the diesel starts from a higher efficiency, the improvement from $\eta_E = 35\%$ to $\eta_E = 50\%$ is not as dramatic as for a gasoline engine. Using the ratio 0.5/0.3, the mpg from a gasoline engine is extrapolated from 30 to 50 mpg (see the curve in Figure 17.9). In the future, the diesel will converge toward the gasoline engine with a 10% better mpg.

Transferring all relevant technology from the R&D pipeline to HEV, a 75 mpg vehicle may be possible. However, rather than focusing on improved mpg, a choice may be made to lower HEV component costs thereby making the HEV more competitive with the CV in the marketplace.

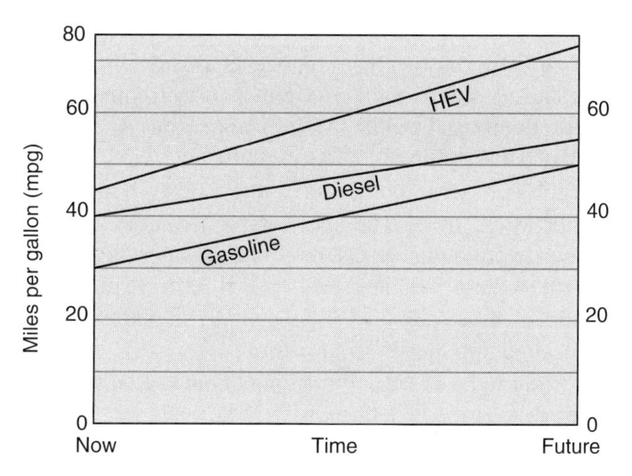

FIGURE 17.9 Trend in mpg due to efficiency improvements in gasoline, diesel, and HEV.

The diesel is more expensive than the gasoline engine. A gasoline–electric hybrid already suffers from a price premium compared to CV. A diesel/electric hybrid has a double whammy of a price premium due to the increment in diesel engine costs. Despite the cost penalty, the diesel/electric hybrid is being taken seriously by Audi, Mercedes Benz, and Volkswagen [25].

SUMMARY

Improvements in ICE advance both the CV and the HEV, which compete in the marketplace. Figures 17.1 and 17.2 give the distribution of losses in a gasoline engine. Table 17.1 lists several new technologies for the gasoline engine. Based on the list, a prediction of 50% gain in gasoline engine efficiency seems reasonable.

Recovery of the thermal energy dumped by the radiator and exhaust is becoming economically viable. Fuel injection provides a means to enhance engine performance. Both direct and dual injection are discussed.

Attention shifts to the diesel engine with Table 17.2 providing a list of promising new technology. The two big issues with regard to diesel emissions are PM and NO_X emissions; both topics are discussed.

A few examples of new designs are presented including engine torque reduction by closing all valves. This action improves energy recovery during regenerative braking. Although not new, the mysteries of the Atkinson cycle are revealed in a thorough discussion. The Alvar engine fits in the spirit of new designs and new directions.

As technology improves, the gasoline engine begins to converge with the diesel engine with regard to fuel economy. This fact may have an influence on development of diesel/electric hybrids.

APPENDIX 17.1: COMBUSTION PROCESSES IN ICE

Gasoline Engines

The processes leading to combustion include

- Intake fuel/air mixture or pure air (direct injection)
- Fuel injection (direct injection)
- Compression
- SI
- Combustion

Carburetion or throttle body injection provides a fuel/air mixture that flows through the intake valves into the cylinder. The *f/a* mixture is (nearly) homogeneous. For direct fuel injection, pure air is sucked into the engine The *f/a* mixture may be either nearly homogeneous or inhomogeneous depending on the injection timing and profile. An inhomogeneous distribution may be described as a stratified charge.

Timing of fuel injection is relative to crankshaft angle (CA). Usually, start of injection occurs before TDC, on the compression stroke. The injection profile is the fuel flow rate as a function of time. The profile, described in colloquial terms, might be a small squirt, a large squirt, and then followed by a smaller squirt. The diesel designed for the Opel Extreme plug-in hybrid uses closed loop fuel injection [17]. Sensors measure pressure in the cylinder in real time. The fuel injection profile is modified to yield the desired time-dependent pressure.

For stoichiometric fuel/air ratio, at the completion of combustion neither fuel nor oxygen (from the air) remain. The equivalence ratio is defined as

$$\phi = \lambda = \frac{\text{Actual } f/a \text{ ratio}}{\text{Stoichiometric } f/a \text{ ratio}} \tag{A.17.1}$$

In the United States, ϕ is used as the symbol for equivalence ratio whereas in Europe, λ is used. For $\phi = \lambda < 1.0$ the mixture is fuel-lean. For $\phi = \lambda = 1.0$, the mixture is stoichiometric. For $\phi = \lambda > 1.0$, the mixture is fuel-rich. For some gasoline engines maximum efficiency occurs for $1.1 < \phi = \lambda < 1.3$.

Fuel injection precedes maximum compression. The spark for ignition also precedes maximum compression, which occurs at or near TDC. From the point of ignition at the spark plug, a flame front (a sheet of flame) progresses through the *f/a* mixture. This is described as surface combustion, that is, combustion occurs crossing a surface (flame front). Release of heat is relatively slow. When the adverb "relatively" is used a reference is needed. This reference is the time for the piston to move one stroke or the crankshaft to move one-half revolution. The time is

$$t_R = \frac{1}{2 \text{ rpm}} \tag{A.17.2}$$

At 3000 rpm, the value of t_R is 10 ms. Combustion extends from just before TDC to about 40° past TDC. The flame speed must be sufficient to consume most of the fuel in as short a time as 10 ms. On occasion, two spark plugs (dual ignition) are used or decrease propagation distance for the flame yielding shorter combustion time.

Engine output, power, is controlled by throttling the mass of *f/a* into the cylinder. A demand for more power causes an increased mass of *f/a* into the engine. Throttling causes the pumping losses (Figure 17.6). Throttling is a characteristic of gasoline (SI) engines.

Knocking or pinging was discussed in Chapter 12 where the advice was given not to buy premium fuel. Frequently, the combustion in an ICE is described as an "explosion," which is erroneous. As stated above, combustion is relatively slow and certainly is not explosive. At some CR, the *f/a* mixture will auto-ignite. When this happens, ignition occurs throughout the volume. Extremely rapid heat release (an explosion) occurs driving shock waves to and fro in the cylinder. These waves cause the audible knocking as well as damage to piston crowns and valves. The temperature increase during the compression stroke is given by

$$\frac{T_2}{T_1} = (\text{CR})^{\frac{\gamma-1}{\gamma}} \tag{A.17.3}$$

In Equation A.17.3, T_2 is the temperature at TDC (without combustion) and T_1 is the value at BDC. The engine CR is CR. Equation A.17.3 applies to isentropic (perfect) compression. CR is the ratio based on the geometry defined by the volumes at TDC and BDC. The ratio of heat capacities is γ. For CR = 10, $T_1 = 200°F = 600°R$, and $\gamma = 1.4$, $T_2 = 614°F = 1274°R$. This temperature may be high enough to cause auto-ignition. For the same initial conditions but with CR = 14, $T_2 = 743°F = 1403°R$. An analysis similar to that above appears in Chapter 22 in "Energy to Compress Hydrogen."

Hot spots may occur in an engine. A little deposit of carbon on the cylinder head may be heated to a glowing red. This hot spot or glow can initiate uncontrolled ignition.

Diesel Engines

The processes leading to combustion include

- Intake pure air (unthrottled)
- Compression (heat the air)
- Fuel injection (typically, direct injection into cylinder shortly before TDC)
- Mixing of air and fuel spray
- Compression ignition (auto-ignition)
- Combustion

Fuel injection is directly into hot air, which has been compressed. Combustion is controlled and initiated by the fuel injection. The processes may overlap. For example, auto-ignition may have occurred in a local region while mixture formation is still ongoing in an adjacent region. Due to auto-ignition, the diesel engine has volume combustion in contrast to surface (flame front) combustion.

One reason for the high efficiency of a diesel is the lack of throttling. Nearly all pumping losses are avoided.

REFERENCES

1. R. El Chammas and D. Clodic, Combined Cycle for Hybrid Vehicles, SP-1973, Society of Automotive Engineers, 2005, R. McGee, M. Duoba, and M. A. Theobald (Ed.), p. 270 (Figures 17.1 and 17.3).
2. U. Spicher, Characteristics, Section 3, *Internal Combustion Engine Handbook: Basics, Components, Systems and Perspective*, R-345, Society of Automotive Engineers, R. van Basshuysen and F. Schäfer (Ed.), 2004, p. 811 (Figure 17.2).
3. R. van Basshuysen, Outlook, Section 29, *Internal Combustion Engine Handbook: Basics, Components, Systems and Perspective*, Society of Automotive Engineers, R-345, R. van Basshuysen and F. Schäfer (Ed.), 2004, p. 811.
4. Editorial Staff, BMW's hybrid vision: Gasoline and steam, *Popular Science*, March 2006, p. 22.
5. F. Markus, Oddball hybrids, *Motor Trend*, June 2006, p. 44.
6. *Fleet Owner*, Available at www.fleetowner.com, December 19, 2005.
7. J. Yamaguchi, Commercial hybrid vehicles from Japan, *Automotive Engineering International*, December 2002, p. 12. (Has data on advances in diesels such as 180 MPa injection.)
8. J. Yamaguchi, Toyota Prius, *Automotive Engineering International*, January 1998, pp. 2110–2114; K. Buchholz, New expedition accents capability, *Automotive Engineering International*, July 2007, p. 10.
9. O. Stenlaas, O. Erlandsson, R. Egnelli, B. Johansson, E. Alm, M. Alakula, and F. Mauss, The effect of unconventional piston movement on SI engine combustion and emissions, SP-1973, Society of Automotive Engineers, 2005, R. McGee, M. Duoba, and M. A. Theobald (Eds.), p. 270 (Figure 17.8).
10. From *Motor Trend* August 2006, p. 148.
11. R. van Basshuysen and F. Schäfer (Eds.), *Modern Engine Technology from A to Z*, R-373, Society of Automotive Engineers, 2007, p. 1047.

12. R. van Basshuysen and F. Schäfer, Eds., *Internal Combustion Engine Handbook: Basics, Components, Systems and Perspective*, R-345, Society of Automotive Engineers, 2004, p. 811.
13. G. Walker, *The Stirling Engine*, Oxford University Press, 1980.
14. F. Marcus, Love child, driving the ultimate bastard engine, *Motor Trend*, December 2007, p. 39.
15. C. Csere, European carmakers protest CO_2 proposals and then parade cars that might meet them, *Car and Driver*, December 2007, p. 11.
16. S. Alceves and D. Flowers, Engine shows diesel efficiency without the emissions, *Science and Technology Review*, Lawrence Livermore National Laboratory, April 2004, pp. 17–19.
17. J. Christoffel, Frankfort concepts, *Automotive Engineering International*, Society of Automotive Engineers, November 2007, pp. 10–12.
18. Staff, Hybrid versus diesel; a 'clean' diesel tops a performance-tuned hybrid, *Consumer Reports*, November 2007, p. 69.
19. S. Ahsley, Diesels come clean, *Scientific American*, March 2007, pp. 80–88.
20. Editorial Staff, Briefly noted, *Mechanical Engineering*, American Society of Mechanical Engineers, September 2006, p. 10.
21. L. Brooke, General motors puts first HCCI prototype on the road, *Automotive Engineering International*, Society of Automotive Engineers, September 2007, pp. 24–25.
22. M. Taylor, When four strokes aren't enough; new engine design doubles power stroke, *AutoWeek*, 3 December 2007, p. 9.
23. S. Birch, Cutting emissions; more answers than questions, *Automotive Engineering International*, Society of Automotive Engineers, June 2007, pp. 26–27.
24. K. Buchholz, Eyes on diesel research, *Automotive Engineering International*, Society of Automotive Engineers, December 2007, p. 42.
25. S. Birch, Powertrain: A4 heralds radical audi technology advances, *Automotive Engineering International*, Society of Automotive Engineers, January 2008, pp. 34–35.

18 Hybrids: Mainstream or Fringe?

INTRODUCTION

The title asks the question whether hybrids will join the mainstream or whether they will remain on the fringe. The answer depends on the type of hybrid in question. Chapter 16 lists many different hybrid possibilities. Energy storage, in addition to batteries, can be ultracapacitor, flywheel, hydraulic/pneumatic, spiral springs, and rubber torsion springs. Energy conversion can be internal combustion engine (ICE) (gasoline, diesel) or fuel cell. For these choices, 15 different hybrids can be considered. In this chapter only hybrid electric vehicle (HEV) and fuel cell vehicle (FCV) are considered. Plug-ins are part of HEV.

HEVs are gasoline/electric or diesel/electric and burn hydrocarbons. The fuel cell/battery and fuel cell/ultracapacitor hybrids typically use hydrogen.

DETERMINING FACTORS

Dominant Factors

Whether or not HEVs or FCVs are on the fringe or mainstream depends on many factors, which form the contents of this chapter. Oil peaking is extremely important, as is discussed in Chapter 2. Oil supply and demand affects oil prices and availability. The status of the global economy affects demand as well as world population. Alternate fuels need to fill the gap between supply and demand that will occur after peaking.

Global warming due to CO_2 emissions is being taken seriously. Governments are restricting the amount of CO_2 emissions. Vehicles burning hydrocarbons are being monitored.

Other Factors

The types of vehicles that will be in future fleets depend on customer desires and preferences. Economics of gas prices and government fees and taxes influence the customer's buying decisions. Those who are environmentally conscious favor hybrids. Fads come and go in automotive features. Chapter 3 discusses objective and subjective items that sway purchase decisions. For the benefit of all society, the customer must have a willingness to accept lesser vehicles. Reference [1] focuses on the numerous and surprising reasons people buy cars.

Alternatives exist to HEV. These include advanced technology gasoline-powered cars, clean diesel cars, and fuel cell-powered vehicles. Also the pure electric vehicle (EV) is being improved and will be an active player.

Government actions affect the mixture of vehicle types to be found on the road in the future. Tax incentives for vehicles with good miles per gallon (mpg) and low emissions are appropriate. Fees on obese vehicles with poor mpg are also effective. The Corporate Average Fuel Economy (CAFE) needs to be increased using valid numbers for mpg. Chapter 12 discusses the details of CAFE.

The government may pass a corporate average CO_2 emissions, which imposes limits. The hybrid is favored once again if the corporate average CO_2 emissions limits are imposed.

Another area where the government can play a role is in citizen energy education. The voters will likely respond favorably to a factual, nonpolitical statement of crises identified in Chapter 2.

Yet another government measure is a gas tax. Annually adding 25 ¢/gal tax spread over 12 years results in gasoline at $6/gal. At this price, which seems outrageously high in the United States, vehicles providing high mpg become much more attractive.

Regulation of Greenhouse Gas

Europe is establishing emission limits that involve a four-year phase-in period from 2012 for fines on manufacturers whose vehicles emit more than an average of 120 grams per kilometer (0.16 lbs/mile) of CO_2. The government in the United States may pass a corporate average CO_2 emissions that imposes limits. Several states in the United States, with California being the gang leader, passed limits on CO_2. These were denied by EPA, which wants to avoid the complexity of state-by-state emissions regulations. See Figure 18.1 for a curve of pounds CO_2 per mile for three vehicle: CV, HEV, and plug-in HEV. The hybrid is favored once again if the corporate average CO_2 emissions become fact.

Consumer Perception of Gas Prices

Most people's view on gas prices occurs at the pump. How often do I need to fill up, and how much does it cost? Chapter 15 discusses the concept of pay at the pump. The overall cost of vehicle ownership is highlighted.

Alternate Fuels

Alternate fuels can be divided into two categories. One category involves hydrocarbons and, hence, CO_2 emissions. The other category includes hydrogen with zero CO_2 emissions. Hydrogen cannot be found in nature as a fuel or source of energy. Hydrogen becomes a fuel that can be transported in a vehicle. Chapter 2 discusses the transportable fuels crisis. Relative to transportation, today's predicament is not solely an energy crisis. It is a transportable (liquid) fuels problem.

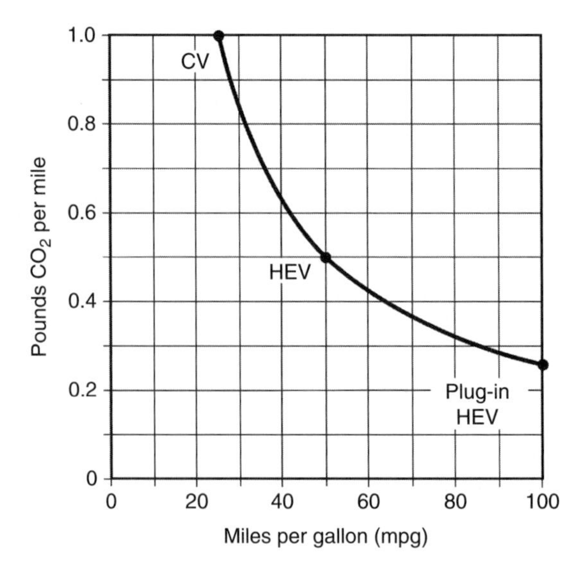

FIGURE 18.1 Level of CO_2 emissions to be expected from CV, HEV, and the PHEV. CO_2 emissions are quantified in terms of the pounds CO_2 per mile of driving.

Uncertainty Concerning Hybrids

One question is long-term reliability of hybrids. The Toyota Prius I was introduced in 1997. The reliability has been good. However, this is a hybrid sold in small numbers from only one car maker. Larger numbers of hybrids from a variety of manufacturers would inspire more confidence.

Maintenance requirements include the added technician skills for hybrids. For service and repair of hybrids additional skills are needed compared to conventional vehicles (CVs). The hourly rate in the repair shop, \$/h of mechanic time, is higher. The rate may be as much as 50% higher. Hence, the cost of service and repair of hybrids will be higher making hybrids less attractive.

Another place where HEVs differ from the CVs is the collision repair of a hybrid. Chapter 20 discusses this facet of hybrid repair.

Mass market acceptability is yet to be demonstrated. Even though hybrids are a hot item on the showroom floor today, the long-term issue remains.

Corporate Average CO_2 Emissions

In response to the dangers of global warming, which is discussed in Appendix B, the government may mandate corporate average CO_2 emissions. The level of the CO_2 emissions from CVs, HEV, and plug-in HEV (PHEV) is shown in Figure 18.1.

The amount of CO_2 emissions is inversely proportional to the fuel economy. This assumes complete combustion. If a gasoline-powered car doubles mpg, the CO_2 emissions drop to one-half. Chapter 22 discusses the fact that the carbon/hydrogen ratio for hydrocarbon fuels has an effect on CO_2 emissions. Not all hydrocarbon fuels have the same carbon/hydrogen ratio.

COMPETITION: HEV VERSUS GAS/DIESEL

Competition to the HEV

Whether or not a hybrid is on the fringe or mainstream depends on the competition. The various technologies that may improve fuel economy for advanced gasoline and clean diesels appear in Chapter 17. The competitors to HEV are

- Vehicles with advanced gasoline engines
- Vehicles with clean diesels
- Fuel cell hybrid
- Advanced pure electrical cars
- Hydrogen-fueled ICE

Figure 17.9 shows the future fuel economy for advanced gasoline (50 mpg), clean diesels (55 mpg), and HEV (75 mpg). Comments are made that the marginal savings do not support fuel economy much above 50 mpg.

Favorable Factors for HEV

The big two factors are superior fuel economy and low emissions. Availability today is also a significant factor.

Unfavorable Factors for HEV

HEVs are complex due to the added battery, M/G, power electronics, and control system. Besides adding cost, the extra parts add weight. The added cost is known as the price premium. The added weight affects the HEV in ways to be described.

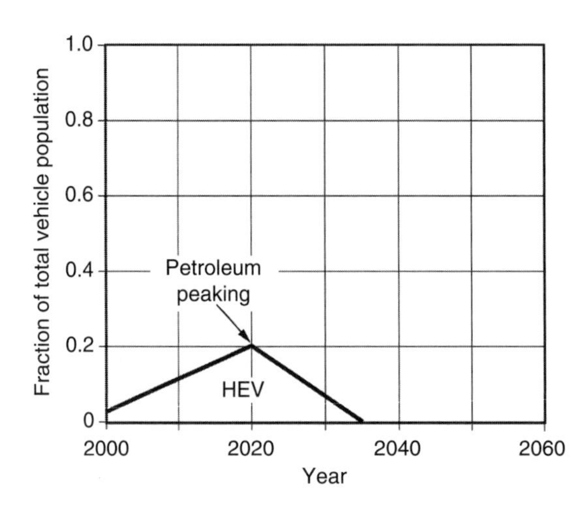

FIGURE 18.2 HEV fraction of the total vehicle population for the years 2000–2060.

FUTURE FOR HEV

The information dealing with the future is speculative. The speculation is based on data that are known today and extrapolated forward into time. The graph in Figure 18.2 shows what fraction the future HEV population will be compared to total vehicle population.

The fraction of the total vehicle population is not the fraction of annual car vehicle sales. For example, the 2020 population fraction is 20% while fraction of annual sales could be 40%. Initially, the gas/diesel vehicles completely dominated the global fleets. Starting from almost 100% gas/diesel in 2000, time is needed for hybrid sales to have an impact. The mixture of cars changes slowly. The next section provides information on the time lag for fleets.

TIME LAG: VEHICLE POPULATIONS

The number of cars that are on the road depends on the number sold in previous years and on the number that are removed from service due to collisions where the car is totaled or where the car is junked because the costs of maintenance were overwhelming. The following analysis sheds some light on this topic. Let us start with a table and then explain each symbol and number in Table 18.1.

TABLE 18.1
Sample Data for Determination of Vehicle Population

Model Year, τ	Age, Years $t - \tau$	Junk Fraction, $f(t - \tau)$	Annual Production Rate, $P(\tau)$		
			Increasing	Constant	Declining
2008	0	0	25	25	5
2007	1	0.2	20	25	10
2006	2	0.5	15	25	15
2005	3	0.8	10	25	20
2004	4	1.0	5	25	25
		Total production	75	125	75

Note: The numbers were chosen for convenience of calculation and not for being typical actual production numbers.

The model year, that is, the year the car was produced, is the symbol, τ. The current year is t and for this example $t = 2008$. The age of a car is simply $t - \tau$. The junk fraction, $f(t - \tau)$, is the fraction of cars that have been sent to the junkyard or, more euphemistically, the recycle center. $f(t - \tau)$ depends on vehicle age. The annual production rate, $P(\tau)$, is the number of cars produced in a single year.

Consider one term from the column for increasing production in Table 18.1. How many cars are on the road in 2008 from cars produced in 2005? This is

$$P(\tau)[1 - f(t - \tau)] = P(2005)[1 - f(2008 - 2005)] = 10(1 - 0.8) = 2 \quad (18.1)$$

For all production years the total number of cars on the road is

$$25(1-0) + 20(1-0.2) + 15(1-0.5) + 10(1-0.8) + 5(1-1) = 50.5$$

Of the 75 cars produced, 50.5/75 = 67% are still on the road in 2008.

Consider the column for declining production in Table 18.1. For all production years the total number of cars on the road is

$$5(1-0) + 10(1-0.2) + 15(1-0.5) + 2(1-0.8) - 25(1-1) = 22.5$$

Of the 75 cars produced, only 22.5/75 = 30% are still on the road in 2008.

The percentages of car on the road are

Increasing production rate	67%
Constant production rate	50%
Declining production rate	30%

An increasing production rate loads the later years where $f(t - \tau)$ is smaller. A declining production rate loads the earlier years (cars are older) where $f(t - \tau)$ is larger.

The concepts can be placed into an equation, which is

$$N(t) = \int_{\tau_0}^{t} P(t)[1 - f(t - \tau)]\mathrm{d}\tau \quad (18.2)$$

$N(t)$ is the number of cars on the road at year, t. The lower limit on the integral is τ_0, which is the first year of production for the cars.

Discussion of Future for HEV

Before discussing the future of HEVs as depicted in Figure 18.2, same facts from Ref. [2] will be presented. Four different vehicle technologies are considered:

- Turbocharged gasoline engines (2016, 2016)
- Low-emissions diesel (2021, 2021)
- Gasoline hybrid HEV (2026, 2021)
- Hydrogen fuel cell hybrid vehicle, FCV (2031, 2026)

The first number in parenthesis is the year that the vehicle will comprise more than one-third of new vehicle production. The second number is the year that major fleet penetration occurs. Major fleet penetration occurs when more than one-third of fleet mileage is driven. These dates can be (somewhat) compared with Figure 18.2.

The increase in HEV fraction from 2000 to 2020 is due to superior attributes. The HEV provides better fuel economy and lesser emissions than either the gasoline or clean diesels. At 2020, oil peaking occurs sending the price of gasoline skyrocketing. In the interval 2000–2020, progress is made on FCV and the necessary hydrogen infrastructure. The FCV has superior traits compared to the HEV. The decrease in HEV fraction from 2020 to 2035 is due, in part, to the superior attributes of the FCV.

Why is 2020 a decisive year? Oil peaking is assumed to occur. Why is the fraction of cars on the road only 20% for HEV and not 30%? In 2000, the fleet starts with almost 100% gasoline or diesel vehicles. As noted above, a time lag exists before the introduction of a new type vehicle has an impact. Also, competition between HEV and gasoline or diesel vehicles continues from 2000 to 2020. Advances in the electrical parts of an HEV are also applicable to the EV. New EVs with greater range between charging will be introduced. Two-car families may decide that one car can be an EV.

Why does the fraction of HEV become zero at 2035? Even low amounts of CO_2 emissions are frowned upon. After peaking, gasoline becomes very expensive. The hydrogen alternative really becomes an alternative. After oil peaking, hydrogen takes off big time.

Home Generation of Electricity

Home generation of electricity is possible using either wind turbines or photovoltaic cells. The electricity can be used to charge a PHEV or an EV. The electricity can also be used to create H_2 by electrolysis. The hydrogen can be used for an FCV. Home generation favors PHEV, EV, and FCV.

Home generation of gasoline is not possible using today's technology. Natural gas may be used from the home gas supply. However, natural gas will have prices in step with gasoline. HEV cannot be supported by home-generated energy. The decline of HEV in the period 2020–2035 is due in part to competition from home generation for PHEV, EV, and FCV.

COMPETITION: HEV VERSUS FCV

HEV is inferior to FCV for two reasons. First, HEV burns hydrocarbon fuels and emits CO_2, which is bad for global warming. Second, after oil peaking, petroleum will be scarce. An FCV will be a simpler vehicle (Figures 18.3 and 18.4).

Figure 18.3 shows the components for an HEV, which include the engine, clutches, and transmission. Not shown are the radiator to cool the engine and the exhaust system. Figure 18.4 illustrates an FCV with the engine-associated parts removed. The fuel cell is considerably lighter than the engine. Due to power transmission by shafts, the engine in HEV is restricted to certain locations in

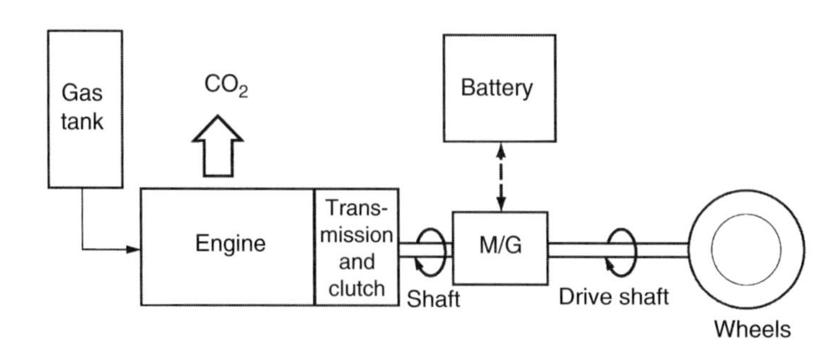

FIGURE 18.3 Heavy, expensive components for an HEV.

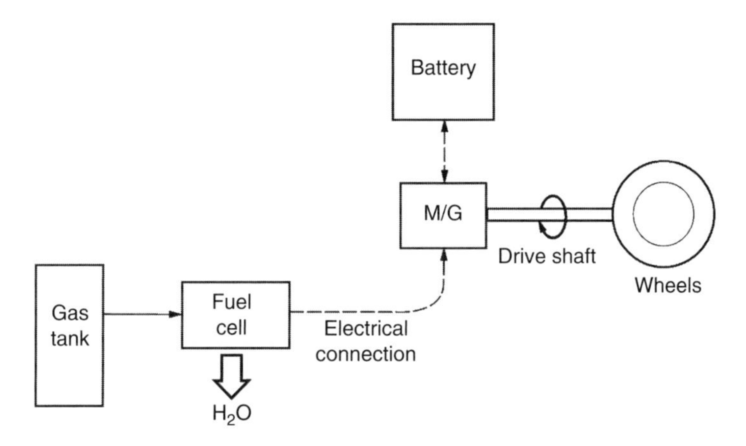

FIGURE 18.4 FCV with the heavy, expensive components of an HEV eliminated.

the vehicle. In both Figures 18.3 and 18.4, the dashed lines are cables for electrical power. The fuel cell can be located almost anywhere in the vehicle. Great design flexibility exists. The FCV has a higher efficiency than an ICE. FCV is an electrochemical device whereas the ICE is a heat engine.

As illustrated, the HEV has CO_2 emissions while the FCV emits water, H_2O. The FCV has fewer components and uses a fuel that is an alternate to petroleum.

FUEL CELL VEHICLE FUTURE

As with the HEV, the future is depicted by a graph of fraction of the on-road fleet as a function of time. This graph is shown in Figure 18.5. The fraction of FCV in the time period 2000–2020 is small. Adaptation of the FCV is limited by the lack of hydrogen service stations alongside the road. The fuel cell costs must be decreased and from 2000 to 2020, cost reductions are anticipated.

After 2020, when oil peaking occurs, a spurt in FCV fraction is shown. Why the spurt? The hydrogen infrastructure was growing from 2000 to 2020 [3]. After oil peaking and scarce petroleum, the FCV becomes much more competitive with HEV and especially with gasoline and diesel. The HEV serves as a bridge between the days of almost 100% gasoline/diesel and the hydrogen mobility

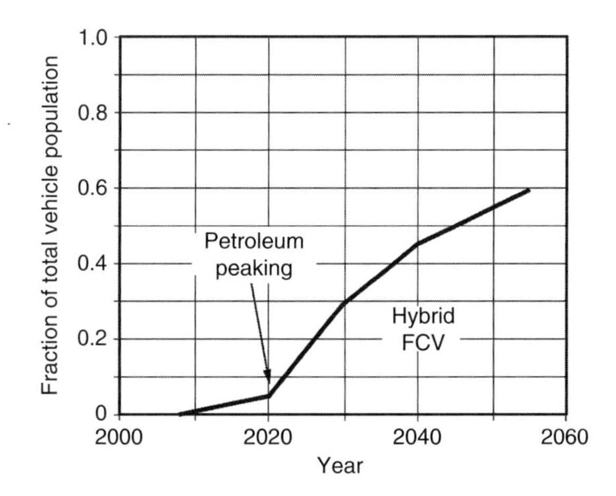

FIGURE 18.5 Fraction of the total vehicle population, which is FCV.

of the FCV. Since the FCV is superior to the HEV, the FCV grows rapidly (Figure 18.5) and the HEV fraction declines slowly (Figure 18.2). Fuel cells concept cars, ca. 2006, are discussed in Ref. [4]. Technical details of FCV are available in Ref. [5].

The diagram below indicates that the gas/diesel was replaced by the HEV due to higher mpg and lower emissions. In turn, the HEV was replaced by the FCV due to the need for an alternate fuel and zero CO_2 emissions by the FCV.

Gas/diesel →→→→→→ HEV →→→→→ FCV

Higher mpg   Alternate fuel

Lower emissions CO_2 emissions

SUMMARY

This chapter starts with the question: Hybrids: Mainstream or Fringe? The prediction is that the HEV is a fringe vehicle while the FCV hybrid will become a mainstream vehicle. In the year 2055, the answer will be known. Until then, the predictions are speculations based on the facts as known at this time.

REFERENCES

1. J. Morgan, *The Distance to the Moon: A Road Trip into the American Dream*, Riverhead Books, New York, 1999, 319 pp.
2. J. B. Heywood, Fueling our transportation future, *Scientific American*, September 2006, pp. 60–63.
3. J. Ogden, High hopes for hydrogen, *Scientific American*, September 2006, pp. 94–101.
4. A. Taylor, III, The new fuel thing, *Fortune,* 2 October 2006, pp. 47–48.
5. R. K. Stobart and S. Kuman (Eds.), *Applications of Fuels Cells in Vehicles 2006*, Society of Automotive Engineers, Warrendale, PA, SP-2006.

19 Factors Influencing Resale Value

The battery dominates the resale value of a hybrid.

INTRODUCTION

Hybrid electric vehicle (HEV) and conventional vehicles (CV) share many of the same depreciation factors. Early in this chapter, common items will be discussed. The HEV is distinctive due to its electrical components. The battery is of prime concern with power electronics and M/G being less important.

Resale value depends on who the buyer is. In order of increasing sales price are trade-in to a dealer, sale to factory dealer, sale to independent car lot, and finally sale to a private individual.

COMMON FACTORS

WARRANTIES

Exposure to future costs is mitigated by a generous warranty. More will be said later about battery warranties.

RELIABILITY AND QUALITY

How many trouble-free miles are yet to be driven? Anticipated repair expenses affect the resale value. The mileage of the vehicle is important. The car maker's reputation for reliability suggests that fewer repair costs will be encountered.

Do not judge hybrids by one car maker. Do not assume all hybrids have the same reliability. However, compared to CVs, hybrids may offer superior reliability in spite of greater complexity.

DESIRABILITY

Human nature, being what it is, makes an item especially desirable when it cannot have it. Translate this to automobiles, and enhanced (inflated) desirability is understandable. Cars with a long waiting list and cars getting more money over manufacturers suggested retail price (MSRP) are desirable. Scarcity creates long waiting lists and raises final sales over MSRP. Caution: car manufacturers can ramp up production. A short spike in desirability occurs.

Other more substantial reasons make a car desirable. These include style, performance, engineering, character, and design. Less substantial reasons involve intangibles, which are nonetheless real.

UNDESIRABILITY

The items that can be defined in terms of numbers (mpg, payload, etc.) are objective factors and less likely to be undesirable. The subjective factors, such as color, can make a car unwanted. An odd color, purple/yellow two tone, will not be popular to the general car-buying public. Overly personalized vehicles will also suffer from low resale value. Low-riders, boom speakers, decals, and tacked-on chrome lower value. Anything that cuts the instrument panel or the exterior body surfaces may greatly degrade value.

Options and Accessories

Options and accessories may or may not help resale value. The initial purchase price of options will depreciate along with the vehicle but in some cases, none of the price will be recovered. Many options are first offered on luxury cars and then work their way down the price ladder. Examples include adaptive cruise control, which uses radar to apply the brakes if an object is close in front. Another example is keyless entry or remote unlocking, which today is very common. Keyless entry is essential. With the advent of global positioning satellite navigation system the onboard LCD screen and voices indicating where to turn are becoming an essential and not a luxury.

Air-conditioning (A/C), which has become standard in all but the cheapest of cars, will be a negative inducement if it is not part of the car. A/C is almost as essential as the engine. A sunroof is neutral; okay if you have it, but not essential. Some years ago the so-called gold kit was popular. The gold kit made all trim items look like polished gold. Extra big wheels (large diameter) are trendy and valuable at trade-in time.

Run-flat tires offer, as the name implies, a certain amount of insurance against a flat tire while on the road. Further, the customer gains storage space since a spare tire is not needed. However, run-flat tires are considerably more expensive and do not wear as long as conventional tires. Run-flat tires may wear out in 30,000 mi, or so, compared to 50,000 mi for the usual tire.

An option offered by some car makers is special paint. At resale time, this special paint is usually a complete loss.

Depreciation

As stated in Chapter 15, depreciation is a stealthy cost of ownership. It takes effect when the owner sells or trades-in the hybrid. Depreciation is important at lease-end for a hybrid. Depreciation determines cost of ownership. Small depreciation yields smaller costs.

Figure 19.1 shows many features of buying a new car and, after a few years, the selling of that car. Your new car becomes a used car as soon as it is registered in your state. A purchase price of $30,000 appears in the figure. Add to that amount typically 10% more for taxes, license, and fees. The out-the-door price (with 10% added in) is $33,000. The instant the key is turned to start the engine to drive off the dealer's lot, the car changes from new to used and loses 15%–20% value. This instantaneous depreciation, which is $5,000 for this example, is shown as the drop from A to B in Figure 19.1. Total cost on the first day of ownership is $8,000.

During the first year, once the $8,000 has been paid, the value remains nearly constant during that year. The line from B to 1 moves time forward by 1 year. At point 1, the exponential decline begins. After 3 years, the typical depreciation is one-half the sticker price, which is (1/2)($30,000) = $15,000. Point 2 is at 3 years and $15,000.

The equation for the exponential decay is

$$V = V_0 \exp\left(-\frac{t}{\tau}\right) \tag{19.1}$$

where

V_0 is the vehicle value at zero time
τ is the characteristic depreciation time

Both V_0 and τ are constants that are determined by the two points that are on the curve. These points are ① and ②. Using the values at these two points, V_0 and τ are found. V_0 is equal to $32,270, which

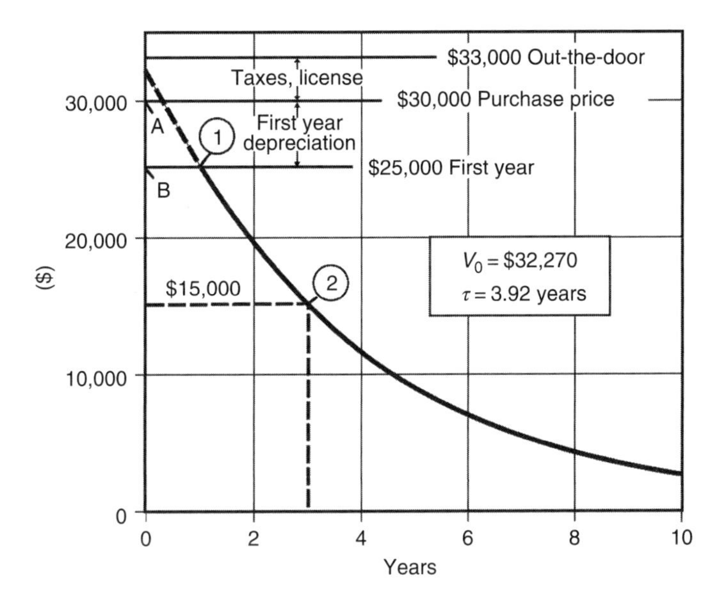

FIGURE 19.1 Depreciation of an automobile showing an exponential decay of value.

is almost the out-the-door price. τ equals 3.92 years. The depreciation curve is calculated using Equation 19.1 and the values for V_0 and τ. At 10 years, the car has depreciated to a value of approximately $2,500.

New Car Rebates

Special deals such as "Buy at employee's prices" and "0% financing" are helpful at the time of buying a new car. Nevertheless, you eventually pay. At resale time the savings now appear as a lower resale value. When you get ready to sell the car or return the car at end of lease, the value will be less due to the initial rebate. This is due, in part, to the fact that the special deals originated because the new car was not very desirable to begin with.

Mileage and Vehicle Condition

Both vehicle mileage and condition have a large effect on resale value. Condition of the automobile has several aspects. Maintenance in accordance with car maker's schedule insures the mechanical components will continue to provide service. Receipts help prove maintenance was actually done. Other aspects include cosmetic measures. Has the car been parked in a garage or has it been out in the sun and the weather on rainy and snowy days? The sun is a powerful destroyer of upholstery. Is the exterior clean without chipped paint? Is the interior clean without food stains on the upholstery?

Mileage was discussed above. Anticipated repair expenses affect the resale value. The mileage of the vehicle is important.

A band of resale values is along the average depreciation curve. This band grows in width as the years go by. During the first year, the band is less than $1000 wide. The cars differ more widely with regard to condition as the years pass. At 5 years, the difference between a cream puff car and a clunker is $4000 with an average resale of about $9000. At about 9 years, the car in poor condition has a zero resale value; the owner must pay to have it towed to the junkyard. At 9 years, a clean,

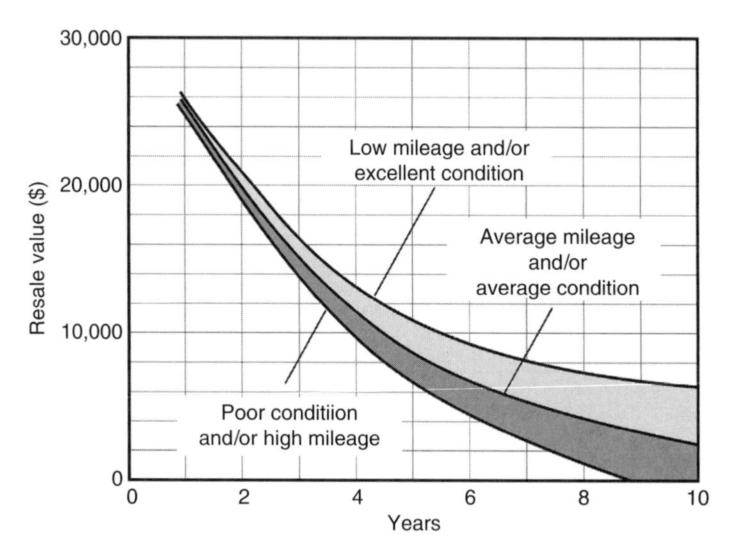

FIGURE 19.2 Effect of car mileage and condition on resale value.

low-mileage car has resale of $8000. For a car with low mileage in excellent condition, after about 6 years, the depreciation is very low. Note the top curve in Figure 19.2.

RESALE FACTORS SPECIFIC TO HYBRIDS

Battery

An expensive battery is unique to the HEV (and EV). As shown in Figure 19.3, when the hybrid is new, the battery is 16% of the value of the hybrid: $3,600/$22,000 = 0.16. When the hybrid is 3 years old, the battery is 28% the value of the depreciated hybrid: $3,600/$12,760 = 0.28. The price of a replacement battery does not depreciate.

If at the end of 3 years, replacement of the battery is necessary, the resale value of the car drops from $12,760 to $9,160. Depreciation increases from 42% (without new battery) to 58% (with new battery), which is a dramatic change.

Warranties

Because of the potential for major repair cost, the warranty on the battery is crucial. In addition to warranty for the first buyer, the warranty must be transferable to the subsequent owners. Otherwise,

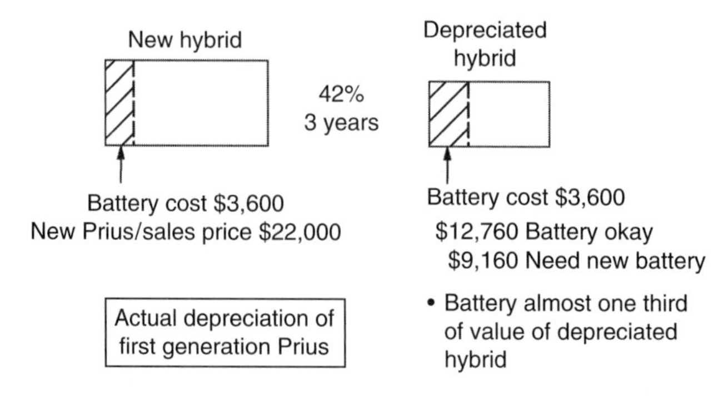

FIGURE 19.3 Battery cost relative to overall vehicle cost when new and after 3 years. The size of the rectangle is proportional to vehicle value.

the hybrid resale value is greatly reduced. The sales brochure for the 2005 Prius II states a battery warranty of 80,000 mi.

Aftermarket Modifications

Enthusiastic owners of HEVs want more mpg. Conversion of an HEV into a plug-in HEV (PHEV) can be done. Chapter 20 discusses the conversion. Avoid voiding the warranty. Some modifications will void the warranty on a new hybrid. This is only an economic danger but can be significant.

At resale time, an HEV that has been converted to a PHEV may be almost impossible to sell. Potential buyers may be afraid of impending difficulties. A PHEV will have a much larger, and more expensive battery. The fraction of vehicle cost that is due to battery will be much larger for a PHEV than for an HEV.

Finger in the Dike: Replacing Individual Cells

Most HEVs monitor the health of every cell in the battery. Many batteries for HEV allow replacement of individual cells. Replacement of individual cells, which might be called putting your finger in the dike, will be much less expensive. Even though cell costs will be relatively low, replacing cells will be high in labor costs. Replacing cells is a stopgap measure. If some cells fail, that implies more cells are on the verge of failure also.

The potential buyer of a hybrid should be aware of replacement of individual cells. If the owner, just before selling the HEV, has replaced only the sick cells, a major battery replacement may just be around the corner.

Depreciation and Pace of Technology

Resale is influenced by the rate of technological advances. The pace of technology does affect depreciation. In a short time, when technology is advancing rapidly, an older hybrid quickly becomes obsolete. Being obsolete reduces the resale value. Is hybrid technology advancing rapidly? The answer is yes, it is. The question of the rate of technology advancements is discussed in Chapter 15.

An example of the effect of very rapid technology advancements is computers. What is the resale value of a 3 year old computer? Next to nothing. What is the resale value of a 5 year old computer? One must pay to have it recycled.

SUMMARY

HEVs and CVs share many common factors that affect resale value. How many trouble-free miles remain in this vehicle? That is a key question all buyers of used cars ask. Reliability and quality are guides. Some automobiles are especially desirable; these cars enjoy high resale. Desirability may be transient. The flip side is undesirability by such things as a purple/yellow paint. Options can help or hurt resale value, and they have been discussed.

In the initial purchase, three prices are important all of which are identified in Figure 19.1. Depreciation tends to follow an exponential decline, and the appropriate equation is Equation 19.1. New car rebates adversely affect resale. Mileage and vehicle condition have a major impact on depreciation. As the years go by, the value gap between the cream puff (low mileage, clean car) and the clunker (high mileage, grimy car) tends to widen.

The battery for an HEV is expensive and lurks as a potential money sink. A generous warranty helps to mitigate battery phobia. PHEVs are particularly vulnerable to battery replacement costs, and all the more so for an aftermarket conversion of an HEV into a plug-in. The stopgap measure is to replace only the sick cells in a battery.

20 Dangers of Aftermarket Add-Ons

The aftermarket has two facets. One is service and repair. The other is performance modification. Aftermarket add-ons are found usually in performance modification.

SERVICE AND REPAIR

PREVALENT BELIEFS CONCERNING HYBRIDS

The beliefs cover a broad range. On one hand, hybrids are viewed as just another car but with more computers and more technologies. On the other hand, some service and repair personnel feel intimidated by hybrids. These people feel that hybrids will require considerable effort to master the high voltage, the batteries, the motor/generators, the power electronics, throttle-by-wire, and brake-by-wire. These thoughts lead to the necessity of the training of the repair and service technicians.

TRAINING

How much does the technician need to know to repair and service a hybrid? The answer to that question is evolving as maintenance data build from hybrids on the road.

SAFETY

A major safety item is the high-voltage battery. The battery is not only dangerous, it is lethal. The attitude cannot be "Oops, I am sorry!" With regard to the battery, only one mistake, that is, electrocution, can cost the technician his/her life.

The trend for the conventional vehicle (CV) electrical system is toward 42 V. Mild hybrids typically use 42 V, which is not dangerous with regard to electrocution. Full hybrids use higher voltage batteries, which are a low impedance source. This means the battery can provide high current at high voltage. High voltage alone is not dangerous. A spark plug has 30,000–40,000 V. The energy in the pulse is very low; consequently, grabbing a spark plug wire is unpleasant but not fatal.

SERVICE AND REPAIR OF HYBRIDS

Some service items for hybrids are common to CV. For example, changing oil, changing antifreeze, and rotating tires. Other items are hybrid specific and are part of hybrid maintenance. Many components, especially on the electrical side, require only minimal maintenance, or do not require routine maintenance for the life of the car.

Since hybrids in large numbers are new, many hybrids are still under warranty. The owner will take any problems to the dealer. Due to newness, the history of which parts fail has not been accumulated. Car part stores do not stock repair parts for hybrids. The logistics for hybrid parts has not been established. For example, a mechanic can buy and install a rebuilt alternator for any CV. However, a rebuilt motor/generator for a hybrid cannot be found. For that matter, even a new motor/generator for a hybrid cannot be found outside a dealership. Status of repair availability determines whether hybrids are on the fringe or in the mainstream.

Repair of Collision Damage

Collision damage is in a separate category of repair. Eventually, hybrids will be involved in collisions, which may damage both body and drive train components. Currently, drive train parts are only available from the appropriate new car dealer. The collision repair can perhaps be done only in a dealer's body shop. Competition from independent shops is reduced or nonexistent; hence, prices are higher.

Auto insurance companies will likely detect the price discrepancy between HEVs and CVs. Insurance rates may increase for hybrids.

PERFORMANCE MODIFICATION

Avoid Voiding the Warranty

Some modifications to be described will void the warranty on a new hybrid. This is only an economic danger but can be significant.

Training

How much does the technician need to know to modify a hybrid? The answer is considerably more than just to service and repair a hybrid. This fact will become apparent in the subsequent discussion.

Safety

As with service and repair of hybrids, a major safety item is the high-voltage battery. Modification of a hybrid may lead to more safety issues for both the personnel doing the modification and the customer buying the modified hybrid. The brave, shade tree, backyard mechanic attempting to install a modification kit may be in particular danger.

Performance Enhancement

Four enhancements with appeal are

- Extended electric-only range
- Extended range until fill-the-gas-tank, that is, total range
- Acceleration; rarely deceleration performance
- More miles per gallon (mpg)

An equally powerful motivation is bragging rights. That is, "My hybrid has more .... than yours!"

Just More Computers

One viewpoint is that hybrids are not really any different from CVs; "they just have more computers." In one sense that is true; however, the numerous electronic control units (ECUs) are very tightly integrated. Part of the theme of Chapter 8 was total interaction of all the parts.

Revisit Figure 8.5, which shows the hybrid ECU, in command of all the subordinate ECUs. The commands from hybrid ECU are based on very extensive, time-consuming, thorough, and expensive optimization.

Tuners are organizations that tune and modify vehicles to enhance performance. These tuners do modify today's CV by reprogramming the engine computer and adding new engine components. The computer that is reprogrammed is the engine ECU of Figure 8.5. Obviously, the CV does not have a hybrid ECU. The tuner has not faced the complete interaction of a hybrid control system.

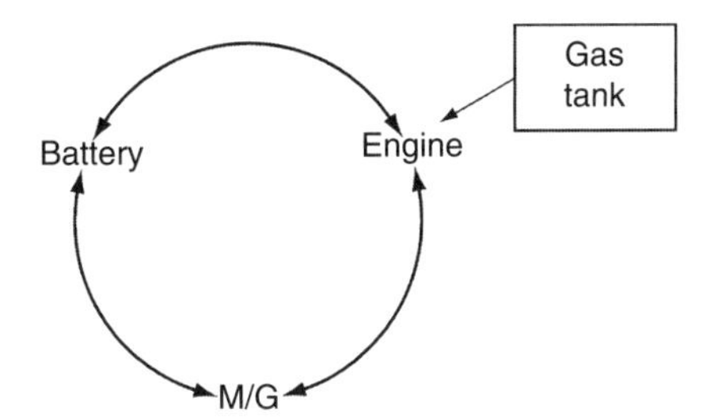

FIGURE 20.1 Major hybrid components are tightly interconnected and integrated through the hybrid ECU. The gas tank is outside the control loop and is a candidate for easy modification.

The analogy of a three-legged stool is appropriate. Suppose a 1 in. extension is made to one leg; this change is equivalent to modifying a sub-ECU within the overall hybrid ECU. The stool is no longer useful for its intended purpose. To restore the utility of the stool, the other two legs must be modified. Likewise, if one sub-ECU is modified, then the other ECUs may need modification to restore optimum operation of the HEV.

Figure 20.1 is intended to provide the feeling of the control network of Figure 8.5. The battery is connected to the engine and the engine is connected to the motor/generator, which in turn is connected to the battery. What happens to one affects operation of all the other components. The gas tank is outside the control loop.

Logic on Complexity of Hybrid Control

One manufacturer may have a 3 year lead in hybrid technology compared to another manufacturer. This lead may not be due solely to hardware but primarily due to hybrid control and integration. In spite of enormous resources, the second company may find it difficult to catch up. If a manufacturer with enormous resources cannot overcome a 3 year lag, how can an aftermarket organization modify a hybrid control to achieve a gain for their customers? With difficulty.

Adding a Second Gas Tank

For some hybrid owners, a performance factor to be increased is range until fill-up-the-gas-tank. Adding another gas tank solves that problem. Further, the gas tank is outside the control loop of Figure 20.1. Hence, no problems are anticipated with an upset control loop. However, now the problem is finding an empty space aboard a closely packed vehicle. Not just any space will be safe or convenient. The fuel tank needs to be connected to the fuel system and meet the needs of the fuel injection system.

The hybrid owner must be strongly motivated to accomplish the task of adding a second gas tank. That addition is easy compared to adding a second battery.

Adding a Second Battery

To add electric-only range with the expectation of increased mpg, the standard battery of a hybrid can be augmented with another. Just as total hybrid range can be extended by adding to the size of the gas tank, the electric-only range is extended by adding battery energy. The expectation of more mpg may not be realized. More mpg will result if the battery is charged by plugging into the local utility. Battery company A123 acquired Hymotion for conversion of HEV into plug-in HEV.

What charges the larger battery? Of course, plug-in is one answer; however, the freedom from plug-in of the normal hybrid is lost. Energy to charge the battery can come from two sources. One is the engine; the other regenerative braking and coasting. Regenerative braking is increased few percent due to added mass of battery. The engine must supply almost all of the added energy.

What is the effect on the hybrid if plug-in is not used? Doubling the battery size does double the electric-only range. Less energy from the engine is available for moving the vehicle. Recall from Chapter 8 a control algorithm, which is

- Control electrical motor and gas engine torque to equal required drive torque
- Operate engine at most efficient points
- Maintain battery state of charge (SOC)

Note the need to maintain battery SOC. In doing so, the ability to operate the engine at most efficient points may fail. Off-optimum operation decreases mpg. Electric-only range is increased but total range is decreased due to loss of mpg.

Making a Home for the Battery

To have a long life, the new battery requires attention to some details. These are

- Cooling
- Temperature monitoring
- Monitor status of each cell
- Installation of safety features such as electrical interlocks
- Monitor SOC
- Attempting to integrate with hybrid control system
- Protection against overtemperature and overcharging

Adding a battery is more than finding a spot and hooking up a couple of wires.

Payload Penalty

For a gain of 5 km in electric-only range, the added battery weighs 50 kg. Assume a hybrid with empty weight of 1900 kg. The payload is 400 kg, which gives a gross weight of 2300 kg. The payload is five passengers with luggage totaling 400 kg. Table 20.1 shows the impact of adding battery weight. See www.iihs.org for details about overloading and safety. Also www.safecar.gov is another source of information.

Performance and Handling Penalties due to Added Battery Weight

Added battery weight has an effect on several of the vehicle handling and performance characteristics including

TABLE 20.1
Payload Penalty due to Added Battery Weight

Electric-Only Range, km	Added Battery Weight, kg	Remaining Payload, kg	Passengers	Luggage, kg
5	0	400	5	90
10	50	350	5	40
15	100	300	4	30
20	150	250	3	50

Note: The vehicle is operated at specified gross weight and is not overloaded.

- Acceleration
- Braking
- Steering
- Depending on location of battery an effect on location of center of gravity (CG)
  - Susceptibility to roll over
  - Weight distribution front-to-rear and directional stability
- Tire rolling friction; more weight, more friction

Performance Penalty

Table 20.2, which was calculated using force, which equals mass times acceleration, shows the loss of performance if a hybrid owner decides to overload the HEV. Braking is stretched out by 11 ft or 8%.

Obesity Penalty due to Added Battery Weight

Gas mileage, mpg, is decreased by added weight. The decrease is given by

$$\frac{\Delta \text{mpg}}{\text{mpg}} = \frac{\Delta F_E}{F_E} = -C\left(\frac{\Delta W}{W}\right) \tag{20.1}$$

where
  W is vehicle weight
  C is a correlation constant which typically has a value $C = 1.2$ for lightweight cars

Using the added battery weight, $\Delta W = 50$ kg, a vehicle weight of 1900 kg, and a mileage of 45 mpg for $W = 1900$ kg, the calculation for mileage loss is

$$\frac{\Delta \text{mpg}}{\text{mpg}} = -1.2\left(\frac{50}{1900}\right) = -0.03 = 3\% \tag{20.2}$$

Table 20.3 was calculated using the above procedure.

TABLE 20.2
Performance Penalty due to Battery Weight

Electric-Only Range, km	Added Battery Mass, kg	Overloaded Mass, kg	Loss of Acceleration, %	Time, s 0 to 60 mph	Braking m	Braking ft
5	0	1900	0	12.0	40	131
10	50	1950	3	12.4	42	135
15	100	2000	5	12.6	43	138
20	150	2050	8	13.0	44	142

Notes: Performance in this case is defined as the increase in 0–60 mph time and increased braking distance. The vehicle is overloaded. The factory fresh, original design HEV yields 5 km electric-only range. This is the top row in the table and serves as the reference vehicle.

TABLE 20.3
Loss of Mileage due to Added Weight of Battery

Electric-Only Range, km	Added Vehicle Mass, kg	Loss of mpg, %	mpg
5	0	0	45
10	50	3	44
15	100	6	42
20	150	9	41

Adding weight to a vehicle creates an mpg hole. A person thinking of modification needs to climb out of the mpg hole before reaping any of the benefits from the modification. For the example of Table 20.3, the hole is 4 mpg for added 150 kg.

Two Classes of Batteries

Table 20.4 provides data for subsequent discussion.

If a second battery is added, one possible benefit is to define two classes of batteries. The first class is the original battery that came with the car. The newly added battery is the second class. Because of life restrictions on depth of discharge, only about 25% of the total battery energy can be used. Since the second battery may only be needed on occasion and not every trip, the life restrictions on depth of discharge become less important. Expanding the allowable range of SOC to 75% decreases the added battery weight by a factor of 3.

$$\frac{75\%}{25\%} = 3$$

Table 20.4 shows the reduced battery weight using 75% range for SOC.

SUMMARY

Since hybrids are relatively new, an aftermarket for service and repair is in its infancy. As market develops, the need for training also develops, especially in the area of safety. The repair of the hybrid drive train differs from collision repair.

TABLE 20.4
Using Allowable SOC of 75%, a Major Savings of Added Battery Weight Is Possible at the Expense of Battery Life and Increased Replacement Costs

	Added Battery Weight, kg, for Different SOC	
Electric-Only Range, km	SOC = 25%	SOC = 75%
5	0	0
10	50	17
15	100	33
20	150	50

Performance modification of hybrids is considerably more complex than that for CVs. Questions about voiding the warranty must be kept in mind. Two examples are discussed. One is adding a second gas tank to extend total range. This example was chosen to compare complexity of adding a second battery. The various penalties due to adding a battery are presented.

21 Safety Issues

Can the occupants of a collision-damaged hybrid be rescued safely in spite of high voltages and possible spilled dangerous chemicals?

OVERVIEW

Automobiles involve tremendous quantities of stored energy. Conventional vehicles (CVs) have a gas tank; the gasoline is a form of stored energy. People have learned to live with whatever danger exists from a tank of gasoline. Hybrid vehicles also have a fuel tank, but in addition have energy stored in batteries. Concentrated energy, if released in a sudden burst, can cause tremendous damage.

To provide some perspective on the magnitude of energy stored in a battery, a calculation was made setting kinetic energy equal to battery energy. A hybrid electrical vehicle (HEV) may have a battery with energy 2 kWh. The same HEV needs a speed equal to 310 km/h (190 mph) to have 2 kWh energy. Think of an HEV hitting a brick wall at 190 mph. Another calculation, which was made setting potential energy equal to 2 kWh, yielded a height of 360 m (1200 ft). Imagine a vehicle dropped from an aircraft flying at 1200 ft. The damage is the same as that caused by sudden, complete release of the battery energy.

Two situations exist. In normal day-to-day driving, the driver and passengers in a hybrid are exposed to the same dangers as a CV. In case of collision-induced damage, the hybrid is different from the CV. The focus of this chapter is on the latter.

WHAT ARE TODAY'S HAZARDS?

One hazard is the nasty chemicals that make a battery work. The lead-acid battery, which has been around for 100 years, has sulfuric acid, which is wicked stuff. Once again, people have learned to live with whatever danger exists from battery acid. Almost every HEV on the road today has a NiMH battery. Each cell of the battery contains potassium hydroxide, which is a strong alkaline chemical that reacts vigorously with several common metals such as aluminum. It is hazardous to all human body tissues. When ruptured, the battery may leak potassium hydroxide. To a minor degree, the hazards exist daily, but an automobile wreck brings these hazards to the forefront. The battery of the future may be lithium-ion (Li-ion). Safety aspects are discussed in Appendix 21.1.

Table 21.1 gives some voltages found in various vehicles. Which parts of an HEV have high voltages? The battery, the cables from battery to engine compartment, the inverter, and the motor/generators (M/Gs) are all at high voltage. More will be stated about design features to mitigate perils of high voltage.

Obviously, a bare cable should not be touched; it could be a high-voltage cable. High-voltage cables are orange. If an HEV is partially submerged in water, the water becomes a conductive path to high-voltage components. Even though the water has high electrical resistance, some current can flow. Seawater has a greater conductivity than freshwater.

The ubiquitous inverter is mentioned in the caption for Table 21.1. Inverters typically have an enormous capacitor to smooth out the electrical ripples. (A capacitor stores electricity.) Even though the battery may be disconnected, all other components listed earlier as being at high voltage may remain at high voltage for a few minutes while the capacitor bleeds down. For your convenience, the

TABLE 21.1
Voltages in Various Vehicles

Vehicle	Voltage, V	Battery
CV	12	Lead acid
Future CV	42	Lead acid
Mild HEV	42	Lead acid
Honda hybrids	144	NiMH
Prius, first generation	300	NiMH
Prius, second generation	500	NiMH
Toyota Highlander,	650	NiMH
Lexus RX400 h, Lexus 450 h	650	NiMH

Note: The voltage shown may be either the battery voltage or the output of the inverter.

list of parts temporarily at high voltage due to the capacitor is repeated; these are the cables from battery to engine compartment, the inverter, and the M/Gs.

In addition to the hazards of high voltage, batteries may exhibit thermal runaway. In any system, thermal runaway occurs when an increase in temperature changes conditions such that a further increase in temperature occurs. Ultimately, this leads to a destructive result. Thermal runaway becomes thermal *ruin*away.

Batteries may be driven into thermal runaway if severely overcharged for a prolonged period at high temperature. Some batteries fail gracefully (lead acid) as a result of thermal runaway. Other batteries (e.g., Ni-Cd) may ignite and catch fire, explode, or create external arcs, which endanger the vehicle.

Some people are concerned with the possible deleterious effect of electromagnetic (EM) fields on health and safety. Examples of concern range from the 60 cycle, high voltage power lines carrying electricity to cities to the EM fields of cell phones. HEVs certainly have far stronger EM fields than CVs. However, conclusive evidence of danger does not exist.

A couple more caution notes apply to HEV. With a CV, when the engine is not running, one can assume that the engine has been turned off. Not so with an HEV. The engine can start unexpectedly to charge the battery. To be certain the engine is off, do just that, turn it off and remove the keyless "key" from the vehicle. Careless handling of a damaged vehicle may cause severe injury or even electrocution.

Many hybrids are smaller cars. A driver switching from a large, heavy car may need to adjust driving style. One evaluation of a hybrid in the popular automobile press described that hybrid as having the handling capability of "… a sack of potatoes!"

WHO HAS AN INTEREST IN HYBRID SAFETY?

Four groups have an interest in hybrid safety. These are

1. Driver and passengers
2. Rescuers
3. Repair and service personnel (see Chapter 20)
4. Car manufacturers

Rescue of Occupants of Hybrid in a Wreck

Can the occupants of a collision-damaged hybrid be rescued safely in spite of high voltages and possible spilled dangerous chemicals? The owners and potential buyers of hybrids want to be reassured and want to have a clear answer. The answer is a conditional yes because

- Safe rescue procedures for HEVs have been developed.
- Special electrically insulated jaws of life are available.

These are effective only if

- Rescue personnel are properly trained for hybrids.
- Special insulated jaws of life have been issued to each rescue unit.

Collisions where jaws of life are necessary are rare compared to a simple fender bender. A fender bender for a hybrid is no different than for any other car. Statistics are on your side.

Agencies Involved in Rescue

Highway patrol, police, or law enforcement officers are usually first on the scene of an accident. They are in overall command at the accident scene and call other rescuers as required.

Fire departments typically operate the jaws of life, clean up minor chemical spills, and control fires if any. Emergency medical personnel and tow trucks complete the rescue brigade.

Guidance is available on the Web to first responders. For Ford products, Toyota and Lexus products, and Honda products, see Refs. [2,3,4], respectively.

Rescue Equipment and Procedures

With proper training and equipment, a rescuer need have no fear about safely removing occupants of a damaged, wrecked car. This is good news for the injured people inside the car.

Because of the dangerous chemicals from a ruptured battery, full chemical-protective suits have been recommended by Toyota and Honda. This includes eye shields, face masks, gloves, etc.

Different gloves serve different purposes. Gloves for protection against chemicals have been mentioned above. Special gloves to electrically insulate the wearer from high voltage are available. High-voltage components can be touched and manipulated. These gloves must be checked for the tiniest pinhole, which may negate the insulation.

The jaws of life have three functions: spreaders, cutters, and hydraulic rams. The cutters, which are like giant tin snips, can remove the section of a roof if necessary. Rams and spreaders can open bent doorways. One fear with cutters is that a high-voltage cable might be cut. Special electrically insulated jaws of life are available; if a high-voltage cable is cut, a significant, life threatening problem is not created.

Built-In Safety

The manufacturers have focused on the safety issue. Serious design and testing of safety has been accomplished to minimize risk. Adverse publicity can squelch sales. In contrast to standard vehicles, which use the chassis as part of the 12V electrical circuit, HEVs use heavily insulated wires for both positive and negative return sides of the high-voltage circuit, or have special conduits to protect the cables. Automatic shutoff systems reduce the potential for postcrash electrical shock from the high-voltage system. Some features built in for safety include

- High-voltage cables and components are color coded orange.
- Battery is located in the safest place; it is between frame rails and ahead of rear axle.
- Redundant safety measures surround high voltages, for example, cutoff relays when an air bag deploys, the power from battery is cut off.
- When an air bag deploys, the power from battery is cut off; see Figure 21.3 for a pyrotechnic cutoff—relay switches may also be used.
- Opening the battery compartment, in some cases, disconnects the battery.

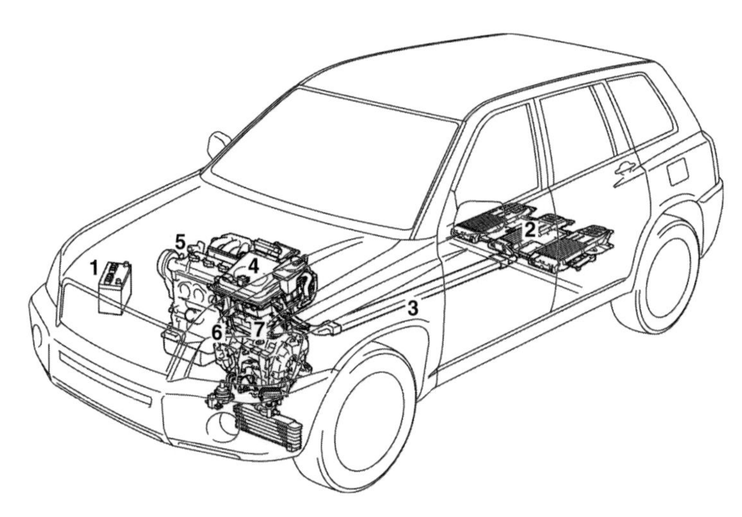

FIGURE 21.1 Toyota Highlander Hybrid components: vehicle view. (Reproduced with permission from *Toyota Highlander Emergency Response Guide*.)

- Orange cables from battery to engine compartment are typically run inside frame rails, which provide protection, or have special conduits to protect the cables.
- Batteries are tested for penetration by sharp objects created in a wreck.

Feel safe about safety items in hybrid design.

An agency of the federal government, National Highway and Traffic Safety Agency, has federal motor vehicle safety standards (FMVSS), which all vehicles must meet. In addition, FMVSS 305 applies to electrically powered vehicles and provides safety standards for electrolyte spillage, electric shock protection, battery retention, and postcrash electrical isolation of the chassis from high-voltage system.

Safety for First Responders and during Dismantling

The first responders need to know the location of components and which are lethal. Likewise, as hybrids go to the graveyard at the end of service life, the junkyard dismantlers need to be aware of the possible dangers. Figures 21.1 and 21.2 are from the *Toyota Highlander Emergency Response Guide*, which also serves as a dismantling manual for the Highlander. Obviously, the information in Figures 21.1 and 21.2 is of great interest to first responders.

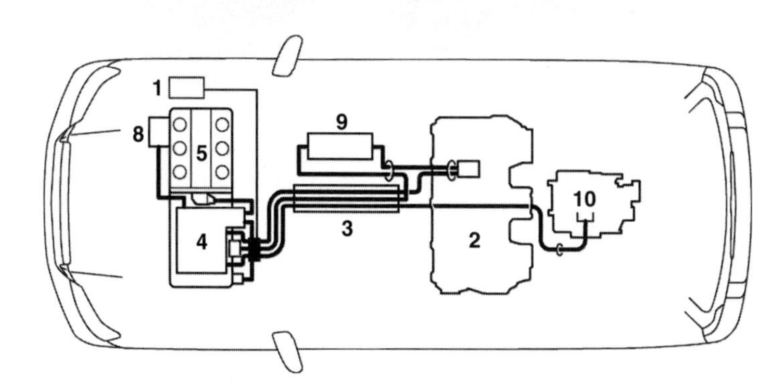

FIGURE 21.2 Toyota Highlander Hybrid components: plan view. For identification of numbers, see the table above with Figure 21.1. (Reproduced with permission from *Toyota Highlander Emergency Response Guide*.)

1. 12 V auxiliary battery; lead-acid battery; supplies headlamps, radio, etc.
2. HEV propulsion battery, 288 V NiMH consisting of 30 modules 9.6 V in series
3. Power cables inside conduit; orange wires high-voltage DC; yellow wires high-voltage AC between inverter/converter and M/G
4. Inverter/converter; inverter boosts DC voltage creating three-phase AC; converter changes AC from M/G to DC for charging propulsion battery; converter operates during regenerative braking
5. Gasoline engine; V-6, 3.3 L giving 155 kW (207 hp)
6. Three-phase AC, permanent magnet, electrical motor to help drive front wheels
7. Three-phase AC, electrical generator contained in transaxle; recharges NiMH battery
8. Air-conditioning (A/C) compressor
9. DC–DC converter (288 V → 42 V yellow wires) for electrical power system for power steering
10. Rear electrical M/G

The transaxle with generator (7) and front motor (6) is similar to the layout shown for the Prius I and II in Figure 3.1. In Figure 3.1, the transaxle with front motor and generator is identified as the hybrid transmission and includes everything within the box defined by dashed lines.

Battery Safety Switch

For both hybrids and CVs, a pyrotechnic, fast-acting safety switch provides very rapid battery isolation during a collision (Figure 21.3). Possible ignition of spilt gasoline or contact with very high voltage is avoided. Arcing, sparking, and vigorous fires or smoldering fires are avoided. The device, which does not have a reset capability, operates similar to the air bag [5].

WHAT ARE FUTURE HAZARDS?

Energy storage falls into two classes. First is energy stored as fuel. Energy conversion is a one-time event. For example in an internal combustion engine, the fuel is burned and is gone. Second is

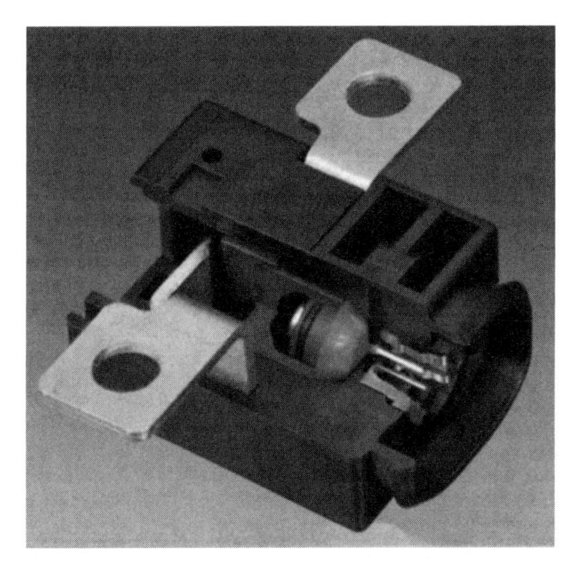

FIGURE 21.3 Battery safety switch. The battery main power cable is disconnected during a collision by pyrotechnic device. (Photograph courtesy of Delphi Corporation; Christoffel, J., Battery safety switch, *Automotive Engineering International*, Society of Automotive Engineers, Warrendale, PA, May 2007, p. 22. With permission.)

TABLE 21.2
Future Safety Hazards Involving HEVs and FCVs

Energy Storage	Type	Status	Safety Hazard	Failure Mode Leading to Rapid Energy Release
Batteries	Recharge	In-service	High-voltage Toxic chemicals	Short circuit Rupture
Supercapacitors	Recharge	R&D	High voltage	—
Flywheels	Recharge	R&D	—	Flywheel rupture
Hydraulic accumulator	Recharge	R&D	—	High-pressure tank rupture
ICE gas/diesel	Fuel	In-service	Fire and explosion	Leaking fuel
ICE hydrogen	Fuel	R&D	Fire and explosion	Leaking fuel
Fuel cells H_2	Fuel	R&D	Fire and explosion	Leaking fuel

Note: Besides being frozen water, ICE is short for internal combustion engine.

energy stored in a rechargeable device such as a battery or a flywheel. Both classes are presented in Table 21.2. In the case of gasoline, diesel fuel, and hydrogen, fire and explosion are major hazards. Of all fuel/air combinations, hydrogen/air has the broadest range of fuel/air mixtures for both deflagration (simple subsonic combustion) and for detonations. Broadest range means from the fuel-lean side to the fuel-rich side.

Hydrogen Safety

Two aspects of hydrogen are relevant to safety. First is the question of combustion or explosion. The Hindenburg passenger airship disaster (May 6, 1937) is always stressed in connection with use of hydrogen. However, NASA and other space agencies have used tons of hydrogen as a fuel without problems. Space applications of hydrogen involve both rocket fuel and fuel cells. Hydrogen mixed with air is explosive but is not toxic. Pure hydrogen is not explosive. Second is the question of hydrogen fuel tanks with very high pressure; pressures up to 10,000 psi (700 atm, 70 MPa) are planned. Possible tank rupture is carefully considered as a safety item.

SUMMARY

To set the stage, perspective is given on damage possible from concentrated stored energy. Some HEVs have battery energy equivalent to a car hitting a brick wall at 190 mph!

The two major hazards in today's HEVs are nasty chemicals and high voltage. For compact electrical motors, high voltage is necessary.

Can the occupants of a hybrid be rescued safely in spite of high voltages and possible spilled dangerous chemicals? The owners and potential buyers of hybrids want to be reassured and want to have a clear answer. The answer was a conditional yes.

The rescuers at a collision need special rescue equipment for their own safety as well as for the injured people. Manufacturers of HEVs have gone the extra mile, so to speak, to reduce risk to minimum.

The future holds new means of storing energy for hybrids. The new technology may reduce hazards. The flywheel does not have high voltages or horrible chemicals.

APPENDIX 21.1: LITHIUM-ION BATTERY SAFETY*

Current Li-ion batteries are not completely stable, but more stable materials are being developed to render them more inherently safe and secure. The acceptable degree of risk must be determined. Significant advances in Li-ion battery safety can be achieved through the use of these new and more stable materials. Graceful failure modes are desired instead of a series of dominos leading to disaster.

The Li-ion battery used in small, mobile appliances (cell phone, laptop, iPod) is not the same as the batteries being developed for HEVs, plug-in HEVs (PHEVs), and EVs. Most commercially available Li-ion batteries utilize a lithiated graphite or carbon negative electrode and a delithiated metal oxide (lithium cobalt oxide [$LiCoO_2$]) positive electrode. There are a couple of exceptions: Sony has introduced a new class of Li-ion batteries that employ intermetallic negative electrodes, and A123 Systems use a lithium metal phosphate positive electrode in their Li-ion cells for power tools. Both these approaches lead to enhanced inherent safety. The intermetallic negative electrodes operate further away from the potential of metallic lithium than do the lithiated graphites and carbons, while the lithium metal phosphate cathodes do not decompose to release oxygen at elevated temperatures, as do many delithiated metal oxides. A slight penalty in cell voltage results from the use of these more stable electrode materials.

There are two main classes of Li-ion batteries: liquid electrolyte cells and polymer electrolyte cells. Both classes of cells employ the same type of liquid organic carbonate solvent systems, but the polymer cells employ a crosslinkable component that immobilizes the liquid solvents. There is another class of solid-state thin-film batteries that employ solid electrolytes, but these batteries are not applicable for use in EVs, HEVs, PHEVs, or other transportation applications. Batteries with solid electrolytes operate at elevated temperatures, which is not an attractive feature.

LI-ION BATTERY PROTECTION IN HEV APPLICATION

Each individual cell of the Li-ion battery will be contained in either a metal can or a flexible pouch container. The overall battery pack will be placed inside a sturdy metal box container with an active thermal management system for controlling the temperature of the cells. Additionally, these batteries will employ some form of a sophisticated electronically controlled battery management system to keep the cells of the battery within safe operating limits. Further, the battery will be installed in the safest location in the vehicle, which is typically between the rear wheels.

CELL-LEVEL SAFETY DEVICES

Commercial cells will employ safety devices that lessen the possibility of harmful failure. These devices may include

- Shutdown separators (overtemperature condition)
- Overpressure control via a rupture disk or pressure relief device (excess internal pressure)
- Thermal interrupt (overcurrent/overcharging)

In most cases, these safety devices permanently and irreversibly disable the cell. The goal is to develop an Li-ion battery that does not require safety devices for safe operation, or at least, requires fewer safety devices.

* Appendix on lithium-ion battery safety was reviewed by Gary Henriksen, manager, Battery Technology Department, Argonne National Laboratory. His suggestions to improve the content have been incorporated. However, errors, if any, are the sole responsibility of the author.

OPENING (RUPTURE) LITHIUM-ION BATTERY

Most of the commonly used electrolyte solvents have high vapor pressures and are flammable. Immobilizing the solvents, via the polymer electrolyte, helps to reduce the vapor pressure and reduce the flammability hazard somewhat. Electrolyte additives are being developed and commercialized to reduce the flammability of these commonly used solvents and alternative (less flammable) solvent systems are being developed. For HEV applications, the battery may be ruptured in a vehicle collision. So minimizing this flammability hazard is important.

CAUSES OF BATTERY ACCIDENTS: TRIGGERS

It is important to control cell temperatures. Once the cell temperature reaches a critical level the cell will go into thermal runaway. Excess temperature can cause a rapid pressure rise inside the cell, cell venting, possible ignition of flammable electrolyt solvents, and even an explosion. Temperature must be monitored and cooling provided. Triggers for battery failure include

- Overcharging
- Overheating
- Short circuits; partial or full
- Contaminants
- Accidental rupture

TRIGGERS: OVERCHARGING OR OVERHEATING: STEPS TO POSSIBLE DISASTER WITH CONVENTIONAL CELL CHEMISTRIES

Step 1: In conventional Li-ion cells, internal self-heating begins with the breakdown of the passivation film on the negative electrode and then lithiated graphite or lithiated carbon reacts with electrolyte in a self-heating reaction.

Step 2: Overheating may generate oxygen from the positive electrode [6]. The oxygen reacts with the electrolyte solvents, creating more heat. This additional heat causes the cell pressure to increase.

Step 3: At a given pressure, the cell case ruptures and flammable electrolyte vents into surrounding air.

Step 4: If an ignition source exists, the flammable solvent gases will ignite.

Overcharging of the cell generates heat that can raise the cell temperature to the level where step 1 initiates. Additionally, overcharging can further destabilize the electrodes and render them more reactive.

TRIGGERS: INTERNAL SHORT CIRCUIT: PARTIAL OR TOTAL

Internal short circuits are especially dangerous. Excessive heat is generated, which raises the cell temperature to levels where the above-mentioned self-heating reactions initiate and this can lead to cell thermal runaway. Metallic contaminants are a possible source of internal short circuits.

POSITIVE ELECTRODE

For an application in an HEV, the typical $LiCoO_2$ will likely be replaced by less expensive and more stable materials, for example, the lithiated metal phosphate. When lithium is removed from most lithium metal oxide positive electrodes (as is the case with charged cells) most of these materials have a tendency to begin releasing oxygen when the cell temperature rises to about 180°C. In the case of delithiated metal phosphate, it does not release oxygen when the cell temperature rises to this level. Another more thermally stable positive electrode material is lithium manganese spinel. These alternative positive electrode materials render Li-ion batteries less likely to go into the thermal runaway state due to little or no oxygen release from the delithiated positive electrode material.

NEGATIVE ELECTRODE

Most Li-ion batteries being developed for HEV applications still employ lithiated graphite or carbon as the negative electrode material. These lithiated graphites and carbons react with the electrolyte solvents to produce heat. During normal operation, these reactions are mitigated via passivation films that form on the electrode surface during the initial charge half cycle. However, conventional passivation films begin to break down below 100°C and then the lithiated graphite/carbon reactions with the electrolyte are initiated. In large cells, these reactions can be sufficient to drive the cells into thermal runaway. Electrolyte additives are being developed that form more stable passivation films on the lithiated graphite and carbon electrodes. Also, more stable alternative negative electrode materials are being developed. These include several types of intermetallic materials and lithium titanate. These alternative negative electrode materials are less reactive with the electrolyte and render Li-ion cells more inherently stable.

REFERENCES

1. National Highway Traffic Safety Administration, Crash and Rollover Rating, www.safecar.gov July 2 2008.
2. Ford First Responder. Available at https://canada.fleet.ford.com/pdfs2006.
3. Toyota and Lexus First Responder. Available at https://techinfo.lexus.com/.
4. Honda First Responder. Available at https://techinfo.honda.com/rjanisis/logon.asp.
5. J. Christoffel, Battery safety switch, *Automotive Engineering International*, Society of Automotive Engineers, Warrendale, PA, May 2007, p. 22.
6. D. Schneider, Who's resuscitating the electric car? *American Scientist*, September–October, 2007, pp. 403–404.

22 Future

The future begins with the past.
Although the automobile of the future will have four wheels,
it will be fundamentally different.

INTRODUCTION

The future in this chapter begins with a review of Chapter 2. The United States is starting with a fleet of vehicles that have poor fuel economy. The mpg is poor compared with the vehicles in the rest of the world. The mpg is poor compared with what might have been possible using the advances in technology.

The future of personal transportation will be dictated by the supply and demand for petroleum. Even without the supply limits of petroleum, the need to control emissions would force a major change from gasoline/diesel cars. Due to the coupling of limited oil supply and greatly restricted emissions, major changes will occur in the future.

Figure 2.1 shows that oil production in the United States is in a state of decline while demand is growing. The result is imported oil. Figure 2.2 shows the switch from positive to negative in 1984. This was the year that the net difference between annual additions to global oil reserves and annual consumption switched from a comfortable positive to a disturbing negative.

Hybrids are a sensible move toward the oil-limited world of tomorrow.

SUPPLY AND DEMAND

Hubbert's curve is a fruitful way to examine the supply and demand for oil. Of course, the supply curve is simply a Hubbert-like curve, which has been drawn in Figure 22.1, with peaking in the year 2025. The demand curves are superimposed. Prior to 2025, the supply nearly equals demand. Near the peak, however, demand begins to exceed supply. The oil markets become erratic and expensive. This is the threshold for possible (likely) economic chaos. See Ref. [8] for one viewpoint on oil and the economy.

The upper demand curve in Figure 22.1 can be decreased by energy conservation. Hybrids decrease demand. A higher Corporate Average Fuel Economy (CAFE) decreases demand. The gap between demand and supply must be filled by alternate fuels. The understanding of future vehicle transportation is entangled with an understanding of alternate fuels.

One approach to alternate fuels is to develop biofuels to replace petroleum fuels while at the same time reducing emissions. For alternate hydrocarbon fuels, the internal combustion engine (ICE) must burn cleaner and have more effective emission controls. Biofuels are a short- as well as long-term solution. As discussed shortly, another approach is to invest heavily in hydrogen as a fuel. H_2 is a long-term solution.

MITIGATION MEASURES

Asymmetric Risk

The peaking of petroleum production poses an asymmetric risk. Little downside exists from being too early. Being too early may result in biomass refineries that are unused for a few years. (Being too early may be an illusion; peaking may be too soon to allow being too early.)

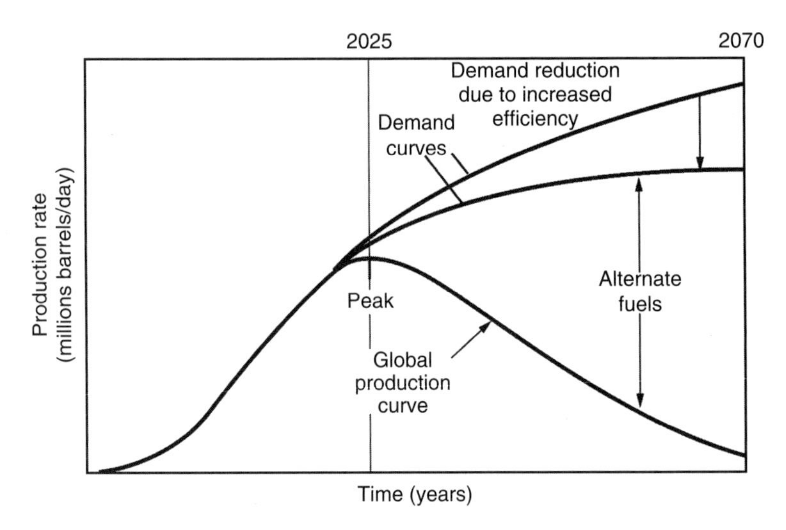

FIGURE 22.1 Global oil production and demand.

Timely actions can be effective and economical. The economy can absorb the changes without major perturbations. Delayed actions could evolve into "mission impossible" relative to creating an alternate energy system. Developing an alternate energy system requires answers to at least two questions. What needs to be done? Who will pay for it?

What Needs to Be Done?

Realist alternate fuels need to be identified. To fully understand all the ramifications of a proposed alternate fuel, prototype refineries providing fuel for prototype vehicles with measurements of emissions and fuel economy are needed. Testing over a period of time in a variety of environments is essential. Favorable results attract investment capital and more government funding. Unfavorable results mean lost time and the need to try something else.

The government must provide leadership so that the population understands the need. The government must provide incentives for attractive designs and penalties for vehicles that will be failures in the future and that impede progress. An enlightened citizenry is needed to elect the government that will be needed. Debate is essential.

Who Will Pay for It?

Funding will be a mixture of private and public monies. Funding for alternate fueling systems requires both private and government investments. New fuels, for example, ethanol, compete with gasoline. The policies of the government can make a new fuel competitive. Added taxes on gasoline and incentives on ethanol encourage private industry to invest in ethanol. Further, the public is encouraged to buy ethanol.

Figure 22.2 illustrates why early funding is important. To develop the alternate fuels as shown in Figure 22.1, $1,000 billion is required. Maybe less, and likely more, will be required. In any event, $1,000 billion puts the funding level in perspective (Figure 22.2).

The abscissa (horizontal axis) in Figure 22.2 is the years-to-go to peaking. The crunch starts near peaking, and the necessary alternatives need to be defined and put in place. If the years-to-go is 20 years, then the annual budget is $50 billion. The amount to be raised by taxes is a small $50 billion. Time exists to allow mistakes without catastrophic consequences. If the years-to-go is 5 years, $200 billion/year must be spent. This amount is difficult to obtain. Even if the $200 billion/year were available, the money cannot be spent wisely. Panic prevails. The economy is in a tailspin. Dead end projects absorb valuable resources.

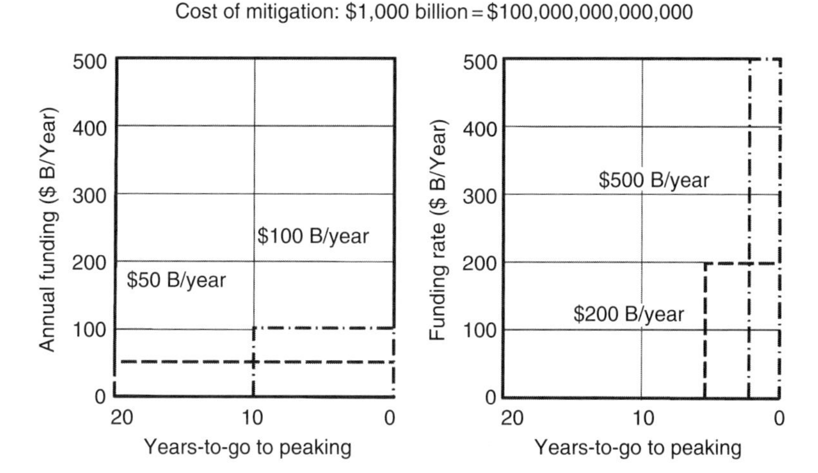

FIGURE 22.2 Mitigation measures and the impact of early funding.

Time is needed to allow mistakes without catastrophic consequences. After 100 years, the petroleum industry should be well understood. If that industry is not fully comprehended, then alternate fuels are a larger mystery. In Europe, diesel is so popular, that a surplus of gasoline exists. A barrel of petroleum contains a certain fraction of diesel fuel, a certain fraction of gasoline, a certain fraction of jet fuel, etc. Since diesel is popular, extra gasoline is refined. The excess gasoline must be exported to the United States and other countries. Changes to the refineries can alter the fraction of gasoline and solve the excess gasoline problem.

Incremental Costs of Mitigation

A new refinery for petroleum costs $10 billion. Perhaps 10 new refineries will be needed if alternate fuels lag. This amounts to $100 billion. However, if conversion to hydrogen and other alternate fuels proceeds at a reasonable pace, the 10 new refineries will not be needed. The $100 billion otherwise spent can be reallocated to hydrogen production, refineries for biomass, and the associated infrastructure. Instead of $1,000 billion for mitigation, the incremental costs might be 50%–60% of that amount.

Painless (Almost) Mitigation Measures

The two proposed measures are called painless because the economy can absorb the changes. Other countries tax motor fuel. The federal gas tax should be increased by 25¢ annually for 12 years at which time gasoline would be $6.00/gal. A vehicle with 50 mpg costs 12¢/mi. An electric vehicle (EV) costing 5¢/mi would be attractive. The revenue from the gas tax can help finance the alternate energy systems.

The other measure is to increase the CAFE. Apply existing CAFE, but use valid numbers for mpg. This should be done immediately. Increase CAFE by 0.5 mpg annually and, in 25 years, the standard is 40 mpg. All vehicles, including sport utility vehicles, minivans, and pick ups, are to be included in the higher CAFE.

Indicator for Alternate Fuels

Just as robins are the harbinger of spring, the announcer for the arrival of alternate fuels will be the significant investment by major oil companies in the production and distribution of alternate fuels.

ABIOTIC OIL

Abiotic oil was introduced and briefly discussed in Chapter 2. One proponent of abiotic oil, Thomas Gold [10], states fossil fuels are a myth. Many geologists, engineers, and scientists believe that abiotic oil is a myth. However, since petroleum is so vital to the world's economies, additional research on abiotic oil seems warranted.

URANIUM ARGUMENT

When the planet Earth was formed, in the swirling soup of atoms and molecules were hydrocarbons such as methane (CH_4). These hydrocarbons form abiotic oil. The arguments against abiotic oil are that origins of petroleum are plant matter processed by millions of years of heat and pressure inside the earth.

Uranium does occur in the earth. Uranium cannot be formed from plant matter processed by millions of years of heat and pressure inside the earth. Uranium cannot be formed from any matter solely by the heat (too low) and pressure (too low) inside the earth. Millions of years are irrelevant in this case. Fossil uranium is impossible. What was the source of the uranium found in the earth? When the planet Earth was formed, in the swirling soup of atoms and molecules were heavy atoms such as uranium.

ALTERNATE FUELS: OVERVIEW

"FUEL" FOR TRANSPORTATION

The word "fuel" is in quotes to indicate that an appropriate definition is needed. A fuel, when properly converted, provides the energy to move a vehicle. The vehicle can be car, truck, train, or bus. For this book, the focus is on land transportation, and mainly on hybrid cars and light trucks.

Consider the boundary between the vehicle and the outside world; think of a vehicle inside a bag. Whatever form of energy crosses that boundary (bag) is fuel. The gas hose at a service station crosses the boundary, and gasoline is the fuel. For an EV, the charging cable crosses the boundary, and the fuel is electricity. For a fuel cell vehicle (FCV), one might be tempted to say the fuel is electricity; however, the fuel is hydrogen. Hydrogen crosses the boundary for storage in the hydrogen tank. Electricity is generated on board the vehicle by the fuel cell. Refer to Figure 22.3, which shows hydrogen produced by electrolysis.

Energy with the arrow "In" depicted in Figure 22.3 can be green energy as will be discussed shortly.

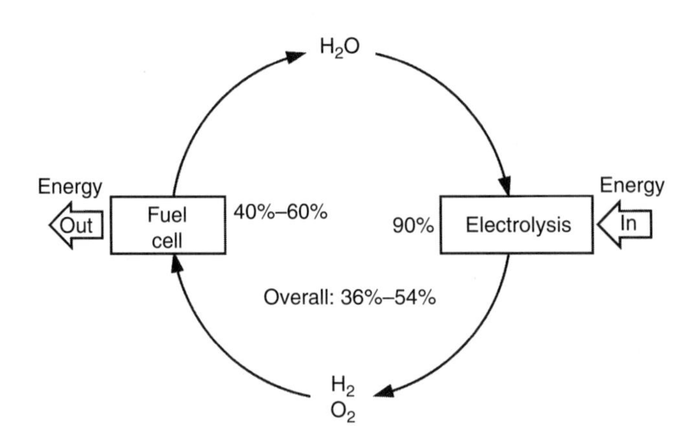

FIGURE 22.3 Hydrogen production by electrolysis.

Some Definitions

A fuel can have three adjectives or modifiers: alternate, renewable, and/or biomass. Table 22.1 assigns the adjectives to the various fuels. When you need values of the properties for the various fuels, consult Table A.22.1.

The easiest definition for alternate fuel is not what it is but what it is not [5,15]. Any fuel that is not a petroleum-based liquid (gasoline, diesel) is an alternate fuel. This definition admits natural gas as an alternate fuel. A fuel that can be produced using available resources and having the ability to replace consumed fuels within a defined time span is considered to be a renewable fuel. Renewable energy includes wind, solar, wave, thermal gradients in the ocean, geothermal, hydropower, and power from the tides. All material from plant and animal life contain energy and have, therefore, the potential for contributing to our energy supply. Biomass fuels or biofuels are energy derived from plants and animals. Biodiesel is produced from plants. Biofuel can be produced from trees. Redwoods, which may grow for 2000 years, are not considered to be a renewable resource whereas pine trees are renewable in 50–60 years of growth. The early steam, wood-burning locomotive discussed in Chapter 2 is an example of propulsion by biomass. Nuclear fuel is not renewable and is the same as petroleum in that it is a source that someday runs out. Energy from municipal refuse is a source of energy. Raw refuse from your community trash truck has about one-half the heating value of coal; this is a surprising number! Animal wastes (manure) and sewage offer another source of energy. In addition to energy, the emission of methane affects global warming. Methane, a powerful greenhouse gas (GHG), receives attention for these two reasons.

Table 22.1 lists the alcohols, methanol and ethanol, along with methane. Methane and natural gas, being gases, are less convenient to use than a liquid. Suggestions have been made to convert

TABLE 22.1
Classification of Alternate Fuels

Fuel	Alternate	Renewable	Biomass
Gasoline (A)	No	No	(A)
Diesel (A)	No	No	(A)
Methanol	Yes	Yes	Yes
Ethanol	Yes	Yes	Yes
Propane	Yes	(B)	(B)
Methane	Yes	(C)	(C)
Natural gas	Yes	No	(A)
Hydrogen	Yes	(D)	No
Biodiesel	Yes	Yes	Yes
Electricity	Yes	(E)	(E)
Coal	Yes	No	(F)
Nuclear fuel	Yes	No	No
Municipal refuse	Yes	Yes	Largely

Notes: (A) From fossils; fuel was biomass millions of years ago. (B) Could be produced from renewable sources; other fuels may be better. (C) The main molecule in natural gas is methane; natural gas is from fossils and is not renewable. However, methane can be made from biomass and is renewable. (D) See Figure 22.3, which shows a closed loop if water is recovered. Consider hydrogen to be renewable. (E) Depends on how the electricity is generated. Wind energy would be renewable but is not biomass. (F) Coal was biomass several 100 million years ago.

methane or natural gas (which are relatively plentiful) to an alcohol. An alternate approach is to convert methane or natural gas to propane. Propane is a gas but can be easily distributed and stored. Propane is more closely related to the feedstock so less energy is lost in the conversion process. Alcohols are used as range extenders in mixtures with gasoline, for example, E85 discussed elsewhere. On the other hand, propane can be used as a dedicated fuel; mixture with gasoline is not needed. Propane is a high-octane fuel allowing higher compression ratios (CRs). Propane does not have an oxygen atom in its chemical structure. This is good relative to energy per gallon (high) but bad relative to emissions. This is not to say propane emits high pollution.

Perspective on Biofuels

Biofuels have been criticized for several side effects. Reference [12] has several arguments against current biofuels. In Ref. [13], the pros and cons of biofuels are discussed. Reference [14] has a statement that captures the thrust of the article. "Everyone seems to think that ethanol is a good way to make cars greener. Everyone is wrong." These arguments have merit. Ethanol from corn has many deficiencies including use of food for fuel. Biofuels can be made from non-food sources. Environmentally, ethanol from corn causes pollution due to the massive amounts of fertilizer that are required.

Carbon/Hydrogen Ratio

When hydrocarbons burn, the carbon atoms in the molecule become CO_2 while the hydrogen atoms yield H_2O, water. From the point of view of emissions and global warming, the amount of CO_2 generated is important, and the amount of water is relatively insignificant. The carbon/hydrogen ratio, R, tells the fraction of combustion products that is CO_2. Figure 22.4 defines the ratio at the top of the figure.

To apply the ratio, the number of atoms is counted. For hydrogen, $C = 0$ and $H = 2$ giving a ratio, R, equal to zero. For methane, $C = 1$ and $H = 4$ with $R = 0.33 = 33\%$. For methane, the main constituent of natural gas, one-third of the combustion products is CO_2. Gasoline is much more complicated than shown in Figure 22.4. Gasoline is a mixture of many hydrocarbon molecules. For the "gasoline" molecule of Figure 22.4, $C = 4$, $H = 10$, and $R = 44\%$. For the "gasoline" molecule of Table 22.2, $C = 8$, $H = 18$, and $R = 48\%$. Relatively, gasoline creates more CO_2 than methane.

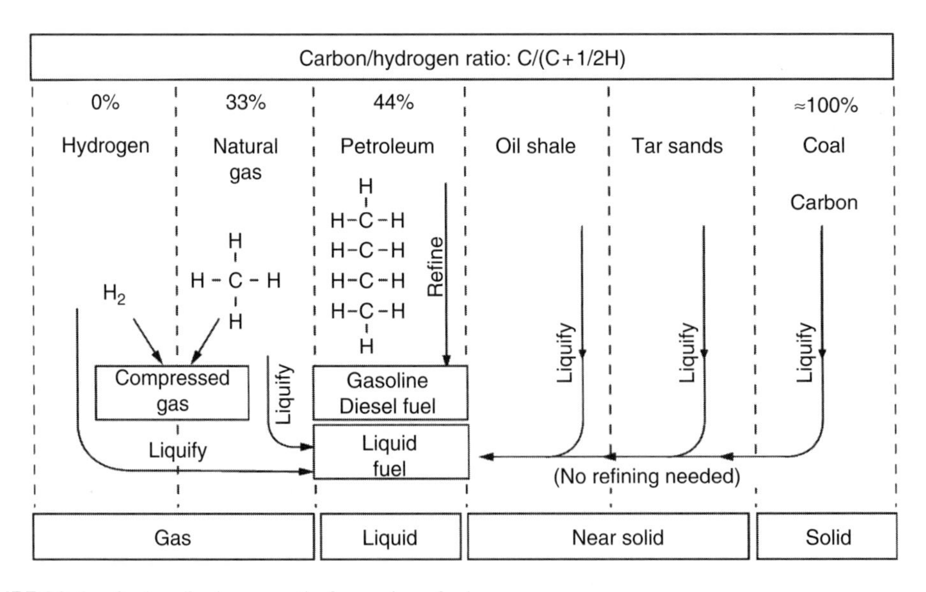

FIGURE 22.4 Carbon/hydrogen ratio for various fuels.

TABLE 22.2
Table of Carbon/Hydrogen Ratio, *R*, for Various Fossil and Alternate Fuels

Fuel	Chemical Symbol	Carbon, C	Hydrogen, H	Ratio, *R*
Hydrogen	H_2	0	1	0
Methane	CH_4	1	4	0.33
Ethane	C_2H_6	2	6	0.40
Propane	C_3H_8	3	8	0.43
Methanol	CH_3OH	1	4	0.33
Ethanol	C_2H_5OH	2	6	0.40
Gasoline	C_8H_{18}	8	18	0.48
Biodiesel	$C_{16}H_{36}$	16	36	0.47
Diesel	$C_{18}H_{38}$	18	38	0.49
Coal	C	1	0	≈1.00

Coal is not a single entity; many different kinds of coal exist. The properties of coal depend on the rank of the coal. Coal contains impurities that are not combustible. These impurities create ash. Anthracite coal is nearly 100% carbon and, hence, when burned has $R \approx 1.0$. Of the major products of combustion, almost all is CO_2.

With regard to global warming, as *R* increases in Table 22.2, the fuel is less and less favorable. Hydrogen is best. Coal is worst. The physical properties of fuels are found in the Appendix 22.1.

Across the bottom of Figure 22.4 are the states of several fuels. Oil shales and tar sands provide near solid material. Oil shale is a fine-grain sedimentary rock containing an organic material called kerogen. Kerogen is not petroleum but can be converted to oil. Rich oil shale yields 50 gal of oil per ton of shale. Vast resources of shale provide <10 gal/ton. "Tar sands" are sands or sandstone impregnated with a heavy oil called bitumen. Bitumen ranges from a viscosity of crankcase oil to a heavy tar. Both oil shale and tar sands can be made liquid. In Figure 22.4, the fact that the common fuels (gasoline, diesel) are liquid indicates the desirability of liquid for storage on board a vehicle. Both hydrogen and natural gas are inconvenient gases. For storage, either can be compressed and stored in a heavy tank or liquefied and stored in a low-temperature, thermally insolated, tank. Likewise, coal can be liquefied to create the desirable liquid fuel. The liquid fuel from oil shales, tar sands, or coal is so heavily processed to obtain the liquid, that further refining is not necessary.

Energy Out/Energy In

One consideration in the selection of a viable alternate fuel is the amount of energy input required for each unit of energy output. Figure 22.5 shows estimated values for a few fuels.

The definition for energy in is the amount of energy required to create the fuel. For example, "energy in" for petroleum involves the amount for exploration, drilling exploratory wells, drilling production wells, pumping oil to the surface, transportation to the refinery, and refining to a useful fuel. "Energy out" is that listed in Table A.22.1. From Table A.22.1, for gasoline the energy out is 18,676 Btu/lb or 116,090 Btu/gal. Conversion efficiencies in an engine are not a factor. From Figure 22.5, petroleum offers 20–100 times more energy to be used than the energy input.

The vertical line at 1.0 is breakeven. At breakeven, as much energy must be input as can be obtained as output. Hydrogen from water is below the breakeven point. This fact will be discussed

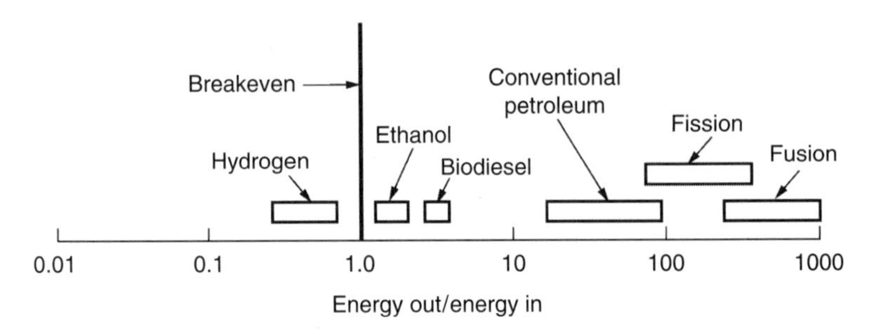

FIGURE 22.5 Energy out/energy in.

later in more detail. Ethanol from corn is only slightly greater than unity. Biodiesel and ethanol from sugarcane are better with a ratio of about three. Nuclear fission has larger values for the ratio, and nuclear fusion is even larger. Developing nuclear fusion, which is cleaner than fission, has been a 60-year-old goal. Fusion (combining) of two nuclei of hydrogen releases energy. One technique to achieve self-sustaining fusion is laser ignition [18]. Another technique is heating of the hydrogen with confinement in a magnetic "bottle."

Coal

Coal is a fossil fuel. Regardless of how dirty or nasty coal may be, enough coal exists to provide the world's energy for 600 years, more or less. Coal as an energy source cannot be ignored. Coal is dirty because of the high CO_2 emitted by combustion. Iron sulfide in coal creates sulfuric acid leading to acid rain. Streams and rivers are polluted with the acid. Strip mining leaves a desolate landscape.

Coal is assigned rank as an indicator of quality. Anthracite coal is ranked highest and is 95% carbon. Lower rank coals have trapped water and impurities. All coals have some hydrogen. Coal was formed in swamp ecosystems 280–345 million years ago. Sulfur in coal poisons the air when burned. Sulfuric acid results from combustion.

Coal is a natural resource that follows the same curves as petroleum follows for production rate and cumulative extraction. Coal will have a peak just as petroleum does. The peak may be decades away. Globally, 40% of electricity is generated in coal-fired plants. For every kWh of electricity generated, 2 lb of CO_2 are typically emitted by the generating plant.

"Btu conversion" of coal implies gasification, methanation, or liquefaction. Exposure of coal to high pressure steam creates syngas, which is a mixture of carbon monoxide, CO, and hydrogen. Syngas can be burned in gas turbines. Combining syngas with steam yields carbon dioxide and hydrogen. In this process, sequestering CO_2 is easier than sequestering CO_2 from the combustion of coal. One research project is to gasify coal underground without mining. Success in this endeavor greatly enhances coal as an energy source.

Methanation is generation of methane gas from coal. Liquefaction of coal, which is shown in Figure 22.4, can be accomplished. Liquefaction of coal is a stopgap measure. The process briefly portrayed is

$$\text{Coal} \rightarrow \text{Syngas} \rightarrow \text{Light H/C} \rightarrow \text{Gasoline/diesel}$$

H/C is shorthand for hydrocarbon. Above approximately $35/barrel for oil, liquefaction of coal becomes economically feasible. Liquid fuel from coal is not viable on a large scale.

The energy density of coal is less than that for gasoline. Gasoline is 11.87 kWh/kg. Coal is 6.67 kWh/kg, which is almost one-half.

Coal Resources

The United States is the "Middle East" of coal. Of all of the world's coal, 26% is in the United States.

ALTERNATE FUELS: DETAILS

Biomass and Alternate Fuels

An alternate name for alternate fuels from biomass is synthetic fuels; this term is used in Europe. The advantages of biomass include capturing of CO_2 from the atmosphere. Growing crops generate oxygen. Biomass allows geographic flexibility with regard to location of growing fields and forests. Production can be continuous. When biofuels are made from what otherwise is waste (cornstalks, corncobs), value is added to the crop. Alcohol fuels and biodiesel are made from biomass.

The disadvantages of biomass include the high cost of production, which requires complicated refineries. The price of gasoline from petroleum must be high to offset biomass production costs. According to Keppler and Röckman [3], growing plants also emit methane. Methane is a powerful GHG.

To encourage private investment in various biomass production and refining facilities, the government must offer incentives in the form of subsidies and tax incentives. In passing, one might question the wisdom of tariffs on ethanol from Brazil.

Cellulosic Materials and Biofuels

Cellulose is one of the three major components of plant cells. Cellulose, which is one of the most abundant biological materials on this planet, is therefore a good source for biofuels. Conversion of corn or sugarcane to ethanol is a mature technology. Conversion of most cellulosic material to fuel is a technology in its infancy. Biological waste such as corncobs and plants such as switch grass can be a source of feedstock provided the technology is mastered. The energy utopia of Figure 22.6 will occur only if cellulosic conversion is made possible.

For any biomass scheme, a life cycle analysis is essential for identifying the winners and losers. Figure 22.6 illustrates the cycle of CO_2 and production of a biofuel from a forest. Absorption of CO_2 and generation of CH_4 are shown in the forest. Solar energy is stored in the trees by photosynthesis. The harvest yields two classes of products: the economically valuable lumber and wood pulp; and the waste material, chips, bark, etc. The waste forms the biomass feedstock for the refinery. The fuel is burned in the vehicle and releases CO_2, which is recycled by the forest. The time span to grow the trees is 50–60 years. The triangular shape of the trees suggests pine.

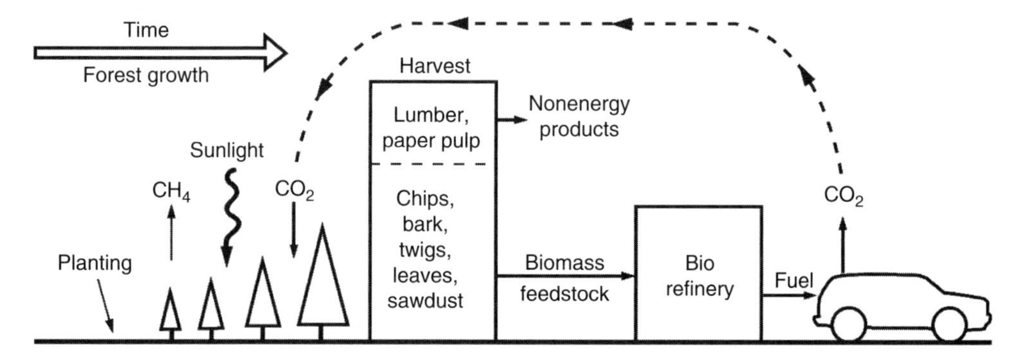

FIGURE 22.6 CO_2 cycle; fuel from biomass.

The trapping of solar energy by trees is not efficient. Of the solar energy falling upon a forest, less than 1% is stored by the tree. This is in contrast to photosynthesis, which is near 100% efficient. Solar cells are much more efficient than trees in capturing solar energy.

Promising candidates for alternate fuels from biomass include switch grass, sugarcane (limited growth potential), bagasse (a sugar by-product), and corn waste such as corncobs and cornstalks. The waste from corn has been assigned the word, *corn stover*. By 2030, combined production of cellulosic ethanol and corn-based ethanol could replace 30% of the petroleum fuels being used today.

In 2004, the major sources of biomass energy were 70% from wood, 20% from waste, and 10% from alcohol. The waste includes manufacturing solid waste, municipal solid waste, and landfill gases. The alcohol is mainly from corn.

Grassoline

Professor Bruce Dale of Michigan State University has been quoted in Ref. [20] as saying:

> I like to call cellulosic ethanol "Grassoline" because it can literally be made from grass. Grassoline is domestically produced, environmentally sound, and helps support rural and regional development.

Cellulosic ethanol does not compete with our food supplies.

Microorganisms in Alternate Fuels

From brewing beer to brewing alternate fuels, microorganisms play an important role. Microorganisms add biology as an important science to production of alternate fuels. The role of microorganisms can be both favorable and unfavorable. Reference [11] discusses one example. The title of Ref. [11] "Microbial deterioration of marine diesel fuel from oil shale" tells the essential aspects of the story. Contamination of fuel can cause problems in storage, distribution, and usage of alternate fuels. The ports in either gasoline or diesel fuel injectors are tiny and are easily plugged. Further, the precise size of the injection ports is important. Contaminated fuel and fuel injectors are a source of misery. A large database exists for microorganisms in petroleum-based fuels. Obviously, the equivalent data do not exist for biodiesel, ethanol, and other fuels.

Favorable aspects and optimism abound for microorganisms and future fuels. Efforts are under way to produce jet fuel with zero net CO_2 production. As discussed elsewhere in this chapter, microorganisms can prepare cellulosic biomass for conversion to alternate fuels such as ethanol and biodiesel. For application in fuels cells, certain microorganisms have the novel ability to assist in the transfer of ions and electrons at the surface of an electrode. Conversion of chemical to electrical energy is thereby enhanced.

Algae, the green scum that grows on the surface of pond water, absorb CO_2 from the atmosphere. Algae produce fat that can be refined into biofuel suitable for diesels. Algae have little in common with other biofuels. It is not a food, it does not require good farmland, and it does not require large amounts of freshwater [19].

Biodiesel Fuel

Biodiesel is typically made from soybeans in the United States and from rapeseed in Europe. Other sources need to be developed; see algae above. Biodiesel from cooking grease has received attention but is unlikely to provide substantial amounts of fuel. An experimental program produces biodiesel from microalgae. This program, if successful, would open up marginal land for production of biodiesel. As shown in Figure 22.7, there are three outputs. One is biodiesel. Another is glycerin, which can be used to make soap. Part of the processed soybeans is useful as animal feed. A commercial biodiesel plant has been built in southern Colorado. Colorado Springs has committed to using B20 in its fleet of diesel vehicles [21].

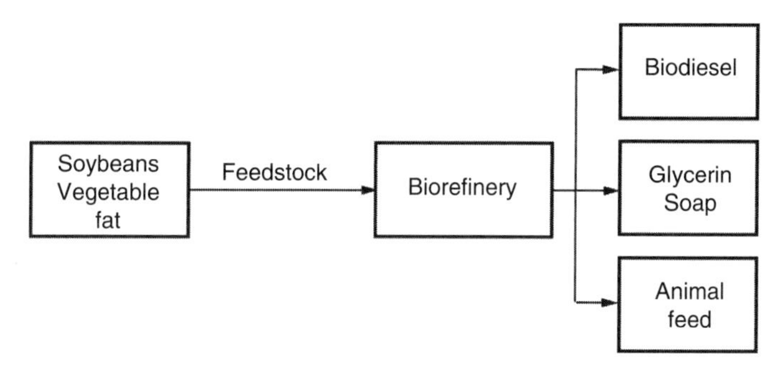

FIGURE 22.7 Biodiesel production with three outputs.

Biodiesel contains zero petroleum although blends, in any proportion, of biodiesel with petrodiesel are possible (see Table 22.3). Biodiesel is available nationwide in the United States. The properties of biodiesel are consistent due to the American Society of Testing and Materials (ASTM), specification ASTM D 6751. Biodiesel has less variability than petrodiesel.

Biodiesel can be used in existing diesel engines without or with only minor modification. This is a huge advantage and was confirmed by extensive tests. B20 has been tested in five buses with an accumulated mileage of 500,000 miles. Bus operations, maintenance, and fuel economy were essentially the same using B20 as petrodiesel [22]. Biodiesel is essentially free of sulfur and aromatics. Greatly reduced sulfur fuels were mandated by EPA. The low sulfur fuels were discussed in Chapter 17.

Biodiesel is similar to Jet A used by the airlines to power jet engines. However, the current biodiesel cannot be used for jet fuel because of the high freezing point. Aircraft operate in the stratosphere where the temperature is at –67°F.

Dimethyl ether (DME) is a clean alternate fuel for diesel engines. DME can be produced from biomass and has properties similar to liquefied natural gas (LNG). Volvo has built a track for testing DME. DME is of particular interest to Sweden since the fuel can be made from a residual product from paper pulp [15].

ALTERNATE FUELS: ETHANOL AS A FUEL

Color Yellow

The color yellow, which is suggestive of corn, is used on many ethanol pumps at service stations. Corn is the major feedstock for ethanol in the United States. Yellow labels are also used. Car makers, producers of ethanol compatible cars, often use yellow as the theme color in advertisements. Ethanol, which is also known as ethyl alcohol or grain alcohol, may also at some future time be produced from corn stover (corn stalks, leaves, cobs, and husks), straw, and sawdust. Ethanol can be produced from three sources. These are, as mentioned above, from corn, from sugarcane, and from cellulosic materials. Production from cellulose is discussed below.

TABLE 22.3
Properties of Biodiesel Compared with Petrodiesel

Fuel	Density (g/cm³)	Energy Content (Btu/gal)	Percent Loss (%)
Diesel #2	0.850	129,500	—
Biodiesel, B100	0.880	118,296	8.65
Biodiesel, B20	0.856	127,259	1.73
Biodiesel, B2	0.851	129,276	0.17

Energy Content

Ethanol has low energy content. The following discussion applies to E100, which is 100% ethanol. Ethanol has less energy per gallon than gasoline. The heating value for ethanol is 76,370 Btu/gal while gasoline has a value of 116,090 Btu/gal. Using these values, ethanol has 34% less energy than gasoline. Fuel economy using ethanol will be one-third less. If a gasoline car yields 36 mpg, the ethanol car will give 24 mpg using pure ethanol. The calculation is

$$[(36 \text{ mi/gal G}) (76{,}370 \text{ Btu/gal E})]/(116{,}090 \text{ Btu/gal G}) = 24 \text{ mpg}$$

In the above equation, E stands for ethanol, and G stands for gasoline. For the same power out, the fuel flow rate must be 152% greater than that for gasoline. Assume equal engine efficiency using ethanol as using gasoline. For the same range, the fuel tank must contain the same energy, Btu. The ratio (116,090 Btu/gal)/(76,370 Btu/gal) gives the 152% value. Fuel tanks must be 152% larger. Fuel pumps and fuel injectors must have 152% greater flow rate. Typical brake-specific fuel consumption (BSFC) for ethanol is 0.75 lb ethanol/hp h. Compare with BSFC values in Chapter 7. The corresponding numbers in terms of energy per mass are: ethanol, 26,800 kJ/kg; and gasoline, 42,700 kJ/kg.

By introducing direct fuel injection, increasing CR, and making other modifications, the ethanol-powered ICE can deliver nearly the same fuel economy as today's gasoline ICE.

Combustion of Ethanol

Looking at the chemical formula for ethanol, C_2H_5OH, the oxygen atom is noted. In the early phases of combustion, oxygen is part of the fuel. Usually fuel and oxygen molecules must collide to initiate combustion. Ethanol burns cleanly with lower emissions. However, the lower energy content of ethanol is due in part to the "onboard" oxygen atom.

Water in the Fuel

Ethanol and water are miscible, which means the two mix readily. Ethanol produced by either ethylene dehydration or brewing (fermentation of sugars) is 95.6% ethanol and 4.4% water. The ethanol must be "dried" for use as a fuel. Even anhydrous ethanol may contain 1.0% water. The word hydrous means with water. Adding the prefix "an," which implies "not," creates a word "anhydrous" meaning "without water."

Water in the ethanol decreases energy content and may adversely affect vehicle operation. A standard is needed that specifies the maximum allowable water in ethanol.

Emissions

Warm-up time for a catalytic converter is longer increasing cold start emissions. Warm-up time is longer because of lower exhaust temperature, which is due to the lower heating value for ethanol. The catalyst should be ethanol specific, that is, designed to work with the exhaust products from the combustion of ethanol.

More ignition energy is needed from the spark plugs for ethanol. Lean misfires may occur. A misfire creates little or no combustion. The exhaust gases due to misfire adversely affect the catalytic converter.

Compression Ratio

Ethanol replaces additives that decrease engine knock or pinging. Lead additives, tetra-ethyl lead (TEL), were used years ago until the toxic nature of lead was recognized. TEL was replaced by methyl tertiary butyl ether (MTBE) with production in 1999 of 200,000 barrels/day. MTBE is also being replaced because of contamination of water supplies.

An E85 blend of gasoline and ethanol, which is termed gasohol, allows higher CR. CR increased to 10–12 is feasible. Using E85, turbocharged engines have been designed with a CR of 15–16 [1].

Fuel Compatibility: Corrosion of Materials

Some materials commonly found in fuel systems are corroded or swell when exposed to pure ethanol or gasoline/ethanol blends with a high proportion of ethanol, for example, E85. These include aluminum, brass, zinc, rubber, PVC, and polyurethane. Stainless steel does not degrade when exposed to ethanol.

Cold Weather Operation

In cold weather, vehicles using E85, or E100, exhibit cold starting problems. Once started, the vehicles exhibit poor drivability while cold. One solution is to heat a small amount of ethanol for the cold start. Another solution is to heat the air supplied to the engine and used for combustion.

Flame Arrestors

When the gas tank or fuel tank is only partially full, in the empty space is a mixture of fuel vapor and air. If only a tiny fraction is fuel, the gas mixture is below the flammability limit. A spark will not ignite the mixture. At the lean flammability limit, sufficient fuel vapor is mixed with the air so that a spark will initiate combustion. The combustion may be deflagration (slow burning) or detonation (very fast burning). Detonation or explosion is very destructive. The amount of fuel in the gaseous mixture is determined by the fuel vapor pressure.

An ethanol blend, E85, has sufficient vapor pressure so that the lean flammability limit may be exceeded at normal ambient temperature. A hazard exists. Sparks or flames must be excluded from the fuel tank. Flame arrestors are installed in the fuel filler pipe to avoid a possible explosion.

Dual-Fuel Vehicles

A vehicle that can operate using pure gasoline or ethanol blends is desirable. If ethanol is not available, gasoline can be substituted. Dual-fuel vehicles have compromises. One compromise is CR. Gasoline demands lower compression while ethanol allows higher compression. Other compromises involve fuel pumps (larger for ethanol) and fuel injectors (also larger for ethanol).

Fortunately, the same air intake can serve both gasoline and ethanol. To analytically show that fact, the heating value of E85 is needed. The heating value for E85 is

$$\mathrm{H}_{\mathrm{E85}} = (0.85)(76{,}370\,\mathrm{Btu/gal}) + (0.15)(116{,}090\,\mathrm{Btu/gal}) = 82{,}330\ \mathrm{Btu/gal} \tag{22.1}$$

The air/fuel ratio for gasoline is 14.7 lb air/lb gasoline. For ethanol, E85, the value is 10.5 lb air/lb ethanol. Assume a gasoline engine consumes 10 lb/h (37.2 mpg at 60 mph) cruising down the freeway. The amount of air is 147 lb/h. Switching to ethanol increases fuel flow rate, M_{E}

$$M_{\mathrm{E}} = M_{\mathrm{G}}\left(\frac{\text{Btu/gal G}}{\text{Btu/gal E}}\right)\left(\frac{\text{Specific gravity E}}{\text{Specific gravity G}}\right) \tag{22.2}$$

As before, E represents ethanol and G, gasoline. Using values for E85 (82,330 Btu/gal) and gasoline, the value of M_E is 14.3 lb E/h. The values for specific gravity are found in Table A.22.1. The air flow is (14.3)(10.5) = 151 lb air/h, which is almost identical to 147 lb air/h for gasoline. This is a very fortunate coincidence involving the properties of the two fuels. The air intake does not need changing when switching between fuels.

Transportation of Ethanol

Ethanol production approaches 6 million barrels per month or, the same amount just in different units, 3 trillion gallons per year. To move the ethanol from the refinery to the user, barges and railroad cars are mainly used. This involves 2500 railroad tank cars per month and 420 barge shipments each month. Tanker trucks play a minor role. Unit trains such as the BNSF *Ethanol Express* are used. Pipelines are much more efficient; however, a number of formidable technical and operational problems need to be addressed.

In 2004, California replaced MTBE with ethanol as an additive creating an oxygenated fuel. An oxygenated fuel is gasoline that has been blended with additives that contain oxygen. The blend is 5.7% ethanol and 94.3% gasoline (E5.7). California is the largest consumer of ethanol in the United States, see Ref. [2].

Rocket Fuel

During World War II, ethanol was used as a fuel for the German V-2 rocket. The oxidizer was liquid oxygen known as LOX.

Can Ethanol Replace Gasoline?

A small fraction of gasoline usage can be replaced by ethanol. Ethanol from corn is unlikely to be competitive without subsidies. Use of subsidies in the short term is desirable. Need for subsidies in the long term means ethanol from corn is not a viable alternate fuel. Corn-derived ethanol can be viewed as taking food from starving peoples in this world. The process of obtaining ethanol from cellulose (corn stover, straw, grasses, sawdust, etc.) must be perfected on a large scale. Wald [4] discusses the viability of ethanol as an alternate fuel.

ALTERNATE FUELS: MORE FUELS

Methanol

Methanol is an alcohol. Methanol is superior to ethanol with regard to CO_2 emission (see Table 22.2). However, methanol (57,250 Btu/gal) lacks energy content compared to ethanol (76,330 Btu/gal) (see Table A.22.1). Energy content is 32% less. Gasoline has 103% more energy than methanol. Methanol is more volatile than ethanol, which may be favorable for combustion.

Natural Gas

Natural gas is a fossil fuel found in the same fields as petroleum or in other fields with only natural gas. Natural gas is mainly methane with small fractions of ethane and propane. Figure 22.4 shows the carbon/hydrogen ratio, R, for natural gas (as methane) with ratio, R, equal to 33%. This value of R is better than that for gasoline with regard to CO_2 emissions. Vehicles powered by natural gas have lower emissions.

Being a gas and not a liquid, natural gas can be stored in a vehicle in three different ways:

1. CNG = Compressed natural gas (3000–3600 psi)
2. LNG = Liquefied natural gas (20–150 psi at −120°C to −170°C)
3. ANG = Adsorbed natural gas (500 psi at ambient temperature)

For applications to cars, trucks, and buses, CNG is more common than LNG. Expensive cryogenic tanks are avoided using CNG rather than LNG. The CNG tank is approximately three times the size of a gasoline tank.

In one case, ANG uses a tank filled with carbon fibers. The gas is adsorbed on the surface of the carbon fibers. According to Burchell and Rogers [6] more than 180 V/V storage has been demonstrated. This means that 180 volumes of natural gas can be stored in one volume of tank. Storage is at 500 psi and ambient temperature. The ANG tank provides about a quarter of the driving range of an equivalent gasoline tank.

Sales of the Honda Civic GX began in 2007. This car, which uses natural gas as the fuel, stores the fuel in the gaseous state. CNG pumps are rare and hard to find. A filling station at home is possible using the gas utility natural gas. To recharge the 8 gal (gasoline equivalent) tank requires 16 h if the charging unit does not use a compressor. For the same amount of energy as a gallon of gasoline, the cost is $1.30. Fuel economy is 29 in the city and 39 on the highway.

Natural Gas/Hydrogen Blends

The blends are identified as HCNG, which is a blend of hydrogen and CNG. HCNG, which is typically 20% hydrogen and 80% natural gas, is one building block on the journey to a hydrogen infrastructure. ICEs can operate using HCNG. Lower emissions of GHG, CO_2, and NO_X result from use of HCNG. The blend has been used in either spark or compression ignition engines. Engine efficiency has no significant change using HCNG compared to CNG.

Electricity as a Fuel

Earlier the distinction was made between hydrogen for fuel cells and electricity for EV. For an EV, the charging cable crosses the boundary, and the fuel is electricity. For an FCV the fuel is hydrogen. To use electricity as a fuel, a battery on board the vehicle is essential. The battery is charged by connection to the power grid. Green energy can be used to generate the electricity.

An EV emits zero emissions and meets the CARB (California Air Resources Board) goal of zero emissions vehicle (ZEV). Appendix A lists the definitions for acronyms. An EV offers lower cost per mile for operation. An EV characteristically has lower maintenance costs except for the battery. The major drawback for an EV is limited range.

ALTERNATE FUELS: HYDROGEN AS A FUEL

Hydrogen: Fuel of the Future

Hydrogen is the fuel of the future. At least that is the opinion of many people. Hydrogen does not occur in nature as the molecule H_2. In the cosmos, H_2 is the dominant molecule. To solve our oil problems, design spacecrafts that are hydrogen tankers. We should send spaceship *Enterprise* to bring back hydrogen. This sentence is said tongue in cheek. Hydrogen as found on earth is not a fuel. Hydrogen is not a primary fuel. Hydrogen is a carrier of energy.

Hydrogen: Fuel of the Past

The world's first turbojet used hydrogen as a fuel. In 1938–1939, Hans Von Ohain was a graduate student in Germany working on the turbojet as his PhD thesis. He selected hydrogen as the fuel. Favorable ignition and ease of combustion were the determining factors. Meanwhile in England, Frank Whittle was also inventing and developing the turbojet.

Critical Milestones

The success of hydrogen as an alternate fuel hinges on the success of fuel cells. Fuel cells provided the power and drinking water for the Apollo astronauts. This application was extremely successful. Fuel cells are also used on the space shuttle. Fuel cells are not new technology. Producing fuel cells at a price suitable for automobile application is new. Getting the cost down is a major hurdle. Hydrogen can be used in ICEs; however, the application in ICE is not as compelling as fuel cells. If FCV cannot provide the same performance, load capacity, costs, and range as ICE vehicles, FCVs will not be commercially viable [16]. Another critical milestone is production of hydrogen, H_2, subject to constraints of cost and minimum level of emissions. The best production method has the least emissions, uses the minimum of green energy, and has the least complexity.

Production of Hydrogen

To produce H_2, start with a molecule or substance containing hydrogen. Water and methane are two candidates. Refer to Figure 22.3, which shows hydrogen production by electrolysis. The input energy for electrolysis can be from a green source. Wind, wave, tidal, photovoltaic, nuclear, and biomass are possible energy sources.

One source of energy for electrolysis of hydrogen is nuclear-powered electricity. Nuclear power does not emit carbon dioxide. Disposal of the radioactive waste is a major problem.

Chapter 6 presents the following possible methods for production of hydrogen:

1. Steam and methane; water gas reaction
2. Partial oxidation
3. Autothermal
4. Electrolysis
5. Methanol reforming
6. Autothermal reforming of gasoline and diesel fuels

Hydrogen Storage

In addition to the success of the fuel cell as a commercially viable option, storage of hydrogen on board the vehicle remains a difficult task. The five options are

1. Gaseous storage.
2. Liquid storage (as was done for the Apollo program).
3. Storage in a hydride or hydrate.
4. Storage in an adsorbed state in carbon, for example.
5. Chemical storage, for example by ammonia, NH_3, which is rich in hydrogen; chemical decomposition on board the vehicle provides the hydrogen.

Gaseous storage is at high pressure typically 5000 psi. Liquid storage is at very low temperature 20° K, which is −423°F or −253°C. Even with the best thermal insulation, a few percent hydrogen is lost each day by boil-off. Hydrides tend to be heavy. Hydrides only work at temperature extremes that are difficult to engineer into vehicles. Research is being focused on hydrides that offer more favorable temperature for storage. The nanotechnology of carbon may offer useful storage. Chemical storage, for example, ammonia, avoids problems of high pressure or low temperature; however, ammonia is expensive.

None of the storage methods are particularly attractive with superior features compared to the other choices. Poor hydrogen storage capability means poor range capability. Range is not as restricted as for an EV using a battery; nonetheless, limited range is a problem to be solved.

Hydrogen Storage: Clathrate Hydrates

At ambient temperature and pressure, when water and hydrocarbon molecules come into contact, little happens. However, at high pressure and low temperature, an ice-like framework of hydrogen bonds can build polyhedral cages around guest molecules. The guest molecules may be methane, nitrogen, and argon. Clathrate hydrates form. The cages are held together by weak van der Waals forces [9]. Think of a bucket of golf balls. Empty spaces exist between the balls. Gases can fill the empty space. The balls are analogous to the polyhedral cages except the cages are of molecular scale thereby invoking the different physics of quantum mechanics.

Clathrate hydrates are a major impediment to the flow in offshore gas and oil pipelines. Pipelines in the arctic region and along the ocean floor can be plugged by a solid mass of clathrate hydrates. While unfortunate for pipelines, under certain conditions, hydrogen can be stored in combination with other molecules. The winner for hydrogen storage is a methane–hydrogen system that has enormous hydrogen (33.4%), H_2, by mass. Counting the H_2 in the methane, 50% of the mass is hydrogen. For comparison, DOE storage goals for 2007, 2010, and 2015 are 4%, 6%, and 9% mass of hydrogen, respectively. Unfortunately, the methane–hydrogen system requires a pressure of 200 MPa (2000 atmospheres) at a low temperature, which are unlikely conditions for a vehicle. However, the technology has been barely scratched.

Energy to Compress Hydrogen

To compress hydrogen gas to 5000 psi requires considerable energy. How much is the energy of compression, E, compared to energy content of hydrogen?

$$E = C_V(T_2 - T_1) = C_V T_1 (T_{21} - 1) \tag{22.3}$$

where T_{21} is shorthand for T_2/T_1. For an isentropic (perfect) compression

$$T_{21} = (p_{21})^{\frac{\gamma-1}{\gamma}} \tag{22.4}$$

Combining the two equations leads to

$$E = C_V T_1 \left[(p_{21})^{\frac{\gamma-1}{\gamma}} - 1 \right] \tag{22.5}$$

The following values are used for a sample calculation:

$$C_V = 2.435\,\text{Btu/lb°F}, \quad T_1 = 70\text{°F} = 530\text{°R}, \quad p_1 = 14.7\,\text{psi}$$

The calculation yields $E = 5530$ Btu/lb, which is 10.6% of the hydrogen energy content of 52,217 Btu/lb. The 10.6% is for a perfect compressor. Compression energy cannot be ignored.

Hydrogen Infrastructure

The distribution of hydrogen to the customers involves three concepts. A large central hydrogen facility, which is 50–300 mi from the roadside pump, allows production on a large scale. The semi-central facility is located 25–100 mi from the customer. A local station is at or near the point of use. Hydrogen is made on the premises. A hydrogen-refueling station may be feasible at home. Photovoltaic cells on an area equivalent to that of two garage doors provide the energy for electrolysis. The hydrogen infrastructure will be expensive [17].

Hydrogen can be piped from the production facility to the customer. Hydrogen gas is prone to leakage. This behavior follows from the kinetic theory of gases, which states the speed of an individual molecule varies inversely as the square root of its mass. Compared to natural gas (methane, which has mass of 16), the hydrogen (which has a mass of 2) is moving $(16/2)^{1/2} = 2.8 \approx 3$ times faster. Also a hydrogen molecule has a smaller collision cross section. A small cross section combined with fast molecules means hydrogen tends to leak. Also, hydrogen causes embrittlement in some metals. The metal becomes brittle instead of remaining ductile. This is a formula for disaster for a high pressure pipe.

Simultaneity

Building the hydrogen infrastructure requires simultaneous production of hydrogen-powered vehicles (not limited to FCVs). Large energy companies, even with government encouragement, will not invest in hydrogen production and distribution systems if the customers are not on the road with near-empty hydrogen tanks. Customers will not buy hydrogen-powered cars if hydrogen filling stations are not handy. It is the classical chicken-and-egg problem. Globally in 2007, there were only 500 H_2-powered cars.

Comparison with Gasoline

The concept of gallon gasoline equivalent (GGE) is needed to compare hydrogen-powered vehicles with gasoline. GGE is determined by comparing the energy in the same volume of fuel. Volume-based GGE for hydrogen is approximately 4 gal of liquid hydrogen that have the same energy as 1 gal of gasoline. An alternate comparison can be based on equal masses of fuel. From Table A.22.1, hydrogen heating value is 52,217 Btu/lb and gasoline is 18,676 Btu/lb. Hence, hydrogen has 2.8 times more energy per pound than gasoline. The low density of liquid hydrogen means large fuel tanks. The density of liquid hydrogen is 0.07 g/cm^3 and gasoline is 0.75 g/cm^3.

The low density of hydrogen and high Btu/lb means low weight. The weight of hydrogen is extremely important for rocket launch vehicles but is largely irrelevant for automobiles. However, the weight of storage tanks for hydrogen is important for automobiles.

PAST AND FUTURE

Past: 1890–1950

Chapter 1 discusses the past for the twentieth century. Figure 22.8 is a graphical summary of the information. The twentieth century starts with three contenders for individual transportation: steam, gasoline, and electric. By 1925, all vehicles except gasoline had disappeared from new car showrooms. Gasoline and diesel were overridingly prevalent. From 1925 to 2000, and almost to 2020, gasoline and diesel remained dominant. In the mid-1910s, hybrids were a tiny, transient blip on the radar.

Future: 2000–2060

Figure 22.9, which covers 2000–2060, is a similar graphical presentation as Figure 22.8. Compare Figure 22.9 with Figure 1.6. One difference is the fuzzy area defined by the years for "gasoline

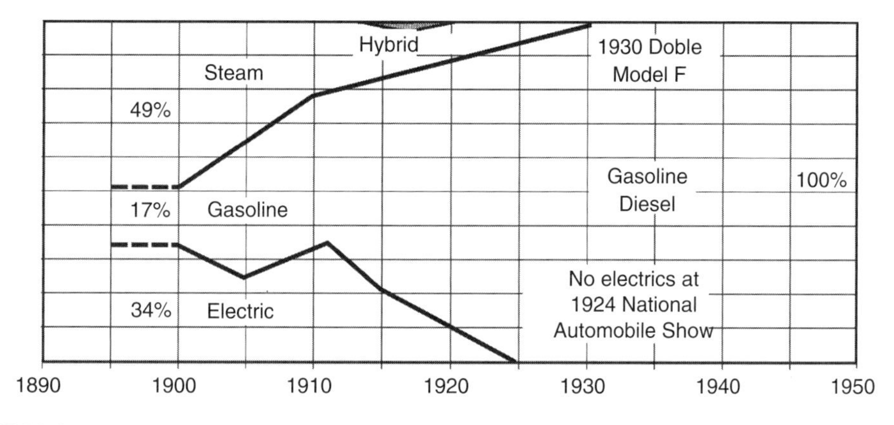

FIGURE 22.8 Past, 1890–1950.

becomes prohibitively expensive" and "the day gasoline essentially runs out." Figure 22.9 takes into account the discussions in Chapter 2, which indicate gasoline production will continue for many decades but will not meet demand. Another difference is the timing of the gasoline crises. Figure 1.6 uses 2037–2042, which was indicated as being somewhat arbitrary. Figure 22.9 shows 2020 as the critical year due to oil peaking.

Conditions at 2000 are similar for both Figures 1.6 and 22.9. The long-term prediction at 2060 is also similar with fewer EV in the latter figure. Both predict an overwhelming abundance of FCVs in 2060.

The different classes of vehicles in Figure 22.9 include gasoline/diesel, pure electric EV_B, hybrid HEV, fuel cell (hybrid, alternate fuel, H_2), alternate fuel, ICE, and kit conversion cars. The rationale for each class of vehicle is now discussed.

Gasoline/Diesel

Gasoline/diesel declines from near 100% in 2000 to 60% in the year of oil peaking. The decline continues to 15% in 2040 and finally to 0% in 2050. The decline is due to stringent emission requirements as well as decreasing petroleum production. As seen in Figure 2.1, some oil production remains even decades after peaking. Initially, the alternate fuels compete with petrofuels. Later, alternate fuels dominate the market, and petrofuels must now compete with the upstart alternate fuels. For the small 15% of gasoline/diesel vehicles, oil will be available. Emissions cause the demise of the gasoline/diesel vehicles.

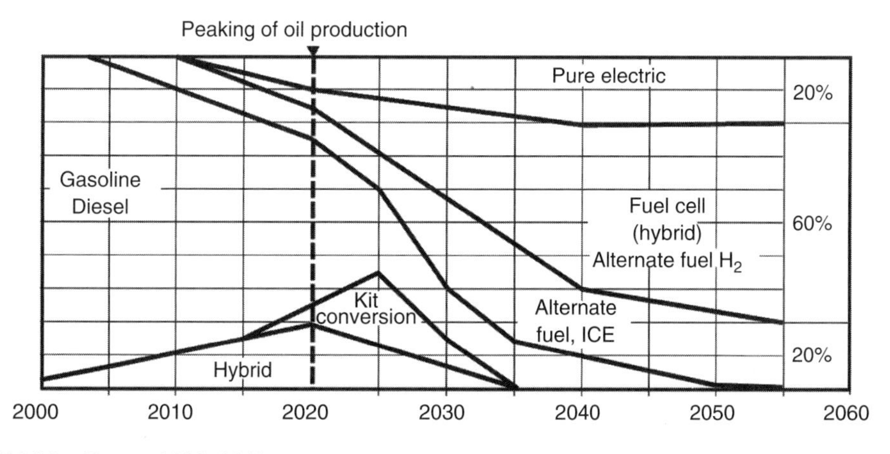

FIGURE 22.9 Future, 2000–2060.

Pure Electric

Over a span of 30 years from 2010 to 2040, the EV with battery storage increases penetration to 20% and maintains that niche for another 15 years up to 2055. EV_B fill the driving needs of many people and offer suitable transportation at competitive rates. In favor of EV_B is zero emissions. Advances in battery technology enhance the attractiveness of EV_B.

Hybrid

The future for HEV has been discussed in connection with Figure 18.2. In fact, Figure 18.2 is superimposed on Figure 22.9. The diagram below, which is reproduced from Chapter 18, indicates that the gas/diesel was replaced by HEV due to higher mpg and lower emissions of the HEV. In turn, the HEV was replaced by FCV due to the need for an alternate fuel and zero CO_2 emissions by the FCV.

Gas/diesel ——————> HEV ——————> FCV

Higher mpg   Alternate fuel

Lower emissions CO_2 emissions

Fuel Cell Vehicle

The FCV provides zero emissions and uses an alternate fuel. These are two major advantages of the FCV. If the fuel cell can be developed to be competitive in price, then the predictions of Figures 22.9 and 1.6 are reasonable. However, serious glitches in the development of the fuel cell make Figures 1.6 and 22.9 immaterial.

Figure 1.6 predicts FCV (or EV_H) having 50% of the market while EV_B has 38% and the niche, solar, photovoltaic recharged EV has 12%. Figure 22.9 gives the FCV 60% of the market with 20% for hydrogen ICEs and 20% for EVs.

Alternate Fuel, ICE

If the alternate fuel is hydrogen, the ICE will be acceptable from emissions point of view. The hydrogen infrastructure will be in place to service the FCV. Use of hydrogen increases the efficiency of the ICE. The hydrogen-fueled ICE provides dynamic performance.

Kit Conversion Cars

White predicts 1,000,000,000 (1 billion cars on the world's roads by 2020 [7]). Note that 2020 is also the predicted year for oil peaking (Figure 22.9). Of the 1 billion cars, an estimated 500,000,000 (that is right, 500 million) will be petroleum-fueled vehicles. These cars with gas tanks will need conversion kits to salvage the body and chassis. The conversion kits will come on the market about 2015. The kit replaces the gas engine with a battery and M/G. Downstream of 2035, the kit cars fade away.

SUMMARY

This book has two discussions of the future. These discussions are found in the first and last chapters. The predictions in the first chapter are based on what the reader may know before reading the book. The predictions in the final chapter are tempered by contents of the book.

The automobile is a symbol of personal freedom; a replacement is unlikely. In the United States, the future starts with a fleet of vehicles that have poor fuel economy. Future actions must not only correct the past but also develop for the oil-limited years. The future is driven by oil supply and demand, technology, and politics. The gap between oil supply and growing demand requires alternate fuels. The ability of engineers and scientists will be strained to meet the technological needs. The technologies of the future for automobiles include biology and electrochemistry. Politicians with the foresight and nerve to implement mitigation measures are desperately required.

The timing for alternate fuels depends on the timing of oil peaking. Demand can be moderated by energy conservation. Hybrids are a step in the right direction. In spite of all the attention, as of today, the sources and chemical nature of viable alternate fuels are not defined. The script for the energy future has not been written.

Mitigation measures must be timely. Funding, of necessity, will be jointly between government and private industry. The level of pain of mitigation actions depends on how seriously the electorate considers the oil-limited events to come. When viewed from the point of view of the possible economic chaos, two painless mitigation actions are increased CAFE (gradually to 40 mpg) and gasoline tax ($6/gal price at pump).

Alternate fuels have been assigned three attributes: being in fact alternate, renewable, and sourced from biomass. Table 22.1 is intended to clarify the classification of fuels.

Besides limitations of oil, emissions play another dominate role. Global warming may be triggered by levels of CO_2 in the atmosphere (see Appendix B). Fuels can be evaluated by the carbon/hydrogen ratio, R, defined in Figure 22.4. The larger the value of R, the larger the relative amount of CO_2 being dumped in the atmosphere.

Coal is so abundant, and the United States holds such an impressive amount, that this energy source cannot be ignored. Regardless of how dirty coal may be, coal will dictate, at least partially, energy considerations and policy. Nuclear energy too is considered a dirty source of energy. Yet nuclear energy will also become an increasingly attractive source.

The major alternate fuels today are biodiesel, ethanol, and hydrogen. Biodiesel has many important advantages. Ethanol, if derived solely from corn, is unlikely to survive critical exposure in the energy market. Hydrogen has many hurdles to clear before being declared a winner.

Finally, the contents of this chapter, as well as that of the book, are summarized by two graphs in Figures 22.8 and 22.9. The graphs are speculative. The graphs are based on information available today. The graphs are intended to stimulate thought and even controversy.

APPENDIX 22.1: PHYSICAL PROPERTIES OF FUELS

TABLE A.22.1
Physical Properties of Alternate and Fossil Fuels

Fuel	Heating Value (Btu/lb)	Heating Value (Btu/gal)	Boiling Temperature (°F)	Chemical Formula	Air/Fuel by Weight	Specific Gravity
Gasoline	18,676	116,090	80–437	C_4 to C_{12}	14.7	0.72–0.78
Diesel #2	18,394	129,050	356–644	C_8 to C_{25}	14.7	0.85
Methanol	8,637	57,250	149	CH_3OH	6.45	0.80
Ethanol	11,585	76,330	172	C_2H_5OH	9.00	0.79
Propane	19,900	84,500	−44	C_3H_8	15.7	0.51
CNG	20,363	19,800	−126.4 to 3.2	(A)	17.2	0.42
Hydrogen	52,217	—	−423	H_2	34.3	0.07
Biodiesel	16,131	118,170	599–662	C_{12} to C_{22}	13.8	0.88
Coal	(B)	—				

Notes: (A) CNG: CH_4 83%–99%; C_2H_6 1%–13%.
(B) Coal is 6.67 kWh/kg; gasoline is 11.87 kWh/kg.

REFERENCES

1. N. M. LaBlanc and R. P. Larsen (Eds.), *2000 Ethanol Vehicle Challenge*, Society of Automotive Engineers, SP-1618, Warrendale, PA, 2001, p. 197.
2. California Energy Commission, Ethanol as a transportation fuel in California, Available at http://www.energy.ca.gov/ethanol/indes.html.
3. F. Keppler and T. Röckman, Methane, plants and climate change, *Scientific American*, February 2007, pp. 52–57.
4. M. L. Wald, Is ethanol for the long haul? *Scientific American*, January 2007, pp. 42–49.
5. S. Birch, Searching for fossil fuel alternatives, *Automotive Engineering International,* Society of Automotive Engineers, March 2007, pp. 56–58.
6. T. Burchell and M. Rogers, Low pressure storage of natural gas for vehicular applications, Paper 2000-01-2205, Society of Automotive Engineers, Warrendale, PA, 2000.
7. J. B. White, One billion cars, *The Wall Street Journal*, April 17, 2006, p. R1.
8. S. Leeb, *The Coming Economic Collapse*, Warner Business Books, New York, 2006.
9. W. L. Mao, C. A. Koh, and E. D. Sloan, Clathrate hydrogen under pressure, Feature Article, *Physics Today*, October 2007, pp. 42–47.
10. F. C. Whittelsey and P. Holey, Bio-hope, bio-hype, *Sierra,* September/October 2007, pp. 50–51.
11. M. E. May and R. A. Neihof, Microbial deterioration of marine diesel fuel from oil shale, Naval Research Laboratory, Washington, DC, April 9, 1981.
12. R. Conniff, Who's fueling whom? *Smithsonian*, November 2007, pp. 109–116.
13. J. K. Bourne, Jr., Green dreams, *National Geographic*, October 2007, pp. 38–59.
14. Staff, Advanced Biofuels; Ethanol, Schmethanol, *The Economist*, September 29, 2007, pp. 84–85.
15. D. Alexander, Fuel for thought, *Automotive Engineering International*, Society of Automotive Engineers, September 2005, pp. 44–48.
16. D. Alexander, Fueling the next generation, *Automotive Engineering International,* Society of Automotive Engineers, September 2005, pp. 73–77.
17. J. Ogden, High hopes for the hydrogen, *Scientific American*, September 2005, pp. 94–101.
18. Staff, Nuclear fusion; fring new shots, using lasers to trigger fusion could prove cheaper than other techniques, *The Economist*, April 21, 2007, pp. 89–90.
19. G. Edmondson, Here comes pond scum power, *Business Week*, December 3, 2007, pp. 65–66.
20. K. Jost, Time for More 'Grassoline', *Automotive Engineering International*, Society of Automotive Engineers, July 2007, p. 6.
21. D. Alexander, Fuel for thought; the demand for clean renewable energy is driving research into different sources of fuel and new, more flexible engines, *Automotive Engineering International*, Society of Automotive Engineers, September 2005, pp. 46–51.
22. K. Proc, R. Barnitt, R. Hayes, M. Ratcliff, R. McCormack, L. Ha, and H. Fang, *100,000-Mile Evaluation of Transit Buses Operated on Biodiesel Blends (B20)*, Society of Automotive Engineers, Paper 2006-01-3253, October 2006.

Appendix A: Acronyms and Relevant Data

ACRONYMS

A/C	Air-conditioning
AFC	Alkaline fuel cell
AFV	Alternate fuel vehicle
ANG	Adsorbed natural gas
API	American Petroleum Institute
AR	Aspect ratio of tire = *H/W*
ASPO	Association for the Study of Peak Oil
ASTM	American Society of Testing and Materials
ATM	Automatic teller machine
AWD	All wheel drive
BDC	Bottom dead center
BMS	Battery management system
BSFC	Brake-specific fuel consumption (pounds fuel/brake horsepower-hour)
CA	Crankshaft angle
CAFE	Corporate Average Fuel Economy
CAT	Compressed air technology
CAV	Compressed air vehicle
C/D	Charge/discharge cycle of a battery
CEC	Computational electrochemistry
CFD	Computational fluid dynamics
CG	Center of gravity
CIDI	Compression ignition (diesel) with direct injection
CNG	Compressed natural gas
CP	Center of pressure for tire patch
CR	Compression ratio
CUV	Crossover utility vehicle
CV	Conventional vehicle
CVT	Continuously variable transmission
DCT	Dual-clutch transmission
DDS	Deep discharge state (Li-ion batteries)
DME	Dimethyl ether (possible diesel fuel)
DoD	Displacement on demand; e.g., deactivating a bank of four cylinders in V8 engine
DoD	Department of Defense, United States
DoD	Depth of discharge for batteries
DOE	Department of Energy
DSG	Direct sequential gearbox
DSG	Direct shift gearbox

DOHC	Dual overhead cam
EDL	Electronic double layer
EDLC	Electrochemical double-layer capacitor
EERE	Energy Efficiency and Renewable Energy (a DOE program)
EGR	Exhaust gas recirculation
EIA	Energy Information Agency (United States)
EMF	Electromotive force (volts)
EPA	Environmental Protection Agency
EPS	Electrical power system
EUR	Estimated ultimate recoverable oil
EVT	Electronically variable transmission
EWB	Electric wedge brakes
f/a	Fuel/air
FC	Fuel cell
FCV	Fuel cell vehicle
FE	Fuel economy
FMVSS	Federal Motor Vehicle Safety Standards
FLC	Fuzzy logic control
F/R	Front/rear (used to denote percentage weight on front and rear wheels)
4WD	Four-wheel drive
FWD	Front wheel drive
GGE	Gallon gasoline equivalent
GHG	Greenhouse gas
GTL	Gas-to-liquid to create diesel fuel from natural gas
GVWR	Gross vehicle weight rating
HC	Hydrocarbon
HCCI	Homogeneous charge compression ignition engine
HCNG	Hydrogen/CNG blend
HDV	Heavy-duty vehicle (i.e., a truck)
HEV	Hybrid electric vehicle
IMA	Integrated motor assist
IPO	Initial public offering of stock in a new company
ISAD	Integrated starter alternator damper
ISG	Integrated starter/generator system (same as M/G)
iVTEC	Intelligent variable valve timing and lift electronic control (Honda)
KE	Kinetic energy (kWh or J)
LDV	Light-duty vehicle
LNG	Liquid natural gas or liquefied natural gas
mbd	million barrels per day
MIEV	Motor in wheel electric vehicle
M/G	A dual-purpose motor/generator
MTBE	Methyl tertiary butyl ether (a fuel additive to prevent pinging)
MY	Model year
NHTSA	National Highway and Traffic Safety Agency
NN	Neural network
NO_X	Emissions of oxides of nitrogen
NRC	National Research Council (part of National Academies)
NREL	National Renewable Energy Laboratory (part of DOE)
OBD	On-board-diagnostics
OCV	Open circuit voltage of a battery or cell
ODAC	Oil Depletion Analysis Center

PbA	Lead-acid battery
PE	Potential energy (kWh or J)
PEM	Proton exchange membrane fuel cell
PHEV	Plug-in HEV
PNGV	Partnership for a New Generation of Vehicles
PM	Particulate matter (black smoke and soot) from diesels
PZEV	Partial zero emissions vehicle
R&D	Research and development
RTOS	Real-time operating system
RWD	Rear wheel drive
SC	Supercapacitor (energy storage)
SEI	Solid electrolyte interphase
SOHC	Single overhead cam
SUV	Sport utility vehicle
TDC	Top dead center
TDI	Turbocharged direct injection
TEL	Tetra-ethyl lead (a fuel additive to prevent pinging)
THS	Toyota hybrid system
TPH	Turbo parallel hybrid
TSD	Toyota synergy drive
TTR	Through-the-road
TV	Tow vehicle
2WD	Two-wheel drive
UC	Ultracapacitor (energy storage)
UFO	Unidentified flying object
ULSD	Ultralow sulfur diesel fuel
UPS	United Parcel Service
VCM	Variable cylinder management
VRLA	Valve-regulated lead-acid battery
VTEC	Variable valve timing and lift electronic control (Honda)
WOT	Wide open throttle
ZEV	Zero emissions vehicle

Gasoline Density

$$\text{Specific gravity} = 0.72$$

$$6.2\ \text{lb/gal} = 0.75\ \text{g/cm}^3 = 0.75\ \text{kg/L}$$

Gasoline Temperature Expansion Coefficient

$$\varepsilon = 0.0006/°\text{F}$$

Gasoline Energy Density

$$33.44\ \text{kWh/gal} = 8.83\ \text{kWh/L}$$

Gasoline-Specific Energy

$$5.39\ \text{kWh/lb} = 11.77\ \text{kWh/kg}$$

Density for Liquid Hydrogen (H_2) and Gasoline

Density of liquid hydrogen is 0.07 g/cm^3 and

Density of gasoline is 0.75 g/cm^3 = 0.75 kg/m^3

Heat Capacity for Hydrogen (H_2)

$$C_P = 3.421 \text{ Btu/lb °F}$$

$$C_V = 2.435 \text{ Btu/lb °F}$$

Heat Capacity for Air

$$C_P = 0.236 \text{ Btu/lb °F}$$

$$C_V = 0.167 \text{ Btu/lb °F}$$

Atmospheric Density at Sea Level

$$\rho = 1.226 \text{ kg/m}^3$$

At 2 km altitude

$$\rho = 1.000 \text{ kg/m}^3$$

Definitions for Specific and Density

As used in this book, the words have the definitions as given. The adjective specific means per unit mass. Thus, specific power has the unit kW/kg where kg is a unit of mass. Specific energy has unit kWh/kg. Specific energy and specific power are used to answer the question, "How much does a component (battery) weigh?" The word density refers to a unit volume. Power density has the unit kW/L, and energy density has the unit kWh/L. Density helps answer the question "How big is a component (battery)?" Liter is a measure of volume. Briefly

Specific = per unit mass

Density = per unit volume

Energy of Fuels

	Energy Density		Specific Energy	
Fuel	**MJ/L**	**kWh/L**	**MJ/kg**	**kWh/kg**
Gasoline	29.0	8.06	45.0	12.50
LPG	22.2	6.16	34.4	9.55
Ethanol	19.6	5.44	30.4	8.44
Methanol	14.6	4.05	22.6	6.28
Gasohol[a] E85	27.6	7.66	42.8	11.89
Diesel	40.9	11.36	63.47	17.63

[a] E85 = 85% gasoline and 15% ethanol.

Gallon = 3.785 L

For gasoline, kWh/gal = (3.785 L/gal)(8.06 kWh/L) = 30.51 kWh/gal

The value of 30.51 kWh/gal does not agree with 33.44 kWh/gal stated above the table.

Energy Required to Propel a Vehicle per Kilometer
For the GM EV-1, the energy/km was 112 W h/km. The energy required to propel a vehicle per kilometer, *E/d*, can also be calculated.

The equation is

$$\frac{E}{d} = \frac{P}{V} \tag{A.1}$$

where
E is the energy to move a vehicle
d is the distance vehicle is moved, 1.0 km
P is the cruise power to move vehicle, 10 kW
V is the velocity of vehicle, 60 mph (26.8 m/s)

A sample calculation with change of units is

$$\frac{E}{d} = \frac{\text{W h}}{\text{km}} = \frac{10\text{ kW}}{26.8\text{ m/s}} = \frac{\text{h}}{3600\text{ s}} = \frac{1000\text{ m}}{\text{km}} = \frac{1000\text{ W}}{\text{kW}} = 104\text{ W h/km} \tag{A.2}$$

The typical value of energy required to propel a vehicle per kilometer is

$$100\text{ W h/km} < \frac{E}{d} < 150\text{ W h/km}$$

This is important for determining the installed energy for the battery. For electric-only operation on a flat road for a 2800 lb vehicle at 60 mph the power is 8 kW. These data are from Ref. 12 of Chapter 6.

Definitions for Devices Converting Electrical Voltage
Definitions for various conversions for electrical quantities

DC → converter → DC
DC → inverter → AC
AC → transformer/rectifier → DC
AC → transformer → AC

Four combinations of DC to AC, etc. are possible as shown above. Accepted names are shown.

Battery Definitions

Specific energy, W h/kg (battery energy)/(mass)
Specific power, W/kg (battery power)/(mass)
Energy density, W h/L (battery energy)/(volume)
Power density, W/L (battery power)/(volume)

Vocabulary and Definitions
Brake-specific fuel consumption (BSFC): Brake comes from the test procedures where the gas engine is not installed in a vehicle; instead, the engine is in a test stand with the crankshaft connected to a brake. The brake, which absorbs the power output of the engine, provides a means to accurately measure horsepower. BSFC has units of

Pounds fuel/brake horsepower-hour

or

Weight of fuel/energy output of engine

SFC can also use mass instead of weight. From Chapter 10, the units for SFC are grams per kilowatt hour (g/kWh).

Waste gate: For turbocharging an engine using a compressor in the intake, which is spun by a turbine in the exhaust, a control of intake pressure is needed. The amount of exhaust gases going through the turbine, and hence the pressure of the intake air, is controlled by a valve called the waste gate.

Feed forward: Some controllers are configured in a feed-forward manner. Everything starts with the driver action. As an example, consider the throttle being pushed down by the driver. The pedal position signal is sent directly to the fuel injector. This allows rapid response. Initially, the fuel injector responds to the pedal command. As the control system does its calculations, fuel injection likely will need adjustment according to the control system.

Air-Conditioning—Required Power
Weissler states 2.8–4.0 kW for a 2005 Cadillac STS (—Ref. [17] of Chapter 12).

Ethanol Facts—Transportation

	Ethanol	Petroleum
Barges: river	420/month	2,800 barges in inventory
Barges: ocean	?	450 barges in inventory
Railroad tank cars	2500/month	211,000 cars in inventory
Pipelines	None	Extensive

Typical Data for a Small Compact Car
Ford Focus Hydrogen-ICE Conventional Vehicle

- State of art (2007)
- Engine 2.3 L, four-cylinder, supercharged, intercooled
- Four-speed automatic
- Mass = 1452 kg (3200 lb)
- Area, $A = 2.06\,\text{m}^2$, $C_D = 0.31$

$$D = \frac{1}{2}\rho V^2 C_D A$$

Rolling resistance: 0.008

- Drag = 0.008 × 3200 lb = 25.6 lb
- Drag = 0.008 × 1452 kg = 114 N
- Power = (114 N)(30 m/s) = 3.4 kW due to rolling resistance

Appendix B: Global Warming

Sunlight enters the atmosphere through a transparent atmosphere. Visible light, that is, sunlight, is transmitted through the atmosphere. The energy from the visible light is absorbed by the earth, and in the process is transformed to longer wavelengths in the infrared. The red end of the visible light blends into the infrared. Infrared light is not visible to humans. Rattlesnakes and pit vipers have infrared sensors in the pits on their nose.

In contrast to visible light, the atmosphere is partially opaque to infrared radiation. The opaqueness is due to certain gases in the atmosphere, which are infrared absorbers. These gases include:

Water vapor (H_2O)	0%–4%
Carbon dioxide (CO_2)	0.035%
Methane (CH_4)	0.0002%
Ozone (O_3)	0.000004%
Nitrous oxide (N_3O)	—
Fluorinated gases	—

The amount of the gases (listed above as %) varies due, in part, to emissions from different sources. Everyone is aware that vehicle emissions affect the level of carbon dioxide, CO_2. Cow manure, melting arctic tundra, and garbage dumps are sources of methane.

Some gases are much more potent infrared absorbers than others. For example, one molecule of methane is 300–400 times more absorbent than one molecule of carbon dioxide. So, why is the focus on carbon dioxide and not on methane? Al Gore, in his movie, *An Inconvenient Truth*, barely mentions methane and the other infrared-active gases.

Carbon dioxide is the infrared-active gas, which is being increased hundredfold due to human activity. The other gases, while not constant, do not change as much as carbon dioxide.

Further, many atmospheric effects are nonlinear. What is meant by nonlinear? An example is the chaotic stock market. Suppose the daily price of a stock is a series as follows:

$41, $42, $43, $44, $___

Since today is Thursday, Friday's price is shown as a blank. If asked what price should appear in the blank on Friday, you would be tempted to fill in $45. Chances are you would be correct; however, the price could just as well be $23 or $89. A graph of the daily price is a straight line (hence, linear) until Friday. A linear variation occurs the first four days and continues on Friday if $45 is correct. Otherwise, the graph has a kink either upward if $89 is correct or downward if $23 is correct. The kink in the curve is characteristic of nonlinear behavior.

Now change the numbers in the above discussion to

41°F, 42°F, 43°F, 44°F, ___°F

Obviously, we are now talking about the weather. The same discussion applies with regard to linear and nonlinear behavior. Suppose the temperature on Friday was a big surprise at 89°F. What caused the change? Or as the climatologists would ask, what triggered the jump to 89°F. The economists ask "What triggered the jump to $89?" What causes the jumps in the stock market, or what triggers for jumps in the weather? This is where all the mystery and uncertainty lies. This is where science and technology struggle to find answers that allow a prediction to be made.

Carbon dioxide is absorbed by the oceans where it forms an acid. Increased CO_2 in the atmosphere translates into increased CO_2 in the oceans. Measurements of the pH in a solution indicate the acidity or alkalinity of that solution. The measured pH values of the oceans have been indicative of increased acidity. The higher acidity level adversely affects coral reefs, shell fish, and other marine life. Computer models for global warming include the oceans as a CO_2 sink. What is favorable for the atmosphere may be disastrous for some creatures living in the ocean.

Carbon dioxide might be such a nonlinear trigger to set off irreversible, unexpected, and catastrophic global temperature changes! Do you bet that growth of carbon dioxide causes a linear or a nonlinear result?

The wild card in the global warming is water in the atmosphere. Water can be in the form of water vapor, as water droplets as in clouds, or as ice crystals, as in high altitude clouds. Water vapor is an infrared absorber. Water vapor, without clouds, contributes to atmospheric warming. On the other hand, in the visible, clouds are both semiopaque (gray, overcast days) and reflective. Sunlight is reflected back into space. Due to reflection, the earth is cooled. The clouds absorb sunlight trapping the energy at higher altitude. The path to increased earth's temperature is less clear.

For evidence on the effect of clouds, see the short article by Christina Reed in Ref. [1]. The contrails formed from aircraft are the result of condensation of water. In the day, these contrails reflect sunlight and help prevent the escape of heat from the earth. Since reflection overwhelms trapped heat, the earth is thereby cooled. In the night, trapping of heat causes temperatures to increase. This is an opposite effect. After 9/11 when aircraft were grounded, analysis of the weather shows daytime temperatures increased while nighttime temperatures decreased. This is consistent with the theory.

In some yet-to-be-discovered way, water in the atmosphere will likely be a dominant factor in global warming.

REFERENCE

1. C. Reed, Hot trails to fight global warming, kiss the red-eye good-bye, Special Issue, *Scientific American*, September, 2006, p. 28.

Index

B

C

D

E

F

G

H

I

J

K

L

M

N

O

S

T

U

V

W

Y

Z